AF386451

Mathematical Sciences Research Institute
Publications

18

Editors

S.S. Chern
I. Kaplansky
C.C. Moore
I.M. Singer

Mathematical Sciences Research Institute
Publications

R.L. Bryant   S.S. Chern   R.B. Gardner
H.L. Goldschmidt   P.A. Griffiths

Exterior Differential Systems

Springer-Verlag
New York   Berlin   Heidelberg   London
Paris   Tokyo   Hong Kong   Barcelona

Robert L. Bryant
Department of Mathematics
Duke University
Durham, NC 27706
USA

S.S. Chern
Mathematical Sciences
   Research Institute
1000 Centennial Drive
Berkeley, CA 94720
USA

Robert B. Gardner
Department of Mathematics
University of North Carolina
Chapel Hill, NC 27514
USA

Hubert L. Goldschmidt
Department of Mathematics
Columbia University
New York, NY 10027
USA

P.A. Griffiths
Department of Mathematics
Duke University
Durham, NC 27706
USA

Mathematical Subject Classifications: 58A15, 58A17, 58G05, 35N10, 53A55

Library of Congress Cataloging-in-Publication Data
Exterior differential forms / Robert L. Bryant . . . [et al.].
     p. cm. — (Mathematical Sciences Research Institute
  publications ; 18)
     Includes bibliographical references and index.
     1. Exterior differential systems. 2. Differential equations,
  Partial. I. Bryant, Robert L. II. Series.
  QA649.E98 1991
  515′.35—dc20   90-47911
                                                         CIP

Printed on acid-free paper.

Photocomposed copy prepared by the Mathematical Sciences Research Institute using AMS-TeX.

9 8 7 6 5 4 3 2 1

ISBN-13: 978-1-4613-9716-8   e-ISBN-13: 978-1-4613-9714-4
DOI: 10.1007/978-1-4613-9714-4

Exterior Differential Systems

TABLE OF CONTENTS

CONTENTS vii

INTRODUCTION

This book gives a treatment of exterior differential systems. It will include both the general theory and various applications.

An *exterior differential system* is a system of equations on a manifold defined by equating to zero a number of exterior differential forms. When all the forms are linear, it is called a pfaffian system. Our object is to study its *integral manifolds*, i.e., submanifolds satisfying all the equations of the system. A fundamental fact is that every equation implies the one obtained by exterior differentiation, so that the complete set of equations associated to an exterior differential system constitutes a *differential ideal* in the algebra of all smooth forms. Thus the theory is coordinate-free and computations typically have an algebraic character; however, even when coordinates are used in intermediate steps, the use of exterior algebra helps to efficiently guide the computations, and as a consequence the treatment adapts well to geometrical and physical problems.

A system of partial differential equations, with any number of independent and dependent variables and involving partial derivatives of any order, can be written as an exterior differential system. In this case we are interested in integral manifolds on which certain coordinates remain independent. The corresponding notion in exterior differential systems is the independence condition: certain pfaffian forms remain linearly independent. Partial differential equations and exterior differential systems with an independence condition are essentially the same object. The latter, however, possess some advantages among which are the facts that the forms themselves often have a geometrical meaning, and that the symmetries of the exterior differential system are larger than those generated simply by changes of independent and dependent variables. Another advantage is that the coordinate-free treatment naturally leads to the intrinsic features of many systems of partial differential equations.

It was Pfaff who pioneered the study of exterior differential systems by his formulation of the Pfaff problem in Pfaff [1814-15]. The exterior derivative of a pfaffian form, called the bilinear covariant, was introduced by Frobenius in 1877 and efficiently used by Darboux in Darboux [1882]. In his book Cartan [1922], Élie Cartan introduced exterior differential forms of higher degree and their exterior derivatives. In 1904–08 he was led to the notion of a pfaffian system in involution through his work in generalizing the Maurer–Cartan forms in Lie groups to infinite Lie pseudogroups. The geometrical concepts introduced in this study apply to general exterior differential systems, as recognized by Goursat. An authoritative account was given in Kähler [1934], culminating in an existence theorem now known as the Cartan–Kähler theorem.

On the side of partial differential equations the basic existence theorem is of course the Cauchy–Kowalewski theorem. More general existence theorems were given by Riquier [1910]. The use of differential operators in studying differential geometry has been traditional and has an extensive literature.

Among our fundamental concepts are *prolongation* and *involutivity*. Intuitively the former is the classical way of adjoining the partial derivatives themselves as new variables, and taking as new equations those obtained by differentiating the old set, while the latter is the property that further prolongations will not give essentially new integrability conditions. Their precise definitions are more subtle, and will be given in Chapters VI and IV respectively. A linear pfaffian system in involution is a "well-behaved" system.

This concept entered in correspondence between Cartan and Einstein [1979] on relativity. While Élie Cartan proved that the Einstein field equations in general relativity based on distant parallelism form an involutive system, Einstein was at first suspicious of the notion. Later he understood it and expressed his satisfaction and appreciation.

Cartan expressed the involutivity condition in terms of certain integers, known as Cartan's test. In modern language this is a homological condition. In fact, Serre proved in 1963 that involutivity is equivalent to the vanishing of certain cohomology groups (see Guillemin–Sternberg [1964]). This makes it possible to use the powerful tool of commutative algebra. At the very beginning one notices the similarity between polynomials and differential operators. It turns out that this relationship goes much deeper, and the theory involves a mixture of both commutative and exterior algebra.

A fundamental problem is whether a given differential system will, after a finite number of prolongations lead to an involutive system. Cartan attempted to answer this question, but it was Kuranishi [1957] who finally proved the Cartan–Kuranishi prolongation theorem. The main tool is homology theory. A slightly weaker version of the theorem will be proved in this book.

We should however emphasize that differential systems not in involution are just as important. In fact, they are probably richer in content. For example, non-generic conditions on a manifold such as isometric embedding in low codimension or the presence of additional geometric structures frequently are expressed by a non-involutive system. The last half of Cartan [1946] and Partie II of his "Œuvres Complètes" (Cartan [1953]) are full of "examples", many of which are topics in their own right. An objective of this book is to call attention to these beautiful results, which have so far been largely ignored.

As the results are coordinate-free, the theory applies well to global problems and to non-linear problems. A guiding problem in the theory is the equivalence problem:

Given two sets of linear differential forms θ^i, θ^{*j} in the coordinates x^k, x^{*l} respectively, $1 \le i, j, k, l \le n$, both linearly independent, and given a Lie group $G \subset GL(n, R)$. To find the conditions that there are functions

$$x^{*i} = x^{*i}(x^1, \ldots, x^n),$$

such that θ^{*j}, after the substitution of these functions, differ from θ^i by a transformation of G. This gives rise to an exterior differential system. Cartan's idea was to introduce the parameters of G as new variables, setting

$$\omega^i = \sum_j g^i_j \theta^j, \quad \omega^{*i} = \sum g^{*i}_j \theta^{*j}, \quad (g^i_j), \quad (g^{*i}_j) \in G.$$

Then in the product of the manifold with G, the condition becomes

$$\omega^i = \omega^{*i},$$

which is symmetrical in both sides. This means we should study the problem in a principal G-bundle and the formulation becomes global.

The equivalence problem gives local Riemannian geometry when $G = O(n)$ and the local invariants of an almost complex structure when $n = 2m$ and $G = GL(m, \mathbb{C})$. Similarly, it gives the local invariants of CR-geometry when $n = 2m - 1$ and G is a suitable subgroup of $GL(m, \mathbb{C})$.

We have stated the equivalence problem because of its importance; it will not be explicitly treated in this book; see Gardner [1989] for a modern exposition.

The subject is so rich that a worker in the field is torn between the devil of the general theory and the angel of geometrical applications, which present all kinds of interesting phenomena. We have attempted to strike a balance. We will develop the general theory both from the standpoint of exterior differential systems and from that of partial differential equations. We will also give a large number of applications. A summary of contents follows:

Chapter I gives a review of exterior algebra, with emphasis on results which are relevant to exterior differential systems. For those who like an intrinsic treatment it includes an introduction to jet bundles.

Chapter II treats some simple exterior differential systems, particularly those which can be put in a normal form by a change of coordinates. They include completely integrable systems (Frobenius theorem) and the pfaffian equation. Cauchy characteristics for exterior differential systems come up naturally. Some arithmetic invariants are introduced for pfaffian systems. Even a pfaffian system of codimension 2, only partially treated in the last section, is a rich subject, with several interesting applications.

Chapter III discusses the generation of integral manifolds through the solution of a succession of initial-value problems. Various basic concepts are introduced. The Cartan–Kähler theorem is given as a generalization of the Cauchy–Kowalewsky theorem; the proof follows that of Kähler. As an application we give a proof of the isometric imbedding theorem of Cartan–Janet.

Chapter IV introduces the important concepts of involution, linear differential systems, tableau and torsion. For linear pfaffian systems the condition of involution, as expressed by Cartan's test, takes a simple form that is useful in computing examples. We also introduce the concept of prolongation, which will be more fully developed in Chapter VI.

As one example we show that the high-dimensional Cauchy–Riemann equations are in involution. We also study the system of q partial differential equations of the second order for one function in n variables and find conditions for their involutivity. A geometrical application is made to the problem of isometric surfaces preserving the lines of curvature. It is an example of an over-determined system which, after several prolongations, leads to a simple and elegant result. In this example the effectiveness of exterior differential systems is manifest.

Chapter V introduces the characteristic variety of a differential system. Intuitively its tangent spaces are hyperplanes of integral elements whose extension fails to be unique. It plays as important a role as the characteristics in classical partial differential equations. For a linear pfaffian system a definition can be given in terms of the symbol and the two agree in the absence of Cauchy characteristics. We discuss in detail the case of surfaces in E^3 and their Darboux frames, as this example illustrates many of the basic notions of exterior differential systems. Some properties of the characteristic variety are given. The deeper ones require the system to be involutive and the use of the complex characteristic variety. Their proofs rely on results of commutative algebra and are postponed to Chapter VIII.

Chapter VI treats prolongation, another well-known process in the case of partial differential equations. The issue is whether any system with an independence condition (I, Ω) can be prolonged to an involutive system in a finite number of steps (Cartan–Kuranishi theorem). With our definition of prolongation we prove that the first prolongation of an involutive linear pfaffian system is involutive, a result that does not seem to appear in the literature. We establish a somewhat weaker version of the Cartan–Kuranishi theorem, thus giving in a sense a positive answer to the above question. As usual the general theory is illustrated by a number of examples.

Chapter VII is devoted to some examples and applications. We give a classification of systems of first-order partial differential equations of two functions in two variables. Other examples include: triply orthogonal systems, finiteness of web rank, isometric imbedding and the characteristic variety.

In Chapter VIII we study the algebra of a linear pfaffian system and its prolongations. The crucial information is contained in the tableau. Its properties are given by the Spencer cohomology groups or the Koszul homology groups, which are dual to each other. Involutive tableau is characterized by the vanishing of certain Spencer cohomology or Koszul homology groups. It is a remarkable coincidence that a regular integral flag and a quasi-regular graded SV-module represent essentially the same object. Homological algebra provides the tools to complete the proofs of the theorems stated in Chapters V and VI, and in particular the Cartan–Kuranishi theorem. As a consequence sheaf theory in commutative algebra and micro-local analysis in partial differential equations become parallel developments.

Chapters IX and X give an introduction to the Spencer theory of overdetermined systems of partial differential equations. While Cartan began his theory in the study of infinite pseudogroups, Spencer had a similar objective, viz., the study of the deformations of pseudogroup structures. His approach is more in the spirit of Lie, with a full use of modern concepts. We see in our account more emphasis on the general theory, although many examples are given. We hope that after the exposition in this book we come to realize that exterior differential systems and partial differential equations are one and the same subject. It is conceivable that different attires are needed for different purposes.

This book grew through our efforts to work through and appreciate Partie II of Cartan's "Œuvres Complètes" (Cartan [1953]), which we found to be full of interesting ideas and details. Hopefully our presentation will help the study of the original work, which we cannot replace. In fact, for readers who have gone through most of this book we propose the following problem as a final examination: Give a report on his famous five-variable paper, "Les systèmes de Pfaff à cinq variables et les équations aux dérivées partielles du second ordre" (Cartan [1910]).

CHAPTER I

PRELIMINARIES

In this chapter we will set up some notations and conventions in exterior algebra, give a description of the basic topic of the book, and introduce the language of jets which allows easy passage between partial differential equations and exterior differential systems. In particular we establish some basic results in exterior algebra Theorems 1.3, 1.5, and 1.7 which will be used in Chapter II.

§1. Review of Exterior Algebra.

Let V be a real vector space of dimension n and V^* its dual space. An element $x \in V$ is called a vector and an element $\omega \in V^*$ a covector. V and V^* have a pairing

$$\langle x, \omega \rangle, \quad x \in V, \quad \omega \in V^*,$$

which is a real number and is linear in each of the arguments, x, ω.

Over V there is the *exterior algebra*, which is a graded algebra:

$$\Lambda(V) = \Lambda^0(V) \oplus \Lambda^1(V) \oplus \cdots \oplus \Lambda^n(V),$$

with

$$\Lambda^0(V) = \mathbb{R}, \quad \Lambda^1(V) = V.$$

An element $\xi \in \Lambda(V)$ can be written in a unique way as

$$\xi = \xi_0 + \xi_1 + \cdots + \xi_n,$$

where $\xi_p \in \Lambda^p(V)$ is called the *p-th component* of ξ. An element

$$\xi = \xi_p \in \Lambda^p(V)$$

is called *homogeneous of degree p* or a *multivector of dimension p*.

Multiplication in $\Lambda(V)$ will be denoted by the wedge sign: $\wedge$. It is associative, distributive, but not commutative. Instead it satisfies the relation

$$\xi \wedge \eta = (-1)^{pq} \eta \wedge \xi, \quad \xi \in \Lambda^p(V), \quad \eta \in \Lambda^q(V).$$

The multivector ξ is called *decomposable*, if it can be written as a monomial

$$(1) \qquad\qquad \xi = x_1 \wedge \cdots \wedge x_p, \quad x_i \in V.$$

We have the following fundamental fact:

Proposition 1.1 (Criterion on linear dependence). *The vectors $x_1, \ldots, x_p$ are linearly dependent if and only if $x_1 \wedge \cdots \wedge x_p = 0$.*

If the decomposable multivector ξ in (1) is not zero, then $x_1, \ldots, x_p$ are linearly independent and span a linear subspace W of dimension p in V. This space can be described by

$$(2) \qquad W = \{x \in V \mid x \wedge \xi = 0\}.$$

Let $x_1', \ldots, x_p'$ be another base in W. Then

$$\xi' = x_1' \wedge \cdots \wedge x_p'$$

is a (non-zero) multiple of ξ. We will call ξ, defined up to a constant factor, the *Grassmann coordinate vector of W*, and write

$$(3) \qquad [\xi] = W,$$

the bracket indicating the class of coordinate vectors differing from each other by a non-zero factor.

In the same way there is over V^* the exterior algebra

$$\Lambda(V^*) = \Lambda^0(V^*) \oplus \Lambda^1(V^*) \oplus \cdots \oplus \Lambda^n(V^*),$$
$$\Lambda^0(V^*) = \mathbb{R}, \quad \Lambda^1(V^*) = V^*.$$

An element of $\Lambda^p(V^*)$ is called a *form of degree p* or simply a *p-form*.

Let e_i be a base of V and ω^k its dual base, so that

$$\langle e_i, \omega^k \rangle = \delta_i^k, \quad 1 \leq i, k \leq n.$$

Then an element $\xi \in \Lambda^p(V)$ can be written

$$(4) \qquad \xi = 1/p! \sum a^{i_1 \cdots i_p} e_{i_1} \wedge \cdots \wedge e_{i_p}$$

and an element $\alpha \in \Lambda^p(V^*)$ as

$$(5) \qquad \alpha = 1/p! \sum b_{i_1 \ldots i_p} \omega^{i_1} \wedge \cdots \wedge \omega^{i_p}.$$

In (4) and (5) the coefficients $a^{i_1 \cdots i_p}$ and $b_{i_1 \ldots i_p}$ are supposed to be anti-symmetric in any two of their indices, so that they are well defined. It follows from (4) that any multivector is a linear combination of decomposable multivectors.

For our applications it is important to establish the explicit duality or pairing of $\Lambda(V)$ and $\Lambda(V^*)$. We require that $\Lambda^p(V)$ and $\Lambda^q(V^*)$, $p \neq q$, annihilate each other. It therefore suffices to define the pairing of $\Lambda^p(V)$

and $\Lambda^p(V^*)$. Since, by the above remark, any multivector is a linear combination of decomposable multivectors, it suffices to have the pairing of

$$\xi = x_1 \wedge \cdots \wedge x_p, \quad x_i \in V,$$

and

$$\alpha = \omega^1 \wedge \cdots \wedge \omega^p, \quad \omega^i \in V^*.$$

We will define

$$(6) \qquad \langle \xi, \alpha \rangle = \det(\langle x_i, \omega^k \rangle), \quad 1 \le i, k \le p.$$

It can be immediately verified that this definition is meaningful, i.e., if ξ (resp. α) is expressed in a different way as a product of vectors (resp. covectors), the right-hand side of (6) remains unchanged.

In terms of the expressions (4) and (5) the pairing is given by

$$(7) \qquad \langle \xi, \alpha \rangle = 1/p! \sum a^{i_1 \ldots i_p} b_{i_1 \ldots i_p}.$$

This is proved by observing that the right-hand side of (7) is linear in the arguments ξ and α and that the right-hand sides of both (6) and (7) are equal when ξ and α are products of the elements of the dual bases.

An endomorphism f of the additive structure of $\Lambda(V)$ is called a *derivation of degree k* if it satisfies the conditions:

(i) $f : \Lambda^p V \to \Lambda^{p+k} V, \ 0 \le p \le n$

(ii) $f(\xi \wedge \eta) = f(\xi) \wedge \eta + (-1)^{kp} \xi \wedge f(\eta)$

for $\xi \in \Lambda^p V, \ \eta \in \Lambda V$.

A derivation of degree -1 is also called an *anti-derivation*.

Given $\xi \in V$, we define the exterior product

$$e(\xi) : \Lambda(V) \to \Lambda(V)$$

by

$$e(\xi)\eta = \xi \wedge \eta \quad \eta \in \Lambda(V).$$

The adjoint operator of $e(\xi)$,

$$\xi \lrcorner : \Lambda(V^*) \to \Lambda(V^*)$$

is called the *interior product*, and is defined by the relation

$$\langle \eta, \xi \lrcorner \alpha \rangle = \langle e(\xi)\eta, \alpha \rangle \quad \eta \in \Lambda(V), \alpha \in \Lambda(V^*).$$

The following result is easily proved:

Proposition 1.2. *If $x \in V$, then $x \lrcorner$ is an anti-derivation.*

Notice that $e(x)$ is neither a derivation nor an anti-derivation.

Definition. A subring $I \subset \Lambda(V^*)$ is called an *ideal*, if:
  a) $\alpha \in I$ implies $\alpha \wedge \beta \in I$ for all $\beta \in \Lambda(V^*)$;
  b) $\alpha \in I$ implies that all its components in $\Lambda(V^*)$ are contained in I.

A subring satisfying the second condition is called *homogeneous*. As a consequence of a) and b) we conclude that $\alpha \in I$ implies $\beta \wedge \alpha \in I$ for all $\beta \in \Lambda(V^*)$. Thus all our ideals are homogeneous and two-sided.

Given an ideal $I \subset \Lambda(V^*)$, we wish to determine the smallest subspace $W^* \subset V^*$ such that I is generated, as an ideal, by a set S of elements of $\Lambda(W^*)$. An element of I is then a sum of elements of the form $\sigma \wedge \beta$, $\sigma \in S$, $\beta \in \Lambda(V^*)$. If $x \in W = (W^*)^\perp$, we have, since the interior product $x \lrcorner$ is an anti-derivation,

$$x \lrcorner \sigma = 0,$$

$$x \lrcorner (\sigma \wedge \beta) = \pm \sigma \wedge (x \lrcorner \beta) \in I.$$

Therefore we define

$$A(I) = \{x \in V \mid x \lrcorner I \subset I\},$$

where the last condition means that $x \lrcorner \alpha \in I$, for all $\alpha \in I$. $A(I)$ is clearly a subspace of V. It will play later an important role in differential systems, for which reason we will call it the *Cauchy characteristic space* of I. Its annihilator

$$C(I) = A(I)^\perp \subset V^*$$

will be called the *retracting subspace* of I.

Theorem 1.3 (Retraction theorem). *Let I be an ideal of $\Lambda(V^*)$. Its retracting subspace $C(I)$ is the smallest subspace of V^* such that $\Lambda(C(I))$ contains a set S of elements generating I as an ideal. The set S also generates an ideal J in $\Lambda(C(I))$, to be called a retracting ideal of I. There exists a mapping*

$$\Delta : \Lambda(V^*) \to \Lambda(C(I))$$

of graded algebras such that $\Delta(I) = J$.

Proof. Suppose $W^* \subset V^*$ be a subspace such that $\Lambda(W^*)$ contains a set S of elements which generate I as an ideal. By the above discussion, if $x \in W = (W^*)^\perp$, we have $x \lrcorner I \subset I$. It follows that $W \subset A(I)$, and consequently, $C(I) = (A(I))^\perp \subset W^*$.

We now choose a complementary space B of $C(I)$ in V^*, so that $V^* = B \oplus C(I)$. Let ω^i, $1 \leq i \leq n$, be a base in V^* with

$$\omega^1, \ldots, \omega^k \in B, \quad \omega^{k+1}, \ldots, \omega^n \in C(I).$$

Its dual base e_i, $1 \leq i \leq n$, then has the property that $A(I) = \{e_1, \ldots, e_k\}$. We define

$$h_j : \Lambda(V^*) \to \Lambda(V^*), \quad 1 \leq j \leq k,$$

by

$$h_j(\alpha) = \alpha - \omega^j \wedge (e_j \lrcorner \alpha), \quad \alpha \in \Lambda(V^*).$$

It is easy to verify that

$$h_j(\alpha \wedge \beta) = h_j(\alpha) \wedge h_j(\beta),$$

so that each h_j is a mapping of graded algebras. The same is therefore true of the composition

$$\Delta = h_k \circ \cdots \circ h_1.$$

Since $e_j \in A(I)$, $1 \leq j \leq k$, we have $h_j(I) \subset I$, from which we get $\Delta(I) \subset I$. Clearly we have the restrictions $\Delta|_B = 0$, $\Delta|_{C(I)} = Id$. Since Δ is a mapping of graded algebras, this implies that $\Lambda(C(I))$ is the image of Δ.

It remains to construct the set S in $\Lambda(C(I))$ which generates I. This is done by induction on the degrees of the elements of I. Let I_p be the set of elements of I of degree p. To exclude the trivial case that $I = \Lambda(V^*)$ itself, we suppose $I_0 = \emptyset$. Using this assmption we have, by the definition of $A(I)$, $x \lrcorner \alpha = \langle x, \alpha \rangle = 0$, $x \in A(I)$, $\alpha \in I_1$. It follows that $A(I) \subset I_1^{\perp}$ of $I_1 \subset C(I)$.

To apply induction suppose that $I_1, \ldots, I_{p-1}$ are generated by elements of $\Lambda(C(I))$. Denote by J_{p-1} the ideal generated by them. Recall that h_j, $1 \leq j \leq k$, are mappings of graded algebras and induce the identity mapping on $C(I)$. They therefore leave J_{p-1} invariant. By the definition of h_1 we have $h_1(\alpha) - \alpha \in J_{p-1}$. Applying $h_2, \ldots, h_k$ successively, we get $\Delta(\alpha) - \alpha \in J_{p-1}$. By replacing α by $\Delta(\alpha)$ as a generator of I, we complete the induction. $\qquad\square$

We wish to make some applications of the retraction theorem (Theorem 1.3). First we recall that, dualizing (2) and (3), the Grassmann coordinate vector $[\alpha]$ of a subspace $W^* \subset V^*$ of dimension p is a non-zero decomposable p-covector such that

$$W^* = \{\omega \in V^* \mid \omega \wedge \alpha = 0\}.$$

This notion can be extended to any p-form α, decomposable or not, by defining

$$L_\alpha = \{\omega \in V^* \mid \omega \wedge \alpha = 0\}.$$

L_α will be called the *space of linear divisors* of α, because of the property given in the following theorem:

Proposition 1.4. *Given a p-form α, let $\omega^1, \ldots, \omega^q$ be a base for L_α. Then α may be written in the form*

$$\alpha = \omega^1 \wedge \cdots \wedge \omega^q \wedge \pi, \ \text{with } \pi \in \Lambda^{p-q}(V^*).$$

Proof. Take first the case $q = 1$. We can suppose ω^1 to be a base element of V^*. By the expression (5) we can write

$$\alpha = \omega^1 \wedge \pi + \alpha_1,$$

where α_1 does not involve ω^1. The hypothesis $\omega^1 \wedge \alpha = 0$ implies $\alpha_1 = 0$, so that the statement is true.

The general case follows by induction on q. $\qquad\qquad\square$

Theorem 1.5. *Let I be an ideal generated by the linearly independent elements $\omega^1, \ldots, \omega^s \in V^*$ and the 2-form $\Omega \in \Lambda^2(V^*)$. Let p be the smallest integer such that*

$$(8) \qquad\qquad \Omega^{p+1} \wedge \omega^1 \wedge \cdots \wedge \omega^s = 0.$$

Then the retracting space $C(I)$ is of dimension $2p+s$ and has the Grassmann coordinate vector

$$\Omega^p \wedge \omega^1 \wedge \cdots \wedge \omega^s.$$

Proof. Consider first the case $s = 0$. An element of I is a linear combination of $\Omega, \Omega^2, \ldots, \Omega^p \neq 0$. Hence by Theorem 1.3, we have $\Omega \in \Lambda(C(I))$, and $\Omega^p \in \Lambda^{2p}(C(I))$. The latter implies

$$\dim C(I) \geq 2p.$$

Let

$$f : V \to V^*$$

be the linear map defined by

$$f(x) = x \lrcorner \Omega, \quad x \in V.$$

Since I does not contain a linear form, we have

$$x \lrcorner \Omega = 0 \text{ if and only if } x \in A(I) = C(I)^\perp.$$

This proves

$$\ker f = A(I),$$

so that

$$(9) \qquad\qquad \dim \ker f = \dim A(I) \leq n - 2p.$$

On the other hand, the equation (8) gives for $s = 0$,

$$x \lrcorner \Omega^{p+1} = (p+1)(x \lrcorner \Omega) \wedge \Omega^p = 0.$$

Hence the space of linear divisors of Ω^p contains the image of f. Since Ω^p is of degree $2p$ and has at most $2p$ linear divisors, we have

$$(10) \qquad\qquad\qquad \dim \mathrm{im}\, f \leq 2p.$$

Now it is an elementary fact that

$$\dim \ker f + \dim \mathrm{im}\, f = n.$$

Therefore the equality signs hold in both (9) and (10). In particular, we have $\dim C(I) = 2p$ and $\Lambda^{2p}(C(I))$ is of dimension one, with Ω^p as a base element, which is thus a Grassmann coordinate vector of $C(I)$.

In the general case, let $W^* = \{\omega^1, \ldots, \omega^s\}$ be the space spanned by the ω's. Then $W = (W^*)^{\perp} \subset V$ and the quotient space V^*/W^* have a pairing induced by that of V and V^*, and are dual vector spaces. We have

$$0 \neq \Omega^p \wedge \omega^1 \wedge \cdots \wedge \omega^s \in \Lambda^{2p+s}(C(I)),$$

so that

$$\dim C(I) \geq 2p + s.$$

Consider the linear map

$$W \xrightarrow{f} V^* \xrightarrow{\pi} V^*/W^*,$$

where π is the projection (into a quotient space) and f is defined by

$$f(x) = x \lrcorner \Omega, x \in W.$$

As above, we wish to find upper bounds for the dimensions of the kernel and image of $f' = \pi \circ f$. The sum of these dimensions is

$$\dim \ker f' + \dim \mathrm{im}\, f' = n - s.$$

A generalization of the above argument gives

$$\dim \ker f' \leq n - 2p - s$$

$$\dim \mathrm{im}\, f' \leq 2p$$

Hence the equality signs hold everywhere and the theorem follows as before. $\qquad\qquad\square$

Proposition 1.6. *Let $\omega^1, \ldots, \omega^s, \pi$ be linearly independent elements of V^* and $\Omega \in \Lambda^2 V^*$; then*

$$(11) \qquad \Omega^p \wedge \omega^1 \wedge \cdots \wedge \omega^s \wedge \pi = 0$$

implies

$$\Omega^{p+1} \wedge \omega^1 \wedge \cdots \wedge \omega^s = 0.$$

Proof. Let $\{\pi\}$ denote the one dimensional space spanned by π and let W^* denote a complement in V^* of $\{\pi\}$ which contains $\omega^1, \ldots, \omega^s$. Then there exist $\alpha \in \Lambda^2 W^*$, $\beta \in W^*$, uniquely determined, such that

$$\Omega = \alpha + \beta \wedge \pi.$$

It follows that

$$\Omega^p = \alpha^p + p\alpha^{p-1} \wedge \beta \wedge \pi$$

and the hypothesis (11) implies

$$\alpha^p \wedge \omega^1 \wedge \cdots \wedge \omega^s \wedge \pi = 0.$$

Since $\alpha^p \wedge \omega^1 \wedge \cdots \wedge \omega^s \in \Lambda(W^*)$, we must have

$$\alpha^p \wedge \omega^1 \wedge \cdots \wedge \omega^s = 0.$$

The conclusion now follows since

$$\Omega^{p+1} = \alpha^{p+1} + (p+1)\alpha^p \wedge \beta \wedge \pi.$$

$\square$

A sequential application of Theorem 1.5 leads to a constructive proof of the algebraic normal form of a two form which is useful for many arguments in the theory of exterior differential systems.

Theorem 1.7. *Let $\Omega \in \Lambda^2(V^*)$ and let r be the smallest integer such that*

$$\Omega^{r+1} = 0.$$

Then there exist $2r$ linearly independent elements $\omega^1, \ldots, \omega^{2r}$ such that

$$(12) \qquad \Omega = \sum_{i=1}^{r} \omega^{r+i} \wedge \omega^i.$$

Proof. The theorem is proved by repeated applications of Theorem 1.5. In fact, from the hypotheses it follows that Ω^r is decomposable and hence has a linear divisor ω^1. Next consider the ideal $l(1) = \{\omega^1, \Omega\}$ generated by ω^1 and Ω. Let r_1 be the smallest integer such that

$$\Omega^{r_1+1} \wedge \omega^1 = 0.$$

Clearly $r_1 + 1 \leq r$.

Then $\Omega^{r_1} \wedge \omega^1$ is the Grassmann coordinate vector of the retraction space $C(l(1))$ and is decomposable and non-zero. Let ω^2 be a linear factor of $\Omega^{r_1} \wedge \omega^1$, which is linearly independent from ω^1. Then

$$\Omega^{r_1} \wedge \omega^1 \wedge \omega^2 = 0.$$

Let r_2 be the smallest integer satisfying

$$\Omega^{r_2+1} \wedge \omega^1 \wedge \omega^2 = 0,$$

so that $r_2 < r_1$.

Continuing this process, we get a sequence of positive integers $r > r_1 > r_2 > \ldots$, which must end with zero. This means that there are linear forms $\omega^1, \ldots, \omega^q$, linearly independent, satisfying

$$\Omega \wedge \omega^1 \wedge \cdots \wedge \omega^q = 0.$$

From this we get
$$\Omega = \sum_{1 \leq i \leq q} \eta_i \wedge \omega^i,$$

where η_i are linear forms. Since $\Omega^r \neq 0$, we must have $q = r$ and η_i, ω^i, $1 \leq i \leq r$ are linearly independent. The theorem is proved by setting

$$\omega^{r+i} = \eta_i.$$

□

Remark. Theorem 1.7 is equivalent to the theorem in linear algebra on the normal form of a skew-symmetric matrix. In fact, in terms of a base ω^i, $1 \leq i \leq n$, of V^* we can write

$$\Omega = 1/2 \sum_{i,j} a_{ij} \omega^i \wedge \omega^j, \qquad a_{ij} + a_{ji} = 0.$$

Let

$$\omega^i = \sum_k s_k^i \omega^{*k}, \qquad 1 \leq i, k \leq n$$

be a change of base. Then

$$\Omega = 1/2 \sum a_{kl}^* \omega^{*k} \wedge \omega^{*l},$$

where

(13)
$$a_{kl}^* = \sum_{i,j} a_{ij} s_k^i s_l^j, \quad 1 \leq i,j,k,l \leq n.$$

If we introduce the matrices

$$A = (a_{ij}), \quad A^* = (a_{ij}^*), \quad S = (s_i^j),$$

of which A and A^* are skew-symmetric, and S is non-singular, then (13) can be written as a matrix equation

$$A^* = SA^t S, \quad {}^t S = \text{ transpose of } S.$$

Theorem 1.7 can be stated as follows: *Given a skew-symmetric matrix A. Its rank is even. There exists a non-singular matrix S, such that*

$$A^* = \begin{pmatrix} 0 & I_p & 0 \\ -I_p & 0 & 0 \\ 0 & 0 & 0 \end{pmatrix},$$

where I_p is the unit matrix of order p.

§2. The Notion of an Exterior Differential System.

Consider a differentiable manifold M of dimension n. Its cotangent bundle, whose fibers are the cotangent spaces $T_x^*(M)$, $x \in M$, we will denote by T^*M. From T^*M we construct the bundle ΛT^*M, whose fibers are

$$\Lambda T_x^* = \sum_{0 \leq p \leq n} \Lambda^p T_x^*,$$

which have the structure of a graded algebra, as discussed in the last section. The bundle ΛT^*M has the subbundles $\Lambda^p T^*M$, whose definition is obvious. Similar definitions are valid for the tangent bundle TM.

A section of the bundle

$$\Lambda^p T^*M = \bigcup_{x \in M} \Lambda^p T_x^* \to M$$

is called an *exterior differential form of degree p*, or a form of degree p or simply a p-form. By abuse of language we will call a differential form a section of the bundle $\Lambda T^* M$; its p-th component is a p-form. All sections are supposed to be sufficiently smooth.

In terms of a system of local coordinates $x^1, \ldots, x^n$ on M, an exterior differential form of degree p has the expression

$$\alpha = 1/p! \sum a_{i_1 \ldots i_p} dx^{i_1} \wedge \cdots \wedge dx^{i_p}, \quad 1 \le i_1, \ldots, i_p \le n,$$

where the coefficients are smooth functions and are anti-symmetric in any two of the indices.

Let $\Omega^p(M) = C^\infty$-sections of $\Lambda^p T^* M$ and let $\Omega^*(M) = \bigoplus \Omega^p(M)$.

Definition. (i) An *exterior differential system* is given by an ideal $\mathcal{I} \subset \Omega^*(M)$ that is closed under exterior differentiation; (ii) an *integral manifold* of the system is given by an immersion $f : N \to M$ such that $f^* \alpha = 0$ for all $\alpha \in \mathcal{I}$.

By our conventions $\mathcal{I} = \bigoplus \mathcal{I}^q$ is a direct sum of its homogeneous pieces $\mathcal{I}^q = \mathcal{I} \cap \Omega^q(M)$, and by differentiation; and by differential closure we have $d\alpha \in \mathcal{I}$ whenever $\alpha \in \mathcal{I}$. We sometimes refer to an ideal $\mathcal{I} \subset \Omega^*(M)$ satisfying $d\mathcal{I} \subseteq \mathcal{I}$ as a *differential ideal*.

In practice, $\mathcal{I}$ will be almost always generated as a differential ideal by a finite collection $\{\alpha_A\}$, $1 \le A \le N$ of differential forms; forms of degree zero, i.e. functions, are not excluded. An integral manifold of $\mathcal{I}$ is given by an immersion

$$f : N \to M$$

satisfying $f^* \alpha = 0$ for $1 \le A \le N$. Then

$$f^*(\beta \wedge \alpha_A) = 0 \quad \text{and} \quad f^*(d\alpha_A) = 0,$$

and so $f^* \alpha = 0$ for all α in the differential ideal generated by the $\{\alpha_A\}$.

The fundamental problem in exterior differential systems is to study the integral manifolds. We may think of these as solutions to the system

$$\alpha_A = 0$$

of exterior equations. When written out in local coordinates, this is a system of P.D.E.'s.

The notion is of such generality that it includes all the ordinary and partial differential equations, as the following examples show:

Example. The second-order differential equations in the (x, y)-plane,

$$\frac{d^2 y}{dx^2} = F(x, y, \frac{dy}{dx})$$

can be written as an exterior differential system

$$dy - y'dx = 0,$$

$$dy' - F(x, y, y')dx = 0$$

in the space of the variables (x, y, y').

Example. Consider the partial differential equation of the first order

$$(14) \qquad F(x^i, z, \frac{\partial z}{\partial x^i}) = 0, \quad 1 \le i \le n.$$

By introducing the partial derivatives as new variables, it can be written as an exterior differential system

$$(15) \qquad \begin{aligned} F(x^i, z, p_i) &= 0, \\ dz - \sum p_i dx^i &= 0 \end{aligned}$$

in the $(2n + 1)$-dimensional space (x^i, z, p_i).

From these examples it is clear that any system of differential equations can be written as an exterior differential system. However, not all exterior differential systems arise in this way. The following example marks the birth of differential systems:

Example. The equation

$$a_1(x)dx^1 + \cdots + a_n(x)dx^n = 0, x = (x^1, \ldots, x^n),$$

is called a *Pfaffian equation.* Pfaff's problem is to determine its integral manifolds of maximal dimension.

From the examples we notice two important concepts. One is an exterior differential system with independence condition (I, Ω) which is given by a closed differential ideal I together with a decomposable p-form

$$\Omega = \omega^1 \wedge \cdots \wedge \omega^p.$$

An integral manifold of (I, Ω) is an integral manifold of I satisfying the additional condition $f^*\Omega \ne 0$. This is the case when we wish to keep some variables independent, as in the case when the system arises from a system of partial differential equations. For instance, in the second example, we take

$$(16) \qquad \Omega = dx^1 \wedge \cdots \wedge dx^n.$$

The partial differential equation (14) is equivalent to the system with independence condition (I, Ω), where I is generated by the left-hand members of (15) and Ω is given by (16). Whether an independence condition should be imposed depends on the particular problem.

The other important concept is that of prolongation, which will be treated in detail later on. In our first and second examples it is necessary to introduce the derivatives as new variables. With more general systems the consideration of higher-order derivatives becomes necessary. Thus we could be forced to introduce higher-dimensional manifolds and related systems, the prolonged systems, whose study is necessary for that of the given system.

§3. Jet Bundles.

A rigorous theory of differential systems depends on a foundation of differentiable manifolds and their differentiable maps. One such foundation is provided by the theory of jets developed by Charles Ehresmann. An introduction will be given below.

We will give a geometric description of the spaces of partial derivatives of maps between two differentiable manifolds. These spaces will be constructed as differentiable manifolds with underlying sets given by equivalence classes of maps. The equivalence relation will be given in a form which is clearly intrinsic by first defining it for normalized functions on the real line and then defining it for general maps by a universal extension. The analytical content of the equivalence relations is then exhibited by a local characterization which is in turn used to provide the differentiable structure.

Historically these ideas were motivated by geometers studying partial differential equations, say

$$F(x^1, \ldots, x^m, z, \partial z/\partial x^1, \ldots, \partial z/\partial x^m) = 0,$$

and their desire to interpret this equation as representing a hypersurface in the space with coordinates

$$x^1, \ldots, x^m, z, \partial z/\partial x^1, \ldots, \partial z/\partial x^m.$$

The idea behind jets is simply to give this a precise formulation.

Let $\mathbb{R}$ denote the real line, with the usual differentiable structure and let t denote a coordinate function in a neighborhood of the origin. If

$$f : \mathbb{R} \to \mathbb{R} \text{ and } g : \mathbb{R} \to \mathbb{R}$$

are two differentiable maps of the real line into itself which map the origin
into the origin, then f and g are said to have the same *r-jet* whenever

$$\frac{df}{dt}(0) = \frac{dg}{dt}(0), \ldots, \frac{d^r f}{dt^r}(0) = \frac{d^r g}{dt^r}(0).$$

Now let N be a differentiable manifold, and let $p \in M$, then *a p-based
parametrized curve u*, written

$$u : (\mathbb{R}, 0) \to (N, p)$$

is a map of the real line into N which takes the origin of $\mathbb{R}$ into p and is
differentiable.

Similarly, a *p-based real valued function v*, written

$$v : (N, p) \to (\mathbb{R}, 0),$$

is a real valued function on N which maps the point p onto the origin of $\mathbb{R}$
and is differentiable.

Let M and N be differentiable manifolds and let

$$f : N \to M \text{ and } g : N \to M$$

be differentiable maps of N into M. Then f and g are said to have the
same *r-jet at a point* $p \in M$ whenever

a) $f(p) = g(p) = q$

and

b) for all p-based parametrized curves $u : (\mathbb{R}, 0) \to (N, p)$,
and for all q-based real valued functions

$$v : (M, q) \to (\mathbb{R}, 0),$$

the differentiable maps

$$v \circ f \circ u \quad \text{and} \quad v \circ g \circ u$$

of the real line into itself mapping the origin into the origin have the same
r-jet.

The relation that two maps have the same r-jet at a point p is an equivalence relation and the equivalence class with the representative

$$f : N \to M$$

will be denoted by

$$j_p^r(f).$$

The point p is called *the source* of $j_p^r(f)$ and the point $f(p)$ is called *the target* of $j_p^r(f)$.

We have given an intrinsic characterization of these equivalence classes. In order to get hold of this notion we express the relation in local coordinates. Given

$$\alpha = (\alpha_1, \ldots, \alpha_m),$$

we define

$$\alpha! = \alpha_1! \ldots \alpha_m! \quad \text{and} \quad |\alpha| = \alpha_1 + \cdots + \alpha_m$$

and given $x = (x^1, \ldots, x^m)$ we define

$$x^\alpha = x^{1^{\alpha_1}} \ldots x^{m^{\alpha_m}}$$

and

$$D_x^\alpha = \frac{\partial^{\alpha_1}}{\partial x^{1^{\alpha_1}}} \cdots \frac{\partial^{\alpha_m}}{\partial x^{m^{\alpha_m}}}$$

with the convention that $D_x^0 f = f(0)$.

Proposition 3.1. *Let f and g be two differentiable maps*

$$f : \mathbb{R}^m \to \mathbb{R}^n \quad \text{and} \quad g : \mathbb{R}^m \to \mathbb{R}^n$$

mapping the origin into the origin. Let $\{x^1, \ldots, x^m\}$ denote coordinates in a neighborhood of the origin of $\mathbb{R}^m$ and $\{z^1, \ldots, z^n\}$ denote coordinates in a neighborhood of the origin of $\mathbb{R}^n$. These coordinates allow us to introduce real valued functions f^i, g^i by

$$f(x) = (f^1(x), \ldots, f^n(x)) \quad \text{and} \quad g(x) = (g^1(x), \ldots, g^n(x)).$$

With these notations f and g have the same r-jet at the origin if and only if

$$(17) \qquad D_x^\alpha f^i(0) = D_x^\alpha g^i(0) \quad (1 \le i \le n, |\alpha| \le r).$$

Proof. Assume that (17) holds and let

$$u : (\mathbb{R}, 0) \to (\mathbb{R}^m, 0)$$

be an arbitrary 0-based curve. Using the $\{x^1, \ldots, x^m\}$ coordinates we may define

$$u(t) = (u^1(t), \ldots, u^m(t)).$$

Next let

$$v : (\mathbb{R}^n, 0) \to (\mathbb{R}, 0)$$

be an arbitrary 0-based real valued function. Then repeated application of
the chain rule and the Leibniz product formula gives rise to an equation

$$\frac{d^k}{dt^k} v \circ f \circ u|_{t=0} = F(D_z^\beta v(0), D_x^\alpha f(0), \frac{d^\gamma u}{dt^\gamma}(0)), \quad |\alpha|, |\beta|, \gamma \le r,$$

where F is a constant coefficient polynomial in the indicated indeterminates. It follows that

$$\begin{aligned}
\frac{d^k}{dt^k} v \circ f \circ u|_{t=0} &= F(D_z^\beta v(0), D_x^\alpha f(0), \frac{d^\gamma u}{dt^\gamma}(0)) \\
&= F(D_z^\beta v(0), D_x^\alpha g(0), \frac{d^\gamma u}{dt^\gamma}(0)) \\
&= \frac{d^k}{dt^k} v \circ g \circ u|_{t=0}
\end{aligned}$$

for $k \le r$, which verifies that f and g have the same r-jet at 0.

Conversely let us assume that

$$j_0^r(f) = j_0^r(g).$$

Then if we take

$$u : (\mathbb{R}, 0) \to (\mathbb{R}^m, 0)$$

to be the 0-based parametrized curve defined by

$$u(t) = (\xi_1 t, \ldots, \xi_m t)$$

with $\xi_i \in \mathbb{R}$, $1 \le i \le m$ and take

$$v^j : (\mathbb{R}^n, 0) \to (\mathbb{R}, 0)$$

to be the 0-based real valued function defined by projection on the j-th
coordinate, i.e.

$$v^j(z^1, \ldots, z^n) = z^j, \quad 1 \le j \le n$$

then by hypothesis

$$\frac{d^k}{dt^k} f^j(\xi_1 t, \ldots, \xi_m t) = \frac{d^k}{dt^k} g^j(\xi_1 t, \ldots, \xi_m t), \quad k \le r,$$

which implies

$$\sum \frac{\partial^k f^j}{\partial x^{1^{i_1}} \ldots \partial x^{m^{i_m}}}(0) \xi_1^{i_1} \ldots \xi_m^{i_m} = \sum \frac{\partial^k g^j}{\partial x^{1^{i_1}} \ldots \partial x^{m^{i_m}}}(0) \xi_1^{i_1} \ldots \xi_m^{i_m},$$

$$i_1 + \cdots + i_m = k.$$

Since this last equation holds for all real $\xi_1, \ldots, \xi_m$, the corresponding co-efficients must be equal, that is

$$D_x^\beta f^j(0) = D_x^\beta g^j(0), \quad 1 \le j \le n, \quad |\beta| \le r$$

as claimed.

In order to carry this last result over to the general situation

$$(N, p) \xrightarrow[g]{f} (M, q)$$

we introduce coordinates h_U with p the origin, and h_V with q the origin and define

$$(\mathbb{R}^m, 0) \xrightarrow[\overline{g}]{\overline{f}} (\mathbb{R}^n, 0)$$

by

$$\overline{f} = h_V \circ f \circ h_U^{-1} \quad \text{and} \quad \overline{g} = h_V \circ g \circ h_U^{-1}.$$

Clearly we have $j_p^r(f) = j_p^r(g)$ if and only if

$$j_0^r(\overline{f}) = j_0^r(\overline{g}).$$

Now let $J_{p,q}^r(N, M)$ denote the set of all r-jets of mappings from N into M with source p and target q. Then define the set

$$J^r(N, M) = \bigcup_{p \in M, q \in N} J_{p,q}^r(N, M).$$

We introduce the natural projections

$$\alpha : J^r(N, M) \to N \quad \text{and} \quad \beta : J^r(N, M) \to N$$

defined by

$$\alpha(j_p^r(f)) = p \quad \text{and} \quad \beta(j_p^r(f)) = f(p).$$

Matters being so, if $\{U^\lambda\}$ denotes a coordinate covering of N and $\{V^\mu\}$ denotes a coordinate covering for M, then we define a topology on the set $J^r(N, M)$ by prescribing a coordinate covering to have underlying open sets

$$W^{\lambda\mu} = \{j_p^r(f) \mid \alpha(j_p^r(f)) \in U^\lambda \quad \text{and} \quad \beta(j_p^r(f)) \in V^\mu\}.$$

Now if $\{x^1, \ldots, x^m\}$ denote the coordinate functions on U^λ and $\{z^1, \ldots, z^n\}$ denote the coordinate functions on V^μ, then Proposition 3.1 implies that we may define a coordinate system on $W^{\lambda\mu}$ by

(18)
$$h(j_p^r(f)) = (x^i(p), z^j(f(p)), D_x^\alpha(z \circ f)(p)), \ 1 \le i \le m, \ 1 \le j \le n, \ 1 \le |\alpha| \le r.$$

We will call these coordinates *the natural coordinates* on the jet space.

The Leibniz product formula together with the chain rule guarantee that a differentiable change of local coordinates in U^λ and in V^μ will induce a differentiable change of local coordinates in $J^r(N, M)$. The fact that this change of local coordinates has non-zero Jacobian determinant follows from the fact that the matrix is block upper triangular with the diagonal blocks given by symmetric powers of the Jacobian of the original coordinate change. Thus we have defined a differentiable structure on $J^r(N, M)$.

The next natural question is to determine the dimension of $J^r(M, N)$ in terms of the dimensions of M and N.

A real valued function on an m-dimensional manifold N

$$f : N \to \mathbb{R}$$

has as many derivatives of order N as there are independent homogeneous polynomials of degree N. This number is

$$\binom{m + N - 1}{m - 1} = \binom{m + N - 1}{N}.$$

The total dimension of $J^r(N, \mathbb{R})$ is thus given by

$$m + 1 + \sum_{i=1}^{r} \binom{m + i - 1}{i} = m + \sum_{i=0}^{r} \binom{m + i - 1}{i} = m + \binom{m + r}{r}.$$

Now each coordinate function in a target space M will give rise to an independent set of derivatives, thus

$$\dim J^r(M, N) = \dim M + \dim N \binom{\dim M + r}{r}.$$

Example. $J^2(\mathbb{R}^2, \mathbb{R})$
Let (x, y) denote coordinates on $\mathbb{R}^2$ and z a coordinate on $\mathbb{R}$. Let

$$p(j_{(x,y)}^2(f)) = \partial f / \partial x \quad \text{and} \quad q(j_{(x,y)}^2(f)) = \partial f / \partial y$$

and

$$r(j_{(x,y)}^2(f)) = \partial^2 f / \partial x^2, \quad s(j_{(x,y)}^2(f)) = \partial^2 f / \partial x \partial y,$$

$$t(j^2_{(x,y)}(f)) = \partial^2 f/\partial y^2.$$

Then

$$h(j^2_{(x,y)}(f)) = (x, y, z, p, q, r, s, t)$$

defines the natural coordinates for $J^1(\mathbb{R}^2, \mathbb{R})$.

If we introduce a change of coordinate on $\mathbb{R}^2$ by

$$S(x, y) = (\xi(x, y), \eta(x, y)),$$

then this induces a transformation of the derivatives. In fact

$$\begin{pmatrix} r \\ s \\ t \\ p \\ q \end{pmatrix} = \left(\begin{array}{c|c} S^2(J(S)) & H(S) \\ \hline 0 & J(S) \end{array} \right) \begin{pmatrix} r' \\ s' \\ t' \\ p' \\ q' \end{pmatrix},$$

where

$$J(S) = \begin{pmatrix} \partial\xi/\partial x & \partial\eta/\partial x \\ \partial\xi/\partial y & \partial\eta/\partial y \end{pmatrix} \qquad H(S) = \begin{pmatrix} \partial^2\xi/\partial x^2 & \partial^2\eta/\partial x^2 \\ \partial^2\xi/\partial x\partial y & \partial^2\eta/\partial x\partial y \\ \partial^2\xi/\partial y^2 & \partial^2\eta/\partial y^2 \end{pmatrix}$$

and

$$S^2(J(S))(S) = \begin{pmatrix} (\partial\xi/\partial x)^2 & 2\partial\xi/\partial x\,\partial\eta/\partial x & (\partial\eta/\partial x^2) \\ \partial\xi/\partial x\,\partial\xi/\partial y & \partial\xi/\partial x\,\partial\eta/\partial y + \partial\xi/\partial y\,\partial\eta/\partial x & \partial\eta/\partial x\,\partial\eta/\partial y \\ (\partial\xi/\partial y)^2 & 2\partial\xi/\partial y\,\partial\eta/\partial y & (\partial\eta/\partial y)^2 \end{pmatrix}.$$

A good viewpoint to keep in mind is that

$$J^r(N, M) \to N \times M,$$

that is, $J^r(N, M)$ sits over $N \times M$, and the coordinate transformations on $N \times M$ induce the action of a linear group on the set of elements in the inverse image of a point.

The notion of jet bundles allows us to formulate general problems in differential geometry. As an example we observe that a partial differential equation for maps

$$f : N \to M$$

can be described by an imbedded submanifold

$$i : \Sigma \to J^r(N, M).$$

A solution is a map $f : N \to M$ such that for all $p \in N$

$$j_p^r(f) \in i(\Sigma).$$

We introduce the *r-graph* of a map f

$$j^r(f) : N \to J^r(N, M)$$

by the definition

$$j^r(f)(p) = j_p^r(f).$$

Then the problem of finding solutions to a partial differential equation is the problem of finding maps whose r-graphs lie on the locus $i(\Sigma)$ of the partial differential equation. This will be illustrated in Chapters IX and X.

Several standard constructions of differential geometry fit into the language of jet bundles. For example the cotangent space $T_p^*(N)$ for $p \in N$ is defined by

$$T_p^*(M) = J_{p,0}^1(N, \mathbb{R})$$

and the differential of a real valued function $f : N \to \mathbb{R}$ at $p \in M$ is defined by

$$df|_p = j_p^1(f - f(p)).$$

The vector space structure on $T_p^*(N)$ is intrinsically induced from the real line by

$$\alpha j_p^1(f) + \beta j_p^1(g) = j_p^1(\alpha f + \beta g).$$

The tangent space $T_p(N)$ for $p \in N$ is defined as the space of linear functionals on $T_p^*(N)$ and is realized by

$$T_p(N) = J_{0,p}^1(\mathbb{R}, N)$$

under the action

$$\langle j_0^1(u), j_p^1(f) \rangle = d/dt(f \circ u)|_{t=0}.$$

In particular the cotangent bundle is defined by

$$T^*(N) = \bigcup_{p \in M} T_p^*(N) \subset J^1(N, \mathbb{R})$$

and the tangent bundle is defined by

$$T(N) = \bigcup_{p \in m} T_p(N) \subset J^1(\mathbb{R}, N).$$

Finally we wish to introduce the contact system $\Omega^r(N, M)$ of a jet bundle $J^r(N, M)$. By a change of notation we can write the natural coordinates in (18) as

$$
\begin{aligned}
&x^i(p), z^\alpha(f(p)), p_i^\alpha, p_{i_1 i_2}^\alpha, \ldots, p_{i_1 \ldots i_r}^\alpha \\
&1 \leq i, i_1, \ldots, i_r \leq m, \quad 1 \leq \alpha \leq n,
\end{aligned}
$$
(19)

where the p's are the partial derivatives with respect to the x^i's, up to the order r inclusive, and are symmetric in their lower indices. The Pfaffian equations

$$
\begin{aligned}
dz^\alpha - \sum p_i^\alpha dx^i &= 0, \\
dp_{i_1}^\alpha - \sum p_{i_1 i_2}^\alpha dx^{i_2} &= 0, \\
dp_{i_1 \ldots i_{r-1}}^\alpha - \sum p_{i_1 \ldots i_{r-1} i_r}^\alpha dx^{i_r} &= 0,
\end{aligned}
$$
(20)

define the *contact system* $\Omega^r(N, M)$. A form in $\Omega^r(N, M)$ is called a *contact form*.

The forms (20) are those that naturally arise when a system of partial differential equations is converted to an exterior differential system. The fundamental property of these systems is contained in the following theorem.

Theorem 3.2. *A section* $\sigma : N \to J^r(N, M)$ *is a* r-graph, *that is* $\sigma(p) = j_p^r(f)$, *if and only if*

$$
\sigma^* \Omega^r(N, M) = 0.
$$

A proof of this theorem and an intrinsic treatment of the contact system can be found in various sources, cf. Gardner and Shadwick [1987] or Goldschmidt and Sternberg [1973].

CHAPTER II

BASIC THEOREMS

In this chapter we consider classical results on simple exterior differential systems that can be established by algebra and ordinary differential equations. In particular these results hold in the C^∞-category. This is to be contrasted with later results that rely on the Cartan–Kähler theorem and hold in the analytic category.

§1. Frobenius Theorem.

Perhaps the simplest exterior differential systems are those whose differential ideal $\mathcal{I}$ is generated algebraically by forms of degree one. Let the generators be

$$\alpha^1, \ldots, \alpha^{n-r},$$

which we suppose to be linearly independent. The condition that $\mathcal{I}$ is closed gives

$$(F) \qquad d\alpha^i \equiv 0, \quad \mod \alpha^1, \ldots, \alpha^{n-r}, \quad 1 \le i \le n - r.$$

This condition (F) is called the *Frobenius condition*. A differential system

$$\alpha^1 = \cdots = \alpha^{n-r} = 0$$

satisfying (F) is called *completely integrable.*

Geometrically the α's span at every point $x \in M$ a subspace W_x of dimension $n - r$ in the cotangent space $T_x^*(M)$ or, what is the same, a subspace $W_x^\perp$ of dimension r in the tangent space T_x. Following Chevalley, such data is known as a *distribution*. Notice that the condition (F) is intrinsic, i.e., independent of local coordinates, and is also invariant under a linear change of the α's with C^∞-coefficients.

The fundamental theorem on completely integrable systems is:

Theorem 1.1 (Frobenius). *Let $\mathcal{I}$ be a differential ideal having as generators the linearly independent forms $\alpha^1, \ldots, \alpha^{n-r}$ of degree one, so that the condition (F) is satisfied. In a sufficiently small neighborhood there is a coordinate system $y^1, \ldots, y^n$ such that I is generated by $dy^{r+1}, \ldots, dy^n$.*

Proof. We will prove the theorem by induction on r. Let $r = 1$. Then the subspace $W_x^\perp \subset T_x$, $x \in M$, is of dimension 1. Relative to a system of local coordinates x^i, $1 \le i \le n$, the equations of the differential system is classically written in the form

$$\frac{dx^1}{X^1(x)} = \cdots = \frac{dx^n}{X^n(x)},$$

where the functions $X^i(x^1, \ldots, x^n)$, not all zero, are the coefficients of a vector field $X = \sum_i X^i(x)\partial/\partial x^i$ spanning $W_x^\perp$. By *the flow box coordinate theorem* (Warner [1971], p. 40), we can choose coordinates $y^1, \ldots, y^n$, such that $W_x^\perp$ is spanned by the vector $\partial/\partial y^1$; then W_x is spanned by $dy^2, \ldots, dy^n$. The latter clearly form a set of generators of $\mathcal{I}$. Notice that in this case the condition (F) is void.

Suppose $r \ge 2$ and the theorem be true for $r - 1$. Let x^i, $1 \le i \le n$, be local coordinates such that

$$\alpha^1, \ldots, \alpha^{n-r}, dx^r$$

are linearly independent. The differential system defined by these $n - r + 1$ forms also satisfies the condition (F). By the induction hypothesis there are coordinates y^i so that

$$dy^r, dy^{r+1}, \ldots, dy^n$$

are a set of generators of the corresponding differential ideal. It follows that dx^r is a linear combination of these forms or that x^r is a function of $y^r, \ldots, y^n$. Without loss of generality we suppose

$$\partial x^r/\partial y^r \ne 0.$$

Since

$$dx^r = \frac{\partial x^r}{\partial y^r}dy^r + \sum_i \frac{\partial x^r}{\partial y^{r+1}}dy^{r+i}, \quad 1 \le i \le n - r$$

we may now solve for dy^r in terms of dx^r and $dy^{r+1}, \ldots, dy^n$. Since $\alpha^1, \ldots, \alpha^{n-r}$ are linear combinations of $dy^r, \ldots, dy^n$ they can now be expressed in the form

$$\alpha^i = \sum_j a_j^i dy^{r+j} + b^i dx^r, \quad 1 \le i, j \le n - r.$$

Since α^i and dx^r are linearly independent, the matrix (a_j^i) must be nonsingular. Hence we can find a new set of generators for I in the form

$$\alpha'^i = dy^{r+i} + p^i dx^r, \quad 1 \le i \le n - r,$$

and the condition (F) remains satisfied. Exterior differentiation gives

$$d\alpha'^i = dp^i \wedge dx^r \equiv \sum_{1 \leq \lambda \leq r-1} \frac{\partial p^i}{\partial y^\lambda} dy^\lambda \wedge dx^r \equiv 0, \quad \mathrm{mod}\ \alpha'^1, \ldots, \alpha'^{n-r}.$$

It follows that

$$\partial p^i / \partial y^\lambda = 0, \quad 1 \leq i \leq n - r, \quad 1 \leq \lambda \leq r - 1,$$

which means that p^i are functions of $y^r, \ldots, y^n$. Hence in the y-coordinates we are studying a system of $n - r$ forms of degree one involving only the $n - r + 1$ coordinates $y^r, \ldots, y^n$. This reduces to the situation settled at the beginning of this proof. Hence the induction is complete. $\qquad\square$

The theorem gives a "normal form" of a completely integrable system, i.e., the system can be written locally as

$$dy^{r+1} = \cdots = dy^n = 0$$

in a suitable coordinate system. The maximal integral manifolds are

$$y^{r+1} = \mathrm{const}, \ldots, y^n = \mathrm{const},$$

and are therefore of dimension r. We say that the system defines a *foliation*, of dimension r and codimension $n - r$, of which these submanifolds are the *leaves*.

The simplest non-trivial case of the Frobenius theorem is the system generated by a single one form in three space. Thus

$$\mathcal{I} = \{Rdx + Sdy + Tdz\}$$

and the condition (F) are the necessary and sufficient conditions that there exist an integrating factor for the one form $\omega = Rdx + Sdy + Tdz$. That is there exists a function μ such that $\mu\omega$ is exact. This example will be considered when condition (F) is not identically satisfied in Example 5.11 of Chapter IV.

The condition (F) has a formulation in terms of vector fields, which is also useful. We add to $\alpha^1, \ldots, \alpha^{n-r}$ the r forms $\alpha^{n-r+1}, \ldots, \alpha^n$, so that α^i, $1 \leq i \leq n$, are linearly independent. Then we have

$$(1) \qquad d\alpha^i = 1/2 \sum_{j,k} c^i_{jk} \alpha^j \wedge \alpha^k, \quad 1 \leq i,j,k \leq n, \quad c^i_{jk} + c^i_{kj} = 0.$$

The condition (F) can be expressed as

$$(2) \qquad c^a_{pq} = 0, \quad 1 \leq a \leq n - r, \quad n - r + 1 \leq p, q \leq n.$$

Let f be a smooth function. The equation

$$(3) \qquad\qquad df = \sum (X_i f)\alpha^i$$

defines n operators or vector fields X_i, which form a dual base to α^i. Exterior differentiation of (3) gives

$$1/2 \sum_{i,j} (X_i(X_j(f)) - X_j(X_i(f)))\alpha^i \wedge \alpha^j + \sum_j X_i(f)d\alpha^i = 0.$$

Substituting (1) into this equation, we get

$$(4) \qquad\qquad [X_i, X_j]f = (X_i X_j - X_j X_i)f = -\sum c_{ij}^k X_k f.$$

It follows that the condition (2) can be written

$$(5) \qquad\qquad [X_p, X_q]f = -\sum c_{pq}^s X_s f, \quad n - r + 1 \le p, q, s \le n.$$

Equation (4) is the dual version of (1). The vectors $X_{n-r+1}, \ldots, X_n$ span at each point $x \in M$ the subspace $W_x^\perp$ of the distribution. Hence the condition (F) or (2) or (5) can be expressed as follows:

Proposition 1.2. *Let a distribution M be defined by the subspace $W_x^\perp \subset T_x$, $\dim W_x^\perp = r$. The condition (F) says that, for any two vector fields X, Y, such that $X_x, Y_x \in W_x^\perp$, their bracket $[X, Y]_x \in W_x^\perp$.*

§2. Cauchy Characteristics.

The Frobenius Theorem shows that a completely integrable system takes a very simple form upon a proper choice of the local coordinates. Given any exterior differential system, one can ask the question whether there is a coordinate system such that the system is generated by forms in a smaller number of these coordinates. This question is answered by the Cauchy characteristics. Its algebraic basis is the retraction theorem (Theorem 1.3 of Chapter I).

Let $\mathcal{I}$ be a differential ideal. A vector field ξ such that $\xi \lrcorner \mathcal{I} \subset \mathcal{I}$ is called a *Cauchy characteristic vector field of* $\mathcal{I}$. At a point $x \in M$ we define

$$A(\mathcal{I})_x = \{\xi_x \in T_x M \mid \xi_x \lrcorner \mathcal{I}_x \subset \mathcal{I}_x\}$$

and $C(\mathcal{I})_x = A(\mathcal{I})_x^\perp \subset T_x^* M$. These concepts reduce to the ones treated in §1, Chapter I. In particular, we will call $C(\mathcal{I})_x$ the retracting space at x and call $\dim C(\mathcal{I})_x$ *the class of* $\mathcal{I}$ at x. We have now a family of ideals $\mathcal{I}_x$ depending on the parameter $x \in M$. When restricting to a point x we have a purely algebraic situation.

Proposition 2.1. *If ξ, η are Cauchy characteristic vector fields of a differential ideal $\mathcal{I}$, so is their bracket $[\xi, \eta]$.*

Proof. Let L_ξ be the Lie derivative defined by ξ. It is well-known

$$L_\xi = d(\xi \lrcorner) + (\xi \lrcorner)d.$$

Since $\mathcal{I}$ is closed, we have $d\mathcal{I} \subset \mathcal{I}$. If ξ is a characteristic vector field, we have $\xi \lrcorner \mathcal{I} \subset \mathcal{I}$. It follows that $L_\xi \mathcal{I} \subset \mathcal{I}$. The lemma follows from the identity

$$(6) \qquad [L_\xi, \eta \lrcorner] \underset{\text{def}}{=} L_\xi \eta \lrcorner - \eta \lrcorner L_\xi = [\xi, \eta] \lrcorner,$$

which is valid for any two vector fields ξ, η.

To prove (6) we observe that L_ξ is a derivation of degree 0 and $\eta \lrcorner$ is a derivation of degree -1, so that $[L_\xi, \eta \lrcorner]$ is also a derivation of degree -1. It therefore suffices to verify (6) when the two sides act on functions f and differentials df. Clearly, when acting on f, both sides give zero. When acting on df, we have

$$[L_\xi, \eta \lrcorner]df = L_\xi(\eta f) - \eta \lrcorner d(\xi f)$$

$$= [\xi, \eta]f = [\xi, \eta] \lrcorner df.$$

This proves (6) and hence the proposition.   □

Theorem 2.2. *Let $\mathcal{I}$ be a finitely generated differential ideal whose retracting space $C(\mathcal{I})$ has constant dimension $s = n - r$. Then there is a neighborhood in which there are coordinates $(x^1, \ldots, x^r;\ y^1, \ldots, y^s)$ such that $\mathcal{I}$ has a set of generators that are forms in $y^1, \ldots, y^s$ and their differentials.*

Proof. By Proposition 1.2 the differential system defined by $C(\mathcal{I})$ (or what is the same, the distribution defined by $A(\mathcal{I})$) is completely integrable. We may choose coordinates $(x^1, \ldots, x^r;\ y^1, \ldots, y^s)$ so that the foliation so defined is given by

$$y^\sigma = \text{const}, \quad 1 \le \sigma \le s.$$

By the retraction theorem, $\mathcal{I}$ has a set of generators which are forms in dy^σ, $1 \le \sigma \le s$. But their coefficients may involve x^ρ, $1 \le \rho \le r$. The theorem follows when we show that we can choose a new set of generators for $\mathcal{I}$ which are forms in the y^σ coordinates in which the x^ρ do not enter. To exclude the trivial case we suppose the $\mathcal{I}$ is a proper ideal, so that it contains no non-zero functions.

Let $\mathcal{I}_q$ be the set of q-forms in $\mathcal{I}$, $q = 1, 2, \ldots$. Let $\varphi^1, \ldots, \varphi^p$ be linearly independent 1-forms in $\mathcal{I}_1$ such that any form in $\mathcal{I}_1$ is their linear combination. Since $\mathcal{I}$ is closed, $d\varphi^i \in \mathcal{I}$, $1 \leq i \leq p$. For a fixed p we have $\frac{\partial}{\partial x^\rho} \in A(\mathcal{I})$, which implies

$$\frac{\partial}{\partial x^\rho} \lrcorner\, d\varphi^i = L_{\partial/\partial x^\rho}\varphi^i \in \mathcal{I}_1,$$

since the left-hand side is of degree 1. It follows that

$$(7) \qquad \frac{\partial \varphi^i}{\partial x^\rho} = L_{\partial/\partial x^\rho}\varphi^i = \sum_j a^i_j \varphi^j, \quad 1 \leq i, j \leq p$$

where the left-hand side stands for the form obtained from φ^i by taking the partial derivatives of the coefficients with respect to x^ρ.

For this fixed ρ we regard x^ρ as the variable and $x^1, \ldots, x^{\rho-1}, x^{\rho+1}, \ldots, x^r, y^1, \ldots, y^s$ as parameters. Consider the system of ordinary differential equations

$$(8) \qquad \frac{dz^i}{dx^\rho} = \sum_j a^i_j z^j, \quad 1 \leq i, j \leq p.$$

Let $z^i_{(k)}$, $1 \leq k \leq p$, be a fundamental system of solutions, so that

$$\det(z^i_{(k)}) \neq 0.$$

We shall replace φ^i by the $\tilde{\varphi}^k$ defined by

$$(9) \qquad \varphi^i = \sum z^i_{(k)} \tilde{\varphi}^k.$$

By differentiating (9) with respect to x^ρ and using (7), (8), we get

$$\frac{\partial \tilde{\varphi}^k}{\partial x^\rho} = 0,$$

so that $\tilde{\varphi}^k$ does not involve x^ρ. Applying the same process to the other x's, we arrive at a set of generators of $\mathcal{I}_1$ which are forms in y^σ.

Suppose this process carried out for $\mathcal{I}_1, \ldots, \mathcal{I}_{q-1}$, so that they consist of forms in y^σ. Let $\mathcal{J}_{q-1}$ be the ideal generated by $\mathcal{I}_1, \ldots, \mathcal{I}_{q-1}$. Let $\psi^\alpha \in \mathcal{I}_q$, $1 \leq \alpha \leq r$, be linearly independent mod $\mathcal{J}_{q-1}$, such that any q-form of $\mathcal{I}_q$ is congruent mod $\mathcal{J}_{q-1}$ to a linear combination of them. By the above argument such forms include

$$\frac{\partial}{\partial x^\rho} \lrcorner\, d\psi^\alpha = L_{\partial/\partial x^\rho}\psi^\alpha.$$

Hence we have

$$\frac{\partial \psi^\alpha}{\partial x^\rho} \equiv \sum b^\alpha_\beta \psi^\beta, \qquad \mod \mathcal{J}_{q-1}, \quad 1 \le \alpha, \beta \le r.$$

By using the above argument, we can replace the ψ^α by $\tilde{\psi}^\beta$ such that

$$\frac{\partial \tilde{\psi}^\alpha}{\partial x^\rho} \in \mathcal{J}_{q-1}.$$

This means that we can write

$$\frac{\partial \tilde{\psi}^\alpha}{\partial x^\rho} = \sum_h \eta^\alpha_h \wedge \omega^\alpha_h,$$

where $\eta^\alpha_h \in \mathcal{I}_1 \cup \cdots \cup \mathcal{I}_{q-1}$ and are therefore forms in y^σ. Let θ^α_h be defined by

$$\frac{\partial \theta^\alpha_h}{\partial x^\rho} = \omega^\alpha_h.$$

Then the forms

$$\overset{\approx}{\psi}{}^\alpha = \tilde{\psi}^\alpha - \sum_h \eta^\alpha_h \wedge \theta^\alpha_h$$

do not involve x^ρ, and can be used to replace ψ^α. Applying this process to all x^ρ, $1 \le \rho \le r$, we find a set of generators for $\mathcal{I}_q$, which are forms in y^σ only. $\qquad \square$

Definition. The leaves defined by the distribution $A(\mathcal{I})$ are called the *Cauchy characteristics.*

Notice that generally r is zero, so that a differential system generally does not have Cauchy characteristics (i.e., they are points). The above theorem allows us to locally reduce a differential ideal to a system in which there are no extraneous variables in the sense that all coordinates are needed to express $\mathcal{I}$ in any coordinate system. Thus the class of $\mathcal{I}$ equals the minimal number of variables needed to describe the system.

An often useful corollary of Theorem 2.2 which illustrates its geometric content is the following:

Corollary 2.3. *Let $f : M \to M'$ be a fibration with vertical distribution $V \subset T(M)$ with connected fibers over $x \in M'$ given by $(\ker f_*)_x$. Then a form α on M is the pull-back $f^*\alpha'$ of a form α' on M' if and only if*

$$v \lrcorner \alpha = 0 \quad \text{and} \quad v \lrcorner d\alpha = 0 \quad \text{for all} \quad v \in V.$$

Proof. By the submersion theorem (Warner [1971], p. 31), there are local coordinates such that

$$f(x^1, \ldots, x^p, x^{p+1}, \ldots, x^N) = (x^1, \ldots, x^p).$$

As such

$$V = \left(\frac{\partial}{\partial x^{p+1}}, \cdots, \frac{\partial}{\partial x^N} \right).$$

Now setting $\mathcal{I} = (\alpha)$, we see that $V \subset A(\mathcal{I})$. Therefore, by Theorem 2.2 there exists a generator for $\mathcal{I}$ independent of $(x^{p+1}, \ldots, x^N)$, and hence of the form $f^*\alpha''$ with $\alpha'' \in M'$. Thus there is a function μ such that

$$\mu\alpha = f^*\alpha''.$$

Since

$$0 = v \lrcorner (d\mu \wedge \alpha'' + \mu d\alpha'') = v(\mu)\alpha'' \quad \text{for all} \quad v \in V,$$

we see that μ is independent of $(x^{p+1}, \ldots, x^N)$ and hence $\mu = \lambda \circ f$ for some function λ defined on M'. Setting $\alpha' = \frac{1}{\lambda}\alpha''$ we have our result that $\alpha = f^*(\alpha')$. $\qquad\qquad\square$

We will apply this theorem to the first order partial differential equation

$$F(x^i, z, \partial z/\partial x^i) = 0, \quad 1 \le i \le n.$$

Following the example starting with §2 equation (14) of Chapter I, the equation can be formulated as the differential system (15), §2, Chapter I. To these equations we add their exterior derivatives to obtain

$$
\begin{aligned}
&F(x^i, z, p_i) = 0 \\[1mm]
&dz - \sum p_i dx^i = 0 \\[1mm]
&\sum (F_{x^i} + F_z p_i) dx^i + \sum F_{p_i} dp_i = 0 \\[1mm]
&\sum dx^i \wedge dp_i = 0.
\end{aligned}
$$

(10)

These equations are in the $(2n + 1)$-dimensional space (x^i, z, p_i). The corresponding differential ideal is generated by the left-hand members of (10).

To determine the space $A(\mathcal{I})$ consider the vector

$$\xi = \sum u^i \partial/\partial x^i + u\partial/\partial z + \sum v_i \partial/\partial p_i$$

and express the condition that the interior product $\xi \lrcorner$ keeps the ideal $\mathcal{I}$ stable. This gives

$$
\begin{aligned}
&u - \sum p_i u^i = 0, \\[1mm]
&\sum (F_{x^i} + F_z p_i) u^i + F_{p_i} v_i = 0, \\[1mm]
&\sum (u^i dp_i - v_i dx^i) = 0.
\end{aligned}
$$

(11)

Comparing the last equation of (11) with the third equation (10) we get

$$(12) \qquad u^i = \lambda F_{p_i}, \quad v_i = -\lambda(F_{x^i} + F_z p_i),$$

and the first equation of (11) then gives

$$(13) \qquad u = \lambda \sum p_i F_{p_i}.$$

The parameter λ being arbitrary, equations (12) and (13) show that $\dim A(\mathcal{I}) = 1$, i.e., the characteristic vectors at each point form a one-dimensional space. This fundamental (and remarkable) fact is the key to the theory of partial differential equations of the first order. The *characteristic curves* in the space (x^i, z, p_i), or *characteristic strips* in the classical terminology, are the integral curves of the differential system

$$(14) \qquad \frac{dx^i}{F_{p_i}} = -\frac{dp_i}{F_{x^i} + F_z p_i} = \frac{dz}{\sum p_i F_{p_i}}.$$

These are the *equations of Charpit and Lagrange*. To construct an integral manifold of dimension n it suffices to take an $(n-1)$-dimensional integral manifold transverse to the Cauchy characteristic vector field (or non-characteristic data in the classical terminology) and draw the characteristic strips through its points. Putting it in another way, an n-dimensional integral manifold is generated by characteristic strips.

We remark that points in (x^i, p_i)-space may be thought of as hyperplanes $\sum p_i dx^i = 0$ in the tangent spaces $T_x(\mathbb{R}^n)$. A curve in (x^i, z, p_i)-space projects to a curve in (x^i, p_i)-space, which is geometrically a 1-parameter family of tangent hyperplanes. This is the meaning of the terminology "strips".

Example. Consider the initial value problem for the partial differential equation

$$z\frac{\partial z}{\partial x} + \frac{\partial z}{\partial y} = 0$$

with initial data given along $y = 0$ by $z(x, 0) = \sqrt{x}$.

Let us introduce natural coordinates in $J^1(2, 1)$ by (x, y, z, p, q). This initial data $D : \mathbb{R} \to \mathbb{R}^2 \times \mathbb{R}$ where $D(x) = (x, 0, \sqrt{x})$ is extended to a map $\delta : \mathbb{R} \to J^2(2, 1)$ where the image satisfies the equation and the strip condition

$$0 = \delta^*(dz - p\,dx - q\,dy) = \frac{1}{2\sqrt{x}}dx - p\,dx$$

here $p = \frac{1}{2\sqrt{x}}$ and $q = 1 - zp = \frac{1}{2}$ and δ is unique. In general there are several choices of δ due to non-linearity of the equation. The extended data becomes

$$\delta(x) = (x, 0, x, \frac{1}{2\sqrt{x}}, 1/2).$$

If we parametrize the equation by $i : \Sigma \to J^1(2,1)$ where $i(x,y,z,p) = (x,y,z,p,1-zp)$, then the data can be pulled back to a map $\Delta : \mathbb{R} \to \Sigma$, where $\Delta(s) = (s,0,\sqrt{s}, \frac{1}{2\sqrt{s}})$.

The Cauchy characteristic vector field is

$$X = z\frac{\partial}{\partial x} + \frac{\partial}{\partial y} - p^2\frac{\partial}{\partial p}$$

and the corresponding flow is given by

$$\frac{dx}{dt} = z, \quad \frac{dy}{dt} = 1, \quad \frac{dz}{dt} = 1, \quad \frac{dp}{dt} = -p^2.$$

The solution for the given data representing the union of characteristic curves along the data is

$$x = \frac{t^2}{2} + (\sqrt{s})t + s, \quad y = t, \quad z = t + \sqrt{s},$$

and eliminating s and t gives an implicit equation $z(x,y)$ by

$$z^2 + zy = x - \frac{y^2}{2}.$$

Next we wish to apply the Cauchy characteristics to prove the following global theorem:

Theorem 2.4. *Consider the eikonal differential equation*

$$(15) \qquad\qquad \sum (\partial z/\partial x^i)^2 = 1 \quad 1 \le i \le n.$$

If $z = z(x^1,\ldots,x^n)$ is a solution valid for all $(x^1,\ldots,x^n) \in E^n$ ($= n$-dimensional euclidean space), then z is a linear function in x^i, i.e.,

$$z = \sum a_i x^i + b,$$

where a_i, b are constants satisfying $\sum a_i^2 = 1$.

Proof. We will denote by E^{n+1} the space of $(x^1,\ldots,x^n,z)$, and identify E^n with the hyperplane $z = 0$. The solution can be interpreted as a graph Γ in E^{n+1} having a one-one projection to E^n. For the equation (15) the denominators in the middle term of (14) are zero, so that the Cauchy characteristics satisfy

$$p_i = \text{const.}$$

The equations (14) can be integrated and the Cauchy characteristic curves, when projected to E^{n+1}, are the straight lines

$$(16) \qquad\qquad x^i = x_0^i + p_i t, \quad z = z_0 + t,$$

where $x_0^i, p_i z_0$ are constants. Hence the graph Γ must have the property that it is generated by the "Cauchy lines" (16), whose projections in E^n form a foliation of E^n.

Changing the notation in the first equation of (16), we write it as

$$x^{*^i} = x^i + \frac{\partial z}{\partial x^i} t,$$

where $z = z(x^1, \ldots, x^n)$ is a solution of (15). For a given $t \in R$ this can be interpreted as a diffeomorphism $f_t : E^n \to E^n$ defined by

$$f_t(x) = x^* = (x^{*^1}, \ldots, x^{*^n}), \quad x, x^* \in E^n.$$

Geometrically it maps $x \in E^n$ to the point x^* at a distance t along the Cauchy line through x; this makes sense, because the Cauchy lines are oriented. Its Jacobian determinant is

$$\mathcal{J}(t) = \det\left(\delta_j^i + \frac{\partial^2 z}{\partial x^i \partial x^j} t \right)$$

and is never zero. But this implies

$$(17) \qquad\qquad \frac{\partial^2 z}{\partial x^i \partial x^j} = 0,$$

and hence that z is linear. For if (17) is not true, then the symmetric matrix $(\partial^2 z / \partial x^i \partial x^j)$ has a real non-zero eigenvalue, say λ, and $\mathcal{J}(-1/\lambda) = 0$, which is a contradiction. $\qquad\square$

Remark. The function

$$z = \left(\sum_i (x^i)^2 \right)^{1/2}$$

satisfies (15), except at $x^i = 0$. Hence Theorem 2.3 needs the hypothesis that (15) is valid for all $x \in E^n$.

§3. Theorems of Pfaff and Darboux.

Another simple exterior differential system is one which consists of a single equation

$$(18) \qquad\qquad\qquad \alpha = 0,$$

where α is a form of degree 1. This problem was studied by Pfaff [1814-15]. The corresponding closed differential ideal $\mathcal{I}$ has the generators $\alpha, d\alpha$. The integer r defined by

$$(19) \qquad (d\alpha)^r \wedge \alpha \neq 0, \quad (d\alpha)^{r+1} \wedge \alpha = 0$$

is called *the rank* of the equation (18). It depends on the point $x \in M$, and is invariant under the change

$$\alpha \to a\alpha, \quad a \neq 0.$$

Putting it in a different way, the two-form $d\alpha \mod \alpha$, has an even rank $2r$ in the sense of linear algebra.

The study of the integral manifolds of (18) is clarified by the normal form, given by the

Theorem 3.1 (The Pfaff problem). *In a neighborhood suppose the equation (18) has constant rank r. Then there exists a coordinate system $w^1, \ldots, w^n$, possibly in a smaller neighborhood, such that the equation becomes*

$$(20) \qquad dw^1 + w^2 dw^3 + \cdots + w^{2r} dw^{2r+1} = 0.$$

Proof. Let $\mathcal{I} = \{\alpha, d\alpha\}$ be the ideal generated by $\alpha, d\alpha$. By Theorem 1.5 of Chapter I and (19), the retraction space $C(\mathcal{I})$ is of dimension $2r + 1$ and has the Grassmann coordinate vector $(d\alpha)^r \wedge \alpha$. By Theorem 2.2 there is a function f_1 such that

$$(d\alpha)^r \wedge \alpha \wedge df_1 = 0.$$

Next let $\mathcal{I}_1$ be the ideal $\{df_1, \alpha, d\alpha\}$. If $r = 0$, our theorem follows from the Frobenius theorem. If $r > 0$, the forms df_1 and α must be linearly independent. Applying Theorem 1.5, Chapter I to $\mathcal{I}_1$, let r_1 be the smallest integer such that
$$(d\alpha)^{r_1+1} \wedge \alpha \wedge df_1 = 0.$$

Clearly $r_1 + 1 \leq r$. The equality sign must hold, as otherwise we get a contradiction to the first equation of (19), by Theorem 1.6, Chapter I. Applying Theorem 2.2 to $\mathcal{I}_1$, there is a function f_2 such that

$$(d\alpha)^{r-1} \wedge \alpha \wedge df_1 \wedge df_2 = 0.$$

Continuing this process, we find r functions $f_1, \ldots, f_r$ satisfying

$$d\alpha \wedge \alpha \wedge df_1 \wedge \cdots \wedge df_r = 0,$$
$$\alpha \wedge df_1 \wedge \cdots \wedge df_r \neq 0.$$

Finally, let $\mathcal{I}_r$ be the ideal $\{df_1, \ldots, df_r, \alpha, d\alpha\}$. Its retraction space $C(\mathcal{L}_r)$ is of dimension $r + 1$. There is a function f_{r+1} such that

$$\alpha \wedge df_1 \wedge \cdots \wedge df_{r+1} = 0,$$
$$df_1 \wedge \cdots \wedge df_{r+1} \neq 0.$$

By modifying α by a factor, we can write

$$\alpha = df_{r+1} + g_1 df_1 + \cdots + g_r df_r.$$

Because of the first equation of (19) the functions $f_1, \ldots, f_{r+1}, g_1, \ldots, g_r$ are independent. Theorem 3.1 follows by setting

$$w^1 = f_{r+1}, \quad w^{2i} = g_i, \quad w^{2i+1} = f_i, \quad 1 \leq i \leq r.$$

$\square$

Corollary 3.2 (Symmetric normal form). *In a neighborhood suppose the equation (18) has constant rank r. Then there exist independent functions $z, y^1, \ldots, y^r, x^1, \ldots, x^r$ such that the equation becomes*

$$(21) \qquad dz + 1/2 \sum_{i=1}^{r}(y^i dx^i - x^i dy^i) = 0.$$

Proof. It suffices to apply the change of coordinates

$$w^1 = z - \frac{1}{2}\sum x^i y^i,$$
$$w^{2i} = y^i, \quad w^{2i+1} = x^i, \quad 1 \leq i \leq r.$$

$\square$

From the normal form (20) we see that the maximal integral manifolds are of dimension r. They are, for instance, given by

$$w^1 = f(w^3, \ldots, w^{2s+1}), \quad s < r$$

$$w^{2t+1} = \text{const}, \quad w^{2t} \text{ arbitrary}, \quad s + 1 \leq t \leq r.$$

Related to the Pfaffian problem are normal forms for the forms themselves and not the ideals generated by them. For one-forms and closed two-forms we have the following theorems.

Theorem 3.3 (Darboux). *Let Ω be a closed two-form satisfying*

$$\Omega^r \neq 0, \quad \Omega^{r+1} = 0, \quad r = \text{const.}$$

Locally there exist coordinates $w^1, \ldots, w^n$ such that

$$(22) \qquad \Omega = dw^1 \wedge dw^2 + \cdots + dw^{2r-1} \wedge dw^{2r}.$$

Proof. We put $\Omega = d\alpha$, where α is a one-form. The argument in the proof of Theorem 3.1 applies, and we can suppose α to be a form in the $2r$ variables $y^1, \ldots, y^{2r}$. In $2r$ variables the Pfaffian equation $\alpha = 0$ must be of rank $\leq r - 1$, and is exactly equal to $r - 1$, because $\Omega^r \neq 0$. Hence we can set

$$\alpha = u(dz^1 + z^2 dz^3 + \cdots + z^{2r-2} dz^{2r-1}),$$

or, by a change of notation

$$\alpha = w^1 dw^2 + \cdots + w^{2r-1} dw^{2r}.$$

This gives the Ω in (22). Since $\Omega^r \neq 0$, the functions $w^1, \ldots, w^{2r}$ are independent and are a part of a local coordinate system. $\qquad \square$

Consider next the case of a one-form α. The *rank r* is defined by the conditions

$$\alpha \wedge (d\alpha)^r \neq 0, \quad \alpha \wedge (d\alpha)^{r+1} = 0.$$

There is a second integer s defined by

$$(d\alpha)^s \neq 0, \quad (d\alpha)^{s+1} = 0.$$

Elementary arguments show that there are two cases:

$$\begin{aligned}
&\text{(i)} \qquad s = r; \\
&\text{(ii)} \qquad s = r + 1.
\end{aligned}$$

Theorem 3.4. *Let α be a one-form. In a neighborhood suppose r and s be constant. Then α has the normal form*

$$(23a) \qquad \alpha = y^0 dy^1 + \cdots + y^{2r} dy^{2r+1}, \quad \text{if } r + 1 = s;$$

$$(23b) \qquad \alpha = dy^1 + y^2 dy^3 + \cdots + y^{2r} dy^{2r+1}, \quad \text{if } r = s.$$

In these expressions, the y's are independent functions and are therefore parts of a local coordinate system.

Proof. Let $\mathcal{I}$ be the differential ideal generated by α and $d\alpha$. By Theorem 3.1 there are coordinates $y^1, \ldots, y^n$ in a neighborhood such that

$$\alpha = u(dy^1 + y^2 dy^3 + \cdots + y^{2r} dy^{2r+1}).$$

A change of notation allows us to write

$$\alpha = z^0 dy^1 + z^2 dy^3 + \cdots + z^{2r} dy^{2r+1}.$$

Then

$$(d\alpha)^{r+1} = c\, dz^0 \wedge dy^1 \wedge dz^2 \wedge dy^3 \wedge \cdots \wedge dz^{2r} \wedge dy^{2r+1}, \quad c = \text{const. } c \neq 0.$$

If $s = r + 1$, this is $\neq 0$, and the functions $z^0, z^2, \ldots, z^{2r}, y^1, y^3, \ldots, y^{2r+1}$ are independent. This proves the normal form (23a).

Consider next the case $r = s$. Then $d\alpha$ is a two-form of rank $2r$. By Theorem 3.3 we can write

$$d\alpha = dw^1 \wedge dw^2 + \cdots + dw^{2r-1} \wedge dw^{2r}$$

$$= d(w^1 dw^2 + \cdots + w^{2r-1} dw^{2r}).$$

Hence the form

$$\alpha - (w^1 dw^2 + \cdots + w^{2r-1} dw^{2r})$$

is closed, and is equal to dv. A change of notation gives (23b). $\square$

Remark. A manifold of dimension $2r + 1$ provided with a one-form α, defined up to a factor, such that

$$\alpha \wedge (d\alpha)^r \neq 0,$$

is called a *contact manifold*. An example is the projectivized cotangent bundle of a manifold, whose points are the non-zero one-forms on the base manifold defined up to a factor. A manifold of dimension $2r$ provided with a closed two-form of maximum rank $2r$ is called a *symplectic manifold*. An example here is the cotangent bundle of a manifold. In terms of local coordinates $x^1, \ldots, x^r$ on an r-dimensional manifold M, points in the projectivized cotangent bundle $\mathbb{P}T^*M$ are non-zero 1-forms

$$\eta = \sum_{i=1}^{r} p_i dx^i,$$

where we identify η and $\lambda\eta$ for $\lambda \neq 0$. In a neighborhood in $\mathbb{P}T^*M$ in which, say, $p_1 \neq 0$ we may normalize by taking $p_1 = -1$. Then $(x^1, \ldots, x^r, p_2, \ldots, p_r)$ are local coordinates on $\mathbb{P}T^*M$ in terms of which

$$\eta = -dx^1 + \sum_{i=2}^{r} p_i dx^i.$$

If we normalize differently on change of local coordinates on M, η changes by a non-zero factor. It then defines the contact structure on $\mathbb{P}T^*M$.

The symplectic structure on T^*M is given locally by

$$\sum_i dp_i \wedge dx^i.$$

It is invariant under changes of coordinates on M. Both contact manifolds and symplectic manifolds play a fundamental role in theoretical mechanics and partial differential equations. Unlike Riemannian manifolds they have no local invariants.

Remark. Darboux's Theorem 3.3 has been generalized in several directions, in particular to Banach manifolds, by Weinstein [1971].

Finally, we wish to make an application of the normal form in Corollary 3.2 to prove a theorem of C. Caratheodory on local accessibility, which played a fundamental role in his "foundations of thermodynamics" (Caratheodory [1909]) and is now of equal importance in control theory. We say that the Pfaffian equation (18) has the *local accessibility property* if every point $x \in M$ has a neighborhood U such that every point $y \in U$ can be joined to x by an integral curve of (18). Then we have

Theorem 3.5 (Caratheodory). *Suppose the rank of the Pfaffian equation*

$$\alpha = 0$$

be constant. It has the local accessibility property if and only if

$$\alpha \wedge d\alpha \neq 0.$$

Proof. The condition is equivalent to saying that the rank r defined in (19) is ≥ 1.

Suppose $r = 0$. This means that the Frobenius condition is satisfied and the equation can locally be written

$$dz = 0.$$

Thus the integral curves are restricted to the leaves $z = $ const, and local accessibility is impossible.

For $r \geq 1$ we use the normal form (21), by supposing that the local coordinates be $z, x^1, y^1, \ldots, x^r, y^r, u^1, \ldots, u^s$ where $2r+s+1 = n = \dim M$. Let x be the origin and let y have the coordinates $(z_0, x_0^1, y_0^1, \ldots, x_0^r, y_0^r, u_0^1, \ldots, u_0^s)$. In the (x^i, y^i)-plane, $1 \leq i \leq r$, let C_i be the curve $(x^i(t), y^i(t))$, $0 \leq t \leq 1$, satisfying

$$x^i(0) = y^i(0) = 0, \quad x^i(1) = x_0^i, \quad y^i(1) = y_0^i.$$

Consider the function

$$z(t) = \frac{1}{2} \int_0^t \sum_{1 \le i \le r} \left(x^i \frac{dy^i}{dt} - y^i \frac{dx^i}{dt} \right) dt.$$

On the curves C_i we impose the further condition

$$z(1) = z_0,$$

which is clearly possible. Geometrically this means that z_0 is the sum of the areas bounded by the curves C_i and the chords joining their end-points. The curve γ in M defined by

$$\left(z(t), x^1(t), y^1(t), \ldots, x^r(t), y^r(t), tu_0^1, \ldots, tu_0^s \right), \quad 0 \le t \le 1,$$

is an integral curve of (21) and joints x to y. $\qquad\square$

Note that the accessibility is by smooth curves, is constructive with an infinite number of solutions, and is valid in the largest domain in which the normal form can be constructed. The theorem was extended by Chow [1940] to fintely generated Pfaffian systems with certain constant rank conditions.

§4. Pfaffian Systems.

A Pfaffian system is a differential system

$$(24) \qquad \alpha^1 = \cdots = \alpha^s = 0,$$

where the α's are one-forms. We suppose them to be linearly independent and $s = $ const. We will denote the Pfaffian system by I and call s its *dimension*. The first properties will be described by the two-forms

$$(25) \qquad d\alpha^i \mod (\alpha^1, \ldots, \alpha^s) \quad 1 \le i \le s.$$

The Frobenius condition is equivalent to saying that they are zero. We shall consider the general case and study their properties.

Geometrically the α's span at every point $x \in M$ a subspace W_x^* of dimension s in the cotangent space T_x^*, or equivalently, a subspace $W_x = (W_x^*)^\perp$ of dimension $n - s$ ($n = \dim M$) in the tangent space T_x. They form a subbundle of the tangent bundle. Already in the case of the Pfaffian problem ($s = 1$), we have shown that there is a local invariant given by the rank. In the general case the local properties could be very complicated. In this and the next sections we shall single out, after a general discussion, some of the simple cases and give some applications.

We can view $I \subset \Omega^1(M)$ as the sub-module over $C^\infty(M)$ of 1-forms

$$\alpha = \sum_i f_i \alpha^i$$

where the f_i are functions. We denote by $\{I\} \subset \Omega^*(M)$ the *algebraic* ideal generated by I. Thus $\beta \in \{I\}$ is of the form

$$\beta = \sum_i \gamma_i \wedge \alpha^i$$

where the γ_i are differential forms. The exterior derivative induces a mapping

$$\delta : I \to \Omega^2(M)/\{I\}$$

that is linear over $C^\infty(M)$. We set

$$I^{(1)} = \ker \delta$$

and call $I^{(1)}$ the *first derived system*. We thus have

$$0 \to I^{(1)} \to I \xrightarrow{\delta} dI/\{I\} \to 0,$$

and $I^{(1)} = I$ exactly in the Frobenius case. Now I is the space of C^∞ sections of a sub-bundle $W \subset T^*M$ with fibres $W_x = \mathrm{span}(\alpha^1(x), \ldots, \alpha^s(x))$. The images of

$$W \otimes \Lambda^q T^*M \to \Lambda^{q+1} T^*M$$

are sub-bundles $W^{q+1} \subset \Lambda^{q+1} T^*M$, and the mapping δ above is induced from a bundle mapping

$$W \xrightarrow{\bar\delta} \Lambda^2 T^*M/W^2.$$

We assume that $\bar\delta$ has constant rank, so that $I^{(1)}$ is the sections of a sub-bundle $W_1 \subset W \subset T^*M$.

Continuing with this construction we arrive at a filtration

$$(26) \qquad\qquad I^{(k)} \subset \cdots \subset I^{(2)} \subset I^{(1)} \subset I^{(0)} = I,$$

defined inductively by

$$I^{(k+1)} = (I^{(k)})_1.$$

We assume that the ranks of mappings $\bar\delta$ are all constant, so that the above filtration corresponds to a flag of bundles

$$W_k \subset \cdots \subset W_2 \subset W_1 \subset W.$$

There will then be a smallest integer N such that $W_{N+1} = W_N$, i.e.

$$I^{(N+1)} = I^{(N)}.$$

We call (26) the *derived flag* of I_0 and N the *derived length*. Note that $I^{(N)}$ is the largest integrable subsystem contained in I. We also define the integers

$$p_0 = \dim I^{(N)}$$
$$(27) \qquad p_{N-i} = \dim I^{(i)}/I^{(i+1)}, \quad 0 \le i \le N-1,$$
$$p_{N+1} = \dim C(I)/I.$$

These are called the *type numbers* of I. Our hypothesis says that they are all constants. The type numbers are not arbitrary; there are inequalities between them. Cf. Gardner [1967].

An integral manifold of I annihilates all the elements of its derived flag, and in particular those of $I^{(N)}$. A function g with differential $dg \in I^{(N)}$ is called a *first integral* of I, since it is constant on all integral manifolds of I.

There are two other integers, which can be defined for a Pfaffian system I. The *wedge length* or the *Engel half-rank* of I is the smallest integer ρ such that

$$(d\alpha)^{\rho+1} \equiv 0 \quad \mod \{I\} \quad \text{for all } \alpha \in I.$$

The *Cartan rank* of I is the smallest integer v such that there exist $\pi^1, \ldots, \pi^v$ in $\Omega^1(M)/I$ with

$$\pi^1 \wedge \cdots \wedge \pi^v \neq 0$$

and

$$d\alpha \wedge \pi^1 \wedge \cdots \wedge \pi^v \equiv 0 \quad \mod \{I\} \quad \text{for all } \alpha \in I.$$

We will suppose that both ρ and v are constants. The following theorems are simple properties concerning the wedge length and the Cartan rank:

Proposition 4.1. *Let I be a Pfaffian system and ρ its wedge length. Then all $(\rho+1)$-fold products of the elements in $dI \mod \{I\}$ are zero.*

Proof. If I is given by the equation (24), an element of the module I is

$$\alpha = t_1 \alpha^1 + \cdots + t_s \alpha^s,$$

where the t's are arbitrary smooth functions. The hypothesis implies

$$(t_1 d\alpha^1 + \cdots + t_s d\alpha^s)^{\rho+1} \equiv 0 \quad \mod \{I\},$$

where the t's can be considered as indeterminates. Expanding the left-hand side of this equation and equating to zero the coefficients of the resulting polynomial in the t's, we prove the proposition (Griffin [1933]).

Proposition 4.2. *Between the wedge length ρ and the Cartan rank v the following inequalities hold:*

$$(28) \qquad \rho \leq v \leq 2\rho.$$

Proof. The condition $(52b)$ can be written

$$d\alpha \equiv 0 \quad \mod \{I, \pi^1, \ldots, \pi^v\}.$$

Hence

$$(d\alpha)^{v+1} \equiv 0, \quad \mod \{I\},$$

so that $\rho \leq v$.

To prove the inequality at the right-hand side we notice that by the definition of ρ there exists $\eta \in I$ such that

$$(d\eta)^\rho \not\equiv 0, \quad \text{and} \quad (d\eta)^{\rho+1} \equiv 0 \quad \mod \{I\}.$$

By Theorem 3.3, $(d\eta)^\rho$ is a monomial of degree 2ρ. Moreover, by Proposition 4.1, we have

$$d\alpha \wedge (d\eta)^\rho \equiv 0 \quad \mod \{I\} \quad \text{for all } \alpha \in I.$$

It follows that $v \leq 2\rho$. $\qquad\qquad \square$

Remark. The bounds for v in (28) are sharp. The lower bound is achieved by a system consisting of a single equation. To achieve the upper bound consider in $R^{3\rho+3}$ with the coordinates $(x_{1k}, x_{2k}, x_{3k}, y^1, y^2, y^3)$, $1 \leq k \leq \rho$, the Pfaffian system

$$\alpha^1 = dy^1 + \sum_k x_{2k} dx_{3k}$$

$$\alpha^2 = dy^2 + \sum_k x_{3k} dx_{1k},$$

$$\alpha^3 = dy^3 + \sum_k x_{1k} dx_{2k}.$$

This system has $v = 2\rho$.

Proposition 4.3. *With our notations the following inequalities hold:*

$$(29) \qquad s + 2\rho \leq \dim C(I) \leq s + \rho + p_N \rho.$$

Proof. We remark that $C(I)$ is the retracting subspace of I. By the definition of ρ the left-hand side inequality is obvious.

To prove the inequality at the right-hand side we recall that by (27)

$$p_N = \dim I/I_1.$$

We choose a basis of I such that

$$(d\alpha^1)^\rho \not\equiv 0 \quad \mod \{I\}.$$

But the left-hand side is a monomial (Theorem 3.3), which we can write as

$$(d\alpha^1)^\rho = \beta^1 \wedge \cdots \wedge \beta^{2\rho} \neq 0 \quad \mod \{I\},$$

when the β's are one-forms. By Proposition 4.1 we have

$$(d\alpha^1)^\rho \wedge d\alpha^j \equiv 0 \quad \mod \{I\}, \quad 2 \leq j \leq p_N$$

or

$$d\alpha^j \in \text{ideal}\{\beta^1, \ldots, \beta^{2\rho}, I\}.$$

Now we can use the proof of Theorem 1.7 on the construction of the canonical form of a two form, by choosing sequentially divisors $\gamma_j^1, \ldots, \gamma_j^{p_j}$ of $d\omega^j$, $2 \leq j \leq p_N$ resulting in

$$\text{Class } I \leq s + 2\rho + p_{N-1}\rho = s + \rho + \rho_N \rho.$$

This proves the right hand side of (29). $\qquad\qquad\square$

Remark. The lower bound for $\dim C(I)$ is achieved by a system consisting of a single equation. To reach the upper bound consider the contact system

$$I = \{dz^\lambda - \sum p_i^\lambda dx^i\}, \quad 1 \leq i \leq m, \quad 1 \leq \lambda \leq n,$$

in the space $(x^i, z^\lambda, p_i^\lambda)$. For this system we have

$$I_1 = 0, \quad s = p_N = n, \quad \rho = v = m$$

and

$$\dim C(I) = mn + m + n.$$

These properties characterize the contact system, as given by the following theorem. For this theorem, and for the rest of this chapter, we shall let $\{\beta^1, \ldots, \beta^s\} \subset \Omega^1(M)$ be the sub-module of 1-forms $\beta - \sum f_i\beta^i$ generated by the set of 1-forms $\beta^1, \ldots, \beta^s$.

Theorem 4.4 (Bryant normal form). *Let $I = \{\alpha^1, \ldots, \alpha^s\}$ be a differential system with $I_1 = 0$. If*

$$\text{(30)} \qquad \dim C(I) = s + vs + v, \quad s \geq 3,$$

there is a local coordinate system containing the coordinates $x^i, z^\lambda, p_i^\lambda$, $1 \leq i \leq v$, $1 \leq \lambda \leq s$, such that

$$I = \{dz^\lambda - \sum p_i^\lambda dx^i\}.$$

Proof. By the definition of v there exist $\pi^1, \ldots, \pi^v$, such that

$$\pi^1 \wedge \cdots \wedge \pi^v \neq 0 \mod I,$$

$$d\alpha^\lambda \wedge \pi^1 \wedge \cdots \wedge \pi^v \equiv 0 \mod I.$$

The last relation can be written

$$d\alpha^\lambda \equiv \sum \eta_i^\lambda \wedge \pi^i \mod I.$$

The hypothesis (30) implies that the forms α^λ, π^i, η_i^λ are linearly independent. By exterior differentiation of the last relation we get

$$\sum \eta_i^\lambda \wedge d\pi^i \equiv 0 \mod \{I, \pi^1, \ldots, \pi^v\},$$

which implies

$$d\pi^i \equiv 0 \mod \{I, \pi^1, \ldots, \pi^v, \eta_i^\lambda\},$$

for every fixed λ. Since $s \geq 3$, this is possible only when

$$d\pi^i \equiv 0 \mod I, \pi^1, \ldots, \pi^v.$$

It follows that the system

$$J = \{\alpha^1, \ldots, \alpha^s, \pi^1, \ldots, \pi^v\}$$

is completely integrable, and we can write

$$J = \{d\xi^1, \ldots, d\xi^{s+v}\},$$

where the ξ's are the first integrals. Then we have

$$\alpha^\lambda = \sum b_A^\lambda d\xi^A, \quad 1 \leq A \leq s + v,$$

in which we can assume that the $(s \times s)$-minor at the left-hand side of the matrix (b_A^λ) is non-zero. Writing

$$\xi^\lambda = z^\lambda, \quad \xi^{s+i} = x^i, \quad 1 \le \lambda \le s, \quad 1 \le i \le v,$$

we can suppose

$$I = \{dz^\lambda - \sum p_i^\lambda dx^i\}.$$

Because of our hypothesis the functions x^i, z^λ, p_i^λ are independent. $\square$

Remark. The theorem is true for $s = 1$, in which case it reduces to the Pfaffian problem. It is not true for $s = 2$. An important counter-example is the following: Consider in R^5 a Pfaffian system

$$I = \{\alpha^1, \alpha^2\},$$

satisfying

$$d\alpha^1 \equiv \alpha^3 \wedge \alpha^4, \quad d\alpha^2 \equiv \alpha^3 \wedge \alpha^5, \qquad \text{mod } I,$$

where $\alpha^1, \ldots, \alpha^5$ are linearly independent one-forms. We have $I_1 = 0$ and

$$s = 2, \quad v = 1, \quad \dim C(I) = 5,$$

so that the hypotheses of Theorem 4.4 are satisfied. But this system has further local invariants; cf. the end of the next section.

Remark. The original Bryant normal form was a deeper theorem proved in his thesis (Bryant [1979]), which can be stated as follows:
The conclusion of Theorem 4.4 remains valid, if the condition (30) is re-placed by

$$\dim C(I) = s + \rho s + \rho.$$

The proof depends on an algebraic argument to show that $\rho = v$.

§5. Pfaffian Systems of Codimension Two.

We follow the notations of the last section and consider a Pfaffian system I defined by (24). If $s = n-1$, the system I is completely integrable. In fact, on the choice of an independent variable, it becomes a system of ordinary differential equations.

In this section we study the case $n = s + 2$. We will show that this case is already a rich subject and the diverse phenomena are present. The case $s - 3$ is the content of Cartan's [1910] paper and the general case has barely been touched. We will also make some applications to ordinary differential equations of the Monge type which have applications to control theory.

To the forms at the left-hand side of (24) we add the forms α^{n-1}, α^n, so that $\alpha^1, \ldots, \alpha^n$ are linearly independent. Then we have

$$d\alpha^i \equiv T^i \alpha^{n-1} \wedge \alpha^n \mod I, \quad 1 \leq i \leq s.$$

If $T^i = 0$, I is completely integrable, and $I^{(1)} = I$. We discard this case and suppose $(T^1, \ldots, T^s) \neq 0$. The α's are defined up to the non-singular linear transformation

$$\begin{pmatrix} \alpha^1 \\ \vdots \\ \alpha^n \end{pmatrix} \longrightarrow \begin{pmatrix} u_1^1 & \cdots & u_s^1 & 0 & 0 \\ \cdots & & & & \\ u_1^s & \cdots & u_s^s & 0 & 0 \\ u_1^{n-1} & \cdots & u_s^{n-1} & u_{n-1}^{n-1} & u_n^{n-1} \\ u_1^n & \cdots & u_s^n & u_{n-1}^n & u_n^n \end{pmatrix} \begin{pmatrix} \alpha^1 \\ \vdots \\ \alpha^n \end{pmatrix}.$$

By choosing the above matrix u properly, we can suppose

$$T^1 = \cdots = T^{s-1} = 0, \quad T^s = 1,$$

i.e.,

$$(31) \qquad d\alpha^1 \equiv \cdots \equiv d\alpha^{s-1} \equiv 0, \quad d\alpha^s \equiv \alpha^{n-1} \wedge \alpha^n, \qquad \mod I.$$

Under this choice $I^{(1)}$ is generated by $\alpha^1, \ldots, \alpha^{s-1}$, and we have $\dim I^{(1)} = s - 1$.

In the case $s = 2$, $n = 4$ we have the theorem:

Theorem 5.1 (Engel's normal form). *Let I be a Pfaffian system of two equations in four variables with derived flag satisfying*

$$\dim I^{(1)} = 1, \quad I^{(2)} = 0.$$

Then locally there are coordinates x, y, y', y'' such that

$$I = \{dy - y'dx, dy' - y''dx\}.$$

Proof. The derived system $I^{(1)}$ is generated by α^1. Since $I^{(2)} = 0$, we have

$$d\alpha^1 \wedge \alpha^1 \neq 0.$$

On the other hand, we have, for dimension reason,

$$(d\alpha^1)^2 \wedge \alpha^1 = 0.$$

By Theorem 3.1 we can therefore suppose

$$\{\alpha^1\} = \{dy - y'dx\}.$$

From (31) we have

$$da^1 \wedge \alpha^1 \wedge \alpha^2 = 0,$$

which gives

$$dx \wedge dy' \wedge \alpha^1 \wedge \alpha^2 = 0,$$

and consequently

$$\alpha^2 \equiv a\,dy' + b\,dx \quad \mathrm{mod}\ \alpha^1.$$

The coefficients a and b are not both zero. If $a \neq 0$, we write

$$\frac{1}{a}\alpha^2 \equiv dy' - y''dx \quad \mathrm{mod}\ \alpha^1.$$

Since

$$d\alpha^2 \wedge \alpha^1 \wedge \alpha^2 = \alpha^1 \wedge \alpha^2 \wedge \alpha^3 \wedge \alpha^4 \neq 0,$$

we have

$$dx \wedge dy \wedge dy' \wedge dy'' \neq 0,$$

so that x, y, y', y'' are independent functions and can serve as local coordinates.

Similarly, if $b \neq 0$, we obtain the form

$$\frac{1}{b}\alpha^2 = dx - y''dy'.$$

The two normal forms

$$I = \{dy - y'dx, dy' - y''dx\}$$

and

$$II = \{dy - y'dx, dx - y''dy'\}$$

are however equivalent since the coordinate change

$$(x, y, y', y'') \rightarrow (y', y - xy', -x, -y'')$$

takes the normal form I into the normal form II. $\square$

If a system is put into Engel normal form then the "general solution" is visibly given by

$$y = f(x), \quad y' = f'(x), \quad y'' = f''(x),$$

where $f(x)$ is an arbitrary function of x. Here general solution means a solution of the Pfaffian system with independence condition: (I, dx) so that $dx \neq 0$.

The Engel normal form is the key tool in the theory of *the Monge equation*

$$(32) \qquad F(x, y, z, y', z') = 0, \quad y' = \frac{dy}{dx}, \quad z' = \frac{dz}{dx},$$

which is an under-determined first order system of one equation for the two unknown functions y and z in the independent variable x.

The Pfaffian system equivalent to this problem is

$$I = \{dy - y'dx, dz - z'dx\}.$$

The manifold in question is the hypersurface (32) in the jet manifold $J^1(R, R^2)$, which is five-dimensional and has the coordinates x, y, z, y', z'. The equation $dF = 0$ gives, when expanded

$$F_{y'}dy' + F_{z'}dz' + F_y(dy - y'dx) + F_z(dz - z'dx) + \frac{dF}{dx}dx = 0,$$

where

$$\frac{dF}{dx} = F_x + F_y y' + F_z z'.$$

To achieve the equations (31), we suppose

$$F_{y'}^2 + F_{z'}^2 \neq 0,$$

and set

$$\alpha^1 = F_{y'}(dy - y'dx) + F_{z'}(dz - z'dx),$$

$$\alpha^2 = -F_{z'}(dy - y'dx) + F_{y'}(dz - z'dx)$$

Then

$$d\alpha^1 \equiv 0 \qquad \mod I$$

$$d\alpha^2 \equiv (F_{z'}dy' - F_{y'}dz') \wedge dx \not\equiv 0 \qquad \mod I.$$

Hence the conditions of Theorem 5.1 are satisfied and we have the following corollary:

Corollary 5.2. *If the Monge equation* (32) *satisfies the condition*

$$F_{y'}^2 + F_{z'}^2 \neq 0,$$

it has a general solution depending upon an arbitrary function in one variable and its first two derivatives.

Example.

$$y'^2 + z'^2 = 1.$$

This can be interpreted either as the equation for unit speed curves in the plane or as null curves in the Lorentzian 3-space with metric $dx^2 - dy^2 - dz^2$.

The equation can be parametrized by

$$y' = \sin\varphi \qquad z' = \cos\varphi$$

and leads to the differential system

$$I = \begin{cases} dy - \sin\varphi\, dx \\ dz - \cos\varphi\, dx. \end{cases}$$

The first derived system is given by

$$I^{(1)} = \{dx - \sin\varphi\, dy - \cos\varphi\, dz\}$$
$$= \{d(x - \sin\varphi\, y - \cos\varphi\, z) + (\cos\varphi\, y - \sin\varphi\, z)d\varphi\}.$$

Following the general theory we set

$$x - \sin\varphi\, y - \cos\varphi\, z = f(\varphi)$$
$$-\cos\varphi\, y + \sin\varphi\, z = f'(\varphi)$$
$$\sin\varphi\, y + \cos\varphi\, z = f''(\varphi)$$

and solve for x, y, z to find

$$x = f''(\varphi) + f(\varphi)$$
$$y = \sin\varphi f''(\varphi) - \cos\varphi f'(\varphi)$$
$$z = \cos\varphi f''(\varphi) + \sin\varphi f'(\varphi),$$

where $f(\varphi)$ is an arbitrary function of φ.

The applications to ordinary differential equations of higher order lead to the Pfaffian system, where

$$\alpha^1 = dy - y'dx,$$

$$\vdots$$

$$\alpha^s = dy^{s-1} - y^{(s)}dx,$$

the space being $(x, y, y', \ldots, y^{(s)})$. This system is of codimension two. It satisfies the relations

$$d\alpha^i = -\alpha^{i+1} \wedge dx, \quad 1 \le i \le s - 1$$
(33)
$$d\alpha^s \ne 0 \mod I.$$

Such a system can be characterized by a set of conditions, as given by the theorem:

Theorem 5.3 (Goursat normal form). *Let*

$$I = \{\alpha^1, \ldots, \alpha^s\}$$

be a Pfaffian system of codimension two in a space of dimension $n = s + 2$. Suppose there exists a Pfaffian form $\pi \neq 0, \mod I$, satisfying
(34)
$$d\alpha^i \equiv -\alpha^{i+1} \wedge \pi \quad \mod \alpha^1, \ldots, \alpha^i, \quad 1 \leq i \leq s-1, \quad d\alpha^s \neq 0 \quad \mod I.$$

Then there is a local coordinate system $x, y, y', \ldots, y^{(s)}$, such that

$$I = \{dy - y'dx, \ldots, dy^{(s-1)} - y^{(s)}dx\}.$$

Proof. The first equation of (33) gives, for $i = 1$,

$$d\alpha^1 \wedge \alpha^1 \neq 0, \quad (d\alpha^1)^2 \wedge \alpha^1 = 0.$$

By Theorem 3.1, we can suppose, by multiplying α^1 by a factor if necessary,

$$\alpha^1 = dy - y'dx.$$

As in the proof of Theorem 5.1, we have

$$d\alpha^1 \wedge \alpha^1 \wedge \alpha^2 = 0.$$

The proof of Theorem 5.1 applies, and we can suppose, by replacing α^2 by a linear combination α^1, α^2 if necessary,

$$\alpha^2 = dy' - y''dx.$$

Equation (34) then gives

$$\alpha^1 \wedge d\alpha^1 = -\alpha^1 \wedge \alpha^2 \wedge \pi = -dx \wedge dy \wedge dy',$$

from which it follows that

$$\pi \wedge dx \wedge dy \wedge dy' = 0,$$

and that we can write

$$\pi = ady' + bdx + cdy.$$

By hypothesis we have

$$\pi \equiv (ay'' + cy' + b)dx \neq 0 \quad \mod \alpha^1, \alpha^2,$$

or

$$ay'' + cy' + b \neq 0.$$

Suppose $s \geq 3$ and suppose, as induction hypothesis,

$$\alpha^3 = dy'' - y'''dx, \ldots, \alpha^{i-1} = dy^{(i-2)} - y^{(i-1)}dx, \quad i \leq s - 1.$$

Equation (34) gives

$$d\alpha^{i-1} = dx \wedge dy^{(i-1)} \equiv -\alpha^i \wedge (ay'' + cy' + b)dx \quad \mathrm{mod} \ \alpha^1, \ldots, \alpha^{i-1}.$$

It follows that, $\mathrm{mod} \ \alpha^1, \ldots, \alpha^{i-1}$ and dx, the form α^i is a non-zero multiple of $dy^{(i-1)}$. We can therefore change α^i to

$$\alpha^i = dy^{(i-1)} - y^{(i)}dx.$$

This completes the induction.

By the second equation of (34) we have

$$\alpha^1 \wedge \cdots \wedge \alpha^s \wedge d\alpha^s \neq 0,$$

giving

$$dx \wedge dy \wedge \cdots \wedge dy^{(s)} \neq 0,$$

so that $x, y, y', \ldots, y^{(s)}$ serve as a local coordinate system. $\qquad \square$

To understand the significance of the Goursat normal form we return to the general case. Suppose the α's be chosen so that the equations (31) are satisfied. They are determined up to the transformation

(35)
$$\begin{pmatrix} \alpha^1 \\ \vdots \\ \alpha^n \end{pmatrix} \longrightarrow \begin{pmatrix} u_1^1 & \cdots & u_{s-1}^1 & 0 & \cdots & 0 \\ \cdots & & & & & \\ u_1^{s-1} & \cdots & u_{s-1}^{s-1} & 0 & \cdots & 0 \\ u_1^s & \cdots & u_{s-1}^s & u_s^s & 0 & \cdots & 0 \\ u_1^{n-1} & \cdots & \cdots & u_s^{n-1} & u_{n-1}^{n-1} & u_n^{n-1} \\ u_1^n & \cdots & \cdots & u_s^n & u_{n-1}^n & u_n^n \end{pmatrix} \begin{pmatrix} \alpha^1 \\ \vdots \\ \alpha^n \end{pmatrix}.$$

Let

$$d\alpha^j \equiv R^j \alpha^s \wedge \alpha^{n-1} + S^j \alpha^s \wedge \alpha^n, \qquad \mathrm{mod} \ \alpha^1, \ldots, \alpha^{s-1}, \quad 1 \leq j \leq s - 1.$$

Under the transformation (35), the rank of the matrix

(36)
$$\begin{pmatrix} R^1 & \cdots & R^{s-1} \\ S^1 & \cdots & S^{s-1} \end{pmatrix}$$

is invariant. In fact, $\dim I^{(2)} = s - 2$ or $s - 3$, according as this rank is 1 or 2. Comparing with (34), we see that a necessary condition for I to be in the Goursat normal form is $\dim I^{(2)} = s - 2$.

Example. The Goursat normal form can be used to study the second-order Monge equation

$$(37) \qquad \frac{dz}{dx} = F(x, y, z, y', y''), \quad y' = \frac{dy}{dx}, \quad y'' = \frac{d^2 y}{dx^2}, \quad F_{y''} \neq 0.$$

This can be studied as a Pfaffian system of codimension two in the space (x, y, z, y', y''), namely, $(s = 3, n = 5)$

$$I = \{dy - y'dx, y' - y''dx, dz - F(x, y, z, y', y'')dx\}.$$

To achieve the equations (31), we set

$$\alpha^1 = dy - y'dx,$$

$$\alpha^2 = dz - Fdx - F_{y''}(dy' - y''dx),$$

$$\alpha^3 = dy' - y''dx,$$

$$\alpha^4 = dx,$$

$$\alpha^5 = dy''.$$

An easy calculation gives

$$\begin{aligned} d\alpha^1 &= dx \wedge dy' = \alpha^4 \wedge \alpha^3 \\ d\alpha^2 &\equiv c\alpha^4 \wedge \alpha^3 + F_{y''y''}\alpha^3 \wedge \alpha^5 \end{aligned} \qquad \mathrm{mod}\ \alpha^1, \alpha^2$$

where c is some function. Hence I can be put in the Goursat normal form only if $F_{y''y''} = 0$, i.e. F is linear in y''.

Consider the system J in the Goursat normal form:

$$\beta^1 = dw - w'dt,$$

$$\beta^2 = dw' - w''dt,$$

$$\beta^3 = dw'' - w'''dt.$$

If $F_{y''y''} \neq 0$, there is no local diffeomorphism

$$x = x(t, w, w', w'', w'''),$$

$$y = y(t, w, w', w'', w'''),$$

$$z = z(t, w, w', w'', w'''),$$

$$y' = y'(t, w, w', w'', w'''),$$

$$y'' = y''(t, w, w', w'', w'''),$$

which maps I into J. In other words, the "general" solution of I or (37) cannot be expressed in terms of an arbitrary function $w(t)$ and its successive derivatives. This was proved by D. Hilbert for the equation

$$\frac{dz}{dx} = \left(\frac{d^2y}{dx^2}\right)^2.$$

On the other hand, for the equation

$$\frac{dz}{dx} = y^m \frac{d^2y}{dx^2},$$

which is linear in y'', É. Cartan gave the solution

$$x = -2tf''(t) - f'(t),$$

$$y^{m+1} = (m+1)^2 t^3 f''^2(t)$$

$$z = (m-1)t^2 f''(t) - mtf'(t) + mf(t),$$

where $f(t)$ is an arbitrary function in t.

We continue with the case $s = 3$, $n = 5$. Its generic situation is when the rank of the matrix (36) is 2. Then the α's can be so chosen that the matrix (30) becomes

$$\begin{pmatrix} 1 & 0 \\ 0 & 1 \end{pmatrix},$$

i.e.

$$d\alpha^1 \equiv \alpha^3 \wedge \alpha^4, \quad d\alpha^2 = \alpha^3 \wedge \alpha^5, \qquad \mod \alpha^1, \alpha^2.$$

By (31) we also have

$$d\alpha^3 \equiv \alpha^4 \wedge \alpha^5, \qquad \mod \alpha^1, \alpha^2, \alpha^3.$$

This generic case is very interesting. A complete system of invariants was determined in Cartan [1910] by the method of equivalence. The fundamental invariant is a ternary quartic (symmetric) differential form. If it vanishes identically, the Pfaffian system is invariant under the exceptional simple Lie group G_2 of 14 dimensions. This is clearly a very natural way that the split real form of G_2 is geometrically realized. The general case involves tensorial invariants. Its treatment has to be divided into cases and is long; cf. Cartan's paper for details.

CHAPTER III

CARTAN–KÄHLER THEORY

In the first two chapters, we have seen how problems in differential geometry and partial differential equations can often be recast as problems about integral manifolds of appropriate exterior differential systems. Moreover, in differential geometry, particularly in the theory and applications of the moving frame and Cartan's method of equivalence, the problems to be studied often appear naturally in the form of an exterior differential system anyway.

This motivates the problem of finding a general method of constructing integral manifolds. When the exterior differential system $\mathcal{I}$ has a particularly simple form, standard differential calculus and the techniques of ordinary differential equations allow a complete (local) description of the integral manifolds of $\mathcal{I}$. Examples of such systems are furnished by the theorems of Frobenius, Pfaff–Darboux, and Goursat (see Chapter II).

However, the differential systems arising in practice are usually more complicated than the ones dealt with in Chapter II. Certainly, one cannot expect to construct the general integral manifold of a differential system $\mathcal{I}$ using *ordinary* differential equation techniques alone. However, at least locally, this problem can be expressed as a problem in *partial* differential equations. It is instructive to see how this can be done.

Let $\mathcal{S} \subset \Omega^*(M)$ be an arbitrary set of differential forms on M. Suppose that we are interested in finding the n-dimensional integral manifolds of the set $\mathcal{S}$. To simplify our notation, we will agree on the index ranges $1 \leq i, j, k \leq n$ and $1 \leq a, b, c \leq m - n$ and make use of the summation convention. We choose local coordinates $x^1, x^2, \ldots, x^n, y^1, \ldots, y^{m-n}$ centered at z on a z-neighborhood $U \subset M$. Let $\Omega = dx^1 \wedge \ldots dx^n$. Let $G_n(TU, \Omega)$ denote the dense open subset of $G_n(TU)$ consisting of the n-planes $P \subset T_w U$ on which Ω restricts to be non-zero. Then there are well defined functions p_i^a on $G_n(TU, \Omega)$ so that, for each $P \in G_n(TU, \Omega)$, the vectors

$$(1) \qquad X_i(P) = (\partial/\partial x^i + p_i^a(P)\partial/\partial y^a)|_w$$

form a basis of P. In fact, the functions x^i, y^a, p_i^a form a coordinate system on $G_n(TU, \Omega)$.

Now, for each q-form φ on U with $q \leq n$ and every multi-index $J = (j_1, j_2, \ldots, j_q)$ with $1 \leq j_1 < j_2 < \cdots < j_q \leq n$ we may define a function $F_{\varphi,J}$ on $G_n(TU, \Omega)$ by setting

$$(2) \qquad F_{\varphi,J}(P) = \varphi(X_{j_1}(P), \ldots, X_{j_q}(P)).$$

(Note that, when $F_{\varphi,J}$ is expressed in the coordinates x^i, y^a, p_i^a, it is linear in the $(k \times k)$-minors of the matrix $p = (p_i^\alpha)$, where $k \leq q$.)

Any submanifold $V \subset U$ of dimension n which passes through $z \in U$ and satisfies $\Omega|_V \neq 0$ can be described in a neighborhood of z as a 'graph' $y^a = u^a(x) = u^a(x^1, \ldots, x^n)$ of a set of $m-n$ functions u^a of the n variables x^i. For each $w = (x, u(x))$ in V, the p-coordinates of $T_w V \in G_n(TU, \Omega)$ are simply the partials $p_i^a = \partial u^a / \partial x^i$ evaluated at x. It follows that V is an integral manifold of $\mathcal{S}$ if and only if the function u satisfies the system of first order partial differential equations

$$(3) \qquad\qquad F_{\varphi,J}(x, u, \partial u/\partial x) = 0$$

for all $\varphi \in \mathcal{S}$ and all J with $\deg(\varphi) = |J| \leq n$.

Thus, constructing integral manifolds of $\mathcal{S}$ is locally equivalent to solving a system of first order partial differential equations of the form (3). Conversely, any first order system of P.D.E. for the functions $u^1, \ldots, u^{m-n}$ as functions of $x^1, \ldots, x^n$ which is linear in the minors of the Jacobian matrix $\partial u/\partial x$ can be expressed as the condition that the graph $(x, u(x))$ in $\mathbb{R}^m$ be an integral manifold of an appropriate set $\mathcal{S}$ of differential forms on $\mathbb{R}^m$.

It is then natural to ask about methods of solving systems of P.D.E. of the form (3). It is rare that the system (3) can be placed in a form to which the classical existence theorems in P.D.E. can be applied directly. In general, even for simple systems $\mathcal{S}$, the corresponding system of equations (3) is overdetermined, meaning that there are more independent equations in (3) than unknowns u. For example, if $m = 2n$ and $\mathcal{S}$ consists of the single differential form $\varphi = dy^1 \wedge dx^1 + dy^2 \wedge dx^2 + \cdots + dy^n \wedge dx^n$, then (3) becomes the system of equations $\partial u^i / \partial x^j = \partial u^j / \partial x^i$, which is overdetermined when $n > 3$. Even when (3) is not overdetermined, it cannot generally be placed in one of the classical forms (e.g., Cauchy–Kowalevski).

Nevertheless, certain systems of equations of the form (3) had been treated successfully (at least, in the real analytic category) in the nineteenth century by methods generalizing the initial value problem (sometimes called the "Cauchy problem" because of Cauchy's work on initial value problems). Let us illustrate such an approach by the following simple example: Consider the following system of first order partial differential equations for one function u of two variables x and y:

$$(4) \qquad\qquad u_x = F(x, y, u), \quad u_y = G(x, y, u).$$

If we seek a solution of (4) which satisfies $u(0, 0) = c$, then we may try to construct such a solution by first solving the initial value problem

$$(5) \qquad\qquad v_x = F(x, 0, v) \quad \text{where} \quad v(0) = c$$

for v as a function of x, and then solving the initial value problem (regarding x as a parameter)

$$(6) \qquad u_y = G(x, y, u) \quad \text{where} \quad u(x, 0) = v(x).$$

Assuming that F and G are smooth in a neighborhood of $(x, y, u) = (0, 0, c)$, standard O.D.E. theory tells us that this process will yield a smooth function $u(x, y)$ defined on a neighborhood of $(x, y) = (0, 0)$. However, the function u may not satisfy the equation $u_x = F(x, y, u)$ except along the line $y = 0$. In fact, if we set $E(x, y) = u_x(x, y) - F(x, y, u)$, then $E(x, 0) = 0$ and we may compute that

$$E_y(x, y) = (G(x, y, u))_x - F_y(x, y, u) - F_u(x, y, u)G(x, y, u)$$
$$= G_u(x, y, u)E(x, y) + T(x, y, u)$$

where

$$T(x, y, u) = F(x, y, u)G_u(x, y, u) - G(x, y, u)F_u(x, y, u) + G_x(x, y, u) - F_y(x, y, u).$$

Suppose that F and G satisfy the identity $T \equiv 0$. Then E satisfies the differential equation with initial condition

$$E_y = G_u(x, y, u)E \quad \text{and} \quad E(x, 0) = 0.$$

By the usual uniqueness theorem in O.D.E., it follows that $E(x, y) \equiv 0$, so u satisfies the system of equations (4). It follows that the condition $T \equiv 0$ is a sufficient condition for the existence of local solutions of (4) where $u(0, 0)$ is allowed to be an arbitrary constant as long as $(0, 0, u(0, 0))$ is in the common domain of F and G.

Note that if we consider the differential system $\mathcal{I}$ (on the domain in $\mathbb{R}^3$ where F and G are both defined) which is generated by the 1-form $\vartheta = du - F dx - G dy$, then $d\vartheta \equiv -T dx \wedge dy \mod \vartheta$, so the condition $T \equiv 0$ is equivalent to the condition that $\mathcal{I}$ be generated algebraically by ϑ. Thus, we recover a special case of the Frobenius theorem. It is an important observation that the process of computing the differential closure of this system uncovers the "compatibility condition" $T \equiv 0$.

Let us pursue the case of first order equations with two independent variables a little further. Given a system of P.D.E. $R(x, y, u, u_x, u_y) = 0$, where u is regarded as a vector-valued function of the independent variables x and y, then, under certain mild constant rank assumptions, it will be possible to place the equations in the following (local) normal form

$$(i) \qquad\qquad u_x^0 = F(x, y, u)$$

$$(ii) \qquad\qquad u_y^0 = G(x, y, u, u_x)$$
$$u_y^1 = H(x, y, u, u_x)$$

by making suitable changes of coordinates in (x, y) and decomposing u into $u = (u^0, u^1, u^2)$ where each of the u^α is a (vector-valued) unknown function of x and y. Note that the original system may thus be (roughly) regarded as being composed of an "overdetermined" part (for u^0), a "determined" part (for u^1), and an "underdetermined" part (for u^2). (This "normal form" generalizes in a straightforward way to the case of n independent variables, in which case the unknown functions u are split into $(n + 1)$ vector-valued components.)

The "Cauchy–Kowalewski approach" to solving this system in the real analytic case can then be described as follows: Suppose that the collection u^α consists of $s_\alpha \geq 0$ unknown functions. For simplicity's sake, we assume that F, G, H are real analytic and well-defined on the entire $\mathbb{R}^k$ (where k has the appropriate dimension). Then we choose s_0 constants, which we write as f^0, s_1 analytic functions of x, which we write as $f^1(x)$, and s_2 analytic functions of x and y, which we write as $f^2(x, y)$. We then first solve the following system of O.D.E. with initial conditions for s_0 functions v^0 of x:

(i')
$$v^0_x = F(x, 0, v^0, f^1(x), f^2(x, 0))$$
$$v^0(0) = f^0$$

and then second solve the following system of P.D.E. with initial conditions for $s_0 + s_1$ functions (u^0, u^1) of x and y:

$$u^0_y = G(x, y, u^0, u^1, f^2(x, y), u^0_x, u^1_x, f^2_x(x, y))$$
$$u^1_y = H(x, y, u^0, u^1, f^2(x, y), u^0_x, u^1_x, f^2_x(x, y))$$

(ii')
$$u^0(x, 0) = v^0(x)$$
$$u^1(x, 0) = f^1(x).$$

This process yields a function $u(x, y) = (u^0(x, y), u^1(x, y), u^2(x, y))$ (where $u^2(x, y)$ is defined to be $f^2(x, y)$) which is uniquely determined by the collection $f = \{f^0, f^1(x), f^2(x, y)\}$. While it is clear that the $u(x, y)$ thus constructed satisfies (ii), it is not at all clear that u satisfies (i). In fact, if we set

$$E(x, y) = u^0_x(x, y) - f(x, y, u(x, y)),$$

then $E(x, 0) \equiv 0$ since $u(x, 0)$ satisfies (i'), but, in general $E(x, y) \neq 0$ for the generic choice of "initial data" f.

In the classical terminology, the system (i), (ii) is said to be "involutive" or "in involution" if, for *arbitrary* analytic initial data f, the unique solution u_f of (i',ii') is also a solution of (i,ii). Because of the nature of the initial conditions f, the classical terminology further described the "general solution" of (i,ii) in the involutive case as "depending on s_0 constants, s_1 functions of one variable, and s_2 functions of two variables".

In the analytic category, the condition of involutivity for the system (i,ii) can be expressed in terms of certain P.D.E., called "compatibility conditions", which must be satisfied by the functions F, G, and H. For example, in the case of (4), the compatibility condition takes the form $T \equiv 0$. Also note that, for the system (4), we have $(s_0, s_1, s_2) = (1, 0, 0)$.

Of course, this notion of involutivity extends to P.D.E. systems with n independent variables.

The condition of involutivity is rather stringent (except in the case $(s_0, \ldots, s_n) = (0, \ldots, 0, s, 0)$, which corresponds to the classical Cauchy problem). Thus, one often must modify the equations in some way in order to reduce to the involutive case.

Let us give an example. Consider the following system of three equations for three unknown functions u^1, u^2, u^3 of three independent variables x^1, x^2, x^3. Here we write ∂_j for $\partial/\partial x^j$ and v^1, v^2, v^3 are some given functions of x^1, x^2, x^3.

$$\partial_2 u^3 - \partial_3 u^2 = u^1 + v^1$$

(7, i,ii,iii)
$$\partial_3 u^1 - \partial_1 u^3 = u^2 + v^2$$

$$\partial_1 u^2 - \partial_2 u^1 = u^3 + v^3$$

The approach to treating (7) as a sequence of Cauchy problems (with $(s_0, s_1, s_2, s_3) = (0, 1, 1, 1)$) is as follows:

(1) Choose three functions $\varphi^1(x^1), \varphi^2(x^1, x^2), \varphi^3(x^1, x^2, x^3)$.

(2) Solve the equation in $\mathbb{R}^2$, $\partial_2 w = -\partial_1 \varphi_2 - \bar{\varphi}^3 - \bar{v}^3$ with the initial condition $w(x^1, 0) = \varphi^1(x^1)$ where $\bar{\varphi}^3(x^1, x^2) = \varphi^3(x^1, x^2, 0)$ and $\bar{v}^3(x^1, x^2) = v^3(x^1, x^2, 0)$.

(3) Solve the pair of equations $\partial_3 u^1 = \partial_1 \varphi^3 + u^2 + v^2$ and $\partial_3 u^2 = \partial_2 \varphi^3 - u^1 - v^1$ with the initial conditions $u^1(x^1, x^2, 0) = w(x^1, x^2)$ and $u^2(x^1, x^2, 0) = \varphi^2(x^1, x^2)$.

(4) Set u^3 equal to φ^3.

However, the resulting set of functions u^a will not generally be a solution to (7, iii). If we set $E = \partial_1 u^2 - \partial_2 u^1 - u^3 - v^3$, then, of course $E(x^1, x^2, 0) = 0$, but if we compute $\partial_3 E = -\{\partial_1(u^1 + v^1) + \partial_2(u^2 + v^2) + \partial_3(u^3 + v^3)\}$, we see that E vanishes identically if and only if the functions u^a satisfy the additional equation

(7, iv) $$0 = \partial_1(u^1 + v^1) + \partial_2(u^2 + v^2) + \partial_3(u^3 + v^3).$$

This suggests modifying our Cauchy sequence by adjoining (7, iv), thus getting a new system with $(s_0, s_1, s_2, s_3) = (0, 1, 2, 0)$ and then proceeding as follows:

(1*) Choose three functions $\varphi^1(x^1), \varphi^2(x^1, x^2), \varphi^3(x^1, x^2)$

(2*) Solve the equation in $\mathbb{R}^2$, $\partial_2 w = -\partial_1 \varphi^2 - \varphi^3 - \bar{v}^3$ with the initial condition $w(x^1, 0) = \varphi^1(x^1)$ where $\bar{v}^3(x^1, x^2) = v^3(x^1, x^2, 0)$.

(3*) Solve the triple of equations with initial conditions

$$\begin{aligned}
\partial_3 u^1 &= \partial_1 u^3 + u^2 + v^2, & u^1(x^1, x^2, 0) &= w(x^1, x^2) \\
\partial_3 u^2 &= \partial_2 u^3 - u^1 - v^1, & u^2(x^1, x^2, 0) &= \varphi^2(x^1, x^2) \\
\partial_3 u^3 &= -\partial_1(u^1 + v^1) - \partial_2(u^2 + v^2) - \partial_3 v^3, & u^3(x^1, x^2, 0) &= \varphi^3(x^1, x^2).
\end{aligned}$$

It then follows easily that the resulting u^a also satisfy $(7, \mathrm{iii})$ (assuming that we are in the analytic category, so that we have existence and uniqueness in the Cauchy problem).

In the example just given, the "compatibility condition" took the form of an extra equation which must be adjoined to the given equations so that the Cauchy sequence approach would work. One can imagine more complicated phenomena. Indeed, in the latter part of the nineteenth century, many examples of systems of P.D.E. were known to be tractable when treated as a sequence of initial value problems, provided that one was able to find a sufficient number of "compatibility conditions."

Around the turn of the century, Riquier [1910] and Cartan [1899] began to make a systematic study of this compatibility problem. Riquier's approach was to work directly with the partial differential equations in question while Cartan, motivated by his research in differential geometry and Lie transformation groups (nowadays called Lie pseudo-groups), sought a coordinate-free approach.

It was Cartan who realized that partial differentiation (which depends on a choice of coordinates) could be replaced by the exterior derivative operator (which does not). His method was to regard a collection of s functions u^a of n variables x^i as defining, via its graph, an n-dimensional submanifold of $\mathbb{R}^{n+s}$. The condition that the collection u^a satisfy a system of first order P.D.E. which was linear in the minors of the Jacobian matrix $\partial u / \partial x$ was then regarded as equivalent to the condition that the graph be an integral of a system $\mathcal{S}$ of differential forms on $\mathbb{R}^{n+s}$. Cartan then let $\mathcal{I}$ be the differential ideal generated by the system $\mathcal{S}$. The problem of constructing n-dimensional integral manifolds of $\mathcal{I}$ by a sequence of Cauchy problems then was reformulated as the problem of "extending" a p-dimensional integral manifold of $\mathcal{I}$ to one of dimension $(p + 1)$.

Cartan's next major insight into this problem was to realize that the condition that a submanifold $N \subset M$ be an integral manifold of a differential ideal $\mathcal{I}$ on M is a condition only on the tangent planes of N. This led him to define the fundamental concept of *integral elements* of a differential system. Namely, the integral elements of dimension p of $\mathcal{I}$ are the p-planes $E \subset T_z M$ on which all of the forms in $\mathcal{I}$ vanish. These form a closed subspace $V_p(\mathcal{I})$ of $G_p(TM)$. Cartan's approach was to study the structure of these subspaces and their interrelationships as p varies. Two important concepts arise which depend only on the structure of $\mathcal{I}$ as an algebraic ideal. These are the notions of *ordinarity* and *regularity* which are treated in detail in Section 1 of this Chapter. Roughly speaking, these concepts describe the smoothness of the spaces $V_p(\mathcal{I})$ and the incidence spaces $V_{p,p+1}(\mathcal{I}) \subset V_p(\mathcal{I}) \times V_{p+1}(\mathcal{I})$. If one thinks intuitively of integral elements as "infinitesimal integral manifolds," then these notions describe the well-posedness of the "infinitesimal Cauchy problem." The main highlight of this section is Theorem 1.11, a version of Cartan's test for an integral element to be ordinary. The version given here is an improvement over Cartan's original version and was suggested to us by the recent work of W. K. Allard [1989].

In Section 2, after stating the classical Cauchy–Kowalevski theorem on the initial value problem for first order P.D.E., we state and prove the fundamental Cartan–Kähler theorem. Roughly speaking, this theorem states that in the real-analytic category, the well-posedness of the initial value problem for an exterior differential ideal $\mathcal{I}$ is determined completely by the infinitesimal (algebraic) properties of the space of integral elements. Here, the condition that the ideal be differentially closed takes the place of the compatibility conditions which one must deal with in the P.D.E. formulation. We also discuss the classical terminology concerning the "generality" of the space of integral manifolds of a differential system, and introduce the important sequence of *Cartan characters*, which generalize the $s_0, s_1, \ldots,$ etc. described above.

In Section 3, we consider a set of examples which demonstrate the use of the Cartan–Kähler theorem in practice. Some of the examples are merely instructive while others are of interest in their own right. One example in particular, the isometric embedding example (Example 3.8), reproduces (with some improvements) Cartan's original proof of the Cartan–Janet isometric embedding theorem.

The following terminology will be used in the remainder of this chapter.

If X is a smooth manifold and $\mathcal{F} \subset C^\infty(X)$ is any set of smooth functions on X, let $Z(\mathcal{F}) \subset X$ denote the set of common zeros of the functions in $\mathcal{F}$. We say that $x \in Z(\mathcal{F})$ is an *ordinary zero* of $\mathcal{F}$ if there exists a neighborhood V of x and a set of functions $f^1, f^2, \ldots, f^q$ in $\mathcal{F}$ whose differentials are independent on V so that

$$Z(\mathcal{F}) \cap V = \{y \in V \mid f^1(y) = f^2(y) = \cdots = f^q(y) = 0\}.$$

By the implicit function theorem, $Z(\mathcal{F}) \cap V$ is then a smooth submanifold of codimension q in V. Note that the set of ordinary zeroes of $\mathcal{F}$ is an open subset of $Z(\mathcal{F})$ (in the relative topology). If we let $Z^0(\mathcal{F})$ denote the set of ordinary zeroes of $\mathcal{F}$, then $Z^0(\mathcal{F})$ is a disjoint union of connected, embedded submanifolds of X. Of course, the components of $Z^0(\mathcal{F})$ do not all have to have the same dimension. By definition, the codimension of $Z^0(\mathcal{F})$ at $x \in Z^0(\mathcal{F})$ is the codimension in X of the component of $Z^0(\mathcal{F})$ which contains x.

A related piece of terminology is the following. If $\mathcal{A} \subset X$ is any subset and $x \in \mathcal{A}$, then we say that $\mathcal{A}$ *has codimension at most (resp., at least)* q *at* x if there exists an open neighborhood V of $x \in X$ so that $\mathcal{A} \cap V$ contains (resp., is contained in) a smooth (embedded) submanifold of V of codimension q which passes through x. Clearly $\mathcal{A}$ has codimension at least q at x and has codimension at most q at x if and only if $\mathcal{A}$ has the structure of a smooth submanifold of codimension q on a neighborhood of x.

§1. Integral Elements.

Throughout this section, M will be a smooth manifold of dimension m and $\mathcal{I} \subset \Omega^*(M)$ will be a differential ideal on M. Recall from Chapter I that an *integral manifold* of $\mathcal{I}$ is a submanifold $\iota : V \to M$ with the property that $\iota^*(\alpha) = 0$ for all $\alpha \in \mathcal{I}$. If $v \in V$ and $E = T_v V \subset T_v M$ is the tangent space to V at v, then $\iota^*(\alpha)|_v = \alpha_E$ where, as usual, α_E denotes the restriction of $\alpha|_v$ to $E \subset T_v M$. It follows that the vanishing of $\iota^*(\alpha)|_v$ for all $\alpha \in \mathcal{I}$ depends only on the tangent space of V at v. This leads to the following fundamental definition.

Definition 1.1. Let M and $\mathcal{I}$ be as above. A linear subspace $E \subset T_x M$ is said to be an *integral element* of $\mathcal{I}$ if $\varphi_E = 0$ for all $\varphi \in \mathcal{I}$. The set of all integral elements of $\mathcal{I}$ of dimension p is denoted $V_p(\mathcal{I})$.

A submanifold of M is an integral manifold of $\mathcal{I}$ if and only if each of its tangent spaces is an integral element of $\mathcal{I}$. Intuitively, one thinks of the integral elements of $\mathcal{I}$ as "infinitesimal integral manifolds" of $\mathcal{I}$.

It is not true, in general, that every integral element of $\mathcal{I}$ is tangent to an integral manifold of $\mathcal{I}$. A simple counterexample is obtained by letting $M = \mathbb{R}^1$ and letting $\mathcal{I}$ be generated by the 1-form $\alpha = x dx$. The space $E = T_0 \mathbb{R}^1$ is an integral element of $\mathcal{I}$, but E is clearly not tangent to any 1-dimensional integral manifold of $\mathcal{I}$.

A more subtle example (which will be used to illustrate several concepts in this section) is the following one.

Example 1.2. Let $M = \mathbb{R}^5$ and let $\mathcal{I}$ be generated by the two 1-forms $\vartheta^1 = dx^1 + (x^3 - x^4 x^5)dx^4$ and $\vartheta^2 = dx^2 + (x^3 + x^4 x^5)dx^5$. Then $\mathcal{I}$ is generated algebraically by the forms $\vartheta^1, \vartheta^2, d\vartheta^1 = \vartheta^3 \wedge dx^4$, and $d\vartheta^2 = \vartheta^3 \wedge dx^5$ where we have written $\vartheta^3 = dx^3 + x^5 dx^4 - x^4 dx^5$. For each $p \in M$, let

$$H_p = \{v \in T_p \mathbb{R}^5 \mid \vartheta^1(v) = \vartheta^2(v) = 0\} \subset T_p \mathbb{R}^5.$$

Then $H \subset T\mathbb{R}^5$ is a rank 3 distribution. A 1-dimensional subspace $E \subset T_p\mathbb{R}^5$ is an integral element of $\mathcal{I}$ if and only if $E \subset H_p$. Thus, $V_1(\mathcal{I}) \cong \mathbb{P}H$ and it is a smooth manifold of dimension 7. Now let

$$K_p = \{v \in T_p\mathbb{R}^5 \mid \vartheta^1(v) = \vartheta^2 v) = \vartheta^3(v) = 0\}.$$

Then $K \subset H$ is a rank 2 distribution on $\mathbb{R}^5$. It is easy to see that, for each $p \in \mathbb{R}^5$, K_p is the unique 2-dimensional integral element of $\mathcal{I}$ based at p. Thus, $V_2(\mathcal{I}) \cong \mathbb{R}^5$. Moreover, $\mathcal{I}$ has no integral elements of dimension greater than 2.

It is not difficult to describe the 1-dimensional integral manifolds of $\mathcal{I}$. Let $f(t) = (f^3(t), f^4(t), f^5(t))$ be an arbitrary smooth immersed curve in $\mathbb{R}^3$. There exist functions $f^1(t), f^2(t)$ (unique up to a choice of 2 constants) which satisfy the differential equations

$$df^1/dt = -(f^3 - f^4 f^5)df^4/dt$$

$$df^2/dt = -(f^3 + f^4 f^5)df^5/dt.$$

Then $F(t) = (f^1(t), f^2(t), f^3(t), f^4(t), f^5(t))$ is an integral manifold of $\mathcal{I}$. Conversely, every 1-dimensional integral manifold of $\mathcal{I}$ is obtained this way. It is now easy to see that there exists an integral manifold of dimension 1 tangent to each element of $V_1(\mathcal{I})$.

On the other hand, by our calculation of $V_2(\mathcal{I})$ above, any 2-dimensional integral manifold of $\mathcal{I}$ is an integral manifold of the differential system $\mathcal{I}_+$ generated by ϑ^1, ϑ^2, and ϑ^3. Using the fact that $d\vartheta^3 = -2\,dx^4 \wedge dx^5$, we see that $\mathcal{I}_+$ is generated algebraically by ϑ^1, ϑ^2, ϑ^3, and $dx^4 \wedge dx^5$. Hence $\mathcal{I}_+$ has no 2-dimensional integral elements, and *a fortiori*, no 2-dimensional integral manifolds. Thus, $\mathcal{I}$ has no 2-dimensional integral manifolds either.

As Example 1.2 shows, the relationship between the integral elements of a differential system and its integral manifolds can be subtle. In general, even the problem of describing the spaces $V_n(\mathcal{I})$ can be complicated. The rest of this section will be devoted to developing basic properties of integral elements of $\mathcal{I}$ and of the subsets $V_n(\mathcal{I})$.

Proposition 1.3. *If E is an n-dimensional integral element of $\mathcal{I}$, then every subspace of E is also an integral element of $\mathcal{I}$.*

Proof. Suppose that $W \subset E$ is a subspace of E. If W were not an integral element of $\mathcal{I}$, then there would be a form φ in $\mathcal{I}$ satisfying $\varphi_W \neq 0$. But then we would clearly have $\varphi_E \neq 0$, contradicting the assumption that E is an integral element of $\mathcal{I}$. $\qquad\square$

Proposition 1.4.

$$V_n(\mathcal{I}) = \{E \in G_n(TM) \mid \vartheta_E = 0 \text{ for all } \vartheta \text{ in } \mathcal{I} \text{ of degree } n\}.$$

Proof. The containment "$\subset$" is clear. Thus, we must prove that if $\vartheta_E = 0$ for all ϑ in $\mathcal{I}$ of degree n then $\varphi_E = 0$ for all φ in $\mathcal{I}$. Suppose that $\varphi_E \neq 0$ for some φ in $\mathcal{I}$ of degree $p < n$. Then there exists η_0 in $\Lambda^{n-p}(E^*)$ so that $\varphi_E \wedge \eta_0$ is a non-zero form in $\Lambda^n(E^*)$. Let η be a smooth $(n-p)$-form on M so that $\eta_E = \eta_0$. Then $\varphi \wedge \eta$ is a form in $\mathcal{I}$ of degree n, but $(\varphi \wedge \eta)_E = \varphi_E \wedge \eta_0 \neq 0$. $\qquad\square$

It follows from Proposition 1.4 that, for each $x \in M$, the set $V_n(\mathcal{I}) \cap G_n(T_xM)$ is an algebraic subvariety of $G_n(T_xM)$. The structure of this algebraic variety can be complicated. Fortunately, it is seldom necessary to confront this problem directly. In practice, the spaces $V_n(\mathcal{I})$ are most often studied by an inductive procedure which uses information about $V_p(\mathcal{I})$ to get information about $V_{p+1}(\mathcal{I})$. The ultimate reason for this approach will be clear in the next section when we prove the Cartan–Kähler theorem, which builds integral manifolds of $\mathcal{I}$ by solving a sequence of initial value P.D.E. problems. A more immediate reason will be furnished by Proposition 1.6 below.

Definition 1.5. Let $e_1, e_2, \ldots, e_p$ be a basis of $E \subset T_xM$. We define the *polar space* of E to be the vector space

$$H(E) = \{v \in T_xM \mid \varphi(v, e_1, e_2, \ldots, e_p) = 0 \text{ for all } \varphi \text{ in } \mathcal{I} \text{ of degree } p+1\}.$$

Note that $E \subset H(E)$. The annihilator of $H(E)$ is denoted $\mathcal{E}(E) \subset T_x^*M$ and is referred to as the space of *polar equations* of E.

The importance of $H(E)$ is explained by the following proposition.

Proposition 1.6. *Let E be an integral element of $\mathcal{I}$ of dimension p. Then a $(p+1)$-plane E^+ containing E is an integral element of $\mathcal{I}$ if and only if it satisfies $E^+ \subset H(E)$.*

Proof. Suppose that $E^+ = E + \mathbb{R}v$ and that $e_1, e_2, \ldots, e_p$ is a basis of E. By Proposition 1.4, E^+ is an integral element of $\mathcal{I}$ if and only if $\varphi_{E^+} = 0$ for all $(p+1)$-forms φ in $\mathcal{I}$. By definition, this latter condition holds if and only if v lies in $H(E)$. $\qquad\square$

Even though the space $V_{p+1}(\mathcal{I}) \cap G_{p+1}(T_xM)$ may be a complicated algebraic variety, for a fixed $E \in V_p(\mathcal{I})$, the space of those $E^+ \in V_{p+1}(\mathcal{I})$ which contain E is a (real) projective space which is canonically isomorphic to $\mathbb{P}(H(E)/E)$. This motivates us to define a function $r : V_p(\mathcal{I}) \to \mathbb{Z}$ by the formula

$$r(E) = \dim H(E) - (p+1).$$

Note that $r(E) \geq -1$ with equality if and only if E lies in no $(p+1)$-dimensional integral element of $\mathcal{I}$, i.e., E is maximal. When $r(E) \geq 0$, the set of $(p+1)$-dimensional integral elements of $\mathcal{I}$ which contain E is then a real projective space of dimension $r(E)$.

This linearization of an exterior algebra problem is related to the linearization process in multi-linear algebra known as "polarization," the most common example being the polarization of a quadratic form on a vector space to produce a bilinear form. We shall not try to make this relationship more precise. We merely offer this comment as a motivation for the name "polar space" for $H(E)$, which was coined by É. Cartan. Other authors have referred to $H(E)$ as the "space of integral enlargements of E" or used similar terminology.

If Ω is any n-form on M, let $G_n(TM, \Omega)$ denote the open set consisting of those E's for which $\Omega_E \neq 0$. If φ is any other n-form on M, we can define a function φ_Ω on $G_n(TM, \Omega)$ by the formula $\varphi_E = \varphi_\Omega(E)\Omega_E$ for all $E \in G_n(TM, \Omega)$. (Since $\Lambda^n(E^*)$ is 1-dimensional with basis Ω_E, this definition makes sense.)

By Proposition 1.4, the set $V_n(\mathcal{I}, \Omega) = V_n(\mathcal{I}) \cap G_n(TM, \Omega)$ is the space of common zeroes of the set of functions

$$\mathcal{F}_\Omega(\mathcal{I}) = \{\varphi_\Omega \mid \varphi \text{ lies in } \mathcal{I} \text{ and has degree } n\}.$$

Definition 1.7. An integral element $E \in V_n(\mathcal{I})$ will be said to be *Kähler-ordinary* if there exists an n-form Ω on M with $\Omega_E \neq 0$ with the property that E is an ordinary zero of the set of functions $\mathcal{F}_\Omega(\mathcal{I})$. We shall use the notation $V_n^o(\mathcal{I}) \subset V_n(\mathcal{I})$ to denote the subspace of Kähler-ordinary points of $V_n(\mathcal{I})$. If E is a Kähler-ordinary integral element and the function r is locally constant on a neighborhood of E in $V_n^o(\mathcal{I})$, then we say that E is *Kähler-regular*. We shall use the notation $V_n^r(\mathcal{I}) \subset V_n^o(\mathcal{I})$ to denote the subspace of Kähler-regular points of $V_n^o(\mathcal{I})$.

The role of Ω in the above definition is not critical. If Ω and Ψ are two n-forms with $\Omega_E \neq 0$ and $\Psi_E \neq 0$, then $E \in G_n(TM, \Omega) \cap G_n(TM, \Psi)$ and the identity $\varphi_\Omega = \varphi_\Psi \cdot \Psi_\Omega$ holds on $G_n(TM, \Omega) \cap G_n(TM, \Psi)$. Since Ψ_Ω never vanishes on $G_n(TM, \Omega) \cap G_n(TM, \Psi)$, it follows that E is an ordinary zero of $\mathcal{F}_\Omega(\mathcal{I})$ if and only if it is an ordinary zero of $\mathcal{F}_\Psi(\mathcal{I})$. Note that $V_n^o(\mathcal{I})$ is an embedded submanifold of $G_n(TM)$ and is an open subset of $V_n(\mathcal{I})$ in the relative topology. Since r is an upper semicontinuous function on $V_n^o(\mathcal{I})$, $V_n^r(\mathcal{I})$ is a open, dense subset of $V_n^o(\mathcal{I})$.

Example 1.2 (continued). We will show that all of the 2-dimensional integral elements of $\mathcal{I}$ are Kähler-regular. Let $\Omega = dx^4 \wedge dx^5$. Then every element $E \in G_2(T\mathbb{R}^5, \Omega)$ has a unique basis of the form

$$X_4(E) = \partial/\partial x^4 + p_4^1(E)\partial/\partial x^1 + p_4^2(E)\partial/\partial x^2 + p_4^3(E)\partial/\partial x^3$$

$$X_5(E) = \partial/\partial x^5 + p_5^1(E)\partial/\partial x^1 + p_5^2(E)\partial/\partial x^2 + p_5^3(E)\partial/\partial x^3.$$

The functions $x^1, \ldots, x^5, p_4^1, \ldots, p_5^3$ form a coordinate system on $G_2(T\mathbb{R}^5, \Omega)$. Computation gives

$$(\vartheta^1 \wedge dx^4)_\Omega = -p_5^1$$

$$(\vartheta^1 \wedge dx^5)_\Omega = p_4^1 - (x^3 - x^4 x^5)$$

$$(\vartheta^2 \wedge dx^4)_\Omega = -p_5^2 - (x^3 + x^4 x^5)$$

$$(\vartheta^2 \wedge dx^5)_\Omega = p_4^2$$

$$(\vartheta^3 \wedge dx^4)_\Omega = -p_5^3 + x^4$$

$$(\vartheta^3 \wedge dx^5)_\Omega = -p_4^3 - x^5.$$

These 6 functions are clearly independent on $G_2(T\mathbb{R}^5, \Omega)$ and their common zeroes are exactly $V_2(\mathcal{I})$. Thus, every point of $V_2(\mathcal{I})$ is Kähler-ordinary. Since none of these elements has any extension to a 3-dimensional integral element, it follows that $r(E) = -1$ for all $E \in V_2(\mathcal{I})$. Thus, every element of $V_2(\mathcal{I})$ is also Kähler-regular.

Similarly, it is easy to see that every $E \in V_1(\mathcal{I})$ is Kähler-ordinary. However, not every element of $V_1(\mathcal{I})$ is Kähler-regular. To see this, note that any $E \in V_1(\mathcal{I})$ on which ϑ^3 does not vanish cannot lie in any 2-dimensional integral element of $\mathcal{I}$. Thus, $r(E) = -1$ for all $E \in V_1(\mathcal{I}, \vartheta^3)$. On the other hand, each $E^+ \in V_1(\mathcal{I})$ on which ϑ^3 *does* vanish lies in a unique $E^+ \in V_2(\mathcal{I})$ and hence has $r(E) = 0$. Since $V_1(\mathcal{I}, \vartheta^3)$ is clearly dense in $V_1(\mathcal{I})$, it follows that $V_1^r(\mathcal{I}) = V_1(\mathcal{I}, \vartheta^3)$.

Returning to the general theory, we shall need to understand the following incidence correspondences:

$$V_{p,p+1}(\mathcal{I}) = \{(E, E^+) \in V_p(\mathcal{I}) \times V_{p+1}(\mathcal{I}) \mid E \subset E^+\}$$

$$V_{p,p+1}^r(\mathcal{I}) = \{(E, E^+) \in V_p^r(\mathcal{I}) \times V_{p+1}(\mathcal{I}) \mid E \subset E^+\}.$$

We let $\pi_p : V_{p,p+1}(\mathcal{I}) \to V_p(\mathcal{I})$ denote the projection onto the first factor and we let $\pi_{p+1} : V_{p,p+1}(\mathcal{I}) \to V_{p+1}(\mathcal{I})$ denote the projection onto the second factor. The fibers of these maps are easy to describe. If $E \in V_p(\mathcal{I})$ has $r(E) \geq 0$, then $(\pi_p)^{-1}(E) \cong \mathbb{P}(H(E)/E) \cong \mathbb{RP}^{r(E)}$. On the other hand, if $E^+ \in V_{p+1}(\mathcal{I})$, then $(\pi_{p+1})^{-1}(E^+) \cong \mathbb{P}(E^+)^*$, the space of hyperplanes in E^+. It is helpful to keep in mind the following diagram.

$$
\begin{array}{ccc}
 & V_{p,p+1}(\mathcal{I}) & \\
\pi_p \swarrow & & \searrow \pi_{p+1} \\
V_p(\mathcal{I}) & & V_{p+1}(\mathcal{I})
\end{array}
$$

This "double fibration" fails, in general, to be surjective or submersive on either base. The next proposition shows that the picture is better for $V_{p,p+1}^r(\mathcal{I})$. Its proof, although technical, is straightforward. Some care is needed to prove that the regularity assumption suffices to guarantee that certain maps have maximal rank.

Proposition 1.8. *If $V^r_{p,p+1}(\mathcal{I})$ is not empty, then it is a smooth manifold. Moreover, the image $\pi_{p+1}(V^r_{p,p+1}(\mathcal{I}))$ is an open subset of $V^o_{p+1}(\mathcal{I})$ and both of the maps $\pi_p : V^r_{p,p+1}(\mathcal{I}) \to V^r_p(\mathcal{I})$ and $\pi_{p+1} : V^r_{p,p+1}(\mathcal{I}) \to \pi_{p+1}(V^r_{p,p+1}(\mathcal{I}))$ are submersions.*

Proof. Let $E \in V^r_p(\mathcal{I})$ have base point $z \in M$ and let $t = \dim M - \dim H(E)$. By hypothesis, there exist t $(p+1)$-forms $\kappa^1, \kappa^2, \ldots, \kappa^t$ in $\mathcal{I}$ so that, for any basis $e_1, e_2, \ldots, e_p$ of E, we have

$$H(E) = \{v \in T_z M \mid \kappa^\tau(v, e_1, e_2, \ldots, e_p) = 0 \text{ for } 1 \leq \tau \leq t\}.$$

Since r is locally constant on a neighborhood of E in $V^r_p(\mathcal{I})$, it follows easily that we must have

$$H(\tilde{E}) = \{v \in T_{\tilde{z}} M \mid \kappa^\tau(v, \tilde{e}_1, \tilde{e}_2, \ldots, \tilde{e}_p) = 0 \text{ for } 1 \leq \tau \leq t\}$$

for all $\tilde{E}$ in $V^r_p(\mathcal{I})$ (based at $\tilde{z}$, with basis $\tilde{e}_1, \tilde{e}_2, \ldots, \tilde{e}_p$) sufficiently near E. From this, it follows that if we set $\mathcal{H} = \{(E, v) \in V_p(\mathcal{I}) \times TM \mid v \in H(E)\}$, then $\mathcal{H}$ is a family of vector spaces over $V_p(\mathcal{I})$ which restricts to each component of $V^r_p(\mathcal{I})$ to be a smooth vector bundle of constant rank. We also conclude that, for each component Z of $V^r_p(\mathcal{I})$ on which r is non-negative, the component $(\pi_p)^{-1}(Z)$ of $V^r_{p,p+1}(\mathcal{I})$ is a smooth bundle over Z.

We now show that the image of π_{p+1} restricted to $V^r_{p,p+1}(\mathcal{I})$ is open in $V_{p+1}(\mathcal{I})$. Let (E, E^+) belong to $V^r_{p,p+1}(\mathcal{I})$. There exists an open neighborhood of E, $U \subset G_p(TM)$, so that $U \cap V_p(\mathcal{I}) \subset V^r_p(\mathcal{I})$ and an open neighborhood of E^+, $U^+ \subset G_{p+1}(TM)$, so that every $\tilde{E}^+$ in U^+ contains a p-plane $\tilde{E}$ in U. Thus, if $\tilde{E}^+ \in U^+ \cap V_{p+1}(\mathcal{I})$, then $\tilde{E} \in U \cap V_p(\mathcal{I}) \subset V^r_p(\mathcal{I})$. It follows that $\pi_{p+1}(V^r_{p,p+1}(\mathcal{I}))$ contains $U^+ \cap V_{p+1}(\mathcal{I})$.

It remains to show that $\pi_{p+1}(V^r_{p,p+1}(\mathcal{I}))$ lies in $V^o_{p+1}(\mathcal{I})$ and that π_{p+1} restricted to $V^r_{p,p+1}(\mathcal{I})$ is a submersion onto its image. To do this, we choose coordinates. Let (E, E^+) belong to $V^r_{p,p+1}(\mathcal{I})$, let $r = r(E) \geq 0$, and let $t = \dim M - \dim H(E) = \dim M - (r + p + 1)$. The cases where either r or t are zero can be handled by obvious simplifications of the following argument, so we assume that r and t are positive.

Choose coordinates $x^1, \ldots, x^p, y, v^1, \ldots, v^r, u^1, \ldots, u^t$ centered on the base point z of E with the properties

(i) E is spanned by the vectors $\partial/\partial x^i$ at z for $1 \leq i \leq p$.

(ii) E^+ is spanned by the vectors in E and the vector $\partial/\partial y$.

(iii) $H(E)$ is spanned by the vectors in E^+ and the vectors $\partial/\partial v^\rho$ at z for $1 \leq \rho \leq r$.

Let $\Omega = dx^1 \wedge \cdots \wedge dx^p$. By hypothesis, there exist a set of p-forms $\{\varphi^1, \ldots, \varphi^q\}$ in $\mathcal{I}$ and an E-neighborhood $U \subset G_n(TM, \Omega)$ so that the functions $f^c = \varphi^c_\Omega$ for $1 \leq c \leq q$ have independent differentials on U and

$$V_p(\mathcal{I}) \cap U = \{\tilde{E} \in U \mid f^c(\tilde{E}) = 0 \text{ for all } c\}.$$

We may also suppose, by shrinking U if necessary, that there are t $(p+1)$-forms $\kappa^1, \kappa^2, \ldots, \kappa^t$ in $\mathcal{I}$ so that

$$H(\tilde{E}) = \{v \in T_{\tilde{z}}M \mid \kappa^\tau(v, \tilde{e}_1, \tilde{e}_2, \ldots, \tilde{e}_p) = 0 \text{ for } 1 \le \tau \le t\}$$

for all $\tilde{E}$ in $V_p(\mathcal{I}) \cap U$ (based at $\tilde{z}$, with basis $\tilde{e}_1, \tilde{e}_2, \ldots, \tilde{e}_p$).

We now want to show that if we set $\Omega^+ = \Omega \wedge dy$ and define $g^c = (\varphi^c \wedge dy)_{\Omega^+}$ and $h^\tau = (\kappa^\tau)_{\Omega^+}$ then there is an open neighborhood $U^+ \subset G_{p+1}(TM, \Omega^+)$ of E^+ so that the set of functions $\{g^c, h^\tau\}$ have independent differentials on U^+ and that

$$V_{p+1}(\mathcal{I}) \cap U^+ = \{\tilde{E}^+ \in U^+ \mid g^c(\tilde{E}^+) = h^\tau(\tilde{E}^+) = 0 \text{ for all } c \text{ and } \tau\}.$$

In particular, we will conclude that $E^+ \in V_{p+1}^o(\mathcal{I})$.

Note that every $\tilde{E}^+ \in G_{p+1}(TM, \Omega^+)$ contains a unique p-plane, which we will denote $\tilde{E} \subset \tilde{E}^+$, on which the differential dy vanishes. Let $U^+ \subset G_{p+1}(TM, \Omega^+)$ be an E^+-neighborhood in $G_{p+1}(TM)$ so that $\tilde{E} \in U$ whenever $\tilde{E}^+ \in U^+$. There exist unique functions A_i^ρ, B_i^τ, a^ρ, and b^τ on $G_{p+1}(TM, \Omega^+)$ so that the $p + 1$ vectors

$$X_i(\tilde{E}^+) = \partial/\partial x^i + A_i^\rho(\tilde{E}^+)\partial/\partial v^\rho + B_i^\tau(\tilde{E}^+)\partial/\partial u^\tau$$

$$Y(\tilde{E}^+) = \partial/\partial y + a^\rho(\tilde{E}^+)\partial/\partial v^\rho + b^\tau(\tilde{E}^+)\partial/\partial u^\tau$$

are a basis of $\tilde{E}^+$. The vectors $X_i(\tilde{E}^+)$ form a basis of $\tilde{E}$. Note that the functions x, y, v, u, A, B, a, and b form a coordinate system on $G_{p+1}(TM, \Omega^+)$.

Since $dy(X_i(\tilde{E}^+)) = 0$, we have the formula $g^c(\tilde{E}^+) = f^c(\tilde{E})$ for all c and all $\tilde{E}^+ \in G_{p+1}(TM, \Omega^+)$. It follows that, if $\tilde{E}^+ \in U^+$ and $g^c(\tilde{E}^+) = 0$ for all c, then $\tilde{E} \in V_p(\mathcal{I}) \cap U$. Since $h^\tau(\tilde{E}^+) = \kappa^\tau(X_1(\tilde{E}^+), \ldots, X_p(\tilde{E}^+), Y(\tilde{E}^+))$, it follows that the equations $h^\tau(\tilde{E}^+) = 0$ imply that $Y(\tilde{E}^+)$ lies in $H(\tilde{E})$ whenever $\tilde{E} \in V_p(\mathcal{I}) \cap U$. It follows that

$$V_{p+1}(\mathcal{I}) \cap U^+ \supset \{\tilde{E}^+ \in U^+ \mid g^c(\tilde{E}^+) = h^\tau(\tilde{E}^+) = 0 \text{ for all } c \text{ and } \tau\}.$$

Since the reverse inclusion is clear, we have proved equality.

It remains to show that the functions $\{g^c, h^\tau\}$ have linearly independent differentials at E^+. To see this, first note that since $h^\tau(\tilde{E}^+) = \kappa^\tau(X_1(\tilde{E}^+), \ldots, X_p(\tilde{E}^+), Y(\tilde{E}^+))$, when we expand the functions h^τ in terms of the coordinates (x, y, v, u, A, B, a, b) they are linear in the functions $\{a^\rho, b^\tau\}$. Thus $h^\tau = N_\rho^\tau + M_\nu^\tau b^\nu$ for some coefficients N and M which depend only on (x, y, v, u, A, B, a). By hypothesis, we have $N_\rho^\tau(E^+) = 0$ and the $t \times t$ matrix $(M_\nu^\tau(E^+))$ is invertible. It follows that, by shrinking U^+

if necessary, we may suppose that $(M_\nu^\tau(\tilde{E}^+))$ is invertible for all $\tilde{E}^+ \in U^+$. Hence we may write $h^\tau = M_\nu^\tau(b^\nu - T^\nu)$ where the functions T^ν depend only on the variables x, y, v, u, A, B and a. It follows that the functions $\{g^c, h^\tau\}$ have independent differentials at E^+ if and only if the functions $\{g^c, b^\tau - T^\tau\}$ have independent differentials at E^+. Since the functions g^c can be expressed in terms of the coordinates x, y, v, u, A, and B alone, it follows that the functions $\{g^c, b^\tau - T^\tau\}$ have independent differentials at E^+ if and only if the functions $\{g^c\}$ have independent differentials at E^+. Let $K \subset U$ be the set of p-planes on which the differential dy vanishes. Then K is clearly a smooth submanifold of U which contains E. Since $\tilde{E}$ lies in K whenever $\tilde{E}^+$ lies in U^+, and since we have the identity $g^c(\tilde{E}^+) = f^c(\tilde{E})$, it follows that functions $\{g^c\}$ have independent differentials at E^+ if and only if the functions $\{f^c\}$ have independent differentials at E after they have been restricted to K. Now every $\tilde{E} \in U$ has a unique basis of the form

$$X_i(\tilde{E}) = \partial/\partial x^i + w_i(\tilde{E})\partial/\partial y + A_i^\rho(\tilde{E})\partial/\partial v^\rho + B_i^\tau(\tilde{E})\partial/\partial u^\tau,$$

and the functions x, y, v, u, A, B, and w form a coordinate system on U centered on E. Also, we have $K = \{\tilde{E} \in U \mid w_i(\tilde{E}) = 0 \text{ for all } i\}$. It follows that the functions $\{f^c\}$ have independent differentials at E after they have been restricted to K if and only if the functions $\{w_i\}$ have independent differentials on the set $V_p(\mathcal{I}) \cap U = \{\tilde{E} \in U \mid f^c(\tilde{E}) = 0 \text{ for all } c\}$. However, since $E^+ \in V_{p+1}(\mathcal{I})$, it follows that, for any vector $\lambda = (\lambda_1, \ldots, \lambda_n)$ where the λ_i are sufficiently small, $V_p(\mathcal{I}) \cap U$ contains the p-plane $E_\lambda \subset E^+$ which is spanned by the vectors $X_i(\lambda) = \partial/\partial x^i + \lambda_i \partial/\partial y$ for i between 1 and p. Since the functions $\{w_i\}$ are independent when restricted to the p-manifold $\mathcal{E} = \{E_\lambda \mid E_\lambda \in U\} \subset V_p(\mathcal{I}) \cap U$, we are done.

Since we have shown that E^+ is a Kähler-ordinary integral element of $\mathcal{I}$, it follows that $\pi_{p+1}(V_{p,p+1}^r(\mathcal{I}))$ is an open subset of $V_{p,p+1}^o(\mathcal{I})$. The fact that π_{p+1} is a submersion when restricted to $V_{p,p+1}^r(\mathcal{I})$ is now elementary. $\quad\square$

The proof of Proposition 1.8 has an important corollary: If $(E, E^+) \in V_{p,p+1}^r(\mathcal{I})$, then the following formula holds

$$(8) \quad \begin{aligned} &(\operatorname{codim} V_{p+1}(\mathcal{I}) \text{ in } G_{p+1}(TM) \text{ at } E^+) \\ &= (\operatorname{codim} V_p(\mathcal{I}) \text{ in } G_p(TM) \text{ at } E) + (\operatorname{codim} H(E) \text{ in } T_zM). \end{aligned}$$

A nested sequence of subspaces $(0)_z \subset E_1 \subset E_2 \subset \cdots \subset E_n \subset T_zM$ where each E_k is of dimension k and E_n is an integral element of $\mathcal{I}$ is called an *integral flag* of $\mathcal{I}$ of length n based at z. If z is an ordinary point of the set of functions $\mathcal{F} = \mathcal{I} \cap \Omega^0(M)$ (i.e., the set of 0-forms in $\mathcal{I}$), and the function r is locally constant on a neighborhood of E_k in $V_k(\mathcal{I})$ for all $k \leq n-1$, then Proposition 1.8 applies inductively to show that each E_k is Kähler-regular for $k \leq n-1$ and that E_n is Kähler-ordinary.

Definition 1.9. Let $\mathcal{I}$ be a differential system on a manifold M. An integral element $E \in V_n(\mathcal{I})$ is said to be *ordinary* if its base point $z \in M$ is an ordinary zero of $\mathcal{I} \cap \Omega^0(M)$ and moreover there exists an integral flag $(0)_z \subset E_1 \subset E_2 \subset \cdots \subset E_n \subset T_z M$ with $E = E_n$ where E_k is Kähler-regular for $k \leq n - 1$. If E is both ordinary and Kähler-regular, then we say that E is *regular*.

In an integral flag $(0)_z \subset E_1 \subset E_2 \subset \cdots \subset E_n \subset T_z M$ where each E_k is Kähler-regular for $k \leq n - 1$, note that each E_k is actually regular for $k \leq n - 1$. Such a flag is called an *ordinary flag*. If, in addition, E_n is also regular, the flag is said to be *regular*. Note that, for integral elements, we have the implications: regular $\Rightarrow$ Kähler-regular, ordinary $\Rightarrow$ Kähler-ordinary, and regular $\Rightarrow$ ordinary. However, these implications are not generally reversible.

Our goal in the remainder of this section is to describe one of the fundamental tests for an integral element to be ordinary. First, we shall introduce a set of constructions which are frequently useful in the study of integral flags of $\mathcal{I}$. It is convenient to assume that the differential ideal $\mathcal{I}$ contains no non-zero forms of degree 0. Since this is usually the case in practice, this restriction is not unreasonable.

Proposition 1.10. *Let $\mathcal{I} \subset \Omega^+(M)$ be a differential ideal which contains no non-zero forms of degree 0. Let $(0)_z \subset E_1 \subset E_2 \subset \cdots \subset E_n \subset T_z M$ be an integral flag of $\mathcal{I}$. Let $e_1, e_2, \ldots, e_n$ be a basis of E_n so that $e_1, e_2, \ldots, e_k$ is a basis of E_k for all $1 \leq k \leq n$. For each $k \leq n$, let c_k be the codimension of $H(E_k)$ in $T_z M$. The numbers c_k satisfy $c_{k-1} \leq c_k$. For each integer a between 1 and c_{n-1}, define the level of a, denoted $\lambda(a)$, to be the smallest integer so that $a \leq c_{\lambda(a)}$. If $c_{n-1} > 0$, then there exists a sequence $\varphi^1, \ldots, \varphi^{c_{n-1}}$ of forms in $\mathcal{I}$ so that φ^a has degree $\lambda(a) + 1$ and so that for all $0 \leq k \leq n - 1$,*

$$(9) \qquad H(E_k) = \{v \in T_z M \mid \varphi^a(v, e_1, \ldots, e_{\lambda(a)}) = 0 \text{ for all } a \leq c_k\}.$$

Proof. Since it is clear that $H(E_{k+1}) \subset H(E_k)$, it follows that $c_{k+1} \geq c_k$. To construct the sequence $\varphi^1, \ldots, \varphi^{c_{n-1}}$, we proceed by induction on the level k. By the very definition of $H(E_0)$, there exist 1-forms $\varphi^1, \ldots, \varphi^{c_0}$ in $\mathcal{I}$ so that (9) holds for $k = 0$. Suppose now that we have constructed a sequence $\varphi^1, \ldots, \varphi^{c_{p-1}}$ so that (9) holds for all $k < p$. Let $\omega^1, \ldots, \omega^n$ be a sequence of 1-forms on M so that their restriction to E is the dual coframe to $e_1, e_2, \ldots, e_n$. Define $\tilde{\varphi}^a \in \mathcal{I}$ by $\tilde{\varphi}^a = \varphi^a \wedge \omega^{\lambda(a)+1} \wedge \omega^{\lambda(a)+2} \wedge \cdots \wedge \omega^p$. Then $\tilde{\varphi}^a$ is a form in $\mathcal{I}$ of degree $p + 1$ and, since φ^a vanishes on E, the identity

$$(10) \qquad \tilde{\varphi}^a(v, e_1, \ldots, e_p) = \varphi^a(v, e_1, \ldots, e_{\lambda(a)})$$

holds for all $v \in T_z M$. If $c_p = c_{p-1}$, then $H(E_p) = H(E_{p-1})$, so (10) shows that (9) already holds for $k = p$. If $c_p > c_{p-1}$, then by the definition of $H(E_p)$, we can choose a set of $(p+1)$-forms in $\mathcal{I}$, $\{\varphi^a \mid c_{p-1} < a \le c_p\}$, so that $H(E_p)$ is the set of vectors v satisfying $\tilde{\varphi}^a(v, e_1, \ldots, e_p) = 0$ for all $a \le c_{p-1}$ as well as $\varphi^a(v, e_1, \ldots, e_p) = 0$ for all $c_{p-1} < a \le c_p$. This completes the induction step. $\qquad\Box$

A sequence $\varphi^1, \ldots, \varphi^{c_{n-1}}$ of forms in $\mathcal{I}$ with the properties given in Propostion 1.10 will be call a *polar sequence* associated to the integral flag $(0)_z \subset E_1 \subset E_2 \subset \cdots \subset E_n \subset T_z M$. Note that the polar sequence does not necessarily carry complete information about $H(E_n)$.

Theorem 1.11 (Cartan's test). *Let $\mathcal{I} \subset \Omega^+(M)$ be an ideal which contains no non-zero forms of degree 0. Let $(0)_z \subset E_1 \subset E_2 \subset \cdots \subset E_n \subset T_z M$ be an integral flag of $\mathcal{I}$ and, for each $k < n$, let c_k be the codimension of $H(E_k)$ in $T_z M$. Then $V_n(\mathcal{I}) \subset G_n(TM)$ is of codimension at least $c_0 + c_1 + \cdots + c_{n-1}$ at E_n. Moreover, each E_k is regular for all $k < n$ (and hence E_n is ordinary) if and only if E_n has a neighborhood U in $G_n(TM)$ so that $V_n(\mathcal{I}) \cap U$ is a smooth manifold of codimension $c_0 + c_1 + \cdots + c_{n-1}$ in U.*

Proof. Set $s = \dim M - n$. There exists a z-centered local coordinate system $x^1, \ldots, x^n, u^1, \ldots, u^s$ with the property that E_k is spanned by the vectors $\{\partial/\partial x^i\}_{i \le k}$ and so that, for all $k < n$,

$$H(E_k) = \{v \in T_z M \mid du^a(v) = 0 \text{ for all } a \le c_k\}.$$

Let $\varphi^1, \ldots, \varphi^{c_{n-1}}$ be a polar sequence for the given flag so that

$$du^a(v) = \varphi^a(v, \partial/\partial x^1, \partial/\partial x^2, \ldots, \partial/\partial x^{\lambda(a)})$$

for all $v \in T_z M$. It follows that

$$\varphi^a = du^a \wedge dx^1 \wedge dx^2 \wedge \cdots \wedge dx^{\lambda(a)} + \psi^a,$$

where ψ^a is a form of degree $\lambda(a)+1$ which is a sum of terms of the following three kinds:

(i) $du^b \wedge dx^J$ where J is a multi-index of degree $\lambda(a)$ which contains at least one index j which is larger than $\lambda(a)$,

(ii) forms which vanish at z,

(iii) forms which are of degree at least 2 in the differentials $\{du^b\}$.

We are now going to show that the forms $\{\varphi^a \mid a \le c_{n-1}\}$ suffice to generate a set of at least $c_0 + c_1 + \cdots + c_{n-1}$ functions on a neighborhood of E_n whose differentials are linearly independent at E_n and whose set of common zeroes contains $V_n(\mathcal{I})$ in a neighborhood of E_n.

Let $\Omega = dx^1 \wedge dx^2 \wedge \cdots \wedge dx^n$. Let $G_n(T\mathcal{U}, \Omega) \subset G_n(TM)$ be the set of n-planes E which are based in the domain $\mathcal{U}$ of the (x, u)-coordinates and for which $\Omega_E \neq 0$. Then there exist functions $\{p_i^a \mid 1 \leq j \leq n$ and $1 \leq a \leq s\}$ on $G_n(T\mathcal{U}, \Omega)$ so that, for each $E \in G_n(T\mathcal{U}, \Omega)$, the vectors $X_i(E) = \partial/\partial x^i + p_i^a(E)\partial/\partial u^a$ form a basis of E. The functions (x, u, p) form an E_n-centered coordinate system on $G_n(T\mathcal{U}, \Omega)$.

For convenience, we define $\lambda(a) = n$ for all $c_{n-1} < a \leq s$. Let us say that a pair of integers (j, a) where $1 \leq j \leq n$ and $1 \leq a \leq s$ is *principal* if it satisfies $j \leq \lambda(a)$ otherwise, we say that the pair is *non-principal*. (For example, there are no principal pairs if $c_0 = s$.) Since, for $j \geq 1$, there are $c_j - c_{j-1}$ values of a in the range $1 \leq a \leq s$ which satisfy $\lambda(a) = j$, it easily follows that the number of principal pairs is $ns - (c_0 + c_1 + \cdots + c_{n-1})$. Hence, the number of non-principal pairs is $c_0 + c_1 + \cdots + c_{n-1}$.

Let (j, a) be a non-principal pair. Define the function F_j^a on $G_n(T\mathcal{U}, \Omega)$ by $F_j^a(E) = \varphi^a(X_j(E), X_1(E), X_2(E), \ldots, X_{\lambda(a)}(E))$. Then we have an expansion

$$F_j^a = p_j^a + P_j^a + Q_j^a$$

where P_j^a is a linear combination (with constant coefficients) of the variables x^i, u^a, and $\{p_i^a \mid (i, a)$ is principal$\}$ and Q_j^a vanishes to second order at E_n. This expansion follows directly from an examination of the terms in ψ^a as described above. It follows that the functions $\{F_j^a \mid (j, a)$ is non-principal$\}$ have linearly independent differentials at E_n. Let $U \subset G_n(T\mathcal{U}, \Omega)$ be a neighborhood of E_n on which these functions have everywhere linearly independent differentials. Then we clearly have

$$V_n(\mathcal{I}) \cap U \subset \{E \in U \mid F_j^a(E) = 0 \text{ for all non-principal } (j, a)\}.$$

It follows that $V_n(\mathcal{I})$ has codimension at least $c_0 + c_1 + \cdots + c_{n-1}$ at E_n, as desired. This proves the first part of the theorem.

In order to prove the second statement of the theorem, we begin by supposing that each E_k is Kähler-regular for all $1 \leq k < n$. Then, by definition, E_n is ordinary. Then (8) shows that we have the following recursion formula for all k between 1 and n:

$$(\text{codim } V_k(\mathcal{I}) \text{ in } G_k(TM) \text{ at } E_k)$$

$$= c_{k-1} + (\text{codim } V_{k-1}(\mathcal{I}) \text{ in } G_{k-1}(TM) \text{ at } E_{k-1}).$$

Since, by hypothesis, $\mathcal{I}$ contains no 0-forms, it follows that $V_0(\mathcal{I}) = G_0(TM) = M$. Thus, by induction, the codimension of $V_n(\mathcal{I})$ in $G_n(TM)$ at E_n has the desired value $c_0 + c_1 + \cdots + c_{n-1}$.

To prove the converse statement, let us now suppose that there is an E_n-neighborhood U in $G_n(TM)$ so that $V_n(\mathcal{I}) \cap U$ is a smooth manifold of codimension $c_0 + c_1 + \cdots + c_{n-1}$ in U. It follows that, by shrinking U if necessary, we may suppose that

$$V_n(\mathcal{I}) \cap U = \{E \in U \mid F_j^a(E) = 0 \text{ for all non-principal } (j,a)\}$$

and that the functions $\mathcal{F} = \{F_j^a \mid (j,a) \text{ is non-principal}\}$ have linearly independent differentials on all of U. If we set $\varphi_j^a = \varphi^a \wedge dx^{K(a,j)}$ where $K(a,j)$ is a multi-index of degree $n - (\lambda(a) + 1)$ with the property that

$$dx^j \wedge dx^1 \wedge dx^2 \wedge \cdots \wedge dx^{\lambda(a)} \wedge dx^{K(a,j)} = dx^1 \wedge dx^2 \wedge \cdots \wedge dx^n,$$

then $F_j^a(E) = \varphi_j^a(X_1(E), X_2(E), \ldots, X_n(E))$ for all $E \in U$. It follows that E_n is Kähler-ordinary. In fact, we can say more. Applying the implicit function theorem to the above expansion of F_j^a, it follows that, by shrinking U if necessary, the submanifold $V_n(\mathcal{I}) \cap U$ in U can be described by equations of the form

$$p_i^a = P_i^a \text{ for all } (i,a) \text{ non-principal,}$$

where the functions P_j^a are functions of the variables x^i, u^a, and $\{p_i^a \mid (i,a)$ is principal$\}$. (For the sake of uniformity, we define the function P_i^a to be p_i^a when the pair (i,a) is principal.) Thus, the variables x^i, u^a, and $\{p_i^a \mid (i,a)$ is principal$\}$ form an E_n-centered coordinate system on $V_n(\mathcal{I}) \cap U$.

By the first part of the proof, we know that $V_{n-1}(\mathcal{I})$ has codimension at least $c_0 + c_1 + \cdots + c_{n-2}$ in $G_{n-1}(TM)$ at E_{n-1}. We will now show that, in fact, $V_{n-1}(\mathcal{I})$ contains a submanifold of codimension $c_0 + c_1 + \cdots + c_{n-2}$ in $G_{n-1}(TM)$ which passes through E_{n-1}. This will imply that E_{n-1} is Kähler-ordinary and that $V_{n-1}(\mathcal{I})$ has codimension $c_0 + c_1 + \cdots + c_{n-2}$ in $G_{n-1}(TM)$ at E_{n-1}. To demonstrate this claim, let $v = (v_1, v_2, \ldots, v_{n-1}) \in \mathbb{R}^{n-1}$, and define a map $\Phi : \mathbb{R}^{n-1} \times V_n(\mathcal{I}) \cap U \to V_{n-1}(\mathcal{I})$ by letting $\Phi(v, E) = E^v$ be the $(n-1)$-plane in E which is spanned by the $n-1$ vectors

$$X_i(E^v) = X_i(E) + v_i X_n(E) \qquad \text{for all } 1 \leq i \leq n-1$$

$$= \partial/\partial x^i + v_i \partial/\partial x^n + (P_i^a(E) + v_i P_n^a(E))\partial/\partial u^a.$$

We claim that Φ has rank $\rho = n + s + (n-1)(s+1) - (c_0 + c_1 + \cdots + c_{n-2})$ at $(0, E_n)$. To see this, note that since $c_0 + c_1 + \cdots + c_{n-2}$ is already known to be a lower bound on the codimension of $V_{n-1}(\mathcal{I})$ in $G_{n-1}(TM)$ at E_{n-1}, the image of Φ must lie in a submanifold of $G_{n-1}(TM)$ whose codimension is at least $c_0 + c_1 + \cdots + c_{n-2}$ and hence the rank of Φ cannot be larger than ρ at any point of some neighborhood of $(0, E_n)$. On the other hand, the rank of Φ at $(0, E_n)$ is equal to ρ, since it is clear that the ρ functions x^i, u^a, $\{v_i \mid 1 \leq i \leq n-1\}$, and $\{P_i^a(E) + v_i P_n^a(E) \mid (i,a) \text{ principal and}$

$i \leq n-1\}$ have linearly independent differentials on a neighborhood of $(0, E_n)$. Thus, the rank of Φ must be identically ρ near $(0, E_n)$.

Moreover, there is a neighborhood $\mathcal{O}$ of $(0, E_n)$ in $\mathbb{R}^{n-1} \times V_n(\mathcal{I}) \cap U$ and a neighborhood U^- of E_{n-1} in $G_{n-1}(TM)$ so that $V_{n-1}(\mathcal{I}) \cap U^-$ is a smooth submanifold of U^- of codimension $c_0 + c_1 + \cdots + c_{n-2}$ and so that $\Phi : \mathcal{O} \to V_{n-1}(\mathcal{I}) \cap U^-$ is a surjective submersion. As noted above, this implies that E_{n-1} is Kähler-ordinary.

We may also conclude that E_{n-1} is Kähler-regular by the following observation. For all $\tilde{E}$ in U^-, the set $\{E \in V_n(\mathcal{I}) \cap U \mid \Phi(v, E) = \tilde{E}$ for some $v\}$ is an open subset of the set $\mathbb{P}(H(\tilde{E})/\tilde{E})$ of n-dimensional integral elements which contain $\tilde{E}$. The dimension of this set is thus $r(\tilde{E})$. However, since Φ is a surjective submersion, this set clearly has the dimension of the fibers of Φ, which is the same as the dimension of the fiber $\Phi^{-1}(E_{n-1})$. It follows that the function r is locally constant on a neighborhood of E_{n-1} in $V_{n-1}(\mathcal{I})$. Thus, E_{n-1} is Kähler-regular, as desired.

By induction, it follows that each E_k is Kähler-regular for all $1 \leq k \leq n-1$. Since $\mathcal{I}$ contains no forms of degree 0, it immediately follows that each E_k is regular for each $1 \leq k \leq n-1$. $\qquad\square$

Example 1.2 (continued). Using Theorem 1.11, we can give a quick proof that none of the elements in $V_2(\mathcal{I})$ are ordinary. For any integral flag $(0)_z \subset E_1 \subset E_2 \subset T_z\mathbb{R}^5$, we know that $c_0 \leq 2$ since there are only 2 independent 1-forms in $\mathcal{I}$. Also, since $E_2 \subset H(E_1)$, it follows that $c_1 \leq 3$. Since there is a unique 2-dimensional integral element at each point of $\mathbb{R}^5$, it follows that $V_2(\mathcal{I})$ has codimension 6 in $G_2(T\mathbb{R}^5)$. Since $6 > c_0 + c_1$, it follows, by Theorem 1.11, that none of the integral flags of length 2 can be ordinary. Hence there are no ordinary integral elements of dimension 2.

Example 1.12. Let $M = \mathbb{R}^6$ with coordinates $x^1, x^2, x^3, u_1, u_2, u_3$. Let $\mathcal{I}$ be the differential system generated by the 2-form

$$\vartheta = d(u_1 dx^1 + u_2 dx^2 + u_3 dx^3) - (u_1 dx^2 \wedge dx^3 + u_2 dx^3 \wedge dx^1 + u_3 dx^1 \wedge dx^2).$$

Of course, $\mathcal{I}$ is generated algebraically by the forms $\{\vartheta, d\vartheta\}$. We have

$$d\vartheta = -(du_1 \wedge dx^2 \wedge dx^3 + du_2 \wedge dx^3 \wedge dx^1 + du_3 \wedge dx^1 \wedge dx^2).$$

We can use Theorem 1.11 to show that all of the 3-dimensional integral elements of $\mathcal{I}$ on which $\Omega = dx^1 \wedge dx^2 \wedge dx^3$ does not vanish are ordinary. Let $E \in V_3(\mathcal{I}, \Omega)$ be fixed with base point $z \in \mathbb{R}^6$. let (e_1, e_2, e_3) be the basis of E which is dual to the basis (dx^1, dx^2, dx^3) of E^*. Let E_1 be the line spanned by e_1, let E_2 be the 2-plane spanned by the pair $\{e_1, e_2\}$, and let E_3 be E. Then $(0)_z \subset E_1 \subset E_2 \subset E_3$ is an integral flag. Since $\mathcal{I}$ is generated by $\{\vartheta, d\vartheta\}$ where ϑ is a 2-form, it follows that $c_0 = 0$. Moreover, since $\vartheta(v, e_1) = \pi_1(v)$ where $\pi_1 \equiv du_1 \mod (dx^1, dx^2, dx^3)$, it follows that

$c_1 = 1$. Note that, since $H(E_2) \supset E_3$, it follows that $c_2 \leq 3$. On the other hand, we have the formula

$$\vartheta(v, e_1) = \pi_1(v)$$

$$\vartheta(v, e_2) = \pi_2(v)$$

$$d\vartheta(v, e_1, e_2) = -\pi_3(v)$$

where in each case, $\pi_k \equiv du_k \mod (dx^1, dx^2, dx^3)$. Since the 1-forms π_k are clearly independent and annihilate $H(E_2)$, it follows that $c_2 \geq 3$. Combined with the previous argument, we have $c_2 = 3$. It follows by Theorem 1.11 that the codimension of $V_3(\mathcal{I})$ in $G_3(T\mathbb{R}^6)$ at E is at least $c_0 + c_1 + c_2 = 4$.

We are now going to show that $V_3(\mathcal{I}, \Omega)$ is a smooth submanifold of $G_3(T\mathbb{R}^6)$ of codimension 4, and thence, by Theorem 1.11, conclude that E is ordinary. To do this, we introduce functions p_{ij} on $G_3(T\mathbb{R}^6, \Omega)$ with the property that, for each $E \in G_3(T\mathbb{R}^6, \Omega)$ based at $z \in \mathbb{R}^6$, the forms $\pi_i = du_i - p_{ij}(E)dx^j \in T_z^*(\mathbb{R}^6)$ are a basis for the 1-forms which annihilate E. Then the functions (x, u, p) form a coordinate system on $G_3(T\mathbb{R}^6, \Omega)$. It is easy to compute that

$$\vartheta_E = (p_{23} - p_{32} - u_1)dx^2 \wedge dx^3 + (p_{31} - p_{13} - u_2)dx^3 \wedge dx^1$$

$$+ (p_{12} - p_{21} - u_3)dx^1 \wedge dx^2$$

$$d\vartheta_E = -(p_{11} + p_{22} + p_{33})dx^1 \wedge dx^2 \wedge dx^3.$$

It follows that the condition that $E \in G_3(T\mathbb{R}^6, \Omega)$ be an integral element of $\mathcal{I}$ is equivalent to the vanishing of 4 functions on $G_3(T\mathbb{R}^6, \Omega)$ whose differentials are independent. Thus $V_3(\mathcal{I}, \Omega)$ is a smooth manifold of codimension 4 in $G_3(T\mathbb{R}^6, \Omega)$, as we desired to show.

The following results will be used in later sections:

Proposition 1.13. *Let $\mathcal{I} \subset \Omega^*(M)$ be a differential ideal which contains no non-zero forms of degree 0. Let $Z \subset V_n(\mathcal{I})$ be a connected component of the space of ordinary integral elements. Then there exists a unique sequence $(c_0, c_1, c_2, \ldots, c_{n-1})$ of integers so that c_k is the codimension of $H(E_k)$ in $T_z M$ for any ordinary integral flag $(0) \subset E_1 \subset \cdots \subset e_n \subset T_z M$ with $E_n \in Z$.*

Proof. Let $\tilde{Z} \subset V_0^r(\mathcal{I}) \times V_1^r(\mathcal{I}) \times \ldots \times V_{n-1}^r(\mathcal{I}) \times Z$ denote the space of ordinary integral flags $F = (E_0, E_1, \ldots, E_n)$ of $\mathcal{I}$ with $E_n \in Z$. We endow $\tilde{Z}$ with the topology and smooth structure it inherits from this product. Note that even though Z is connected, $\tilde{Z}$ may not be connected. However, if we define $c_k(F) = \dim M - \dim H(E_k)$, then the functions c_k for $k < n$

are clearly locally constant on $\tilde{Z}$. We must show that these functions are actually constant on $\tilde{Z}$.

To do this, suppose that for some $p < n$, c_p were not constant on $\tilde{Z}$. Then there would exist non-empty open sets $\tilde{Z}_1, \tilde{Z}_2$ so that $c_p \equiv p$ on $\tilde{Z}_1$ and $c_p \neq p$ on $\tilde{Z}_2$. The images Z_1, Z_2 of these two sets under the submersion $\tilde{Z} \to Z$ would then be an open cover of Z. By the connectedness of Z, they would have to intersect non-trivially. In particular, there would exist an $E \in Z$ and two p-planes $E^1, E^2 \in V_p^r(\mathcal{I}) \cap G_p(E)$ for which $r(E^1) = p \neq r(E^2)$.

We shall now show that this is impossible. Since $E \subset T_z M$ is an integral element, it follows that $G_p(E) \subset V_p(\mathcal{I})$ and hence that $V_p^r(\mathcal{I}) \cap G_p(E)$ is an open subset of $G_p(E)$. Moreover, since the function r is locally constant on $V_p^r(\mathcal{I})$, it follows that $V_p^r(\mathcal{I}) \cap G_p(E)$ is a subset of the open set $G_p^*(E) \subset G_p(E)$ on which r is locally constant. Thus, it suffices to show that r is constant on $G_p^*(E)$.

Let $\varphi^1, \ldots, \varphi^q$ be a set of $(p+1)$-forms in $\mathcal{I}$ with the property that a $(p+1)$-plane $E_{p+1} \subset T_z M$ is an integral element of $\mathcal{I}$ if and only if each of the forms $\varphi^1, \ldots, \varphi^q$ vanish on E_{p+1}. (Since we are only considering planes based at z, such a finite collection of forms exists.) Then for any $E_p \in G_p(E)$ with basis $e_1, e_2, \ldots, e_p$,

$$H(E_p) = \{v \in T_z M \mid \varphi^a(v, e_1, e_2, \ldots, e_p) = 0, \ 1 \le a \le q\}.$$

By the usual argument involving the ranks of linear equations whose coefficients involve parameters, it follows that $\dim H(E_p)$ is locally constant on $G_p(E)$ only on the open set where it reaches its minimum. Thus, r is constant on $G_p^*(E)$, as we wished to show. $\qquad\square$

Proposition 1.14. *Let $\mathcal{I} \subset \Omega^*(M)$ be a differential ideal which contains no non-zero forms of degree 0. If $\varphi^1, \ldots, \varphi^{c_{n-1}}$ is a polar sequence for the ordinary integral flag $(0) \subset E_1 \subset \cdots \subset E_n \subset T_z M$, then it is also a polar sequence for all nearby integral flags.*

Proof. Obvious. $\qquad\square$

We conclude this section by proving a technical proposition which provides an effective method of computing the numbers c_i which are associated to an integral flag. We need the following terminology: If $J = (j_1, j_2, \ldots, j_p)$ is a multi-index of degree p taken from the set $\{1, 2, \ldots, n\}$, then we define $\sup J$ to be the largest of the integers $\{j_1, j_2, \ldots, j_p\}$. If $J = \emptyset$ is the (unique) multi-index of degree 0, we define $\sup J = 0$.

Proposition 1.15. *Let $\mathcal{I} \subset \Omega^+(M)$ be an ideal which contains no non-zero forms of degree 0. Let $E \in V_n(\mathcal{I})$ be based at $z \in M$. Let $\omega^1, \ldots, \omega^n, \pi^1, \ldots, \pi^s$ (where $s = \dim M - n$) be a coframing on a z-neighborhood so* that $E = \{v \in T_z M \mid \pi^a(v) = 0 \text{ for all } a\}$. For each $p \leq n$, define $E_p = \{v \in E \mid \omega^k(v) = 0 \text{ for all } k > p\}$. Let $\{\varphi^1, \varphi^2, \ldots, \varphi^r\}$ be a set of forms which generate $\mathcal{I}$ algebraically where φ^ρ has degree $d_\rho + 1$.

Then, for each ρ, there exists an expansion

$$\varphi^\rho = \sum_{|J| = d_\rho} \pi^\rho_J \wedge \omega^J + \tilde{\varphi}^\rho$$

where the 1-forms π^ρ_J are linear combinations of the π's and the terms in $\tilde{\varphi}^\rho$ are either of degree 2 or more in the π's or else vanish at z.

Moreover, we have the formula

$$H(E_p) = \{v \in T_z M \mid \pi^\rho_J(v) = 0 \text{ for all } \rho \text{ and } \sup J \leq p\}.$$

In particular, for the integral flag $(0)_z \subset E_1 \subset E_2 \subset \cdots \subset E_n \subset T_z M$ of $\mathcal{I}$, c_p is the number of linearly independent 1-forms in the set $\{\pi^\rho_J|_z \mid \sup J \leq p\}$.

Proof. The existence of the expansion cited for φ^ρ is an elementary exercise in exterior algebra using the fact that E is an integral element of $\mathcal{I}$. The "remainder term" $\tilde{\varphi}^\rho$ has the property that $\tilde{\varphi}^\rho(v, e_1, e_2, \ldots, e_{d_\rho}) = 0$ for all $v \in T_z M$ and all $\{e_1, e_2, \ldots, e_{d_\rho}\} \subset E$. If $e_1, e_2, \ldots, e_n$ is the basis of E which is dual to the coframing $\omega^1, \omega^2, \ldots, \omega^n$ and $K = (k_1, k_2, \ldots, k_{d_\rho})$ is a multi-index with $\deg K = d_\rho$, we have the formula $\varphi^\rho(v \wedge e_K) = \pi^\rho_K(v)$. The stated formulas for $H(E_p)$ and c_p follow immediately. $\square$

To see the utility of Proposition 1.15, consider Example 1.12. Here, $\mathcal{I}$ is generated by two forms $\varphi^1 = \vartheta$ and $\varphi^2 = -d\vartheta$, of degrees 2 and 3 respectively. If $E \in G_3(T\mathbb{R}^6, \Omega)$ is any integral element, then the annihilator of E is spanned by 1-forms $\pi_i = du_i - p_{ij} dx^j$ for some numbers p_{ij}. It is clear that we have expansions of the form

$$\varphi^1 = \vartheta = \pi_1 \wedge dx^1 + \pi_2 \wedge dx^2 + \pi_3 \wedge dx^3 + \tilde{\varphi}^1$$

$$\varphi^2 = -d\vartheta = \pi_3 \wedge dx^1 \wedge dx^2 + \pi_2 \wedge dx^3 \wedge dx^1 + \pi_1 \wedge dx^2 \wedge dx^3 + \tilde{\varphi}^2,$$

on a neighborhood of the base point of E. By Proposition 1.15, it follows that the annihilator of $H(E_1)$ is spanned by $\{\pi_1\}$ and the annihilator of $H(E_2)$ is spanned by $\{\pi_1, \pi_2, \pi_3\}$. Thus, we must have $c_1 = 1$ and $c_2 = 3$, as we computed before.

§2. The Cartan–Kähler Theorem.

In this section, we prove the Cartan–Kähler theorem, which is the fundamental existence result for integral manifolds of a real-analytic differential system. This theorem is a coordinate-free, geometric generalization of the classical Cauchy–Kowalevski theorem, which we now state.

We shall adopt the index ranges $1 \leq i, j \leq n$ and $1 \leq a, b \leq s$.

Theorem 2.1 (Cauchy–Kowalevski). *Let y be a coordinate on $\mathbb{R}$, let $x = (x^i)$ be coordinates on $\mathbb{R}^n$, let $z = (z^a)$ be coordinates on $\mathbb{R}^s$, and let $p = (p_i^a)$ be coordinates on $\mathbb{R}^{ns}$. Let $\mathcal{D} \subset \mathbb{R}^n \times \mathbb{R} \times \mathbb{R}^s \times \mathbb{R}^{ns}$ be an open domain, and let $G : \mathcal{D} \to \mathbb{R}^s$ be a real analytic mapping. Let $\mathcal{D}_0 \subset \mathbb{R}^n$ be an open domain and let $f : \mathcal{D}_0 \to \mathbb{R}^s$ be a real analytic mapping so that the "1-graph"*

$$(11) \qquad \Gamma_f = \{(x, y_0, f(x), Df(x)) \mid x \in \mathcal{D}_0\}$$

lies in $\mathcal{D}$ for some constant y_0. (Here, $Df(x) \in \mathbb{R}^{ns}$, the Jacobian of f, is described by the condition that $p_i^a(Df(x)) = \partial f^a(x)/\partial x_i$.)

Then there exists an open neighborhood $\mathcal{D}_1 \subset \mathcal{D}_0 \times \mathbb{R}$ of $\mathcal{D}_0 \times \{y_0\}$ and a real analytic mapping $F : \mathcal{D}_1 \to \mathbb{R}^s$ which satisfies the P.D.E. with initial condition

$$(12) \qquad \begin{aligned} \partial F/\partial y &= G(x, y, F, \partial F/\partial x) \\ F(x, y_0) &= f(x) \text{ for all } x \in \mathcal{D}_0. \end{aligned}$$

Moreover, F is unique in the sense that any other real-analytic solution of (12) agrees with F on some neighborhood of $\mathcal{D}_0 \times \{y_0\}$.

We shall not prove the Cauchy–Kowalevski theorem here, but refer the reader to other sources, such as Trêves [1975] or Spivak [1979]. We remark, however, that the assumption of real analyticity is necessary in both the function G (which defines the system of P.D.E.) and the initial condition f. In the smooth category, there are examples where the existence part of the above statement fails and there are other examples where the uniqueness part of the above statement fails.

We now turn to the statement of the Cartan–Kähler theorem. If $\mathcal{I} \subset \Omega^*(M)$ is a differential ideal, we shall say that an integral manifold of $\mathcal{I}$, $V \subset M$, is a *Kähler-regular* integral manifold if the tangent space $T_v V$ is a Kähler-regular integral element of $\mathcal{I}$ for all $v \in V$. If V is a connected, Kähler-regular integral manifold of $\mathcal{I}$, then we define $r(V)$ to be $r(T_v V)$ where v is any element of V.

Theorem 2.2 (Cartan–Kähler). *Let $\mathcal{I} \subset \Omega^*(M)$ be a real analytic differential ideal. Let $P \subset M$ be a connected, p-dimensional, real analytic, Kähler-regular integral manifold of $\mathcal{I}$.*

Suppose that $r = r(P)$ is a non-negative integer. Let $R \subset M$ be a real analytic submanifold of M which is of codimension r, which contains P, and which satisfies the condition that $T_x R$ and $H(T_x P)$ are transverse in $T_x M$ for all $x \in P$.

Then there exists a real analytic integral manifold of $\mathcal{I}$, X, which is connected and $(p + 1)$-dimensional and which satisfies $P \subset X \subset R$. This *manifold is unique in the sense that any other real analytic integral manifold of $\mathcal{I}$ with these properties agrees with X on an open neighborhood of P.*

Proof. The theorem is local, so it suffices to prove existence and uniqueness in a neighborhood of a single point $x_0 \in P$. Let $s = \dim M - (r + p + 1)$. (The following proof holds with the obvious simplifications if any of p, r, or s are zero. For simplicity of notation, we assume that they are all positive.)

Our hypothesis implies that the vector space $T_x R \cap H(T_x P)$ has dimension $p+1$ for all $x \in P$. It follows that we may choose a local (real analytic) system of coordinates centered on x_0 of the form $x^1, \ldots, x^p, y, u^1, \ldots, u^s, v^1, \ldots, v^r$ so that P is given in this neighborhood by the equations $y = u = v = 0$, R is given in this neighborhood by the equations $v = 0$, and, for all $x \in P$, the polar space $H(T_x P)$ is spanned by the vectors $\{\partial/\partial x^j\}_{1 \le j \le p} \cup \{\partial/\partial y\} \cup \{\partial/\partial v^\rho\}_{1 \le \rho \le r}$.

Now, there exists a neighborhood U of $T_{x_0} P$ in $G_p(TM)$ so that every $E \in U$ with base point $z \in M$ has a basis of the form

$$X_i(E) = (\partial/\partial x^i + q_i(E)\partial/\partial y + p_i^\sigma(E)\partial/\partial u^\sigma + w_i^\rho(E)\partial/\partial v^\rho)|_z.$$

The functions x, y, u, v, q, p, w form a coordinate system on U centered on $T_{x_0} P$. By the definition of $H(T_{x_0} P)$, there exist s real analytic $(p+1)$-forms $\kappa^1, \ldots, \kappa^s$ in $\mathcal{I}$ with the property that

$$H(T_{x_0} P) = \{v \in T_{x_0} M \mid \kappa^\sigma(v, \partial/\partial x^1, \partial/\partial x^2, \ldots, \partial/\partial x^p) = 0 \text{ for } 1 \le \sigma \le s\}.$$

In fact, we may even assume that $\kappa^\sigma(v, \partial/\partial x^1, \partial/\partial x^2, \ldots, \partial/\partial x^p) = du^\sigma(v)$ for $1 \le \sigma \le s$ and all $v \in T_{x_0} M$. By the Kähler-regularity of $T_{x_0} P$, we may assume, by shrinking U if necessary, that, for all $E \in V_p(\mathcal{I}) \cap U$ with base point $z \in M$, we have

$$H(E) = \{v \in T_z M \mid \kappa^\sigma(v, X_1(E), X_2(E), \ldots, W_p(E)) = 0 \text{ for } 1 \le \sigma \le s\}.$$

If we seek $v \in H(E)$ of the form

$$v = (a\partial/\partial y + b^\sigma \partial/\partial u^\sigma + c^\rho \partial/\partial v^\rho)|_z,$$

then the s equations $\kappa^\sigma(v, X_1(E), X_2(E), \ldots, X_p(E)) = 0$ are, of course, linear equations for the quantities a, b, and c of the form

$$A^\sigma(E)a + B_\tau^\sigma(E)b^\tau + C_\rho^\sigma(E)c^\rho = 0.$$

Again, by hypothesis, when $E = T_{x_0} P$ these s equations are linearly independent and reduce to the equations $b^\sigma = 0$. Thus, by shrinking U if necessary, we may assume that the $s \times s$ matrix $B(E) = (B_\tau^\sigma(E))$ is invertible for all $E \in U$. It follows that there exist unique real analytic functions G^σ on U so that, for each $E \in U$ based at $z \in M$, the vector $Y(E) = (\partial/\partial y + G^\sigma(E)\partial/\partial u^\sigma)|_z$ satisfies

$$\kappa^\sigma(Y(E), X_1(E), X_2(E), \ldots, X_p(E)) = 0.$$

Since the functions x, y, u, v, q, p, w form a coordinate system on U centered on $T_{x_0}P$, we may regard the functions G^σ as functions of these variables.

We first show that there exists a real analytic submanifold of R of the form $v = 0$, $u = F(x, y)$ on which the forms κ^σ vanish. Note that the following vectors would be a basis of the tangent space to such a submanifold at the point $z(x, y) = (x, y, F(x, y), 0)$:

$$X_i(x, y) = (\partial/\partial x^i + \partial_i F^\sigma(x, y)\partial/\partial u^\sigma)|_{z(x,y)}$$

$$Y(x, y) = (\partial/\partial y + \partial_y F^\sigma(x, y)\partial/\partial u^\sigma)|_{z(x,y)}.$$

It follows that the function F would have to be a solution to the system of P.D.E. given by

$$(13) \qquad\qquad \partial_y F^\sigma = G^\sigma(x, y, F, 0, 0, \partial_x F, 0).$$

Moreover, in order that the submanifold contain P (which is given by the equations $y = u = v = 0$), it is necessary that the function F satisfy the initial condition

$$(14) \qquad\qquad\qquad F(x, 0) = 0.$$

Conversely, if F satisfies (13) and (14), then the submanifold of R given by $v = 0$ and $u = F(x, y)$ will both contain P and be an integral of the set of forms $\{\kappa^\sigma\}_{1 \le \sigma \le s}$.

By the Cauchy–Kowalevski theorem, there exists a unique real analytic solution F of (13) and (14). We let $X \subset R$ denote the (unique) submanifold of dimension $p + 1$ constructed by this method. Replacing the functions u in our coordinate system by the functions $u - F(x, y)$ will not disturb any of our normalizations so far and allows us to suppose, as we shall for the remainder of the proof, that X is described by the equations $u = v = 0$.

We must now show that X is an integral manifold of $\mathcal{I}$. We have already seen that X is the unique connected real analytic submanifold of dimension $p + 1$ which satisfies $P \subset X \subset R$ and is an integral of the forms $\{\kappa^\sigma\}_{1 \le \sigma \le s}$. We now show that all of the p-forms in $\mathcal{I}$ vanish on X.

Again using the Kähler-regularity of $T_{x_0}P$, let $\beta^1, \ldots, \beta^a$ be a set of real analytic p-forms in $\mathcal{I}$ so that the functions $f^c(E) = \beta^c(X_1(E), \ldots, X_p(E))$ for $1 \le c \le a$ have linearly independent differentials on U and have the locus $V_p(\mathcal{I}) \cap U$ as their set of common zeros. (We may have to shrink U once more to do this.) Since $T_{x_0}X$ lies in $V_{p+1}(\mathcal{I})$ by construction,

Proposition 1.8 shows that $T_{x_0}X$ is Kähler-ordinary. In fact, the proof of Proposition 1.8 shows that the $(p+1)$-forms $\{\beta^c \wedge dy\}_{1 \leq c \leq a} \cup \{\kappa^\sigma\}_{1 \leq \sigma \leq s}$ have $V_{p+1}(\mathcal{I}) \cap U^+$ as their set of ordinary common zeros in some neighborhood U^+ of $T_{x_0}X$ in $G_{p+1}(TM)$. Thus, in order to show that X is an integral manifold of $\mathcal{I}$, it suffices to show that the forms $\{\beta^c \wedge dy\}_{1 \leq c \leq a}$ vanish on X.

Shrinking U^+ if necessary, we may suppose that every $E^+ \in U^+$ has a basis $X_1(E^+), X_2(E^+), \ldots, X_p(E^+), Y(E^+)$ which is dual to the basis of 1-forms $dx^1, dx^2, \ldots, dx^p, dy$. If we set

$$B^c(E^+) = \beta^c \wedge dy(X_1(E^+), X_2(E^+), \ldots, X_p(E^+), Y(E^+))$$

$$K^\sigma(E^+) = \kappa^\sigma(X_1(E^+), X_2(E^+), \ldots, X_p(E^+), Y(E^+)),$$

then we have

$$V_{p+1}(\mathcal{I}) \cap U^+ = \{E^+ \in U^+ \mid B^c(E^+) = K^\sigma(E^+) = 0\}.$$

Since $\mathcal{I}$ is an ideal, the forms $\beta^c \wedge dx^i$ are also in $\mathcal{I}$ and hence vanish on $V_{p+1}(\mathcal{I}) \cap U^+$. Thus, if we set

$$B^{ci}(E^+) = \beta^c \wedge dx^i(X_1(E^+), X_2(E^+), \ldots, X_p(E^+), Y(E^+)),$$

then the functions B^{ci} are in the ideal generated by the functions B^c and K^σ. It follows that there exist real analytic functions A and L on U^+ so that

$$B^{ci} = A^{ci}_b B^b + L^{ci}_\sigma K^\sigma.$$

Since $K^\sigma(T_zX) = 0$ for all $z \in X$ by construction, it follows that

$$B^{ci}(T_zX) = A^{ci}_b(T_zX)B^b(T_zX)$$

for all $z \in X$.

Since $\mathcal{I}$ is differentially closed, the forms $d\beta^c$ are in $\mathcal{I}$. Thus, if we set

$$D^c(E^+) = d\beta^c(X_1(E^+), X_2(E^+), \ldots, X_p(E^+), Y(E^+)),$$

there must exist functions G and H on U^+ so that

$$D^c = G^c_b B^b + H^c_\sigma K^\sigma.$$

Again, since $K^\sigma(T_zX) = 0$, we must have

$$D^c(T_zX) = G^c_b(T_zX)B^b(T_zX)$$

for all $z \in X$.

Now, if we restrict the forms β^c to X, then we have an expansion of the form

$$\beta^c|_X = B^c(x,y)dx^1 \wedge \cdots \wedge dx^p$$

$$+ \sum_i (-1)^{p-i+1} B^{ci}(x,y)dx^1 \wedge \cdots \wedge dx^{i-1} \wedge dx^{i+1} \wedge \cdots \wedge dx^p \wedge dy,$$

where, for $z = (x,y,0,0) \in X$, we have set $B^c(x,y) = B^c(T_z X)$ and $B^{ci}(x,y) = B^{ci}(T_z X)$. We also have the formula

$$d\beta^c|_X = (-1)^p \left(\partial_y B^c(x,y) + \sum_i \partial_i B^{ci}(x,y) \right) dx^1 \wedge \cdots \wedge dx^p \wedge dy$$

$$= D^c(x,y)dx^1 \wedge \cdots \wedge dx^p \wedge dy,$$

where we have written $D^c(x,y) = D^c(T_z X)$ for z as before. Using the formulas

$$D^c(x,y) = G^c_b(x,y)B^b(x,y)$$

$$B^{ci}(x,y) = A^{ci}_b(x,y)B^b(x,y),$$

we see that the functions $B^c(x,y)$ satisfy a linear system of P.D.E. of the form

$$\partial_y B^c(x,y) = \tilde{A}^{ci}_b(x,y)\partial_i B^b(x,y) + \tilde{G}^c_b(x,y)B^b(x,y)$$

for some functions $\tilde{A}$ and $\tilde{G}$ on X. Moreover, since $T_z X$ is an integral element of $\mathcal{I}$ when $y = 0$, we have the initial conditions $B^c(x,0) = 0$. By the uniqueness part of the Cauchy-Kowalevski theorem and the fact that all of the functions involved are real-analytic, it follows that the functions $B^c(x,y)$ must vanish identically. In turn, this implies that the forms β^c vanish on X. Hence X is an integral manifold of $\mathcal{I}$, as we wished to show. Since we have already established uniqueness, we are done. $\square$

The role of the "restraining manifold" R in the Cartan–Kähler theorem is to convert the "underdetermined" Cauchy problem one would otherwise encounter in extending P to a $(p+1)$-dimensional integral to a determined problem. In the coordinate system we introduced in the proof, we could have taken, instead of R, which was defined by the equations $v^\rho = 0$, the submanifold $\tilde{R}$, defined by the equations $v^\rho = f^\rho(x,y)$ where the functions f^ρ are "small" but otherwise arbitrary real analytic functions of the $p+1$ variables $(x^1, x^2, \ldots, x^p, y)$. The construction in the above proof would then have lead to an integral manifold $\tilde{X}$ of dimension $p+1$ defined locally by equations $u^\sigma = g^\sigma(x,y)$ and $v^\rho = f^\rho(x,y)$. In this sense, the $(p+1)$-dimensional extensions of a given p-dimensional, Kähler-regular integral manifold P of $\mathcal{I}$ depend on $r(P)$ functions of $p+1$ variables.

The Cartan–Kähler theorem has the following extremely useful corollary. (In fact, this corollary is used more often than Theorem 2.2 and is often called the Cartan–Kähler theorem, even in this book.)

Corollary 2.3. *Let $\mathcal{I}$ be an analytic differential ideal on a manifold M. Let $E \subset T_x M$ be an ordinary integral element of $\mathcal{I}$. Then there exists an integral manifold of $\mathcal{I}$ which passes through x and whose tangent space at x is E.*

Proof. Assume that the dimension of E is n and let $(0)_x = E_0 \subset E_1 \subset E_2 \subset \cdots \subset E_n = E \subset T_x M$ be an ordinary integral flag. Suppose that, for some $p < n$, we have found a p-dimensional, regular, real analytic integral manifold X_p of $\mathcal{I}$ which passes through x and which satisfies $T_x X_p = E_p$. Then it is easy to see that there exists a real analytic manifold $R_p \subset M$ which contains X_p, is of codimension $r(E_p)$, and satisfies $T_x R_p \cap H(E_p) = E_{p+1}$. Shrinking X_p if necessary, we may assume that $T_z R_p$ is transverse to $H(T_z X_p)$ for all $z \in X_p$. Applying Theorem 2.2, we see that there exists a real analytic integral manifold of $\mathcal{I}$ of dimension $(p+1)$, X_{p+1}, with the property that $T_x X_{p+1} = E_{p+1}$. If $p+1 < n$, then E_{p+1} is a Kähler-regular integral element of $\mathcal{I}$ and hence, by shrinking X_{p+1} if necessary, we may assume that X_{p+1} is a $(p+1)$-dimensional, Kähler-regular, real analytic integral manifold of $\mathcal{I}$. If $p+1 = n$, then X_{p+1} is the desired integral manifold. $\qquad\square$

We conclude this section by explaining some classical terminology regarding the "generality" of the space of ordinary integral manifolds of an analytic differential system $\mathcal{I}$. For simplicity, let us suppose that $\mathcal{I}$ is a real analytic differential system on a manifold M and that $\mathcal{I}$ contains no non-zero 0-forms. Let $(0)_z = E_0 \subset E_1 \subset \cdots \subset E_n \subset T_z M$ be an ordinary integral flag of $\mathcal{I}$. As usual, for $0 \leq k \leq n-1$, let c_k be the codimension of $H(E_k)$ in $T_z M$. For convenience of notation, let us set $c_{-1} = 0$ and $c_n = s = \dim M - n$. Then we may choose a z-centered coodinate system on some z-neighborhood of the form $x^1, x^2, \ldots, x^n, u^1, u^2, \ldots, u^s$ so that E_k is spanned by the vectors $\{\partial/\partial x^j\}_{1 \leq j \leq k}$ and so that, for $k < n$,

$$H(E_k) = \{v \in T_z M \mid du^a(v) = 0 \text{ for all } a \leq c_k\}.$$

For any integer a between 1 and s, let $\lambda(a)$, the *level* of a, be the integer k between 0 and n which satisfies $c_{k-1} < a \leq c_k$. The number of integers of level k is clearly $c_k - c_{k-1}$. The (non-negative) number $s_k = c_k - c_{k-1}$ is called the kth *Cartan character* of the given integral flag.

Let $\Omega = dx^1 \wedge dx^2 \wedge \cdots \wedge dx^n$ and let $V_n(\mathcal{I}, \Omega)$ denote the space of n-dimensional integral elements of $\mathcal{I}$ on which Ω does not vanish. For each $\tilde{E} \in V_n(\mathcal{I}, \Omega)$, let us define $\tilde{E}_k = \{v \in \tilde{E} \mid dx^j(v) = 0 \text{ for all } j > k\}$. Then for each k, the map $\tilde{E} \mapsto \tilde{E}_k$ is a continuous mapping from $V_n(\mathcal{I}, \Omega)$ to $V_k(\mathcal{I})$. It follows that there exists a (connected) neighborhood U of E_n in $V_n(\mathcal{I}, \Omega)$ with the property that $\tilde{E}_k$ is Kähler-regular for all $k < n$ and all $\tilde{E} \in U$. By shrinking U if necessary, we may even suppose that the 1-forms

$\{dx^j\}_{1 \le j \le n}$ and $\{du^a\}_{a > c_k}$ are linearly independent on $H(\tilde{E}_k)$ for all $k < n$ and all $\tilde{E} \in U$.

We now want to give a description of the collection $\mathcal{C}$ of real analytic n-dimensional integral manifolds of $\mathcal{I}$ whose tangent spaces all belong to U and which intersect the locus $x = 0$. By Corollary 2.3, we know that $\mathcal{C}$ is non-empty. If X belongs to $\mathcal{C}$, then locally, we may describe X by equations of the form $u^a = F^a(x^1, x^2, \ldots, x^n)$. If the index a has level k, let us define f^a to be the function of k variables given by $f^a(x^1, x^2, \ldots, x^k) = F^a(x^1, x^2, \ldots, x^k, 0, 0, \ldots, 0)$. By convention, for level 0 we speak of "functions of 0 variables" as "constants".) Then the collection $\{f^a\}_{1 \le a \le s}$ is a set of s_0 constants, s_1 functions of 1 variable, s_2 functions of 2 variables, $\ldots$, and s_n functions of n variables.

We now claim that the collection $\{f^a\}_{1 \le a \le s}$ characterizes X in the sense that any $\tilde{X}$ in $\mathcal{C}$ which gives rise to the same collection of functions $\{f^a\}_{1 \le a \le s}$ agrees with X on a neighborhood of the point $(x, u) = (0, F(0))$. Moreover, the functions in the collection $\{f^a\}_{1 \le a \le s}$ are required to be "small", but are otherwise arbitrary. It is in this sense that $\mathcal{C}$ is parametrized by s_0 constants, s_1 functions of 1 variable, s_2 functions of 2 variables, $\ldots$, and s_n functions of n variables. It is common to interpret this as meaning that the local n-dimensional integrals of $\mathcal{I}$ depend on s_0 constants, s_1 functions of 1 variable, s_2 functions of 2 variables, $\ldots$, and s_n functions of n variables.

To demonstrate our claim, let $\{f^a\}_{1 \le a \le s}$ be a collection of real analytic functions which are suitably "small" and where f^a is a function of the variables $x^1, x^2, \ldots, x^{\lambda(a)}$. For $1 \le k \le n$, define the manifold R_k to be the locus of the equations $x^{k+1} = x^{k+2} = \cdots = x^n = 0$ and $u^a = f^a(x^1, \ldots, x^k, 0, \ldots, 0)$ where a ranges over all indices of level greater than or equal to k. The codimension of R_k is $n - k + (s - c_{k-1}) = r(E_{k-1})$. Define X_0 to be the point $(x, u) = (0, f(0))$. By sucessive applications of the Cartan–Kähler theorem, we may construct a unique nested sequence of integral manifolds of $\mathcal{I}$, $\{X_k\}_{0 \le k \le n}$, which also satisfy the conditions $X_k \subset R_k$. (It is at this stage that we use the assumption of "smallness" in order to guarantee that the necessary transversality conditions hold.) This clearly demonstrates our claim.

The following proposition shows that the sequence of Cartan characters has an invariant meaning.

Proposition 2.4. *Let $\mathcal{I} \subset \Omega^*(M)$ be a smooth differential system which contains no 0-forms. Let $Z \subset V_n(\mathcal{I})$ be a component of the space of ordinary integral elements. Then the sequence of Cartan characters $(s_0, s_1, \ldots, s_n)$ is the same for all ordinary integral flags $(0)_z \subset E_1 \subset \cdots \subset \tilde{E}_n$ with $E_n \in Z$.*

Proof. This follows immediately from Proposition 1.13 and the definitions of the s_k, namely: $s_0 = c_0$, $s_k = c_k - c_{k-1}$ for $1 \le k < n$, and $s_n = s - c_{n-1}$.

Usually, the component Z must be understood from context when such statements as "The system $\mathcal{I}$ has Cartan characters $(s_0, s_1, \ldots, s_n)$." are made. In fact, in most cases of interest, the space of ordinary, n-dimensional integral elements of $\mathcal{I}$ has only one component anyway.

§3. Examples.

In this section, we give some applications of the Cartan–Kähler theorem. Some of the examples are included merely to demonstrate techniques for calculating the quantities one must calculate in order to apply the Cartan–Kähler theorem, while others are more substantial. The most important example in this section is the application of the Cartan–Kähler theorem to the problem of isometric embedding (see Example 3.8).

Example 3.1 (The Frobenius theorem). Let M be a manifold of dimension $m = n + s$ and let $\mathcal{I}$ be a differential system which is generated algebraically in degree 1 by a Pfaffian system $I \subset T^*M$ of rank s. Then at each $x \in M$, there is a unique integral element of dimension n, namely $I_x^\perp \in T_x M$. In fact, every integral element of $\mathcal{I}$ based at x must be a subspace of $I_x^\perp$, since $H((0)_x) = I_x^\perp$. Thus, if $(0)_x \subset E_1 \subset \cdots \subset E_n = I_x^\perp$ is an integral flag, then we have $H(E_p) = I_x^\perp$ for all $0 \leq p \leq n$. Thus $c_p = s$ for all p. It follows by Theorem 1.11 that $V_n(\mathcal{I})$ must have codimension at least ns in $G_n(TM)$. On the other hand, since there is a unique integral element of $\mathcal{I}$ at each point of M, it follows that $V_n(\mathcal{I})$ is a smooth manifold of dimension $n + s$ while $G_n(TM)$ has dimension $n + s + ns$. Thus, $V_n(\mathcal{I})$ is a smooth submanifold of codimension ns in $G_n(TM)$. By Theorem 1.11, it follows that all of the elements of $V_n(\mathcal{I})$ are ordinary. If we now assume that $\mathcal{I}$ is real analytic, then the Cartan–Kähler theorem applies (in the form of Corollary 2.3) to show that there exists an n-dimensional integral manifold of $\mathcal{I}$ passing through each point of M. The characters are $s_0 = s$ and $s_p = 0$ for all $p > 0$. Thus, according to our discussion at the end of Section 2, the local integral manifolds of $\mathcal{I}$ of dimension n depend on s constants. This is in accordance with the usual theory of foliations.

The assumption of analyticity is, of course, not necessary. We have already proved the Frobenius theorem in the smooth category in Chapter II. The reader might find it helpful to compare this proof with the proof given there.

Example 3.2 (Orthogonal coordinates). Let g be a Riemannian metric on a manifold N of dimension n. We wish to know when there exist local coordinates $x^1, x^2, \ldots, x^n$ on N so that the metric takes the diagonal form

$$g = g_{11}(dx^1)^2 + g_{22}(dx^2)^2 + \cdots + g_{nn}(dx^n)^2.$$

Equivalently, we wish the coordinate vector fields $\{\partial/\partial x^i\}_{1\leq i\leq n}$ to be orthogonal with respect to the metric. Such a local coordinate system is said to be *orthogonal*.

If $x^1, x^2, \ldots, x^n$ is such an orthogonal coordinate system, then the set of 1-forms $\eta_i = \sqrt{g_{ii}}dx^i$ is a local orthonormal coframing of N. These forms clearly satisfy the equations $\eta_i \wedge d\eta_i = 0$. Conversely, if $\eta_1, \eta_2, \ldots, \eta_n$ is a local orthonormal coframing on M which satisfies the equations $\eta_i \wedge d\eta_i = 0$, then the Frobenius theorem implies that there exist local functions $x^1, x^2, \ldots, x^n$ on N so that $\eta_i = f_i dx^i$ (no summation) for some non-zero functions f_i. It follows that the functions $x^1, x^2, \ldots, x^n$ form a local orthogonal coordinate system on N. Thus, our problem is essentially equivalent to the problem of finding local orthonormal coframes $\eta_1, \eta_2, \ldots, \eta_n$ which satisfy the equations $\eta_i \wedge d\eta_i = 0$.

Let $\mathcal{F} \to N$ denote the bundle of orthonormal coframes for the metric g on N. Thus, for each $x \in N$, the fiber $\mathcal{F}_x$ consists of the set of all orthonormal coframings of the tangent space $T_x N$. The bundle $\mathcal{F}$ has a canonical coframing $\omega_i, \omega_{ij} = -\omega_{ji}$ which satisfies the structure equations of É. Cartan:

$$dw_i = -\sum_j \omega_{ij} \wedge \omega_j$$

$$dw_{ij} = -\sum_k \omega_{ik} \wedge \omega_{kj} + \tfrac{1}{2} \sum_{k,l} R_{ijkl}\, \omega_k \wedge \omega_l.$$

The forms ω_i have the "reproducing property" described as follows: If $\eta = (\eta_1, \eta_2, \ldots, \eta_n)$ is any local orthonormal coframing defined on an open set $U \subset M$, then η may be regarded as a local section $\eta : U \to \mathcal{F}$ of $\mathcal{F}$. Then the formula $\eta^*(\omega_i) = \eta_i$ holds.

We set $\Omega = \omega_1 \wedge \omega_2 \wedge \cdots \wedge \omega_n$. Notice that Ω does not vanish on the submanifold $\eta(U) \subset \mathcal{F}$ since $\eta^*(\Omega) = \eta_1 \wedge \eta_2 \wedge \cdots \wedge \eta_n \neq 0$. Conversely, it is clear that any n-dimensional submanifold $X \subset \mathcal{F}$ on which Ω does not vanish is locally of the form $\eta(U)$ for some section η. Let $\mathcal{I}$ denote the differential system on $\mathcal{F}$ generated by the n 3-forms $\Theta_i = \omega_i \wedge d\omega_i$. If $X \subset \mathcal{F}$ is an integral of the system $\mathcal{I}$ on which Ω does not vanish, then clearly X is locally of the form $\eta(U)$ where η is a local orthonormal coframing satisfying our desired equations $\eta_i \wedge d\eta_i = 0$. Conversely, a local orthonormal coframing η satisfying $\eta_i \wedge d\eta_i = 0$ has the property that $\eta(U)$ is an n-dimensional integral manifold of $\mathcal{I}$ on which Ω does not vanish.

We proceed to analyse the n-dimensional integral manifolds of $\mathcal{I}$ on which Ω does not vanish. Note that $\mathcal{I}$ is generated algebraically by the 3-forms Θ_i and the 4-forms $\Psi_i = d\Theta_i$. Let $E \in V_n(\mathcal{I}, \Omega)$ be based at $f \in \mathcal{F}$. When we restrict the forms ω_i, ω_{ij} to E, the forms ω_i remain linearly independent, and we have relations of the form

$$\Theta_i = -\left(\sum_j \omega_{ij} \wedge \omega_j\right) \wedge \omega_i = 0.$$

It follows that there exist 1-forms $\lambda_i = \sum_j L_{ij}\omega_j$ so that $\left(\sum_j \omega_{ij} \wedge \omega_j\right) = \lambda_i \wedge \omega_i$. Collecting terms, we have the equation

$$\sum_j (\omega_{ij} + L_{ij}\omega_i - L_{ji}\omega_j) \wedge \omega_j = 0.$$

Since the forms $\varphi_{ij} = (\omega_{ij} + L_{ij}\omega_i - L_{ji}\omega_j)$ are skew-symmetric in their indices, it follows that the above equations can only hold if we have

$$\omega_{ij} + L_{ij}\omega_i - L_{ji}\omega_j = 0.$$

Conversely, we claim that if $\{L_{ij}\}_{i \neq j}$ is any set of $n^2 - n$ numbers, then the n-plane $E \subset T_f\mathcal{F}$ annihilated by the 1-forms $\varphi_{ij} = (\omega_{ij} + L_{ij}\omega_i - L_{ji}\omega_j)$ is an integral element of $\mathcal{I}$ on which Ω does not vanish. To see this, note that for such E, we have the identity $-d\omega_i = \left(\sum_j \omega_{ij} \wedge \omega_j\right) = \lambda_i \wedge \omega_i$. It immediately follows that $\Theta_i = \omega_i \wedge d\omega_i$ and $\Psi_i = d\omega_i \wedge d\omega_i$ must vanish on E.

It follows that the space of integral elements of $\mathcal{I}$ which are based at a point of $\mathcal{F}$ is naturally a smooth manifold of dimension $n^2 - n$. Moreover, the space $V_n(\mathcal{I}, \Omega)$ is a smooth manifold of dimension $\dim \mathcal{F} + (n^2 - n)$. Thus, the codimension of $V_n(\mathcal{I}, \Omega)$ in $G_n(\mathcal{F})$ is $(n - 2)\binom{n}{2}$.

When $n = 2$, we are looking for integrals of dimension 2. However, $\mathcal{I}$ has no non-zero forms of degree less than 3. It follows that any surface in $\mathcal{F}$ is an integral of $\mathcal{I}$.

From now on, we assume that $n \geq 3$. Since $\dim \mathcal{F} = \frac{1}{2}(n^2 + n)$, if $(0)_f \subset E_1 \subset \cdots \subset E_n$ is an integral flag, it follows that $c_p \leq \binom{n}{2}$ for all $0 \leq p \leq n$. However, since $\mathcal{I}$ contains no non-zero forms of degree less than 3, it follows that $c_0 = c_1 = 0$.

Moreover, since $\mathcal{I}$ contains only n 3-forms, it follows that $c_2 \leq n$. Thus, we have the inequality

$$c_0 + c_1 + c_2 + \cdots + c_{n-1} \leq n + (n - 3)\binom{n}{2}$$

It follows, by Theorem 1.11, that for $n \geq 4$, none of the elements of $V_n(\mathcal{I}, \Omega)$ are ordinary. Thus, the Cartan–Kähler theorem cannot be directly applied in the case where $n \geq 4$. This is to be expected since a Riemannian metric in n variables has $\binom{n}{2}$ "off diagonal" components in a general coordinate system and a choice of coordinates depends on only n

functions of n variables. Thus, if $n < \binom{n}{2}$, (which holds when $n \geq 4$) we do not expect to be able to diagonalize the "generic" Riemannian metric in n variables by a change of coordinates.

Let us now specialize to the case $n = 3$. By Theorem 1.11 and the calculation above, an integral element $E \in V_3(\mathcal{I}, \Omega)$ is an ordinary integral element if and only if it contains a 2-dimensional integral element E_2 whose polar space has codimension 3. Now a basis for the 3-forms in $\mathcal{I}$ can be taken to be $\{\omega_{23} \wedge \omega_2 \wedge \omega_3, \omega_{31} \wedge \omega_3 \wedge \omega_1, \omega_{12} \wedge \omega_1 \wedge \omega_2\}$. If $E \in V_3(\mathcal{I}, \Omega)$ is given, then let $v_1, v_2 \in E$ be two vectors which span a 2-plane E_2 on which none of the 2-forms $\{\omega_2 \wedge \omega_3, \omega_3 \wedge \omega_1, \omega_1 \wedge \omega_2\}$ vanish. Then it immediately follows that the polar equations of E_2 have rank 3. Thus E is ordinary. This yields the following theorem.

Theorem 3.3 (Cartan). *Let (M^3, g) be a real analytic Riemannian metric. Let S be a real analytic surface in M and let $\eta : S \to \mathcal{F}$ be a real analytic coframing along S so that none of the 2-forms $\{\eta_2 \wedge \eta_3, \eta_3 \wedge \eta_1, \eta_1 \wedge \eta_2\}$ vanishes when restricted to S. Then there is an open neighborhood U of S in M and a unique real analytic extension of η to U so that the equations $\eta_i \wedge d\eta_i = 0$ hold.*

Proof. The surface $\eta(S) \subset \mathcal{F}$ is a Kähler-regular integral of $\mathcal{I}$ by the above calculation. The rest follows from the Cartan–Kähler theorem.

Corollary 3.4. *Let (M^3, g) be a real analytic Riemannian metric. Then every point of M lies in a neighborhood on which there exists a real analytic orthogonal coordinate system for g.*

Remark. Theorem 3.3 has been proved in the smooth category by DeTurck and Yang [1984]. Their proof relies on the theory of the characteristic variety of an exterior differential system, and in that context, this example will be revisited in Chapter V. For example, the conditions on the $\eta_i \wedge \eta_j$ in the above theorem say exactly that S is non-characteristic.

Example 3.5 (Special Lagrangian geometry). This example is due to Harvey and Lawson [1982]. Let $M = \mathbb{C}^n$ with complex coordinates $z_1, z_2, \ldots, z_n$. Let $\mathcal{I}$ be the ideal generated by the 2-form Φ and the n-form Ψ where

$$\Phi = (\sqrt{-1}/2)(dz_1 \wedge d\bar{z}_1 + dz_2 \wedge d\bar{z}_2 + \cdots + dz_n \wedge d\bar{z}_n)$$

and

$$\Psi = \mathrm{Re}(dz_1 \wedge dz_2 \wedge \cdots \wedge dz_n)$$

$$= \tfrac{1}{2}(dz_1 \wedge dz_2 \wedge \cdots \wedge dz_n + d\bar{z}_1 \wedge d\bar{z}_2 \wedge \cdots \wedge d\bar{z}_n).$$

Note that $\mathcal{I}$ is invariant under the group of motions of $\mathbb{C}^n$ generated by the translations and the rotations by elements of $SU(n)$.

We want to examine the set $V_p(\mathcal{I})$ for all p. First assume that $E \in V_p(\mathcal{I})$ where p is less than n. Let $e_1, e_2, \ldots, e_p$ be an orthonormal basis for E, where we use the standard inner product on $\mathbb{C}^n$. Since, for any two vectors $v, w \in \mathbb{C}^n$, the formula $\Phi(v, w) = \langle \sqrt{-1}\, v, w \rangle$ holds, it follows that $\langle \sqrt{-1}\, e_j, e_k \rangle = 0$ for all j and k. Thus, the vectors $e_1, e_2, \ldots, e_p$ are Hermitian orthogonal as well as Euclidean orthogonal. Since $p < n$, it follows that by applying a rotation from $SU(n)$, we may assume that $e_k = \partial/\partial x^k$ where we define the usual real coordinates on $\mathbb{C}^n$ by the equation $z^k = x^k + \sqrt{-1}\, y^k$. It follows that the group of motions of $\mathbb{C}^n$ which preserve $\mathcal{I}$ acts transitively on the space $V_p(\mathcal{I})$ for all $p < n$. In particular, the polar spaces of all of the elements of $V_p(\mathcal{I})$ have the same dimension. Since $\mathcal{I}$ contains no non-zero 0-forms, Proposition 1.10 now shows that every integral element of $\mathcal{I}$ of dimension less than n is Kähler-regular. Thus, every integral element of $\mathcal{I}$ of dimension n is ordinary.

For each $p < n$, let E_p be spanned by the vectors $\{\partial/\partial x^k\}_{k \leq p}$. Then it is easy to compute that, for $p < n - 1$,

$$H(E_p) = \{v \in T_z\mathbb{C}^n \mid dy^k(v) = 0 \text{ for all } k \leq p\}.$$

On the other hand, we have

$$H(E_{n-1}) = \{v \in T_z\mathbb{C}^n \mid dy^k(v) = 0 \text{ for all } k \leq n - 1 \text{ and } dx^n(v) = 0\}.$$

Thus, $c_p = p$ for all $p < n - 1$ and $c_{n-1} = n$. In particular, note that there are no integral elements of dimension $n + 1$ or greater, and each $E \in V_n(\mathcal{I})$ is the polar space of any of its $(n - 1)$-dimensional subspaces. It follows that the group of motions of $\mathbb{C}^n$ which preserve $\mathcal{I}$ acts transitively on $V_n(\mathcal{I})$ as well.

Using the technique of calibrations, Harvey and Lawson show that any n-dimensional integral manifold $N^n \subset \mathbb{C}^n$ of $\mathcal{I}$ is absolutely area minimizing with respect to compact variations. They call such manifolds *special Lagrangian*. We may now combine our discussion of the integral elements of $\mathcal{I}$ with the Cartan–Kähler theorem to prove one of their results:

Theorem 3.6. *Every $(n - 1)$-dimensional real analytic submanifold $P \subset \mathbb{C}^n$ on which Φ vanishes lies in a unique real analytic n-dimensional integral manifold of $\mathcal{I}$.*

Remark. Because of the area minimizing property of the n-dimensional integrals of $\mathcal{I}$, it follows that every integral of $\mathcal{I}$ is real analytic. Thus, if $P \subset \mathbb{C}^{n-1}$ is an integral of Φ which is not real analytic, then there may be no extension of P to an n-dimensional integral of $\mathcal{I}$ in $\mathbb{C}^n$. As an example of such a P, consider the submanifold defined by the equations

$$z^n = y^1 = y^2 = \cdots = y^{n-2} = y^{n-1} - f(x^{n-1}) = 0$$

where f is a smooth function of x^{n-1} which is not real analytic. This shows that the assumption of real analyticity in the Cartan–Kähler theorem cannot be omitted in general.

Example 3.7 (An equation with degenerate symbol). This example is a generalization of Example 1.12. Let a vector field V be given in $\mathbb{R}^3$. Let λ be a fixed constant. We wish to determine whether there exists a vector field U in $\mathbb{R}^3$ which satisfies the system of 3 equations

$$\operatorname{curl} U + \lambda U = V.$$

Note that even though this is three equations for three unknowns, this set of equations cannot be put in Cauchy–Kowalevski form. In fact, computing the divergence of both sides of the given equation, we see that U must satisfy a fourth equation

$$\lambda \operatorname{div} U = \operatorname{div} V.$$

Of course, if $\lambda = 0$, then a necessary and sufficient condition for the existence of such a vector field U is that $\operatorname{div} V = 0$. If $\lambda \neq 0$, then the situation is more subtle.

We shall set up a differential system whose 3-dimensional integrals correspond to the solutions of our problem. Let $\mathbb{R}^6$ be given coordinates $x^1, x^2, x^3, u^1, u^2, u^3$. We regard x^1, x^2, x^3 as coordinates on $\mathbb{R}^3$. If the components of the vector field V are (v^1, v^2, v^3), let us define the forms

$$\alpha = u^1 dx^1 + u^2 dx^2 + u^3 dx^3$$

$$\beta = u^1 dx^2 \wedge dx^3 + u^2 dx^3 \wedge dx^1 + u^3 dx^1 \wedge dx^2$$

$$\gamma = v^1 dx^2 \wedge dx^3 + v^2 dx^3 \wedge dx^1 + v^3 dx^1 \wedge dx^2.$$

Now let $\mathcal{I}$ be the differential system on $\mathbb{R}^6$ generated by the 2-form $\Theta = d\alpha + \lambda\beta - \gamma$. Any 3-dimensional integral manifold of Θ on which the form $\Omega = dx^1 \wedge dx^2 \wedge dx^3$ does not vanish is locally a graph of the form $(x, u(x))$ where the components of $u(x)$ determine a vector field U on $\mathbb{R}^3$ which satisfies the equation $\operatorname{curl} U + \lambda U = V$. Conversely, any local solution of this P.D.E. gives rise to an integral of $\mathcal{I}$ by reversing this process.

We now turn to an analysis of the integral elements of $\mathcal{I}$. The pair of forms $\Theta, d\Theta$ clearly suffice to generate $\mathcal{I}$ algebraically. The cases where λ vanishes or does not vanish are markedly different.

If $\lambda = 0$, then $d\Theta = -d\gamma = -(\operatorname{div} V)\Omega$. Thus, if $\operatorname{div} V \neq 0$, then there cannot be any integral of $\mathcal{I}$ on which Ω does not vanish.

On the other hand, suppose $\operatorname{div} V = 0$. Then $\mathcal{I}$ is generated by Θ alone since then $d\Theta \equiv 0$. If $E \in V_3(\mathcal{I}, \Omega)$, then the annihilator of E is spanned by three 1-forms of the form $\pi^i = du^i - \sum_j A^i_j dx^j$ for some numbers A. It follows that, at the base point of E, the form Θ can be written in the form

$$\Theta = \pi^1 \wedge dx^1 + \pi^2 \wedge dx^2 + \pi^3 \wedge dx^3.$$

Setting $\omega^i = dx^i$, we may apply Proposition 1.15 to show that the characters of the associated integral flag satisfy $c_p = p$ for all $0 \leq p \leq 3$. On the other hand, it is clear from this formula for Θ that there exists a 6-parameter family of integral elements of Θ at each point of $\mathbb{R}^6$. By Theorem 1.11, it follows that all of the elements of $V_3(\mathcal{I}, \Omega)$ are ordinary. The character sequence is given by $s_0 = 0$ and $s_p = 1$ for $p = 1, 2, 3$. At this point, if we assumed that V were real analytic, we could apply the Cartan–Kähler theorem to show that there exist local solutions to our original problem. However, in this case, an application of the Poincaré lemma will suffice even without the assumption of real analyticity.

Now let us turn to the case where $\lambda \neq 0$. If, for any (i, j, k) which is an even permutation of $(1, 2, 3)$, we set

$$\pi_i = du^i + \tfrac{1}{2}[(\lambda u^j - v^j)dx^k - (\lambda u^k - v^k)dx^j] - \lambda^{-1}\partial_i v^i dx^i;$$

then

$$\Theta = \pi_1 \wedge dx^1 + \pi_2 \wedge dx^2 + \pi_3 \wedge dx^3$$

$$\lambda^{-1}d\Theta = \pi_1 \wedge dx^2 \wedge dx^3 + \pi_2 \wedge dx^3 \wedge dx^1 + \pi_3 \wedge dx^1 \wedge dx^2.$$

It follows that any $E \in V_3(\mathcal{I}, \Omega)$ is annihilated by 1-forms of the form $\vartheta_i = \pi_i - \sum_j p_{ij} dx_j$ where, in order to have $\Theta_E = 0$, we must have $p_{ij} = p_{ji}$ and, in order to ahve $(d\Theta)_E = 0$, we must have $p_{11} + p_{22} + p_{33} = 0$. Thus $V_3(\mathcal{I}, \Omega)$ is a smooth manifold of codimension 4 in $G_3(T\mathbb{R}^6)$. On the other hand, by Proposition 1.15, it follows that $c_0 = 0$, $c_1 = 1$, and $c_2 = 3$ for the integral flag associated to the choice $\omega^i = dx^i$. Since $c_0 + c_1 + c_2 = 4$, it follows that all of the elements of $V_3(\mathcal{I}, \Omega)$ are ordinary.

By the Cartan–Kähler theorem, it follows that, if V is real analytic, then there exist (analytic) local solutions to the equation $\operatorname{curl} U + \lambda U = V$ for all non-zero constants λ.

Example 3.8 (Isometric embedding). We now wish to consider the problem of locally isometrically embedding an n-dimensional manifold M with a given Riemannian metric g into Euclidean space $\mathbb{E}^N$ where N is some integer yet to be specified. Note that the condition that a map $u : M \to \mathbb{E}^N$ be an isometric embedding is a set of non-linear, first-order partial differential equations for u. Precisely, if $x^1, x^2, \ldots, x^n$ is a set of local coordinates on M and $g = \sum g_{ij} dx^i \circ dx^j$, then the equations that u must satisfy are $g_{ij} = \partial_i u \cdot \partial_j u$. This is $\tfrac{1}{2}n(n+1)$ for the N unknown components of u. Thus, we do not expect to have any solution if $N < \tfrac{1}{2}n(n+1)$. It is the contention of the Cartan–Janet isometric embedding theorem (which we prove below) that such a local isometric embedding is possible if the metric g is real analytic and $N = \tfrac{1}{2}n(n+1)$. Note that even though the isometric embedding system is determined if $N = \frac{1}{2}n(n+1)$, it cannot be put in Cauchy–Kowalevski form for $n > 1$.

We begin by writing down the structure equations for g. Since our results will be local, we may as well assume that we can choose an orthonormal coframing $\eta_1, \eta_2, \ldots, \eta_n$ on M so that the equation $g = (\eta_1)^2 + (\eta_2)^2 + \cdots + (\eta_n)^2$ holds. By the fundamental lemma of Riemannian geometry, there exist unique 1-forms on M, $\eta_{ij} = -\eta_{ji}$, so that the first structure equations of É. Cartan hold:

$$d\eta_i = -\sum_j \eta_{ij} \wedge \eta_j.$$

The second structure equations of É. Cartan also hold:

$$d\eta_{ij} + -\sum_k \eta_{ik} \wedge \eta_{kj} + \tfrac{1}{2} \sum_{k,l} R_{ijkl}\, \eta_k \wedge \eta_l.$$

Here, the functions R_{ijkl} are the components of the Riemann curvature tensor and satisfy the usual symmetries

$$R_{ijkl} = -R_{jikl} = -R_{ijlk}$$

$$R_{ijkl} + R_{iklj} + R_{iljk} = 0.$$

Let $\mathcal{F}_n(\mathbb{E}^N)$ denote the bundle over $\mathbb{E}^N$ whose elements consist of the $(n+1)$-tuples $(x; e_1, e_2, \ldots, e_n)$ where $x \in \mathbb{E}^N$ and $e_1, e_2, \ldots, e_n$ are an orthonormal set of vectors in $\mathbb{E}^N$. Note that $\mathcal{F}_n(\mathbb{E}^N)$ is diffeomorphic to $\mathbb{E}^N \times SO(N)/SO(N-n)$. For several reasons, it is more convenient to work on $\mathcal{F}_n(\mathbb{E}^N)$ than on the full frame bundle of $\mathbb{E}^N$. We shall adopt the index ranges $1 \le i, j, k, l \le n < a, b, c \le N$. Let $\mathcal{U} \subset \mathcal{F}_n(\mathbb{E}^N)$ be an open set on which there exist real analytic vector-valued functions $e_a : \mathcal{U} \to \mathbb{E}^N$ with the property that for all $f = (x; e_1, e_2, \ldots, e_n) \in \mathcal{U}$, the vectors $e_1, e_2, \ldots, e_n, e_{n+1}(f), \ldots, e_N(f)$ form an orthonormal basis of $\mathbb{E}^N$. We also regard the components of f as giving vector-valued functions $x, e_i : \mathcal{U} \to \mathbb{E}^N$. It follows that we may define a set of 1-forms on $\mathcal{U}$ by the formulae

$$\omega_i = e_i \cdot dx$$

$$\omega_a = e_a \cdot dx$$

$$\omega_{ij} = e_i \cdot de_j = -\omega_{ji}$$

$$\omega_{ia} = e_i \cdot de_a = -\omega_{ai} = -e_a \cdot de_i$$

$$\omega_{ab} = e_a \cdot de_b = -\omega_{ba}.$$

Of these forms, the set $\{\omega_i\} \cup \{\omega_a\} \cup \{\omega_{ij}\}_{i<j} \cup \{\omega_{ai}\}$ forms a coframing of $\mathcal{U}$. We shall have need of the following structure equations

$$d\omega_i = -\sum_j \omega_{ij} \wedge \omega_j - \sum_b \omega_{ib} \wedge \omega_b$$

$$d\omega_a = -\sum_j \omega_{aj} \wedge \omega_j - \sum_b \omega_{ab} \wedge \omega_b$$

$$d\omega_{ij} = -\sum_k \omega_{ik} \wedge \omega_{kj} - \sum_b \omega_{ib} \wedge \omega_{bj}.$$

Now, on $M \times \mathcal{U}$, consider the differential system $\mathcal{I}_-$ generated by the 1-forms $\{\omega_i - \eta_i\}_{i \leq n} \cup \{\omega_a\}_{a > n}$. Let $\Omega = \omega_1 \wedge \omega_2 \wedge \cdots \wedge \omega_n$.

Proposition 3.9. *Any n-dimensional integral of $\mathcal{I}_-$ on which Ω does not vanish is locally the graph of a function $f : M \to \mathcal{U}$ with the property that the composition $x \circ f : M \to \mathbb{E}^N$ is a local isometric embedding. Conversely, every local isometric embedding $u : M \to \mathbb{E}^N$ arises in a unique way from this construction.*

Proof. First suppose that we have an isometric embedding $u : M \to \mathbb{E}^N$. Let $\{E_i\}$ be the orthonormal frame field on M which is dual to the coframing $\{\eta_i\}$. For each $z \in M$, define $f(z) = (u(z); du(E_1(z)), \ldots, du(E_n(z)))$. Now consider the graph $\Gamma_u = \{(z, f(z)) \mid z \in M\} \subset M \times \mathcal{U}$. Clearly, Γ_u is an integral of $\mathcal{I}_-$ if and only if f satisfies $f^*(\omega_i) = \eta_i$ and $f^*(\omega_a) = 0$. However, since $e_a(f(z))$ is normal to the vectors $du(E_i(z))$ by construction, we have $f^*(\omega_a) = (e_a \circ f) \cdot d(x \circ f) = (e_a \circ f) \cdot du = 0$. Also, since u is an isometric embedding, we have, for all $v \in T_z M$, $f^*(\omega_i)(v) = (e_i \circ f) \cdot du(v) = du(E_i(z)) \cdot du(v) = E_i(z) \cdot v = \eta_i(v)$. Note also that, on Γ_u, we have $\Omega = \omega_1 \wedge \omega_2 \wedge \cdots \wedge \omega_n = \eta_1 \wedge \eta_2 \wedge \cdots \wedge \eta_n$. Since the latter form is non-zero when projected onto the factor M, it follows that Ω is non-zero on Γ_u.

Now suppose that $X \subset M \times \mathcal{U}$ is an n-dimensional integral manifold of $\mathcal{I}_-$ on which Ω does not vanish. Then since $\omega_i = \eta_i$ on X, it follows that the form $\eta_1 \wedge \eta_2 \wedge \cdots \wedge \eta_n$ also does not vanish on X. It follows that the projection $X \to M$ onto the first factor is a local diffeomorphism. Thus we may regard X locally as the graph of a function $f{:}M \to \mathcal{U}$. Now let $u = x \circ f$. We claim that $u{:}M \to \mathbb{E}^N$ is an isometry and moreover that $e_i \circ f = du(E_i)$. This will establish both parts of the proposition. To see these claims, note that we have $f^*(\omega_i) = \eta_i$ and $f^*(\omega_a) = 0$. Since $f^*(\omega_a) = f^*(e_a \cdot dx) = (e_a \circ f) \cdot du = 0$, it follows that $(e_a \circ f)(z)$ is normal to $du(E_i(z))$ for all i and a and $z \in M$. Thus, the vectors $du(E_i(z))$ are linear combinations of the vectors $\{(e_j \circ f)(z)\}_{j \leq n}$. On the other hand, for any $v \in T_z M$, we have $E_i(z) \cdot v = \eta_i(v) = f^*(\omega_i)(v) = f^*(e_i \cdot dx)(v) = (e_i \circ f)(z) \cdot du(v)$. Using the fact that $\{E_i(z)\}$ is an orthonormal basis for

$T_z M$ and that $\{(e_i \circ f)(z)\}$ is an orthonormal basis for $du(T_z M)$, we see that du must be an isometry and that $e_i \circ f = du(E_i)$, as claimed. $\qquad\square$

We are now going to show that any integral of $\mathcal{I}_-$ on which Ω does not vanish is actually an integral of a larger system $\mathcal{I}$ (defined below). Suppose that X is such an n-dimensional integral. Then let us compute, on X,

$$0 = d(\omega_i - \eta_i) = -\sum_j (\omega_{ij} - \eta_{ij}) \wedge \eta_j.$$

Since the forms η_i are linearly independent on X and since the forms $\vartheta_{ij} = (\omega_{ij} - \eta_{ij})$ are skew-symmetric in their lower indices, this implies that the forms ϑ_{ij} must vanish on X. The geometric meaning of this fact is that the Levi–Civita connection of a Riemannian metric is the same as the connection induced by any isometric embedding into Euclidean space.

Let us now consider the differential system $\mathcal{I}$ on $M \times \mathcal{U}$ which is generated by the set of 1-forms $\{\omega_i - \eta_i\}_{i \leq n} \cup \{\omega_a\}_{n < a} \cup \{\omega_{ij} - \eta_{ij}\}_{i < j < n}$. We are going to show that if $N \geq \frac{1}{2} n(n + 1)$, then there is an ordinary integral element of $\mathcal{I}$ at every point of $M \times \mathcal{U}$. We begin by describing a set of forms which generate $\mathcal{I}$ algebraically. Let I denote the Pfaffian system generated by the 1-forms in $\mathcal{I}$. We compute that

$$d(\omega_i - \eta_i) \equiv 0 \mod I$$

$$d\omega_a \equiv -\sum_i \omega_{ai} \wedge \omega_i \mod I$$

$$d(\omega_{ij} - \eta_{ij}) \equiv \sum_a \omega_{ai} \wedge \omega_{aj} - \frac{1}{2} \sum_{k,l} R_{ijkl} \omega_k \wedge \omega_l \mod I.$$

Thus, $\mathcal{I}$ is generated algebraically by the 1-forms in I and the 2-forms $\Theta_a = \sum_i \omega_{ai} \wedge \omega_i$ and $\Theta_{ij} = \sum_a \omega_{ai} \wedge \omega_{aj} - \frac{1}{2} \sum_{k,l} R_{ijkl} \omega_k \wedge \omega_l$. Let $E \subset T_{(x,f)}(M \times \mathcal{U})$ be an n-dimensional integral element of $\mathcal{I}$ on which the form Ω does not vanish. Then, in addition to annihilating the 1-forms in I, E must annihilate some 1-forms of the form $\pi_{ai} = \omega_{ai} - \sum_j h_{aij} \omega_j$ for some numbers h_{aij}. The condition that Θ_a vanish on E is the condition that $h_{aij} = h_{aji}$ for all a, i, and j. Using this information, the condition that Θ_{ij} also vanish on E becomes the quadratic equations on h_{aij}:

$$\sum_a (h_{aik} h_{ajl} - h_{ail} h_{ajk}) = R_{ijkl}(x).$$

These equations represent the Gauss equations.

Let W be the Euclidean vector space of dimension $r = N - n$. We can interpret the numbers $h_{aij} = h_{aji}$ as a collection of $\binom{n+1}{2}$ vectors $h_{ij} = (h_{aij})$ in W. In fact, we may interpret $h = (h_{aij})$ as an element of the vector space $W \otimes S^2(\mathbb{R}^n)$ in the obvious way. If we let $\mathcal{K}_n \subset \Lambda^2(\mathbb{R}^n) \otimes \Lambda^2(\mathbb{R}^n)$ denote the space of Riemann curvature tensors in dimension n, then there is a well defined quadratic map $\gamma : W \otimes S^2(\mathbb{R}^n) \to \mathcal{K}_n$ defined for $h = (h_{aij}) \in W \otimes S^2(\mathbb{R}^n)$ by

$$\gamma(h)_{ijkl} = \sum_a (h_{aik}h_{ajl} - h_{ail}h_{ajk}).$$

We shall need the following algebraic lemma, whose proof we postpone until the end of our discussion.

Lemma 3.10. *Suppose that* $r = N - n \geq \binom{n}{2}$. *Let* $\mathcal{H} \subset W \otimes S^2(\mathbb{R}^n)$ *be the open set consisting of those elements* $h = (h_{aij})$ *so that the vectors* $\{h_{ij} \mid i \leq j < n\}$ *are linearly independent as elements of* W. *Then* $\gamma : \mathcal{H} \to \mathcal{K}_n$ *is a surjective submersion.*

Assume this lemma for the moment, we now state our main result for local isometric embedding.

Theorem 3.11 (Cartan-Janet). *Suppose that* $N \geq \frac{1}{2}n(n+1)$. *If the Riemannian metric* g *on* M *is real analytic, then every point of* M *has a neighborhood which has a real analytic isometric embedding into* $\mathbb{E}^N$.

Proof. By virtue of the Cartan–Kähler theorem, it suffices to show that, for every $(x, f) \in M \times \mathcal{U}$, there exists an ordinary integral element $E \in V_n(\mathcal{I}, \Omega)$ based at (x, f). Let $Z \subset M \times \mathcal{U} \times \mathcal{H}$ denote the set of triples (x, f, h) so that the equation $\gamma(h) = R(x)$ holds where $R(x) = (R_{ijkl}(x)) \in \mathcal{K}_n$ is the Riemann curvature tensor at $x \in M$. By Lemma 3.10 and the implicit function theorem, Z is a smooth submanifold of $M \times \mathcal{U} \times \mathcal{H}$ of codimension $n^2(n^2 - 1)/12$ $(= \dim \mathcal{K}_n)$ and the projection onto the first two factors, $Z \to M \times \mathcal{U}$ is surjective. In particular, note that the dimension of Z is

$$\dim Z = \dim(M \times \mathcal{U}) + (N - n) \cdot \tfrac{1}{2}n(n + 1) - n^2(n^2 - 1)/12.$$

We define a map $\varepsilon : Z \to V_n(\mathcal{I}, \Omega)$ by letting $\varepsilon(x, f, h)$ be the n-plane based at (x, f) which is annihilated by the 1-forms

$$\{\omega_i - \eta_i\}_{i \leq n} \cup \{\omega_a\}_{n < a} \cup \{\omega_{ij} - \eta_{ij}\}_{i < j \leq n} \cup \{\pi_{ai} = \omega_{ai} - \sum_j h_{aij}\omega_j\}_{i \leq n, n < a}.$$

It is clear that the map ε is an embedding. By our previous discussion, it maps onto an open submanifold of $V_n(\mathcal{I}, \Omega)$. We are now going to show that the image $\varepsilon(Z)$ consists entirely of ordinary integral elements.

Let $E = \varepsilon(x, f, h)$ with $(x, f, h) \in Z$. Let $E_p \subset E$ be the subspace annihilated by the 1-forms ω_i where $i > p$. We want to compute the codimension of $H(E_p)$ for all $p < n$. To do this, we will apply Proposition 1.15.

Of course, all of the 1-forms in $\mathcal{I}$ lie in the polar equations of E_p for all p. We may express the 2-forms in terms of $\{\pi_{ai}, \omega_i\}$ as follows:

$$\sum_i \omega_{ai} \wedge \omega_i \equiv \sum_i \pi_{ai} \wedge \omega_i$$

$$\sum_a \omega_{ai} \wedge \omega_{aj} - \tfrac{1}{2} \sum_{k,l} R_{ijkl}\, \omega_k \wedge \omega_l \equiv \sum_{a,k}(h_{ajk}\pi_{ai} - h_{aik}\pi_{ai}) \wedge \omega_k + Q_{ij}$$

where Q_{ij} is a 2-form whose terms are either quadratic in π or else vanish at the base point (x, f). It follows by Proposition 1.15 that the polar equations of E_p are spanned by the 1-forms

$$\{\omega_i - \eta_i\} \text{ for } i \le n$$

$$\{\omega_a\} \text{ for } a > n$$

$$\{\omega_{ij} - \eta_{ij}\} \text{ for } i < j \le n$$

$$\{\pi_{ai}\} \text{ for } i \le p \text{ and } a > n$$

$$\{(h_{aik}\pi_{aj} - h_{ajk}\pi_{ai})\} \text{ for } k \le p \text{ and } i < j \le n.$$

The first 3 types of terms are the same for all $p \ge 0$ so they contribute $N + \tfrac{1}{2}n(n-1)$ forms for all $p \ge 0$. The fourth type of term contributes $pr = p(N - n)$ terms which are clearly linearly independent from the previous terms. In the fifth type of term, the cases where the i, j indices are both less than or equal to p are obviously linear combinations of terms of the fourth kind. The remaining terms of the fifth kind can be broken into the subcases where either 1) $i \le p < j \le n$ or else 2) $p < i < j \le n$.

In case 1), in view of the terms of the fourth kind, we may replace these terms by the simplified expressions $\{h_{ik} \cdot \pi_j \mid i, k \le p < j \le n\}$ where we have written $\pi_j = (\pi_{aj})$ and regard π_j as a W-valued 1-form (all of whose components are linearly independent). These terms are clearly linearly independent from the terms of the fourth kind due to the assumption that the vectors $\{h_{ij} \mid i, j < n\}$ are linearly independent. They contribute $(n - p) \cdot \tfrac{1}{2}p(p + 1)$ more terms to the polar equations.

In case 2) the remaining expressions are given by the collection $\{h_{ik} \cdot \pi_j - h_{jk} \cdot \pi_i \mid k \le p < i < j \le n\}$. Again, the assumption that the vectors $\{h_{ij} \mid i, j < n\}$ are linearly independent shows that these $\tfrac{1}{2}p(n - p)(n - p - 1)$ terms are linearly independent from all of the previous terms. It follows that the rank of the polar equations for E_p is equal to

$$c_p = N + \tfrac{1}{2}n(n - 1) + rp + (n - 1) \cdot \tfrac{1}{2}p(p + 1) + \tfrac{1}{2}p(n - p)(n - p - 1)$$

$$= N + \tfrac{1}{2}n(n - 1) + rp + \tfrac{1}{2}pn(n - p).$$

We may now compute

$$c_0 + c_1 + \cdots + c_{n-1} = Nn(n+1)/2 + n^2(n^2-1)/12.$$

However, this is precisely the codimension of $\varepsilon(Z)$ in $G_n(T(M \times \mathcal{U}))$. It now follows from Theorem 1.11 that E is ordinary. $\qquad\square$

Proof of Lemma 3.10. Throughout this argument, whenever $p < n$, we identify $\mathbb{R}^p$ with the subspace of $\mathbb{R}^n$ consisting of those elements of $\mathbb{R}^n$ whose last $n - p$ coordinates are zero. As above, we let $\mathcal{K}_p \subset \Lambda^2(\mathbb{R}^p) \otimes \Lambda^2(\mathbb{R}^p)$ denote the space of elements $R = (R_{ijkl})$ which satisfy the relations

$$R_{ijkl} = -R_{jikl} = -R_{ijlk}$$

$$R_{ijkl} + R_{iklj} + R_{iljk} = 0.$$

Note that if $p < n$ then $\mathcal{K}_p \subset \mathcal{K}_n$. It is well known that the dimension of $\mathcal{K}_p$ is $p^2(p^2-1)/12$ for all $p \geq 0$. (Actually, our calculations will contain a proof of this result.)

Let W be an Euclidean vector space of dimension $r \geq \frac{1}{2}n(n-1)$. Then, as we defined γ before, note that $\gamma(W \otimes S^2(\mathbb{R}^p)) \subset \mathcal{K}_p$. We are going to prove Lemma 3.10 by induction on p between the values 1 and n. Fix an element $R = (R_{ijkl}) \in \mathcal{K}_n$. For each $p \leq n$, we let R^p denote the element of $\mathcal{K}_p$ got from R by setting all of the components with an index greater than p equal to zero.

First note that since $\mathcal{K}_1 = (0)$, the lemma is trivially true for $p = 1$. Suppose now that, for some $p < n$, we have shown that there is an element $h^p = (h^p_{ij})$ in $W \otimes S^2(\mathbb{R}^p)$ with $\gamma(h^p) = R^p$ and with all of the vectors $\{h^p_{ij} \in W \mid i \leq j \leq p\}$ linearly independent and that the differential of the mapping $\gamma : W \otimes S^2(\mathbb{R}^p) \to \mathcal{K}_p$ is surjective at any such h^p. We now try to construct a corresponding extension h^{p+1}. Let $v_1, \ldots, v_p$ be p vectors in W. Consider the equations

$$h^p_{ik} \cdot v_j - h^p_{ij} \cdot v_k = R_{(p+1)ijk}$$

where i, j, k run over all choices of indices less than or equal to p. We want to show that there exist vectors $v_1, \ldots, v_p$ so that these equations hold. To see this, note that the tensor $L_{ijk} = R_{(p+1)ijk}$ in $\mathbb{R}^p \otimes \Lambda^2(\mathbb{R}^p)$ lies in the kernel of the skew-symmetrizing map $\mathbb{R}^p \otimes \Lambda^2(\mathbb{R}^p) \to \Lambda^3(\mathbb{R}^p)$ by the symmetries of the Riemann curvature tensor. It follows, by the exactness of the sequence

$$0 \to S^3(\mathbb{R}^p) \to S^2(\mathbb{R}^p) \otimes \mathbb{R}^p \to \mathbb{R}^p \otimes \Lambda^2(\mathbb{R}^p) \to \Lambda^3(\mathbb{R}^p) \to 0$$

that there exists an element $r \in S^2(\mathbb{R}^p) \otimes \mathbb{R}^p$ so that $r_{ikj} - r_{ijk} = L_{ijk}$. Thus, it suffices to find the vectors v_i so that $h^p_{ij} \cdot v_k = r_{ijk} = r_{jik}$. By the independence assumption on h^p, such vectors v_i exist. If $p < n - 1$ then there is even room to choose the vectors v_i so that they and the vectors h^p_{ij} are linearly independent. Once the v_i have been chosen, we choose a vector w so that the following equations hold:

$$w \cdot h^p_{ij} - v_i \cdot v_j = R_{(p+1)i(p+1)j}.$$

Again, by the independence assumption on h^p and the fact that the Riemann curvature tensor has the well-known symmetry $R_{ijkl} = R_{klij}$, this can be done. Also, if $p < n - 1$, we have room to choose w so that the vectors h^p_{ij}, v_i, w are all linearly independent in W.

We can now define an element h^{p+1} of $W \otimes S^2(\mathbb{R}^{p+1})$ by letting

$$h^{p+1}_{ij} = h^p_{ij} \quad \text{when } i, j \le p$$

$$h^{p+1}_{(p+1)i} = h^{p+1}_{i(p+1)} = v_i \quad \text{when } i \le p$$

$$h^{p+1}_{(p+1)(p+1)} = w.$$

It is clear that $\gamma(h^{p+1}) = R^{p+1}$. Moreover, using the assumption of surjectivity of the differential of $\gamma : W \otimes S^2(\mathbb{R}^p) \to \mathcal{K}_p$ at h^p, and the explicit formula for the equations defining the extension, it is clear that the differential of $\gamma : W \otimes S^2(\mathbb{R}^{p+1}) \to \mathcal{K}_{p+1}$ is surjective at h^{p+1}. Finally, note that if h^{p+1} is any element of $W \otimes S^2(\mathbb{R}^{p+1})$ where the vectors h^{p+1}_{ij} with $i \le j \le p$ are linearly independent in W, then the induction hypothesis implies that the differential of $\gamma : W \otimes S^2(\mathbb{R}^p) \to \mathcal{K}_p$ is surjective at the corresponding restricted element h^p. Thus, by the above argument, the differential of $\gamma : W \otimes S^2(\mathbb{R}^{p+1}) \to \mathcal{K}_{p+1}$ is surjective at h^{p+1}. $\qquad\square$

CHAPTER IV

LINEAR DIFFERENTIAL SYSTEMS

The goal of this chapter is to develop the formalism of linear Pfaffian differential systems in a form that will facilitate the computation of examples.

Let $\mathcal{I}$ be a differential ideal on a manifold M. In practice we usually seek integral manifolds of $\mathcal{I}$ that satisfy a transversality condition, and this then leads to the concept of a *differential system with independence condition* $(\mathcal{I}, \Omega)$ to be explained in Section 1. There we also introduce the fundamental concept of *involution* for such systems. Recall from the proof of the Cartan–Kähler theorem in Chapter III that integral manifolds are constructed by solving a succession of Cauchy initial value problems. Roughly speaking, to be involutive means, according to that proof, that the solutions to the $(k+1)^{st}$ initial value problem remain solutions to the family of kth initial value problems depending on x^{k+1}.[1] On the other hand, intuitively a system is involutive when all of the integrability conditions implicit in the system are satisfied. It is not obvious that these two viewpoints coincide. Although it is relatively simple to define, the concept of involution is subtle and gaining an understanding of it will be one of the main goals of this chapter.

In Section 2 we introduce the important concept of *linearity* for a differential system with independence condition. We also introduce the *linearization* of an arbitrary differential system at an integral element. This is a linear differential system with constant coefficients that, roughly speaking, corresponds to linearizing and freezing the coefficients of an arbitrary P.D.E. system. Both of these concepts play a fundamental role in developing the theory.

Section 3 introduces the purely algebraic concept of a *tableau*. The motivation arises from trying to extend the concept of the *symbol* of a P.D.E. system to general exterior differential systems. The purely algebraic notion of *involutivity* of a tableau is also defined, and we explain how this arises naturally from the consideration of involutive differential systems.

In Section 4 we introduce the definition of the *tableau A_E of a differential ideal $\mathcal{I}$* at an integral element E. This tableau appears naturally both as the tangent space to the variations of an integral element over a point and as the homogeneous 1-jets of integral manifolds to the linearization of $\mathcal{I}$ at E. A non-trivial theorem is that A_E is involutive in case E is an ordinary

---

[1] In this regard we refer to the introduction to Chapter III. Involutivity implies that solutions to (ii$'$) and (i$'$) are also solutions to (i) and (ii).

integral element.

Section 5 takes up the very important class of *linear Pfaffian systems*. These include most examples and will be the systems mainly used throughout the rest of Chapters V–VII. Associated to a linear Pfaffian system are two invariants, its *tableau* and *torsion* (or integrability conditions). These are discussed in some detail, and Cartan's test for involution is seen to have a very simple and computable form using the tableau and torsion.

In Section 6 we introduce the concept of the *prolongation* $(\mathcal{I}^{(1)}, \Omega)$ of an exterior differential system $\mathcal{I}$. This is a linear Pfaffian system that is defined on the space of integral elements of $\mathcal{I}$. Intuitively, $(\mathcal{I}^{(1)}, \Omega)$ is obtained by introducing the first derivatives as new variables, and its effect is to impose the first order integrability conditions in the original system. This section is preparatory to Chapter VI, where the main results will be proved; it is put here so that the concept of prolongation is available for computation of examples.

In Section 7 we give a number of examples, including the conditions that a pair of 2nd order P.D.E.'s for one unknown function be in involution. Finally, in section 8 we give a non-trivial and natural example from surface geometry of an overdetermined, non-involutive system requiring prolongation.

In this chapter we will let $\{\sharp\}$ denote the *algebraic* ideal in $\Omega^* M$ generated by a set of differential forms $\sharp$ (for example, $\sharp$ may be the sections of a subbundle $I \subset T^* M$). We shall also denote by φ_E the restriction of a form φ on M to an n-plane $E \subset T_x M$. If $\mathcal{I} \subset \Omega^*(M)$ is a differential ideal, we denote by $I \subset T^* M$ the sub-bundle spanned by the values of the 1-forms in $\mathcal{I}$ (assuming, of course, the obvious constant rank condition). Finally we will use the summation convention.

§1. Independence Condition and Involution.

We suppose given a closed differential system $\mathcal{I}$ on a manifold M. Many problems require the existence of integral manifolds of $\mathcal{I}$ satisfying a transversality condition given by the following:

Definition 1.1. A *differential system with independence condition*, denoted by $(\mathcal{I}, \Omega)$, is given by a closed differential ideal $\mathcal{I}$ together with an equivalence class of n-forms Ω where the following conditions are satisfied:

(i) Ω and Ω' are equivalent if

$$\Omega \equiv f\Omega' \text{ modulo } \mathcal{I}$$

where f is a non-zero function;

(ii) locally Ω may be represented by a decomposable n-form

$$(1) \qquad\qquad \Omega = \omega^1 \wedge \cdots \wedge \omega^n$$

where the ω^i are 1-forms; and

(iii) $\Omega_x \notin \mathcal{I}_x$ for any $x \in M$.

In intrinsic terms, under suitable constant rank assumptions the degree one piece, I, of $\mathcal{I}$ is given by the sections of a sub-bundle $I \subset T^*M$. There should be an additional sub-bundle $J \subset T^*M$ with

$$\begin{cases} I \subset J \subset T^*M \\ \operatorname{rank} J/I = n. \end{cases}$$

The ω^i above give local sections of J that induce a framing of J/I and Ω represents a non-vanishing section of $\Lambda^n(J/I)$. We shall usually work locally and write $\Omega = \omega^1 \wedge \cdots \wedge \omega^n$ as in (1) above.

Definition 1.2. (i) An *integral element for* $(\mathcal{I}, \Omega)$ is an n-dimensional integral element for $\mathcal{I}$ on which Ω is non-zero; and (ii) an *integral manifold for* $(\mathcal{I}, \Omega)$ is given by an n-dimensional integral manifold

$$f : N \to M$$

for $\mathcal{I}$ such that each $f_*(T_y N)$ is an integral element of $(\mathcal{I}, \Omega)$.

Integral elements of $(\mathcal{I}, \Omega)$ are thus given by the n-planes $E \in G_n(TM)$ that satisfy

$$(2) \qquad \begin{cases} \Theta_E = 0, \quad \text{for all } \Theta \in \mathcal{I} \\ \Omega_E \neq 0 \end{cases},$$

where we recall our notation φ_E for the restriction of a differential form φ to E. In intrinsic terms, the first equation in (2) implies that $E \subset T_x M$ lies in $I_x^\perp$; thus the restriction mapping $J_x/I_x \to E^*$ is well-defined, and the second equation in (2) says that this mapping should be an isomorphism. We denote by

$$G(\mathcal{I}, \Omega) \subset G_n(TM)$$

the set of integral elements of $(\mathcal{I}, \Omega)$. If we think of the set $G_n(\mathcal{I})$ of all n-dimensional integral elements of $\mathcal{I}$ as being a subvariety of $G_n(TM)$ (say, in the real-analytic case), then for each irreducible component Z of $G_n(\mathcal{I})$ the intersection $G(\mathcal{I}, \Omega) \cap Z$ is either empty or is a dense open subset.

Example 1.3. Any P.D.E. system

$$(3) \qquad F^\lambda(x^i, z^a, \partial z^a/\partial x^i, \ldots, \partial^k z^a/\partial x^I) = 0, \qquad \partial x^I = \partial x^{i_1} \cdots \partial x^{i_k},$$

may be written as a differential system with independence condition. For instance, in the 2nd order case ($k = 2$) we introduce variables

$$p_i^a, p_{ij}^a = p_{ji}^a$$

and then the system is defined on the space with coordinates $(x^i, z^a, p_i^a, p_{ij}^a)$ and is generated by the equations

$$\begin{cases} F^\lambda(x^i, z^a, p_i^a, p_{ij}^a) = 0 \\ dz^a - p_i^a dx^i = 0 \\ dp_i^a - p_{ij}^a dx^j = 0, \end{cases}$$

and their exterior derivatives, with the independence condition given by $\Omega = dx^1 \wedge \cdots \wedge dx^n$. An integral manifold of the differential system with independence condition is locally the same as a solution to the P.D.E. system.

This example may be expressed in coordinate free terms by thinking of a P.D.E. system as defined by a submanifold M of a suitable jet manifold $J^k(X, Y)$ and by restricting the contact system on $J^k(X, Y)$ to M (cf. Chapters I and IX, X).

It is clear that any P.D.E. system may be written as a differential system $(\mathcal{I}, \Omega)$ on a manifold M. However, the diffeomorphisms f of M that preserve the structure $(\mathcal{I}, \Omega)$ may be strictly larger than those induced by changes of dependent and independent variables separately.[2] In addition, we may utilize non-integrable co-framings of M adapted to the structure of $(\mathcal{I}, \Omega)$ in order to isolate the geometry of the P.D.E. These points of view will be extensively illustrated by examples below and in Chapters V and VII.

Example 1.4. Suppose we are given two manifolds X, Y and a set of geometric conditions on immersions

$$f : X \to Y$$

that are expressed in local coordinates by a P.D.E. system. An example is when X and Y are Riemannian manifolds and f is an isometric immersion, as discussed in Chapter III. We may then set up a differential system with independence condition $(\mathcal{I}, \Omega)$ on a suitable manifold $M \subset J^k(X, Y)$ whose integral manifolds are locally k-jets of mappings f satisfying the given geometric conditions. The independence or transversality condition simply reflects the fact that a submanifold $N \subset X \times Y$ with $\dim N = \dim X$ is locally the graph of an immersion f if, and only if, $\pi^* \Omega \neq 0$ where Ω is any volume form on X and $\pi : N \to X$ is the projection.

Example 1.5. Let M be a manifold and consider the Grassmann bundle

$$\pi : G_n(TM) \to M$$

---

[2] We have seen one instance of this given by the local normal form (i) and (ii) of an arbitrary P.D.E. system in the introduction to Chapter III.

whose fiber $G_n(T_x M)$ over any point $x \in M$ is the Grassmann manifold of all n-planes in $T_x M$. Given any n-dimensional manifold N and immersion $f : N \to M$, there is a *canonical lifting*

$$
(4) \qquad
\begin{array}{c}
G_n(TM) \\
\hat{f}_* \nearrow \quad \downarrow \pi \\
N \xrightarrow{\quad f \quad} M
\end{array}
$$

where $\hat{f}_*(y) = f_*(T_y N) \subset T_{f(y)} M$. We will define a differential system with independence condition $(\mathcal{L}, \Phi)$ on $G_n(TM)$ whose integral manifolds are locally the liftings $\hat{f}_*$ in (4) above. $\mathcal{L}$ will be a Pfaffian system and we will define $(\mathcal{L}, \Phi)$ by giving the sub-bundles

$$
I \subset J \subset T^* G_n(TM)
$$

as explained above.

Points of $G_n(TM)$ will be written as (x, E) where $E \subset T_x M$ is an n-plane, and we then set

$$
I_{(x,E)} = \pi^*(E^\perp)
$$

$$
J_{(x,E)} = \pi^*(T_x^* M).
$$

Let us see what this means in local coordinates. Setting $\dim M = m = n+s$, relative to a local coordinate system $(x^1, \ldots, x^n, y^1, \ldots, y^s)$ on M an open set U in $G_n(TM)$ is given by tangent n-planes to M on which

$$
(5) \qquad dx^1 \wedge \cdots \wedge dx^n \neq 0.
$$

In this open set tangent planes are defined by equations

$$
(6) \qquad dy^\sigma - p_i^\sigma dx^i = 0 \qquad 1 \leq i \leq n,\ 1 \leq \sigma \leq s,
$$

and $(x^i, y^\sigma, p_i^\sigma)$ forms a local coordinate system on $G_n(TM)$. The canonical system $(\mathcal{L}, \Phi)$ is locally generated by the tautological 1-forms

$$
\theta^\sigma = dy^\sigma - p_i^\sigma dx^i
$$

with the independence condition $\Phi = dx^1 \wedge \cdots \wedge dx^n$. Given an integral manifold

$$
g : Y \to G_n(TM)
$$

satisfying the conditions (5) and (6), we set $f = \pi \circ g$ and then $f^*(dx^1 \wedge \cdots \wedge dx^n) \neq 0$. We may then take $x^1, \ldots, x^n$ as local coordinates on Y in terms of which g is given by

$$
x^i \to (x^i, y^\sigma(x), p_i^\sigma(x)).
$$

From (6) we conclude that

$$p_i^\sigma(x) = \frac{\partial y^\sigma(x)}{\partial x^i}$$

as claimed.

This construction will be used below to define the prolongation of a differential system $\mathcal{I}$ on the manifold M.

We now come to one of the main concepts in the theory:

Definition 1.6. The differential system with independence condition $(\mathcal{I}, \Omega)$ is *in involution at* $x \in M$ if there exists an ordinary integral element $E \subset T_x M$ for $(\mathcal{I}, \Omega)$.

We sometimes say that $(\mathcal{I}, \Omega)$ is *involutive,* and we shall usually drop reference to the point $x \in M$, it being understood that the system is involution at each point of M.

When $(\mathcal{I}, \Omega)$ is in involution and we are in the real analytic case, the Cartan–Kähler theorem may be applied to conclude the existence of local integral manifolds of $(\mathcal{I}, \Omega)$ passing through $x \in M$. Conversely, the Cartan–Kuranishi prolongation states roughly that any local integral manifold of $(\mathcal{I}, \Omega)$ is an integral manifold of a suitable involutive prolongation $(\mathcal{I}^{(q)}, \Omega)$ of $(\mathcal{I}, \Omega)$—this will be explained in section 6 below and more fully in Chapter VI.

Definition 1.7. A P.D.E. *system* (3) *is involutive* if the corresponding exterior differential system with independence condition is involutive.

To make this precise, we should include reference to the point x on the manifold M, but we shall omit this. Of course, the definition is valid for P.D.E. systems of any order.

Example 1.8. On a 6-dimensional manifold with basis $\theta^1, \theta^2, \omega^1, \omega^2, \pi^1, \pi^2$ for the 1-forms, we consider a Pfaffian system $\theta^1 = \theta^2 = 0$ with independence condition $\omega^1 \wedge \omega^2 \neq 0$ and structure equations

$$(7) \qquad \begin{aligned} d\theta^1 &\equiv \pi^1 \wedge \omega^1 \mod \mathcal{I} \\[2mm] d\theta^2 &\equiv \pi^1 \wedge \omega^2 \mod \mathcal{I}. \end{aligned}$$

We shall show that this system is not in involution.

For this we denote by $\partial/\partial\theta^1$, $\partial/\partial\theta^2$, $\partial/\partial\omega^1$, $\partial/\partial\omega^2$, $\partial/\partial\pi^1$, $\partial/\partial\pi^2$ the basis of tangent vectors dual to the above basis of forms. A one-dimensional integral element of $\mathcal{I}$, i.e., a general vector in the space $\theta^1 = \theta^2 = 0$, is

$$(8) \qquad \xi = \xi^0 \frac{\partial}{\partial \pi^1} + \xi^1 \frac{\partial}{\partial \pi^2} + \xi^2 \frac{\partial}{\partial \omega^1} + \xi^3 \frac{\partial}{\partial \omega^2}.$$

Using self-evident notation, the polar equations

$$\langle d\theta^1, \xi \wedge \tilde{\xi} \rangle = 0 = \langle d\theta^2, \xi \wedge \tilde{\xi} \rangle$$

of the vector ξ in (8) are

(9)
$$\begin{cases} \xi^0 \tilde{\xi}^2 - \xi^2 \tilde{\xi}^0 = 0 \\ \xi^0 \tilde{\xi}^3 - \xi^3 \tilde{\xi}^0 = 0. \end{cases}$$

This linear system has rank 2 if $\xi^0 \neq 0$. The latter is therefore the condition for ξ to be regular.

On the other hand, any 2-plane E^2 on which $\theta^1 = \theta^2 = 0$ and $\omega^1 \wedge \omega^2 \neq 0$ is given by linear equations in the tangent space

$$\begin{cases} \pi^1 = p_1^1 \omega^1 + p_2^1 \omega^2 \\ \pi^2 = p_1^2 \omega^1 + p_2^2 \omega^2. \end{cases}$$

The condition that this 2-plane be integral is $p_1^1 = p_2^1 = 0$. Thus, any $E^1 \subset E^2$ will have a basis vector

$$\eta = \xi^1 \frac{\partial}{\partial \pi^2} + \xi^2 \frac{\partial}{\partial \omega^1} + \xi^3 \frac{\partial}{\partial \omega^2}.$$

Comparing with the above remark, we see that E^2 contains no regular one-dimensional integral element, and is therefore not ordinary.

The situation can perhaps be explained more intuitively as follows: From (7) we find, as a consequence of $\theta^1 = \theta^2 = 0$, that

$$\pi^1 \wedge \omega^1 = 0, \qquad \pi^1 \wedge \omega^2 = 0$$

an any integral manifold of $\mathcal{I}$. Using the transversality condition $\omega^1 \wedge \omega^2 \neq 0$, the first equation says that π^1 is a multiple of ω^1 and the second equation says that it is also a multiple of ω^2 on any integral manifold of $(\mathcal{I}, \Omega)$. Combining these two conclusions, we get $\pi^1 = 0$ on any integral manifold. This last equation and its exterior derivative must be added to the system. Thus the integral manifolds must satisfy additional equations which result through differentiations and not just through algebraic operations. This is one of the simplest phenomena for "failure" of involution.

We remark that with the independence condition given by $\pi = \pi^1 \wedge \pi^2 \neq 0$ the system $(\mathcal{I}, \pi)$ is in involution.

Although it is relatively simple to define, the concept of involution is one of the most difficult in the theory. Gaining both a computational and a theoretical understanding of it will be one of the main goals of this chapter.

§2. Linear Differential Systems.

The concept of a linear exterior differential system with an independence condition is an extremely useful one. In order to define it we will first show that the set of n-planes in a fixed vector space that satisfy a set of linear equations with a transversality condition has a natural affine linear structure. More precisely, we will prove that:

(10)   *On a vector space T for which we have a filtration*

   $I \subset J \subset T^*$ with $\dim J/I = n$, the n-planes $E \subset T$

   which satisfy

   $$E \subset I^\perp$$

   $$J/I \cong E^* \text{ (i.e. the restriction } J/I \to E^* \text{ is}$$
   $$\text{an isomorphism)}$$

   form a subset of $G_n(T)$ on which there is a natural

   affine linear structure.

To establish (10) we shall first treat the case when $I = 0$. For this we use coordinates (x^i, y^σ) in $\mathbb{R}^{n+s} \cong \mathbb{R}^n \oplus \mathbb{R}^s$ where $1 \leq i, j \leq n$ and $1 \leq \sigma, \rho \leq s$. The n-planes on which $dx^1 \wedge \cdots \wedge dx^n \neq 0$, i.e., n-planes that project isomorphically onto the $\mathbb{R}^n$ factor, are given by equations

$$y^\sigma = p_i^\sigma x^i.$$

Under an invertible linear change

$$x'^i = A_j^i x^j$$

$$y'^\sigma = B_\rho^\sigma y^\rho + C_i^\sigma x^i$$

we have

(11) $$p_i'^\sigma A_j^i = B_\rho^\sigma p_j^\rho + B_\rho^\sigma C_j^\rho$$

or, in obvious matrix notation,

$$p' = BpA^{-1} + BCA^{-1}.$$

It follows that the p's transform affine linearly.

To treat the general case when $I \neq 0$, we consider $\mathbb{R}^{h+n+s}$ with coordinates (u^a, x^i, y^σ) where $1 \leq a \leq h$. Then n-planes on which

$$du^a = 0$$

$$dx^1 \wedge \cdots \wedge dx^n \neq 0$$

are given by linear equations

$$(12) \qquad \begin{aligned} u^a &= p_i^a x^i \\ y^\sigma &= p_i^\sigma x^i \\ p_i^a &= 0. \end{aligned}$$

Under an invertible linear change

$$u'^a = D_b^a u^b$$

$$x'^i = A_j^i x^j + E_a^i u^a$$

$$u'^\sigma = B_\rho^\sigma y^\rho + C_i^\sigma x^i + F_a^\sigma u^a$$

it is easy to check that the p_i^a and p_i^α defined by the first two equations in (12) transform *quadratically*. However, when we impose the third equation the remaining non-zero p_i^σ's transform by (11). Taking I to be spanned by the u^a's and J/I by the x^i's, by virtue of $u^a = 0$ on E the map $J/I \to E^*$ is well defined and the condition that this be an isomorphism is $dx^1 \wedge \cdots \wedge dx^n \neq 0$ on E. From this we conclude (10).

Now let $(\mathcal{I}, \Omega)$ be a differential system with independence condition over a manifold M. Applying this construction fibrewise where $T = T_x M$, $I = I_x$, $J = J_x$ we conclude that

(13)   *The subset $G(I, \Omega)$ of tangent n-planes E satisfying*

$$\begin{cases} \theta_E = 0, & \text{for all } \theta \in I \\ \Omega_E \neq 0 \end{cases}$$

forms in a natural way a bundle of affine linear spaces

over M.

In the future we shall ususally write the above equations more simply as

$$\begin{cases} \theta = 0, & \theta \in I \\ \Omega \neq 0 \end{cases}.$$

Clearly the set $G(\mathcal{I}, \Omega)$ of integral elements of $(\mathcal{I}, \Omega)$ is a subset of $G(I, \Omega)$.

Definition 2.1. The differential system $(\mathcal{I}, \Omega)$ is *linear* if the fibres of $G(\mathcal{I}, \Omega) \to M$ are affine linear subspaces of the fibres of $G(I, \Omega) \to M$.

Implicit in the above discussion is that the definition of linearity requires an independence condition.

Roughly speaking, a partial differential equation system is linear when its solutions may be linearly superimposed. For a differential system on a manifold the concept of linearity only makes sense infinitesimally. The integral elements of $(\mathcal{I}, \Omega)$ are the infinitesimal solutions and the above definition is the corresponding concept of linear. We will see that many but not all differential systems are linear. Moreover, given any differential system $\mathcal{I}$ and an integral element E, we will define its linearization $(\mathcal{I}_E, \Omega_E)$ at E, which will correspond to linearizing an arbitrary P.D.E. at a solution. Before doing this we need to develop conditions that will allow us to recognize when $(\mathcal{I}, \Omega)$ is linear. We mention that both these conditions and the linearization $(\mathcal{I}_E, \Omega_E)$ are implicit in the proofs of the results in Chapter III above, (cf. Proposition 1.15 in that chapter).

We let $(\mathcal{I}, \Omega)$ be a differential system with independence condition. Locally we choose a coframe

$$\theta^1, \ldots, \theta^h; \omega^1, \ldots, \omega^n; \pi^1, \ldots, \pi^s$$

adapted to the filtration

$$I \subset J \subset T^*(M).$$

Differential forms on M may then be locally written as

$$\psi = f_{\Sigma K A}\pi^\Sigma \wedge \omega^K \wedge \theta^A \tag{14}$$

where $\Sigma = (\sigma_1, \ldots, \sigma_l)$, $K = (k_1, \ldots, k_p)$ and $A = (a_1, \ldots, a_q)$ are increasing multi-indices and $\pi^\Sigma = \pi^{\sigma_1} \wedge \cdots \wedge \pi^{\sigma_l}$, etc.

Definition 2.2. We will say that $(\mathcal{I}, \Omega)$ is *linearly generated* if locally it is generated algebraically by forms ψ which are of combined total degree one in the θ^a's and π^σ's.

It follows that $(\mathcal{I}, \Omega)$ is algebraically generated by forms

$$\tag{15} \left\{ \begin{array}{l} \theta^a \\ \psi = f_{\sigma K}\pi^\sigma \wedge \omega^K, \end{array} \right.$$

and it is clear that this condition is intrinsic. Integral elements of $(\mathcal{I}, \Omega)$ are then defined by the equations

$$\tag{16} \left\{ \begin{array}{l} \theta^a = 0 \\ \pi^\sigma = p_i^\sigma \omega^i \qquad\qquad \text{where} \\ \sum_{\sigma, K} f_{\sigma K} p_i^\sigma \omega^i \wedge \omega^K = 0. \end{array} \right.$$

Since these equations are linear in the p_i^σ's we conclude that:

If $(\mathcal{I}, \Omega)$ is linearly generated, then it is linear.

We shall now give some examples of systems that are linearly generated. These will all be generated in degrees $p \geq 2$; the very important case of linear Pfaffian systems will be treated below. In these examples we will denote by $\Omega^{*,1}(M) \subset \Omega^*(M)$ the forms (14) that are at most linear in the π^σ's, i.e., that have $|\Sigma| \le 1$. The system $(\mathcal{I}, \Omega)$ is then linearly generated if it is algebraically generated by $\Omega^{*,1}(M) \cap \mathcal{I}$.

Example 2.3 (The third fundamental theorem of Lie). Let $\mathbb{R}^n$ be endowed with a Lie algebra structure $[\ ,\ ] : \mathbb{R}^n \times \mathbb{R}^n \to \mathbb{R}^n$. One version of the third fundamental theorem of Lie is that there exists a neighborhood U of $0 \in \mathbb{R}^n$ and an $\mathbb{R}^n$-valued 1-form η on U so that $\eta|_0 : T_0\mathbb{R}^n \to \mathbb{R}^n$ is an isomorphism and so that $d\eta = -1/2[\eta, \eta]$. To establish this, let x^i be linear coordinates on $\mathbb{R}^n$ and let (p^i_j) be the usual coordinates on $GL(n, \mathbb{R})$. We are seeking functions $p^i_j(x)$ so that the forms $\eta^i = p^i_j(x)dx^j$ satisfy both $\det(p^i_j(0)) \ne 0$ and the differential equation $d(pdx) = -1/2[pdx, pdx]$. Thus, let $M = GL(n, \mathbb{R}) \times \mathbb{R}^n$ and let

$$\theta = d(pdx) + 1/2[pdx, pdx].$$

Let $\mathcal{I}$ be the ideal in $\Omega^*(M)$ generated by the n 2-form components of θ. We easily compute that

$$d\theta = 1/2[\theta, pdx] - 1/2[pdx, \theta] - 1/2[[pdx, pdx], pdx]$$

$$\equiv -1/2[[pdx, pdx], pdx] \mod \mathcal{I}$$

$$\equiv 0 \mod \mathcal{I}$$

since $[[pdx, pdx], pdx] = 0$ by the Jacobi identity. Thus $\mathcal{I}$ is differentially closed. As independence condition, we take J so that its sections are spanned by $dx^1, \ldots, dx^n$. Clearly an integral manifold of $(\mathcal{I}, \Omega)$ is locally of the form $(x^i, p^i_j(x))$ where $\eta = (\eta^i) = (p^i_j(x)dx^j)$ satisfies our conditions. Note that $\mathcal{I}$ is linearly generated: We have $\theta^i = dp^i_j \wedge dx^j +$ (terms quadratic in $dx^i) \in \Omega^{*,1}(M)$. More explicitly, we may write $\theta^i = dp^i_j \wedge dx^j + 1/2T^i_{jk}dx^k \wedge dx^j = \pi^i_j \wedge dx^j$ where $\pi^i_j = dp^i_j + 1/2T^i_{jk}dx^k$, and the $\{dx^i, \pi^i_j\}$ form a coframing of M.

An integral element of $(\mathcal{I}, \Omega)$ is obviously described by a set of equations of the form $\pi^i_j - p^i_{jk}dx^k = 0$ where $p^i_{jk} = p^i_{kj}$. Thus the dimension S of the space of integral elements over a point is given by $S = n \binom{n+1}{2} = n^2(n+1)/2$. On the other hand, it may be easily seen that the characters s_k are given by $s_0 = 0$, $s_1 = s_2 = \cdots = s_n = n$. Since $s_1 + 2s_2 + \cdots + ns_n = n(1 + 2 + \cdots + n) = n^2(n+1)/2$, from Theorem 1.10 in Chapter III it follows that $(\mathcal{I}, \Omega)$ is involutive. Since $\mathcal{I}$ is clearly analytic, an application of the Cartan–Kähler theorem yields Lie's theorem.

This proof is not the most elementary, of course, but it is perhaps the simplest conceptually. Note that once existence is proved, the Frobenius theorem suffices to prove that any two solutions η_1 on U_1 and η_2 on U_2 are locally equivalent via diffeomorphism $U_1 \simeq U_2$.

Example 2.4 (Closed self-dual forms on four-manifolds). Let X^4 denote an oriented Riemannian 4-manifold. Let $M^7 = \Lambda^2_+(X)$ denote the bundle of self-dual 2-forms on X. Let $\varphi \in \Omega^2(M)$ denote the tautological 2-form on M which satisfies $\varphi(v_1, v_2) = \alpha(\pi_*(v_1), \pi_*(v_2))$ where $v_i \in T_\alpha M$, and $\pi : M \to X$ is the projection. Thus $\varphi|_{T_\alpha M} = \pi^*(\alpha)$. This form φ has the "reproducing" property: If $\beta = *\beta$ is a 2-form on X, then when we regard β as a section $\beta : X \to M$, we have $\beta^*(\varphi) = \beta$. Moreover $\beta^*(d\varphi) = d\beta$. Let $\mathcal{I}$ be the system algebraically generated by $d\varphi \in \Omega^3(M)$. Clearly $\mathcal{I}$ is differentially closed. Let $\Omega = \pi^*(\text{vol}) \in \Omega^4 M$. The integrals of $(\mathcal{I}, \Omega)$ are locally sections of $M \to X$ which are the graphs of local closed self-dual 2-forms on X. We claim that $(\mathcal{I}, \Omega)$ is linearly generated and involutive. To see this, it suffices to work locally, so let $\omega^1, \omega^2, \omega^3, \omega^4$ be an oriented orthonormal coframing on $U \subseteq X$. Of course $M|_U \cong U \times \mathbb{R}^3$ and there exist unique (linear) coordinates p_2, p_3, p_4 on the $\mathbb{R}^3$ factor so that, on $M|_U$, we have

$$\varphi = p_2(\omega^1 \wedge \omega^2 + \omega^3 \wedge \omega^4) + p_3(\omega^1 \wedge \omega^3 + \omega^4 \wedge \omega^2) + p_4(\omega^1 \wedge \omega^4 + \omega^2 \wedge \omega^3).$$

Now $\Omega = \omega^1 \wedge \omega^2 \wedge \omega^3 \wedge \omega^4$, and we have

$$\begin{aligned}
d\varphi = dp_2 \wedge (\omega^1 \wedge \omega^2 + \omega^3 \wedge \omega^4) &+ dp_3 \wedge (\omega^1 \wedge \omega^3 + \omega^4 \wedge \omega^2) \\
&+ dp_4 \wedge (\omega^1 \wedge \omega^4 + \omega^2 \wedge \omega^3) + T
\end{aligned}$$

where T is a 3-form which is cubic in the $\{\omega^i\}$. Clearly $d\varphi \in \Omega^{*,1}(M)$, so $(\mathcal{I}, \Omega)$ is linearly generated. It is not difficult to show that there exist forms π_2, π_3, π_4 on $M|_U$ so that

$$d\varphi = \pi_2 \wedge (\omega^1 \wedge \omega^2 + \omega^3 \wedge \omega^4) + \pi_3 \wedge (\omega^1 \wedge \omega^3 + \omega^4 \wedge \omega^2) + \pi_4 \wedge (\omega^1 \wedge \omega^4 + \omega^2 \wedge \omega^3)$$

where $\pi_i \equiv dp_i \mod \omega^1, \ldots, \omega^4$. Given this, and keeping the notations from the proceeding example, we easily compute that $S \equiv 8$ on M, and that for any integral flag, we have $s_0 = s_1 = 0$, $s_1 = 1$, $s_3 = 2$, $s_4 = 0$. Since $8 = s_1 + 2s_2 + 3s_3 + 4s_4$, from Proposition 1.15 in Chapter III we again see that Cartan's Test is satisfied and so the system is involutive. Applying the Cartan–Kähler theorem then yields the following result: *Suppose that X^4 is an analytic 4-manifold with an orientation and an analytic Riemannian metric. Let $H^3 \subset X^4$ be an analytic imbedded hypersurface and let $\alpha \in \Omega^2(H^3)$ be a closed analytic 2-form on H^3. Then there exists an open set $U \supset H^3$ and a closed self-dual 2-form β on U so that $\beta|_{H^3} = \alpha$.* We leave details to the reader. This extension theorem is easily seen to be false if α is not assumed to be analytic. We note in closing that, as a P.D.E. system, this is four equations for the three unknown coefficients p_2, p_3, p_4.

In order to further motivate our concept of linearity and for later use, we shall define the linearization $\mathcal{I}_E$ of an arbitrary differential system $\mathcal{I}$ at an n-dimensional integral element $E \subset T_{x_0} M$ lying over $x_0 \in M$. The linearization will have the following properties:

i) it is a differential system $(\mathcal{I}_E, \Omega_E)$ with independence condition defined on the vector space $M_E = E \oplus Q$ where $Q = T_{x_0} M / E$;

ii) $(\mathcal{I}_E, \Omega_E)$ is a constant coefficient, linearly generated (and therefore linear) exterior differential system.

(N.B.: A *constant coefficient differential system* is an exterior differential system defined on a vector space and which is generated as a differential system by translation-invariant differential forms. If there is an independence condition, then this should also be translation invariant. Strictly speaking, in order to construct the full differential system we should take the differential ideal in the set of all smooth (or real-analytic) forms generated by our constant coefficient forms. However, this enlargement will not effect the calculation of such quantities as polar equations or the integral manifolds, and so we shall not insist on it.)

iii) if E is an ordinary integral element of $\mathcal{I}$, then $(\mathcal{I}_E, \Omega_E)$ is involutive and has the same Cartan characters as does E.

(Implicit in (iii) is the assertion that *all* integral elements of $(\mathcal{I}_E, \Omega_E)$ have the same Cartan characters s_k.)

To define $\mathcal{I}_E$, we let $E^\perp \subset T_{x_0} M$ be the space of 1-forms that annihilate E and we denote by

$$\{E^\perp\} \subset \Lambda^*(T_{x_0}^* M)$$

the exterior ideal generated by $E^\perp$. Then, because E is an integral element of $\mathcal{I}$ we have $\mathcal{I}_{x_0} \subset \{E^\perp\}$. There is a canonical exact sequence

$$0 \to \{\Lambda^2 E^\perp\} \to \{E^\perp\} \to E^\perp \oplus \Lambda^* E^* \to 0$$
$$\|$$
$$Q^* \otimes \Lambda^* E^*.$$

We let $P_E \subset Q^* \otimes \Lambda^* E^* \subset \Lambda^*(Q^* \oplus E^*)$ denote the image of $\mathcal{I}_{x_0}$ and define $\mathcal{I}_E$ to be the ideal in $\Lambda^*(Q^* \oplus E^*)$ generated by P_E. Then $\mathcal{I}_E$ is an ideal of exterior forms on $E \oplus Q$. We let Ω_E be a volume form on E.

Definition 2.5. $(\mathcal{I}_E, \Omega_E)$ is the *linearization* of $\mathcal{I}$ at E.

When expressed in a set of linear coordinates on $E \oplus Q$, the elements of $\mathcal{I}_E$ have constant coefficients and hence are closed differential forms. Thus, $(\mathcal{I}_E, \Omega_E)$ is a constant coefficient differential system with independence condition. In order to see that it is linearly generated, it is instructive to see what this construction means in coordinates. For this we choose a local coframe $\omega^1(x), \ldots, \omega^n(x), \pi^1(x), \ldots, \pi^t(x)$ on M so that the forms $\pi^\sigma(x_0)$ span $E^\perp$. We then choose linear coordinates $x^1, \ldots, x^n$ on E and

$y^1, \ldots, y^t$ on Q such that $\omega^i(x_0)|_E = dx^i$ and $\pi^\sigma(x_0)|_Q = dy^\sigma$. Finally, we set $f = f(x_0)$ for any locally defined function $f(x)$ on M. Let $\psi \in \mathcal{I}$ and write

$$\psi(x) = f_I(x)\omega^I(x) + f_{\sigma J}(x)\pi^\sigma(x) \wedge \omega^J(x) + f_{\sigma\rho K}(x)\pi^\sigma(x) \wedge \pi^\rho(x) \wedge \omega^K(x) + \ldots .$$

We note that, because of $\psi|_E = 0$, $f_I = 0$ and we define

$$\overline{\psi} = f_{\sigma J}\, dy^\sigma \wedge dx^J \in Q^* \otimes \Lambda^* E^*$$

(17)
$$\cap$$

$$\Lambda^*(Q^* \oplus E^*).$$

Intuitively, $\overline{\psi}$ is obtained by setting $x = x_0$—i.e., by freezing coefficients— and by ignoring quadratic terms in the $\pi^\sigma(x_0)$—i.e., terms that vanish to second order on E. It is clear that

(18)
$$P_E = \{\overline{\psi} : \psi \in \mathcal{I}\}$$

is the above set of algebraic generators of $\mathcal{I}_E$. Thus $(\mathcal{I}_E, \Omega_E)$ is linearly generated. Moreover, the following proposition is an immediate consequence of Proposition 1.13 of Chapter III.

Proposition 2.6. *Let $E^p \subset E$ be any p-plane, and let $H(E^p) \subset T_{x_0}M$ be the polar space of E^p as an integral element of $\mathcal{I}$. Then, as an integral element of $\mathcal{I}_E$ the polar space of E^p is $E \oplus (H(E^p)/E) \subset E \oplus Q$.*

We will establish the third property mentioned above of $(\mathcal{I}_E, \Omega_E)$ following a general discussion of the concept of tableau in the next section.

§3. Tableaux.

One of the most important concepts in the theory of exterior differential systems is that of a tableau. This is a purely algebraic concept defined as follows:

Definition 3.1. A *tableau* is given by a linear subspace

$$A \subset \mathrm{Hom}(V, W)$$

where V, W are vector spaces.

We let $v_1, \ldots, v_n$ and $w_1, \ldots, w_s$ be bases for V, W respectively and choose a basis

$$A_\varepsilon = A^a_{\varepsilon i} w_a \otimes v_i^*$$

for A (we have chosen to use v_i^* instead of v^i for the dual basis to v_i). Then a general element of A

$$A(\zeta) = A_\varepsilon \zeta^\varepsilon$$
$$= A^a_{\varepsilon i} \zeta^\varepsilon w_a \otimes v_i^*$$

may be thought of as a matrix

$$A(\zeta) = \| A^a_{\varepsilon i} \zeta^\varepsilon \|$$

whose entries are linear functions of the coordinates ζ^ε on A. Therefore, from a linear algebra point of view, the study of tableaux is the same as studying matrices whose entries are linear functions. This will be apparent when we introduce the symbol associated to A.

Example 3.2. Let V and W be vector spaces with coordinates $x^1, \ldots, x^n$ and $y^1, \ldots, y^s$ dual to bases $v_1, \ldots, v_n$ and $w_1, \ldots, w_s$ for V and W respectively. We consider a first order linear homogeneous, constant coefficient P.D.E. system

$$(19) \qquad B^{\lambda i}_a \frac{\partial y^a(x)}{\partial x^i} = 0 \qquad \lambda = 1, \ldots, r.$$

The linear solutions

$$y^a(x) = A^a_j x^j$$

to (19) form a tableau $A \subset \mathrm{Hom}(V, W)$. We shall call A *the tableau associated to* P.D.E. *system* (19). It is clear, conversely, that every tableau is uniquely associated to such a P.D.E. system.

Definition 3.3. Given a tableau $A \subset \mathrm{Hom}(V, W) = W \otimes V^*$, the associated *symbol* is given by the annihilator

$$B = A^\perp \subset W^* \otimes V.$$

Example 3.2 (continued). Assuming that the forms

$$B^\lambda = B^{\lambda i}_a w_a^* \otimes v_i \in W^* \otimes V$$

are linearly independent, the classical definition of the symbol associated to (19) assigns to each covector $\xi = \xi_i dx^i \in V^*$ the matrix

$$(20) \qquad \sigma(\xi) = \| B^{\lambda i}_a \xi_i \|.$$

In coordinate-free terms

$$\sigma(\xi) : W \to W \otimes V^* / A \cong A^{\perp *}$$

is given by

$$\text{(21)} \qquad\qquad \sigma(\xi)(w) = w \otimes \xi \quad \text{mod } A.$$

From (20) it is clear that giving the symbol of the P.D.E. system (19) is equivalent to giving the symbol B of the tableau A.

We now consider a tableau $A \subset W \otimes V^*$. The $(q+1)^{st}$ symmetric product $S^{q+1}V^*$ may be considered as the space of homogeneous polynomials of degree $q+1$ on V, and we have the usual differentiation operators

$$\frac{\partial}{\partial x^i} : S^{q+1}V^* \to S^q V^*.$$

We extend $\partial/\partial x^i$ to $W \otimes S^{q+1}V^*$ by treating W as constants, so that by definition

$$\partial/\partial x^i(w_a \otimes P^a(x)) = w_a \otimes \frac{\partial P^a}{\partial x^i}$$

where the $P^a(x)$ are homogeneous polynomials in $x^1, \ldots, x^n$.

Definition 3.4. Given a tableau $A \subset W \otimes V^*$, the qth *prolongation*

$$A^{(q)} \subset W \otimes S^{q+1}V^*$$

is defined inductively by $A^{(0)} = A$ and, for $q \geq 1$,

$$A^{(q)} = \{P : \frac{\partial P}{\partial x^i} \in A^{(q-1)} \text{ for all } i\}.$$

It is clear that $A^{(q)}$ is the subspace consisting of all $P \in W \otimes S^{q+1}V^*$ satisfying

$$\text{(22)} \qquad\qquad \frac{\partial^q P(x)}{\partial x^{i_1} \ldots \partial x^{i_q}} \in A$$

for all $i_1, \ldots, i_q$.

In case A is the tableau associated to the constant coefficient, linear homogeneous P.D.E. system (19) it is clear that: $A^{(q)}$ *is the set of homogeneous polynomial solutions of degree $q+1$ to the* P.D.E. *system* (19). What will turn out to be a more profound interpretation of the first prolongation $A^{(1)}$ follows.

First, we consider the exterior differential system $(\mathcal{I}_A, \Omega_A)$ associated to the P.D.E. system (19) corresponding to the tableau $A \subset W \otimes V^*$. To describe $(\mathcal{I}_A, \Omega_A)$ we consider as usual the space $J^1(V, W)$ of 1-jets of mappings from V to W and let (x^i, y^a, p_i^a) be the standard coordinates induced from the coordinate systems x^i, y^a on V, W respectively. We then define

$$M \subset J^1(V, W)$$

by the equations $B_a^{\lambda i} p_i^a = 0$. Then $(\mathcal{I}_A, \Omega_A)$ is the exterior differential system with independence condition $\Omega_A = dx^1 \wedge \cdots \wedge dx^n \neq 0$ obtained by restricting the contact system on $J^1(V, W)$ to M. We recall that the contact system is generated algebraically by the differential forms

$$(23) \qquad \begin{cases} \theta^a = dy^a - p_i^a dx^i \\ d\theta^a = -dp_i^a \wedge dx^i. \end{cases}$$

The restrictions to M of the 1-forms dx^i, θ^a, dp_i^a span the cotangent spaces and are subject to the relations

$$(24) \qquad B_a^{\lambda i} dp_i^a = 0 \qquad \lambda = 1, \ldots, r$$

that define $T_q M \subset T_q J^1(V, W)$ for $q \in M$.

If $E \subset T_q M$ is any integral element of $(\mathcal{I}_A, \Omega_A)$, then since $dx^1 \wedge \cdots \wedge dx^n|_E \neq 0$ it follows that the $dx^i|_E$ form a basis for E^*, and consequently E is defined by a set of linear equations

$$(25) \qquad \begin{cases} \theta^q = 0 \\ dp_i^a = p_{ij}^a dx^j \end{cases}$$

subject to the conditions that $d\theta^a|_E = 0$ and that the linear relations (24) are satisfied on E. Substituting dp_i^a from (25) into the second equation in (23) gives

$$p_{ij}^a = p_{ji}^a,$$

and then the linear relations (24) give

$$B_a^{\lambda i} p_{ij}^a = 0.$$

Taken together these two equations are equivalent to the condition that

$$P = p_{ij}^a w_a \otimes x^i x^j \in A^{(1)}.$$

In summary:

$$(26) \qquad \textit{For the exterior differential system associated to the}$$
$$\textit{P.D.E. system (19), the space of integral elements over}$$
$$\textit{any fixed point is naturally identified with the } 1^{st}$$
$$\textit{prolongation } A^{(1)} \textit{ of the tableau associated to (19).}$$

This result will be extended to general linear Pfaffian systems in section 5 below.

We now want to explain the concept of involutivity for a tableau A. Although it is a purely algebraic concept, it will turn out to be equivalent to the condition that the exterior differential system $(\mathcal{I}_A, \Omega_A)$ associated to the P.D.E. system (19) with tableau A should be involutive. From Theorem 1.11 of Chapter III we see that this in turn is expressed by the condition that an inequality between the dimension of the space of integral elements of $(\mathcal{I}_A, \Omega_A)$ and an expression involving the ranks of the polar equations should be an equality. This together with (26) above should help to motivate the following discussion leading up to the definition of involutivity for a tableau A.

First we need two definitions. If $U \subset W \otimes S^q V^*$ is any subspace we set

$$(27) \qquad U_k = \left\{ P \in U : \frac{\partial P}{\partial x^1} = \cdots = \frac{\partial P}{\partial x^k} = 0 \right\}.$$

We note that

$$(A^{(1)})_k = (A_k)^{(1)}$$

since both sides are equal to $\{P \in W \otimes S^2 V^* : \partial P/\partial x^i \in A$ and $\partial^2 A/\partial x^i \partial x^j = 0$ for all i and $1 \le j \le k\}$. We denote either side of this equality by $A_k^{(1)}$. Clearly we have

$$(28) \qquad \partial/\partial x^k : A_{k-1}^{(1)} \to A_{k-1}.$$

We observe that the subspaces A_k give a filtration

$$(29) \qquad 0 = A_n \subset A_{n-1} \subset \cdots \subset A_1 \subset A_0 = A,$$

and that the numbers $\dim A_k$ are upper-semi-continuous and constant on a dense open set of coordinate systems for V^*.

Definition 3.5. Let $A \subset W \otimes V^*$ be a tableau and $x^1, \ldots, x^n \in V^*$ a generic coordinate system for which the $\dim A_k$ are a minimum. We then define the *characters $s'_1, \ldots, s'_n$ of the tableau A* inductively by

$$(30) \qquad s'_1 + \cdots + s'_k = \dim A - \dim A_k.$$

Although it is not immediately obvious, it can be shown that

$$s'_1 \ge s'_2 \ge \cdots \ge s'_n$$

(cf. the normal form (90) below).

The following gives an algebraic analogue of the inequality in Cartan's test:

Proposition 3.6. *We have*

$$(31) \qquad \dim A^{(1)} \le s_1' + 2s_2' + \cdots + n s_n'$$

with equality holding if, and only if, the mappings (28) are surjective.

Proof. We note that

$$(32) \qquad \begin{cases} \dim A = s_1' + \cdots + s_n' \\ \dim A_k = s_{k+1}' + \cdots + s_n'. \end{cases}$$

From the exact sequence

$$0 \to A_k^{(1)} \to A_{k-1}^{(1)} \xrightarrow{\ \partial/\partial x^k\ } A_{k-1}$$

we have

$$\dim A_{k-1}^{(1)} - \dim A_k^{(1)} \le \dim A_{k-1}.$$

Adding these up and using $A_0^{(1)} = A^{(1)}$ gives

$$\dim A^{(1)} \le \dim A + \dim A_1 + \cdots + \dim A_{n-1}.$$

Substituting (32) into the right hand side gives the result. $\qquad\square$

Definition 3.7. The *tableau A is involutive* in case the equality

$$\dim A^{(1)} = s_1' + 2s_2' + \cdots + n s_n'$$

holds in (31).[3]

Proposition 3.8. *The involutivity of the tableau A is equivalent to the involutivity of the P.D.E. system (19) associated to A.*

Proof. It follows from the discussion preceeding (26) that the space $G(\mathcal{I}_A, \Omega_A)$ of integral elements of $(\mathcal{I}_A, \Omega_A)$ fibers over M with the fibres each being a linear space naturally isomorphic to the first prolongation $A^{(1)}$. In order to apply Theorem 1.11 in Chapter III it will thus suffice to work in the space of integral elements lying over the origin in M.

We next set

$$(33) \qquad \begin{cases} \omega^i = dx^i|_M \\ \pi_i^a = -dp_i^a|_M \end{cases}$$

so that the structure equations (23) and (24) of $(\mathcal{I}_A, \Omega_A)$ become

$$(34) \qquad \begin{cases} d\theta^a = \pi_i^a \wedge \omega^i \\ B_a^{\lambda i} \pi_i^a = 0. \end{cases}$$

---

[3]This formulation of involutivity is due to Matsushima [1954-55]

A substitution

$$\begin{cases} \pi_i^a \to \pi_i^a - p_{ij}^a \omega^j \\ p_{ij}^a w_a \otimes x^i x^j \in A^{(1)} \end{cases}$$

leaves these structure equations unchanged. By means of such a substitution the integral element (25) in $G(\mathcal{I}_A, \Omega_A)$ is now defined by the equations

$$(35) \qquad \begin{cases} \theta^a = 0 \\ \pi_i^a = 0 \end{cases}$$

and is subject to the requirement that the $d\theta^a = 0$ on this n-plane. Calling this n-plane E, we will determine the circumstances such that E satisfies the conditions in Theorem 1.11 of Chapter III.

For this we let $e_1, \ldots, e_n \in T_0 M$ be the basis for E defined by the equations

$$\begin{cases} \theta^a(e_k) = \pi_i^a(e_k) = 0 \\ \omega^i(e_k) = \delta_k^i. \end{cases}$$

Then $e_1, \ldots, e_k$ spans a subspace $E_k \subset E$ and we claim that:

(36)   *The rank of the polar equations associated to E_k is*

$$s + s_1' + \cdots + s_k' \text{ where } s = \dim W.$$

Proof of (36). The cotangent space $T_0^* M$ is spanned by the 1-forms ω^i, θ^a, π_i^a, of which the ω^i and θ^a are linearly independent and the π_i^a are subject exactly to the second equations in (34). If we define subspaces R, S of $T_0 M$ by

$$R = \{v \in T_0 M : \pi_i^a(v) = 0\}$$

$$S = \{v \in T_0 M : \omega^i(v) = 0 = \theta^a(v)\},$$

then we have $T_0 M = R \oplus S$. Moreover, the mapping $S \to W \otimes V^*$ defined by

$$(37) \qquad v \to \pi_i^a(v) w_a \otimes x^i$$

is injective, and by the second equation in (34) the image of this mapping is the tableau $A \subset W \otimes V^*$. We shall identify S with A and denote by S_k the subspace of S corresponding to A_k. Thus

$$S_k = \{v \in S : \pi_i^a(v) = 0 \text{ for } 1 \le i \le k\}.$$

We now shall show that the polar equations of $E_k = \text{span}\{e_1, \ldots, e_k\}$ are given by

$$(38) \qquad \begin{cases} \theta^a = 0 \\ \pi_i^a = 0 \text{ for } 1 \le i \le k. \end{cases}$$

This is immediate from (34), since for $v \in T_0 M$ and $1 \leq i \leq k$

$$d\theta^a(e_i, v) = (\pi^a_j \wedge \omega^j)(e_i, v)$$
$$= -\pi^a_i(v).$$

Since the rank of the equations $\pi^a_i(v) = 0$ clearly depends only on the projection of $v \in R \oplus S$ to S, we see that the rank of the equations (38) is given by

$$s + \dim(S/S_k) = s + \dim A - \dim A_k$$
$$= s + s'_1 + \cdots + s'_k.$$

by the definition (39). This completes the proof of (36).

We may now complete the proof of Proposition 3.10. In fact, using (32) and (36) the inequality in Theorem 1.11 in Chapter III is just (31); moreover, the condition for equality in (31) is just the condition that the integral element defined by (35) be ordinary. In fact the inequality there is $\mathrm{codim}\{G_n(\mathcal{I}) \subset G_n(TM)\} \geq c_0 + c_1 + \cdots + c_{n-1}$. We have shown that

$$c_k = s + s'_1 + \cdots + s'_k,$$
$$\dim A^{(1)} = \dim G_{n,x}(\mathcal{I}),$$

and combining these three relations and unwinding the arithmetic gives (31). $\square$

§4. Tableaux Associated to an Integral Element.

Let $\mathcal{I}$ be a differential system on a manifold M. Let $E \subset T_{x_0} M$ be an n-dimensional integral element of $\mathcal{I}$ and set $Q = T_{x_0} M / E$. We will canonically associate to E a tableau

$$A_E \subset \mathrm{Hom}(E, Q)$$

that has a number of important properties.

For this we use coordinates x^i, y^σ on E, Q as in the discussion following Definition 2.5 of the linearization $(\mathcal{I}_E, \Omega_E)$ of $\mathcal{I}$ at E. Recall that $\mathcal{I}_E$ is defined on $M_E = E \oplus Q$ and is algebraically generated by constant coefficient differential forms

$$(39) \qquad\qquad \overline{\psi} = f_{\sigma J} dy^\sigma \wedge dx^J$$

that are linear in the dy^σ's, and that the independence condition is given by $\Omega_E = dx^1 \wedge \cdots \wedge dx^n$. It follows that integral elements $\tilde{E}$ of $(\mathcal{I}_E, \Omega_E)$ lying over the origin are given by graphs of linear mappings

$$p : E \to Q$$

satisfying the following conditions:

$$(40) \qquad \begin{cases} dy^\sigma - p_i^\sigma\, dx^i = 0 \\ f_{\sigma J} p_i^\sigma\, dx^i \wedge dx^J = 0. \end{cases}$$

Here the first equation expresses p in coordinates, and the second equation expresses the condition that $\overline{\psi}|_{\tilde{E}} = 0$ for all $\overline{\psi} \in P_E$, where P_E given by equation (18).

Definition 4.1. The *tableau* $A_E \subset \operatorname{Hom}(E, Q)$ *associated to* $E \in G_n(\mathcal{I})$ is the linear subspace of $\operatorname{Hom}(E, Q)$ defined by the equations (40).

It is clear that A_E is canonically associated to E. One geometric interpretation is that by definition A_E is canonically identified with the set of integral elements lying over the origin of the linearization of $\mathcal{I}$ at E. Another geometric interpretation of A_E is as follows: We set $T = T_{x_0} M$ and consider the n-dimensional integral elements of $\mathcal{I}$ lying over x_0 as a subset

$$G_{n,x_0}(\mathcal{I}) \subset G_n(T).$$

It is well known that there is a canonical isomorphism

$$T_E(G_n(T)) \cong \operatorname{Hom}(E, Q),$$

and we will show that:

$(41) \qquad$ *If $E(t) \subset G_n(T)$ is a smooth arc of integral elements*

$\qquad\qquad$ *of $\mathcal{I}$ lying over x_0 with $E(0) = E$, then*

$$E'(0) \in A_E.$$

Proof. We let $v_1, \ldots, v_n, w_1, \ldots, w_s$ be the basis for T dual to $dx^1, \ldots, dx^n$, $dy^1, \ldots, dy^s$, and we extend the v_i to a smoothly varying basis $v_i(t)$ for $E(t)$. Setting

$$v_i'(0) = \alpha_i^j v_j + \beta_i^\sigma w_\sigma,$$

then by definition

$$E'(0) = \beta_i^\sigma [w_\sigma] \otimes dx^i \in Q \otimes E^*$$

where $[w_\sigma] \in Q$ is the equivalence class defined by $w_\sigma \in T$. On the other hand, as in the discussion following the Definition 2.3 of the linearization, we consider $\psi \in \mathcal{I}$ of degree n and set

$$\psi(x_0) = f_{\sigma J} dy^\sigma \wedge dx^J + f_{\sigma \rho K} dy^\sigma \wedge dy^\rho \wedge dx^K + \dots .$$

Then, setting $v(t) = v_1(t) \wedge \dots \wedge v_n(t)$ and using that $E(t) \in G_{n,x_0}(\mathcal{I})$, we have

$$0 = \langle \psi(x_0), v(t) \rangle$$

$$= f_{\sigma J} \langle dy^\sigma \wedge dx^J, v(t) \rangle + f_{\sigma \rho K} \langle dy^\sigma \wedge dy^\rho \wedge dx^K, v(t) \rangle + \dots .$$

Taking the derivative of this equation at $t = 0$ gives

$$f_{\sigma J} \beta_i^\sigma \, dx^i \wedge dx^J = 0,$$

and comparing with (40) gives our assertion. $\qquad\qquad\square$

When we defined the linearization $(\mathcal{I}_E, \Omega_E)$ of a differential ideal $\mathcal{I}$ at $E \in G_{n,x_0}(\mathcal{I})$, we said that $\mathcal{I}_E$ was obtained by setting $x = x_0$ (freezing coefficients) and by throwing out forms in $\mathcal{I}$ that vanish to second order or higher on E. This is now explained by the proof of (41).

From (41) we have the following geometric interpretation of the tableau A_E:

Proposition 4.2. *If the set $G_{n,x_0}(\mathcal{I})$ of n-dimensional integral elements of $\mathcal{I}$ lying over $x_0 \in M$ is a smooth manifold near E, then its tangent space is the tableau A_E associated to E.*

In general, $G_{n,x_0}(\mathcal{I})$ is an algebraic subvariety of $G_n(T_{x_0}M)$ and A_E is its *Zariski tangent space* at E. This is because the Zariski tangent space to any algebraic variety is the span of tangent vectors to smooth arcs lying in the variety.

Theorem 4.3. *If E is an ordinary integral element of $\mathcal{I}$, then the linearization $(\mathcal{I}_E, \Omega_E)$ is involutive and has the same Cartan characters.*

Proof. Let $0 \subset E_1 \subset \dots E_n = E$ be an ordinary integral flag of $E \subset T_{x_0}M$ as an integral element of $\mathcal{I}$. Let c_k be the codimension of the polar space of E_k. By Proposition 2.6 above, this number is the same whether we regard E_k as an integral element of $\mathcal{I}$ or $\mathcal{I}_E$. By the proof of Theorem 1.11 of Chapter III, the fact that E is ordinary implies that $G_{n,x_0}(\mathcal{I})$ is a smooth submanifold of $G_n(T_{x_0}M)$ of codimension $c_0 + c_1 + \dots + c_{n-1}$. By Proposition 4.2 above, the vector space $T_E(G_{n,x_0}(\mathcal{I}))$ is isomorphic to A_E. Thus A_E has codimension $c_0 + c_1 + \dots + c_{n-1}$ in $T_E(G_n(E \oplus Q)) \cong Q \otimes E^*$. Since by the argument given above, $G(\mathcal{I}_E, \Omega_E) \cong (E \oplus Q) \times A_E$, it follows that

$G(\mathcal{I}_E, \Omega_E)$ is a smooth submanifold of codimension $c_0 + c_1 + \cdots + c_{n-1}$ in the space of all n-dimensional tangent planes at points of $E \oplus Q$. By Theorem 1.11 of Chapter III, it follows that E is an ordinary integral element of $\mathcal{I}_E$. The equality of characters is now obvious. $\qquad\square$

Another interpretation of the tableau arises by considering the constant coefficient, linear homogeneous P.D.E. system associated to the differential system $(\mathcal{I}_E, \Omega_E)$. By this we mean the following: Using the above notations, an integral manifold of $(\mathcal{I}_E, \Omega_E)$ is locally given by a graph

$$x^i \to (x^i, y^\sigma(x))$$

on which all the generators of $\mathcal{I}_E$ restrict to zero. This means that

$$(42) \qquad f_{\sigma J} \frac{\partial y^\sigma(x)}{\partial x^i} dx^i \wedge dx^J = 0.$$

It is then clear that: *the space of homogeneous 1-jets of solutions to (42) is the tableau A_E; more generally, the space of degree $q + 1$ homogeneous polynomial solutions to (42) is the qth prolongation $A_E^{(q)}$ of the tableau A_E* (cf. the discussion below Definition 3.4).

This observation leads to an interesting point. Suppose we clear out the exterior algebra and write (42) as a constant coefficient, linear homogeneous P.D.E. system

$$(43) \qquad B_\sigma^{\lambda i} \frac{\partial y^\sigma(x)}{\partial x^i} = 0.$$

As we have seen in Proposition 3.7 above the condition that the tableau A_E be involutive is that the exterior differential system in $(x^i, y^\sigma, p_i^\sigma)$ space

$$(44) \qquad \begin{cases} dy^\sigma - p_i^\sigma dx^i = 0 \\ B_\sigma^{\lambda i} p_i^\sigma = 0 \\ dx^1 \wedge \cdots \wedge dx^n \neq 0 \end{cases}$$

associated to (43) be involutive. As we have also seen in section 3 above (cf. (26) there), integral elements of this system are given by the equations

$$dp_i^\sigma - p_{ij}^\sigma dx^j = 0$$

where $p = p_{ij}^\sigma w_\sigma \otimes x^i x^j \in A_E^{(1)}$. Thus, the following result, which by Theorem 4.3 relates a property of the integral elements of $(\mathcal{I}_E, \Omega_E)$ lying over the origin (these are just A_E) to a property of the integral elements of (44) lying over the origin (these are just the 1st prolongation $A_E^{(1)}$) is by no means obvious:

Theorem 4.4. *If $E \in G_n(\mathcal{I})$ is an ordinary integral element, then the tableau A_E is involutive.*

Proof. We retain the notations in §§2 and 3 above. We are given that E is an ordinary integral element of $\mathcal{I}$, and we want to show that equality holds in the inequality for $\dim A_E^{(1)}$ in (31).

Now, by Theorem 4.3, the condition that E be ordinary for $\mathcal{I}$ implies that it is ordinary for the linearization $(\mathcal{I}_E, \Omega_E)$, and we shall prove that

$$(45) \qquad \left\{ \begin{array}{c} E \text{ ordinary for} \\ (\mathcal{I}_E, \Omega_E) \end{array} \right\} \Rightarrow \dim A_E^{(1)} = s_1' + 2s_2' + \cdots + ns_n'.$$

This is a purely algebraic statement, and although it is possible to give a purely algebraic argument, here we will use the Cartan–Kähler theorem, which is an analytic result. Recalling that $A_E^{(1)}$ may be identified with the homogeneous 2-jets of integral manifolds of $(\mathcal{I}_E, \Omega_E)$, in outline the analytic proof goes as follows:

$$(\mathcal{I}_E, \Omega_E) \text{ involutive} \quad \Rightarrow \quad \left\{ \begin{array}{c} \text{there are ``enough'' integral} \\ \text{manifolds for } (\mathcal{I}_E, \Omega_E) \end{array} \right\}$$

$$\Downarrow$$

$$\text{a lower bound on } \dim A_E^{(1)}$$

Here is the formal argument.

Proof. As noted above, we may assume that $E \in G(\mathcal{I}_E, \Omega_E)$ is an ordinary integral element of $\mathcal{I}_E$. For the rest of the proof, we shall work with the differential system $\mathcal{I}_E$ on $E \oplus Q$. We may assume that our linear coordinate systems $x^1, \ldots, x^n$ on E and $y^1, y^2, \ldots, y^s$ on Q have been chosen so that the subspaces $E_p = \{v \in E \mid x^j(v) = 0 \text{ for all } j > p\}$ form an ordinary flag, and so that $H(E_p) = E \oplus \{w \in Q \mid y^\sigma(w) = 0 \text{ for all } \sigma > s - c_p\}$ for $p < n$. Here we are using the notations from the proof of Proposition 1.10 in Chapter III, and we recall our convention that $c_n = s$ and $c_{-1} = 0$.

By definition, there exist r forms $\varphi^1, \varphi^2, \ldots, \varphi^r$ in $\mathcal{I}$ so that the forms $\overline{\varphi}^\rho$ for $1 \leq \rho \leq r$ generate $\mathcal{I}_E$ algebraically. Here, $\overline{\varphi}^\rho$ refers to the construction given by (26). These forms have expansions

$$\overline{\varphi}^\rho = \sum_{\sigma, |J| = p_\rho} f_{\sigma J}^\rho dy^\sigma \wedge dx^J$$

for $1 \leq \rho \leq r$ where $\overline{\varphi}^\rho$ has degree $p_\rho + 1$, and the $f_{\sigma J}^\rho$ are some constants. Clearly, $\mathcal{I}_E$ is real analytic and moreover, as a differential form on $E \oplus Q$ each $\overline{\varphi}^\rho$ is closed. By Theorem 4.3, the Cartan–Kähler theorem applies.

We now refer to Chapter III. Recall that there we defined the *level* of an index σ in the range $1 \leq \sigma \leq s$ to be the integer k (in the range $0 \leq k \leq n$)

so that $s - c_k < \sigma \leq s - c_{k-1}$. We also recall from these that the characters $s_0, s_2, \ldots, s_n$ of $\mathcal{I}$ in a neighborhood of E are defined by

$$s_k = \text{ number of } \sigma\text{'s of level } k.$$

By the discussion following the proof of the Cartan–Kähler theorem, the real analytic integral manifolds of $(\mathcal{I}_E, \Omega_E)$ are given in a neighborhood of $x = 0$ by equations of the form $y^\sigma = F^\sigma(x^1, x^2, \ldots, x^n)$ where the F^σ are real analytic and moreover are uniquely specified by knowing the following data:

the s_0 constants $\qquad\qquad f^\sigma = F^\sigma(0, 0, \ldots, 0)$ $\qquad$ when σ has level 0
the s_1 functions $\qquad\quad f^\sigma(x^1) = F^\sigma(x^1, 0, \ldots, 0)$ $\qquad$ when σ has level 1
the s_2 functions $\qquad f^\sigma(x^1, x^2) = F^\sigma(x^1, x^2, 0, \ldots, 0)$ $\qquad$ when σ has level 2

$$\vdots \qquad\qquad\qquad\qquad \vdots \qquad\qquad \vdots$$

the s_n functions $\quad f^\sigma(x^1, x^2, \ldots, x^n) = F^\sigma(x^1, x^2, \ldots, x^n)$ $\quad$ when σ has level n.

Note that due to the fact that the forms $\overline{\varphi}^\rho$ have constant coefficients, it follows that if $y^\sigma = F^\sigma(x^1, x^2, \ldots, x^n)$ is a real analytic solution in a neighborhood of $x = 0$ and we let F_k^σ be the homogeneous term of degree k in the power series expansion of F^σ, then $y^\sigma = F_k^\sigma$ is also an integral manifold of $\mathcal{I}_E$. It follows also that if the f^σ are each chosen to be homogeneous polynomials of degree k in the appropriate variables, then the corresponding F^σ will also be homogeneous polynomials of degree k. If we regard the collection $F_k = (F_k^\sigma)$ as a Q-valued polynomial on E of degree k, then we see that the subspace $\mathcal{S}^k \subset Q \otimes S^k(E^*)$ consisting of those polynomial maps of degree k whose graphs in $E \oplus Q$ are integral manifolds of $\mathcal{I}_E$ is a vector space whose dimension is given, for $k > 0$, by the formula

$$\dim \mathcal{S}^k = s_1 + s_2 \binom{k+1}{1} + s_3 \binom{k+2}{2} + \cdots + s_n \binom{k+n-1}{n-1}.$$

Note that $\mathcal{S}^1 = A_E$ by definition. Moreover, we plainly have that $\mathcal{S}^k = (A_E)^{(k-1)}$ for all $k \geq 1$.

We now claim that the characters $s_1, s_2, \ldots, s_n$ of $\mathcal{I}$ in a neighborhood of E and $s_1', s_2', \ldots, s_n'$ of the tableau A_E are related by

$$(*) \qquad\qquad\qquad s_k' = s_k + \cdots + s_n.$$

Once we have established then we are done, since it follows that the dimension of $(A_E)^{(1)} \cong \mathcal{S}^2$ is given by

$$s_1 + s_2 \binom{3}{1} + s_3 \binom{4}{2} + \cdots + s_n \binom{n+1}{n-1}$$

$$= (s_1 + \cdots + s_n) + 2(s_2 + \cdots + s_n) + \cdots + n s_n$$

$$= s_1' + 2s_2' + \cdots + n s_n'$$

by $(*)$.

To establish $(*)$, we have from Definition 3.5 that

$$\dim(A_E)_k = s'_{k+1} + \cdots + s'_n,$$

and we also have from the definition that $(A_E)_k$ consists of the Q-valued linear functions on E that lie in $A_E \subset Q \otimes E^*$ and that do not depend on $x^1, \ldots, x^k$. Thus, $(A_E)_k$ is isomorphic to the space of linear integral manifolds $y^\sigma = F^\sigma(x^{k+1}, \ldots, x^n)$ of $(\mathcal{I}_E, \Omega_E)$ as described above and which do not depend on $x^1, \ldots, x^k$. By the count of the number of such solutions there are (starting from the top)

$(n-k)s_n$ $-$ dimensions worth coming from an arbitrary linear function $f^\sigma(x^{k+1}, \ldots, x^n)$ where σ has level n

$(n-k-1)s_{n-1}$ $-$ dimensions worth coming from an arbitrary linear function $f^\sigma(x^{k+1}, \ldots, x^{n-1})$ where σ has level $n-1$

$$\vdots$$

s_{k+1} $-$ dimensions worth coming from an arbitrary linear function $f^\sigma(x^{k+1})$ where σ has level $k+1$.

Thus

$$\dim(A_E)_k = s_{k+1} + 2s_{k+2} + \cdots + (n-k)s_n.$$

This gives the equations

$$s'_{k+1} + \cdots + s'_n = s_{k+1} + 2s_{k+2} + \cdots + (n-k)s_n \quad \text{for} \quad 0 \le k \le n-1,$$

which may then be solved to give $(*)$. $\qquad\qquad\qquad\qquad\qquad\qquad\square$

§5. Linear Pfaffian Systems.

The general theory takes a concrete and simple form for Pfaffian systems.

Definition 5.1. A *Pfaffian system* is an exterior differential system with independence condition $(\mathcal{I}, \Omega)$ such that $\mathcal{I}$ is generated as an exterior differential system in degrees zero and one.

Locally $\mathcal{I}$ is algebraically generated by a set $f_1, \ldots, f_r$ of functions and $\theta^1, \ldots, \theta^{s_0}$ of 1-forms together with the exterior derivatives $df_1, \ldots, df_r$ and

$d\theta^1, \ldots, d\theta^{s_0}$. Equating the functions f_i to zero effectively means restricting to submanifolds, and we shall not carry this step along explicitly in our theoretical developments. However, in practice it is obviously important; for example, in imposing integrability conditions during the process of prolongation (cf. §6 below). Thus, unless mentioned to the contrary, we shall assume that $\mathcal{I}$ is generated as a differential ideal by the sections of a sub-bundle $I \subset T^*M$.[4] Moreover, as explained following Definition 1.1 above, we shall assume that the independence condition corresponds to a sub-bundle $J \subset T^*M$ with $I \subset J$. Denoting by $\{J\} \subset \Omega^*(M)$ the algebraic ideal generated by the C^∞ sections of J, we shall prove the following

Proposition 5.2. *The necessary and sufficient condition that $(\mathcal{I}, \Omega)$ be linear is that*

$$(46) \qquad\qquad dI \equiv 0 \mod \{J\}.$$

Proof. We will use the proof as an opportunity to derive the local structure equations of a Pfaffian system $(\mathcal{I}, \Omega)$. Choose a set of 1-forms

$$\theta^1, \ldots, \theta^{s_0}; \omega^1, \ldots, \omega^n; \pi^1, \ldots, \pi^t$$

that is adapted to the filtration

$$I \subset J \subset T^*M$$

and that gives a local coframing on M. Throughout this section we shall use the ranges of indices

$$1 \le a, b \le s_0$$
$$1 \le i, j \le n$$
$$1 \le \varepsilon, \delta \le t.$$

The above local coframing is defined up to invertible linear substitutions

$$\tilde{\theta}^a = A^a_b \theta^b$$
$$(47) \qquad\qquad \tilde{\omega}^i = B^i_j \omega^j + B^i_a \theta^a$$
$$\tilde{\pi}^\varepsilon = C^\varepsilon_\delta \pi^\delta + C^\varepsilon_i \omega^i + C^\varepsilon_a \theta^a,$$

reflecting the filtration $I \subset J \subset T^*M$.

---

[4] Although it is somewhat cumbersome, we shall use s_0 to denote the rank of I; this s_0 is the first of the Cartan characters $s_0, s_1, \ldots, s_n$.

The behavior of the 2-forms $d\theta^a$ is fundamental to the differential system. In terms of them we will determine the conditions that $(\mathcal{I}, \Omega)$ be linear. For this we write

$$(48) \qquad d\theta^a \equiv A^a_{\varepsilon i}\pi^\varepsilon \wedge \omega^i + \frac{1}{2}c^a_{ij}\omega^i \wedge \omega^j + \frac{1}{2}e^a_{\varepsilon\delta}\pi^\varepsilon \wedge \pi^\delta \quad \text{mod } \{I\}$$

where we can suppose that

$$c^a_{ij} + c^a_{ji} = 0 = e^a_{\varepsilon\delta} + e^a_{\delta\varepsilon}.$$

Here we recall our notation that $\{I\}$ is the algebraic ideal generated by the θ^a's. Integral elements of $(\mathcal{I}, \Omega)$ are defined by $\theta^a = 0$ together with linear equations

$$(49) \qquad\qquad\qquad\qquad \pi^\varepsilon - p^\varepsilon_i \omega^i = 0$$

where by (48)

$$(50) \qquad\qquad (A^a_{\varepsilon i}p^\varepsilon_j - A^a_{\varepsilon j}p^\varepsilon_i) + c^a_{ij} + e^a_{\varepsilon\delta}(p^\varepsilon_i p^\delta_j - p^\varepsilon_j p^\delta_i) = 0.$$

These equations are linear in p^ε_i if, and only if,

$$e^a_{\varepsilon\delta} = 0.$$

This is equivalent to

$$(51) \qquad\qquad d\theta^a \equiv A^a_{\varepsilon i}\pi^\varepsilon \wedge \omega^i + \frac{1}{2}c^a_{ij}\omega^i \wedge \omega^j \quad \text{mod } \{I\},$$

which is also clearly the condition that $(\mathcal{I}, \Omega)$ be linearly generated. It is also clear that (51) is just (46) written out in terms of bases. $\qquad\square$

The proof shows that a Pfaffian system is linear if, and only if, it is linearly generated.

Definition 5.3. We shall say that the Pfaffian system $(\mathcal{I}, \Omega)$ is *linear* if either of the equivalent conditions (46) or (51) is satisfied.

We want to comment on the equation (51). Assume that $(\mathcal{I}, \Omega)$ is linear and write

$$(52) \qquad\qquad\qquad\qquad d\theta^a \equiv \pi^a_i \wedge \omega^i \quad \text{mod } \{I\}$$

where the π^a_i are 1-forms. It is clear from (47) that

$(53) \qquad$ *the π^a_i are well-defined as sections of T^*M/J,*

$\qquad\qquad$ *and under a change of coframe (47) the π^a_i transform*

$\qquad\qquad$ *like the components, relative to our chosen coframe, of*

$\qquad\qquad$ *a section of $I^* \otimes J/I$.*

The relation between (51) and (52) is

$$(54) \qquad\qquad \pi_i^a \equiv A_{\varepsilon i}^a \pi^\varepsilon \quad \mathrm{mod}\ \{\theta^a, \omega^i\},$$

and the most general 1-forms π_i^a satisfying (52) are given by

$$\pi_i^a = A_{\varepsilon i}^a \pi^\varepsilon + c_{ji}^a \omega^j + p_{ji}^a \omega^j$$

where the $A_{\varepsilon i}^a$ and c_{ji}^a are as above and the p_{ij}^a are free subject to $p_{ij}^a = p_{ji}^a$.

In intrinsic terms, for linear Pfaffian systems the exterior derivative induces a bundle mapping

$$(55) \qquad\qquad \overline{\delta} : I \to (T^*M/J) \otimes J/I$$

given locally by

$$(56) \qquad\qquad \overline{\delta}(\theta^a) = A_{\varepsilon i}^a \pi^\varepsilon \otimes \omega^i$$

where the ω^i are viewed as sections of J/I and the π^ε are viewed as sections of T^*M/J.[5]

Example 5.4. Referring to Example 1.3, we consider a partial differential equation of second order

$$F(x^i, z, \partial z/\partial x^i, \partial^2 z/\partial x^i \partial x^j) = 0.$$

This is equivalent to the Pfaffian differential system

$$(57) \qquad \begin{cases} F(x^i, z, p_i, p_{ij}) = 0 \\ \theta = dz - p_i dx^i = 0 \\ \theta_i = dp_i - p_{ij} dx^j = 0 \end{cases}$$

with independence condition $dx^1 \wedge \cdots \wedge dx^n \neq 0$ in the space of variables $x^i, z, p_i, p_{ij} = p_{ji}$.

The exterior derivatives of θ and θ_i are clearly in the algebraic ideal generated by dx^i, θ, θ_i. Hence the Pfaffian system is linear.

(58) *Remark.* In general, it is true that (i) the contact systems on the jet spaces $J^k(\mathbb{R}^n, \mathbb{R}^{s_0})$ are linear Pfaffian systems, and (ii) the restriction of a linear Pfaffian system to a submanifold is again a linear Pfaffian system (assuming that the independence form Ω is non-zero modulo the ideal $\mathcal{I}$ on the submanifold). Hence this example is valid for a P.D.E. system of any order.

---

[5] This $\overline{\delta}$ is closely related to the maps $\delta = d \bmod \mathcal{I}$ encountered in the discussion of the derived flag.

Example 5.5. Referring to Example 1.5, the canonical system $(\mathcal{L}, \Phi)$ on $G_n(TM)$ is a linear Pfaffian system.

We next want to define the important concept of the tableau bundle associated to a linear Pfaffian system $(\mathcal{I}, \Omega)$ satisfying suitable constant rank conditions. This will be a sub-bundle

$$A \subset I^* \otimes J/I$$

given for each $x \in M$ by a tableau

$$A_x \subset I_x^* \otimes J_x/I_x$$

as defined in §3 above, and with the property we always assume that $\dim A_x$ is locally constant. To define A_x we let

$$J_x^\perp \subset T_x M$$

be given as usual by $J_x^\perp = \{v \in T_x M : \eta(v) = 0 \text{ for all } \eta \in J\} = \{v \in T_x M : \theta^a(v) = \omega^i(v) = 0\}$. Then, referring to (53) above, the quantities

$$\pi_i^a(v) \in I_x^* \otimes J_x/I_x, \qquad v \in J_x^\perp$$

are well defined, and we set

$$(59) \qquad\qquad A_x = \{\pi_i^a(v) : v \in J_x^\perp\}.$$

More precisely, the choice of framing θ^a for I and ω^i for J/I induce bases w_α and x^i for I_x^* and J_x/I_x, respectively. Then A_x is spanned by the quantities

$$\pi(v) = \pi_i^a(v) w_a \otimes x^i$$

for $v \in J_x^\perp$.

Definition 5.6. Assuming that $\dim A_x$ is locally constant on M, we define the *tableau bundle* $A \subset I^* \otimes J/I$ by the condition that its fibres be given by (59).

Remark. We observe from (52) that *the mapping*

$$J_x^\perp \to I_x^* \otimes J_x/I_x$$

given by

$$v \to \|\pi_i^a(v)\|$$

is injective if, and only if, there are no vectors $v \in J_x^\perp$ *satisfying*

$$v \lrcorner \mathcal{I}_x \subset \mathcal{I}_x.$$

In particular, this is the case if there are no Cauchy characteristic vectors for $\mathcal{I}$, and in this situation the tableau A_x has as basis the matrices

$$A_\varepsilon = \|A^a_{\varepsilon i}\|, \qquad \varepsilon = 1, \ldots, t.$$

In general, these matrices A_ε *span* A_x but may not give a basis.

In intrinsic terms, by dualizing (55) with respect to T^*M/J and I we have a bundle mapping

$$(60) \qquad\qquad \pi : J^\perp \to I^* \otimes J/I,$$

and with our constant rank assumption the tableau bundle is the image $\pi(J^\perp)$. The mapping π is given in the above coordinates by $\pi(v) = \pi^a_i(v) w_a \otimes x^i$.

From (51) it is clear that *for linear Pfaffian systems the tableau bundle encodes what we might call the "principal part" of the behavior of the 2-forms $d\theta^a$ mod $\{I\}$.* Here, principal part refers to the term $A^a_{\varepsilon i} \pi^\varepsilon \wedge \omega^i$; the other term $\frac{1}{2} c^a_{ij} \omega^i \wedge \omega^j$ will also be discussed below.

For each $x \in M$, the characters $s'_i(x)$ of the tableau A_x are defined; we shall assume that these are locally constant and shall call them the *reduced characters* of the linear Pfaffian system. In the discussion of examples it will frequently be convenient to let $x \in M$ be a typical point, set

$$\begin{cases} W = I^*_x \\ V^* = J_x/I_x \\ A = A_x \text{ and } s'_i = s'_i(x), \end{cases}$$

and speak of $A \subset W \otimes V^*$ as *the tableau of $(\mathcal{I}, \Omega)$* without reference to the particular point $x \in M$.

In Definition 3.4 above, we introduced the prolongation of a tableau and in (26) we gave an interpretation of the prolongation. This interpretation may be extended as follows:

Proposition 5.7. *Assume that $(\mathcal{I}, \Omega)$ is a linear Pfaffian system and that the set of integral elements $G_x(\mathcal{I}, \Omega)$ of $(\mathcal{I}, \Omega)$ lying over $x \in M$ is non-empty. Then $G_x(\mathcal{I}, \Omega)$ is an affine linear space whose associated vector space may be naturally identified with the prolongation $A^{(1)}_x$ of A_x.*

Proof. We work over a fixed point $x \in M$ and omit reference to it. Referring to the proof of Proposition 5.2, the equations of integral elements are

$$\begin{cases} \theta^a = 0 \\ \pi^\varepsilon - p^\varepsilon_i \omega^i = 0 \end{cases}$$

where

$$(A^a_{\varepsilon i} p^\varepsilon_j - A^a_{\varepsilon j} p^\varepsilon_i) + c^a_{ij} = 0.$$

These equations define an affine linear space, and assuming that this is non-empty (a point we shall take up next) the associated vector space is defined by the homogeneous linear equations

$$(61) \qquad A^a_{\varepsilon i} p^\varepsilon_j = A^a_{\varepsilon j} p^\varepsilon_i.$$

Given a solution p^ε_j to these equations, then we set $P^a_{ij} = A^a_{\varepsilon i} p^\varepsilon_j = P^a_{ji}$ and see that

$$P = P^a_{ij} w_a \otimes x^i x^j \in W \otimes S^2 V^*$$

satisfies the relations

$$B^{\lambda i}_a P^a_{ij} = 0$$

that define A; hence $P \in A^{(1)}$. Conversely, if

$$P = P^a_{ij} w_a \otimes x^i x^j \in A^{(1)},$$

then each $\partial P / \partial x^j \in A$ and so is a linear combination

$$\partial P / \partial x^j = A^a_{\varepsilon i} p^\varepsilon_j w_a \otimes x^i$$

of a spanning set of matrices $A_\varepsilon = \| A^a_{\varepsilon i} \|$ of A. The condition $\partial^2 P / \partial x^i \partial x^j = \partial^2 P / \partial x^j \partial x^i$ is then equivalent to (61). $\qquad\square$

In §3, we have defined the symbol associated to a tableau, and here we have the corresponding

Definition 5.8. Let $(\mathcal{I}, \Omega)$ be a linear Pfaffian system with tableau bundle $A \subset I^* \otimes J/I$. Then the *symbol bundle* is defined to be

$$B = A^\perp \subset I \otimes (J/I)^*.$$

As explained above, we shall frequently omit reference to the point $x \in M$ and simply refer to B as the *symbol of the Pfaffian differential system*.

Example 5.9. We consider a 1st order P.D.E. system

$$(62) \qquad F^\lambda(x^i, y^a, \partial y^a / \partial x^i) = 0.$$

We write this as the Pfaffian differential system

$$(63) \qquad \begin{cases} \theta^a = dy^a - p^a_i dx^i = 0 \\ \Omega = dx^1 \wedge \cdots \wedge dx^n \neq 0 \end{cases}$$

in the submanifold M of (x^i, y^a, p^a_i) space defined by the equations

$$(64) \qquad F^\lambda(x^i, y^a, p^a_i) = 0$$

(we assume that these define a submanifold). The structure equations of (63) are

$$d\theta^a \equiv \pi_i^a \wedge \omega^i \quad \mod (I)$$

where $\pi_i^a = -dp_i^a|_M$ and $\omega^i = dx^i|_M$. From (64) the π_i^a are subject to the relations

$$\frac{\partial F^\lambda}{\partial p_i^a}\pi_i^a \equiv 0 \quad \mod \{J\}.$$

It follows from the above discussion that at each point $q = (x^i, y^a, p_i^a)$ of M the fibre of the symbol bundle is spanned by the matrices

$$B^\lambda = \|(\partial F^\lambda/\partial p_i^a)(q)\|.$$

Thus, *the symbol of the Pfaffian system* (63) *associated to the* P.D.E. *system* (62) *agrees with the classical definition of the symbol of such a system.*

In general, we have chosen our notations for a linear Pfaffian system so that their structure equations look like the structure equations of the special system (63). In this regard, we offer without proof the following easy

Proposition 5.10. *A linear Pfaffian system is locally equivalent to the Pfaffian system* (63) *arising from a* P.D.E. *system* (62) *if, and only if, the Frobenius condition*

$$dJ \equiv 0 \quad \mod \{J\}$$

is satisfied.

It is well known that a P.D.E. system may have compatibility conditions obtained from the equality of mixed partials, and we shall find the expression of these conditions for a general linear Pfaffian system. More specifically, we consider the compatibility conditions for the affine linear equations

$$(65) \qquad (A_{\varepsilon i}^a(x)p_j^\varepsilon - A_{\varepsilon j}^a(x)p_i^\varepsilon) + c_{ij}^a(x) = 0$$

whose solutions give the integral elements $G_x(\mathcal{I}, \Omega)$ lying over a point $x \in M$. The compatibility conditions for this system of linear equations in the p_i^ε may lead to relations on the $A_{\varepsilon i}^a(x)$ and $c_{ij}^a(x)$. These are called *integrability conditions*. Their presence means that the set $G(\mathcal{I}, \Omega)$ of integral elements of $(\mathcal{I}, \Omega)$ projects onto a *proper subset* of M, and we should restrict our consideration to this subset.

As the following simple example shows, the presence of integrability conditions is an important phenomenon for "over-determined" systems of partial differential equations, and usually imposes strong restrictions on the solution.

Example 5.11. In the (x, y, z)-space consider the system of P.D.E.'s of the first order:

$$z_x = A(x, y, z), \qquad z_y = B(x, y, z).$$

This is equivalent to the differential system

$$(66) \qquad\qquad \theta = dz - A\,dx - B\,dy = 0,$$

also in (x, y, z) space and with the independence condition

$$(67) \qquad\qquad dx \wedge dy \neq 0.$$

In the above notation we have rank $I = 1$, rank $J = 3$ and thus $J = T^*M$, and at each point of M there is a unique 2-plane (66) satisfying the independence condition (67). The condition that this 2-plane be an integral element is that

$$d\theta = -dA \wedge dx - dB \wedge dy$$

restrict to zero on it. Working this out gives

$$(68) \qquad\qquad A_y + A_z B = B_x + B_z A,$$

which is the usual integrability condition. If it is not identically satisfied, there are two cases: a) The relation (68) does not involve z and is therefore a relation between x, y, so that the system has no integral manifold satisfying the independence condition (67); b) The relation (68) gives z as a function of x, y, which is then the only possible solution, and thus the equation has a solution or not depending on whether it is satisfied or not by this function.

Example 5.12. In an open set $U \subset \mathbb{R}^n (n \geq 2)$ with coordinates $x^1, \ldots, x^n$, we assume given a smooth 2-form $\varphi = \frac{1}{2}\varphi_{ij}dx^i \wedge dx^j$, $\varphi_{ij} + \varphi_{ji} = 0$, and consider the equation

$$d\eta + \varphi = 0$$

for a 1-form η. When written out, this equation becomes the P.D.E. system

$$\partial\eta_i/\partial x^j - \partial\eta_j/\partial x^i + \varphi_{ij} = 0.$$

The associated exterior differential system is defined on the submanifold M of (x^i, η_i, p_{ij}) space by the equations

$$(69) \qquad\qquad \begin{cases} p_{ij} - p_{ji} + \varphi_{ij} = 0 \\ \theta_i = d\eta_i - p_{ij}dx^j = 0 \\ dx^1 \wedge \cdots \wedge dx^n \neq 0. \end{cases}$$

We seek an integral element $E \subset T_q M$ defined by

$$(70) \qquad dp_{ij} - p_{ijk} dx^k = 0$$

together with

$$\begin{cases} d\theta_i|_E = 0 \\ (dp_{ij} - dp_{ji} + d\varphi_{ij})|_E = 0. \end{cases}$$

The first of these equations gives

$$(71) \qquad p_{ijk} = p_{ikj},$$

and using (70) the second equations give

$$(72) \qquad p_{ijk} - p_{jik} + \partial \varphi_{ij}/\partial x^k = 0.$$

It is an elementary consequence of (71) and (72) that $d\varphi = 0$. In other words, *the necessary and sufficient condition that the Pfaffian system* (69) *have an integral element lying over each point of M is that $d\varphi = 0$.*

This again illustrates our assertion that the compatibility conditions for the equations (65) are integrability conditions; more precisely, they are *first order integrability conditions.*

We will now see how these integrability conditions are reflected in the structure equations of a linear Pfaffian differential system. Referring to the proof of Proposition 5.2 above, we assume that $\theta^1, \ldots, \theta^{s_0}, \omega^1, \ldots, \omega^n,$ $\pi^1, \ldots, \pi^t$ is a local coframe for M adapted to the filtration $I \subset J \subset T^* M$. This coframe is defined up to an invertible linear transformation (47). The linearity of the Pfaffian system is expressed by the equation (51), i.e., by the absence of $\pi^\delta \wedge \pi^\varepsilon$ terms in the $d\theta^a$'s. By abuse of notation we shall write the system as

$$(73) \qquad \begin{cases} \theta^a = 0 \\ d\theta^a \equiv A^a_{\varepsilon i} \pi^\varepsilon \wedge \omega^i + \tfrac{1}{2} c^a_{ij} \omega^i \wedge \omega^j \mod \{I\} \\ \Omega = \omega^1 \wedge \cdots \wedge \omega^n \neq 0. \end{cases}$$

Under a substitution (47) with only the diagonal blocks being non-zero—i.e., with $\|C^\varepsilon_i\| = 0$ (the θ-terms don't matter because of the congruence in the 2nd equation above)—we see that c^a_{ij} transforms like a section of $I^* \otimes \Lambda^2(J/I)$. Under a substitution (47) with the diagonal blocks being the identity, i.e., given by

$$(74) \qquad \pi^\varepsilon \to \pi^\varepsilon + p^\varepsilon_i \omega^i,$$

we have that

$$(75) \qquad c^a_{ij} \to c^a_{ij} + (A^a_{\varepsilon j} p^\varepsilon_i - A^a_{\varepsilon i} p^\varepsilon_j).$$

In intrinsic terms, we have a mapping

$$(76) \qquad \bar{\pi} : J^{\perp} \otimes J/I \to I^* \otimes \Lambda^2(J/I)$$

induced by (60) and given in the above bases by

$$p_i^{\varepsilon} \to (A_{\varepsilon j}^a p_i^{\varepsilon} - A_{\varepsilon i}^a p_j^{\varepsilon}),$$

and the c_{ij}^a give a section of $I^* \otimes \Lambda^2(J/I)/\text{image }\bar{\pi}$, where we now assume that the mapping (76) has locally constant rank.

Definition 5.13. We denote by $[c]$ the section of the bundle $I^* \otimes \Lambda^2(J/I)/\text{image }\bar{\pi}$, given in bases by $\{c_{ij}^a\}$ modulo the equivalence relation (75). Then $[c]$ is called the *torsion* of the linear Pfaffian system (73).

From the proof of Proposition 5.5 we have the

Proposition 5.14. *The necessary and sufficient condition that there exist an integral element of* (73) *over a point* $x \in M$ *is that the torsion* $[c](x) = 0$.

By our discussion above, we see that the torsion reflects the 1st order integrability conditions in the Pfaffian system (73). For this reason, we shall sometimes say that *the integrability conditions are satisfied* rather than *the torsion vanishes*. On the other hand, assuming always that the mapping (76) has constant rank, we see that the vanishing of the torsion in an open neighborhood U of a point $x \in M$ is equivalent to being able to make a smooth substitution (74) in U such that the $c_{ij}^a = 0$ in (73). For this reason, we shall sometimes say that *the torsion may be absorbed* (by a substitution (74)) rather than *the torsion vanishes*. In summary, we have that *the satisfaction of the integrability conditions for* (73) *is expressed by being able to absorb the torsion*.

There is an alternate way of writing (73), called the *dual form*, that is especially useful in computing examples. To explain it we set

$$(77) \qquad \pi_i^a = A_{\varepsilon i}^a \pi^{\varepsilon} + c_{ji}^a \omega^j$$

so that the second equation in (73) becomes

$$(78) \qquad d\theta^a \equiv \pi_i^a \wedge \omega^i \quad \mod \{I\}.$$

The 1-forms π_i^a are not linearly independent modulo J, but are subject to the relations

$$(79) \qquad B_a^{\lambda i} \pi_i^a \equiv C_j^{\lambda} \omega^j \quad \mod \{I\},$$

where by (77)

$$(80) \qquad C_j^{\lambda} = B_a^{\lambda i} c_{ji}^a.$$

Here, we are working over an open set $U \subset M$ and omitting reference to the point $x \in M$, and the $B_a^{\lambda i}$ give a basis for symbol bundle over U. Summarizing, the dual form of the structure equations (73) is

$$(81) \qquad \begin{cases} \theta^a = 0 \\ d\theta^a \equiv \pi_i^a \wedge \omega^i \mod \{I\} \\ B_a^{\lambda i} \pi_i^a \equiv C_j^\lambda \omega^j \mod \{I\} \\ \Omega = \omega^1 \wedge \cdots \wedge \omega^n \neq 0. \end{cases}$$

Proposition 5.15. *Assuming that the operator $\bar{\pi}$ in (76) has constant rank, the following are equivalent:*

(i) *the space $G(\mathcal{I}, \Omega)$ of integral elements surjects onto M;*

(ii) *locally, we may choose the π^ε so that $c_{ij}^a = 0$ in (73);*

(iii) *locally, we may choose the π_i^a so that $C_j^\lambda = 0$ in (81).*

Proof. We have proved that (i) $\Rightarrow$ (ii) $\Rightarrow$ (iii) above (see (80) for (ii) $\Rightarrow$ (iii)). Assuming (iii), we consider the family of n-planes

$$\begin{cases} \theta^a = 0 \\ \pi_i^a = 0. \end{cases}$$

By the third equation in (81), these n-planes are well-defined, and by the first and second equations there they are integral elements. $\qquad \square$

If any of the equivalent conditions in the proposition are satisfied, we shall say that *the integrability conditions are satisfied* or that *the torsion may be absorbed.*

Suppose that (i) in the proposition is satisfied. Then we may choose our coframe $\theta^1, \ldots, \theta^{s_0}$, $\omega^1, \ldots, \omega^n$, $\pi^1, \ldots, \pi^t$ so that the structure equations (73), (81) became respectively

$$(82) \qquad \begin{cases} \theta^a = 0 \\ d\theta^a \equiv A_{\varepsilon i}^a \pi^\varepsilon \wedge \omega^i \mod \{I\} \\ \Omega = \omega^1 \wedge \cdots \wedge \omega^n \neq 0 \end{cases}$$

$$(83) \qquad \begin{cases} \theta^a = 0 \\ d\theta^a \equiv \pi_i^a \wedge \omega^i \mod \{I\} \\ B_a^{\lambda i} \pi_i^a \equiv 0 \mod \{I\} \\ \Omega = \omega^1 \wedge \cdots \wedge \omega^n \neq 0. \end{cases}$$

These are the forms of the structure equations that we shall use in examples where the torsion is absorbed.

Referring to structure equations (73), we have discussed the tableau and torsion of a linear Pfaffian system. We will now express Cartan's test for involution in terms of these invariants. For this we work in a neighborhood U of a point $x \in M$ and assume that the quantities $\dim A_x$, $s_i'(x)$, $\dim A_x^{(1)}$, and rank $\bar{\pi}_x$ in (76) all are constant.

Theorem 5.16. *The linear Pfaffian system $(\mathcal{I}, \Omega)$ is involutive at $x \in M$ if, and only if,*
  (i) *the torsion vanishes in U*
  (ii) *the tableau A_x is involutive.*

Proof. We may replace U by M, and then by Proposition 5.14 the vanishing of the torsion is equivalent to the surjectivity of the mapping $G(\mathcal{I}, \Omega) \to M$. By Proposition 5.7 we then have that

$$\dim G(\mathcal{I}, \Omega) = \dim M + \dim A_x^{(1)}.$$

On the other hand, using the structure equations (83) we see first that the equations

$$\begin{cases} \theta^a(x) = 0 \\ \pi_i^a(x) = 0 \end{cases}$$

define an integral element $E \subset T_x M$ having a basis e_i where $\langle \omega^j(x), e_i \rangle = \delta_i^j$, and secondly the proof of (36) in section 3 above shows that the rank of the polar equations of $E_k = \text{span}\{e_1, \ldots, e_k\}$ is given by

$$s_0 + s_1' + \cdots + s_k'.$$

On the other hand, the Cartan characters r_k and s_k associated to E in terms of the dimensions of the polar spaces $H(E^k)$ are given for $k \geq 0$ by the relations

$$(84) \qquad \begin{cases} \dim H(E^k) = r_{k+1} + k + 1 \\ s_k = r_k - r_{k+1} - 1 \geq 0. \end{cases}$$

We set $r_0 = m = \dim M$, reflecting the assumption that there are integral elements over each point. Then $r_0 - r_1 - 1 = \text{rank}\, I_x = s_0$, so that our notations are consistent. From (84) we infer that, for $1 \leq k \leq n - 1$,

$$(85) \qquad r_{k+1} + k + 1 = n + t - (s_0 + s_1' + \cdots + s_k').$$

Subtracting these equations for k and $k - 1$ and using the second equation in (84) gives

$$(86) \qquad s_k = s_k'$$

for $k = 1, \ldots, n-1$. This proves: *For any integral element E, the characters $s_k = s_k(E)$ are equal to the reduced characters s_k'.* In particular, the s_k are the same for all integral elements E lying over a fixed point $x \in M$. The inequality in Cartan's test given by Theorem 1.11 in Chapter III is then

$$(87) \qquad \dim A_x^{(1)} \leq s_1' + 2s_2' + \cdots + ns_n',$$

and, by the Definition 3.7 above, equality holds if, and only if, A_x is involutive. $\qquad\square$

To conclude this section, we want to give a practical method for computing the s_i' so that one can, with relative ease, check for equality in (87). For a 1-form φ we set

$$\overline{\varphi} = \varphi(x) \text{ modulo } J_x$$
$$= \varphi(x) \text{ modulo } \{\theta^a(x), \omega^i(x)\},$$

and we shall omit reference to the point $x \in M$. We define the *tableau matrix* by

$$(88) \qquad \pi = \begin{bmatrix} \overline{\pi}_1^1 & \cdots & \overline{\pi}_n^1 \\ \vdots & & \vdots \\ \overline{\pi}_1^{s_0} & \cdots & \overline{\pi}_n^{s_0} \end{bmatrix}.$$

Then, assuming that the ω^i are chosen generically, we have again from the proof of Proposition 3.8 above that

$$(89) \qquad s_1' + \cdots + s_k' = \left\{ \begin{array}{l} \text{number of independent 1-forms } \overline{\pi}_i^a \\ \text{in the first } k \text{ columns of } \pi \end{array} \right\}.$$

In practice, this equality will allow us to determine the s_i' by "eyeballing" the tableau matrix (88).

For an illustration of the use of the tableau matrix, we shall put it in a normal form and use this to give an especially transparent proof of Cartan's test.

We assume that the torsion has been absorbed so that the structure equations (83) and relations (86), (89) hold. Then, amongst the 1-forms π_1^a exactly s_1' are linearly independent. (We remind the reader that we are omitting reference to the point x.) We may then assume that $\theta^1, \ldots, \theta^{s_0}$ are chosen so that $\overline{\pi}_1^1, \ldots, \overline{\pi}_1^{s_1'}$ are independent. Then all of the forms π_i^a for $a > s_1'$ are linear combinations of $\overline{\pi}_1^1, \ldots, \overline{\pi}_1^{s_1'}$, since otherwise we could choose $\omega^1, \ldots, \omega^n$ so that at least $s_1' + 1$ of the forms $\overline{\pi}_1^a$ were linearly independent in contradiction to (89). Having said this, among the forms $\overline{\pi}_2^1, \ldots, \overline{\pi}_2^{s_1'}$ exactly s_2' are independent modulo $\overline{\pi}_1^1, \ldots, \overline{\pi}_1^{s_1'}$. We may assume that $\theta^1, \ldots, \theta^{s_1'}$ have been chosen so that $\overline{\pi}_2^1, \ldots, \overline{\pi}_2^{s_2'}$ are linearly independent modulo $\overline{\pi}_1^1, \ldots, \overline{\pi}_1^{s_1'}$. We note that $s_2' \le s_1'$, since otherwise s_1' would not be the number of independent $\overline{\pi}_1^a$ for a *generic* choice of $\omega^1, \ldots, \omega^n$.

Continuing in this way, we may choose $\theta^1, \ldots, \theta^{s_0}$ so that the tableau matrix looks like

$$
(90) \qquad
\left\|
\begin{array}{cccc}
\bar{\pi}^1_1 & \bar{\pi}^1_2 & \cdots & \bar{\pi}^1_n \\[4pt]
\vdots & \vdots & & \bar{\pi}^{s'_n}_n \\[4pt]
\vdots & \bar{\pi}^{s'_2}_2 & & * \\[4pt]
\bar{\pi}^{s'_1}_1 & * & & * \\[4pt]
* & * & & * \\[4pt]
* & * & \cdots & *
\end{array}
\right\|
$$

with the property that: *for $b > s'_k$ the form $\bar{\pi}^b_k$ is a linear combination of the forms $\bar{\pi}^a_i$ where $i \leq k$, $a \leq s'_i$.*[6] *We also note that $s'_1 \geq s'_2 \geq \cdots \geq s'_n$.*

Definition 5.17. The forms $\bar{\pi}^a_i$ where $a \leq s'_i$ are called the *principal components*. Thus the principal components are independent and span the space of the $\bar{\pi}^a_i$'s.

We will now derive Cartan's test. Integral elements are defined by linear equations

$$\theta^a = 0$$
$$\pi^a_i - p^a_{ij}\omega^j = 0$$

where $p^a_{ij} = p^a_{ji}$ and $B^{\lambda i}_a p^a_{ij} = 0$ by (83). Clearly it suffices to consider only those linear equations

$$(91) \qquad \pi^a_i - p^a_{ij}\omega^j = 0, \qquad a \leq s'_i$$

corresponding to the principal components, as the remaining linear equations are consequences of these by writing

$$\pi^b_k = \left\{ \begin{array}{l} \text{linear combination of the } \pi^a_i \text{ for } i \leq k \text{ and } a \leq s'_i \\ \text{together with the } \omega^j \text{ and } \theta^b. \end{array} \right\}$$

Since $p^a_{ij} = p^a_{ji}$ the integral element (46) is determined by the quantities

$$p^a_{ij}, a \leq \min s'_i, s'_j$$

or equivalently by the quantities

$$(92) \qquad p^a_{ij}, \quad i \leq j, \quad a \leq s'_j.$$

---

[6]Actually, we have proved that the $\bar{\pi}^b_k$ for $b > s'_1$ are linear combinations of $\bar{\pi}^1_1, \ldots, \bar{\pi}^{s'_1}_1$; the $\bar{\pi}^b_k$ for $s'_2 < b \leq s'_1$ are linear combinations of $\bar{\pi}^1_1, \ldots, \bar{\pi}^{s'_1}_1, \bar{\pi}^1_2, \ldots, \bar{\pi}^{s'_2}_2$; the $\bar{\pi}^b_k$ for $s_3 < b \leq s'_2$ are linear combinations of $\bar{\pi}^1_1, \ldots, \bar{\pi}^{s'_1}_1, \bar{\pi}^1_2, \ldots, \bar{\pi}^{s'_2}_2, \bar{\pi}^1_3, \ldots, \bar{\pi}^{s'_3}_3$; and so forth.

For $j = 1$ we have the p_{11}^a for $a \leq s_1'$. For $j = 2$ we have the p_{12}^a and p_{22}^a for $a \leq s_2'$. For $j = 3$ we have the p_{13}^a, p_{23}^a, and p_{33}^a for $a \leq s_3'$. Continuing in this way, we see that there are at most

$$\tag{93} s_1' + 2s_2' + 3s_3' + \cdots + ns_n'$$

independent quantities (92), and this is the inequality in Cartan's test.

Suppose that equality holds, and consider integral elements of dimension p that satisfy the additional equations $\omega^{p+1} = \cdots = \omega^n = 0$. These integral elements are given by equations (91) where we set $\omega^{p+1} = \cdots = \omega^n = 0$, from which it follows that they are uniquely determined by the quantities (92) where $j \leq p$. If the space of integral elements is of dimension equal to (93), the p_{ij}^a in (92) can be freely specified. From this it follows that every p-dimensional integral element given by $\omega^{p+1} = \cdots = \omega^n = 0$ extends to a $(p+1)$-dimensional integral element given by $\omega^{p+2} = \cdots = \omega^n = 0$, and in fact does so in $(p+1)s_p'$-dimensional ways. Thus, we have a C-regular flag, which means that the system is involutive. $\qquad\square$

Remark. To understand the relation between this discussion and the proof of Cartan's test in Chapter III, we assume as above that E^n is defined by the equations $\theta^a = \pi_i^b = 0$ and $E^k \subset E^n$ by the additional equations $\omega^{k+1} = \cdots = \omega^n = 0$. In the filtration

$$\tag{94} \begin{aligned} E_0 \subset E_1 \subset E_2 \subset \cdots \subset E_n \subset H(E_n) \subset \ldots \\ \subset H(E_2) \subset H(E_1) \subset H(E_0) \subset T \end{aligned}$$

we have

$$H(E_0) = \{\theta^b = 0\}$$

$$H(E_1) = \{\theta^b = 0,\ \pi_1^a = 0 \text{ for } a \leq s_1'\}$$

$$H(E_2) = \{\theta^b = 0,\ \pi_1^a = 0 \text{ for } a \leq s_1' \text{ and } \pi_2^b = 0 \text{ for } b \leq s_2'\}$$

$$\vdots \qquad \vdots$$

$$H(E_n) = \{\theta^b = 0,\ \pi_i^a = 0 \text{ for } a \leq s_i'\}.$$

From this it is clear that the normal form (90) is simply the implication on the tableau of chosing the π_i^a adapted to the filtration (94) in the manner just explained.

A very useful insight into involutivity is to express its consequences on the symbol relations when the tableau matrix is in the normal form (90). The general case is called the *Guillemin normal form*, which will be further discussed at the end of Chapter VIII. Here we will first take up the special case when the system is involutive and

$$\tag{95} s_1' = s_0, \qquad s_2' = \cdots = s_n' = 0.$$

Thus the principal components are the π_1^a, and using the additional index range

$$2 \leq \rho, \ \sigma \leq n$$

we will have relations

$$(96) \qquad \pi_\rho^a \equiv C_{\rho b}^a \pi_1^b,$$

where the congruence is modulo the θ^a's and ω^i's. These are a complete set of symbol relations, and setting

$$C_\rho = \|C_{\rho b}^a\|$$

we will prove that:

(97)   *A tableau satisfying* (95) *with symbol relations* (96) *is involutive if, and only if, the commutation relations*

$$(98) \qquad [C_\rho, C_\sigma] = 0$$

are satisfied.

Proof. Among the equations

$$\pi_i^a - p_{ij}^a \omega^j = 0$$

that define integral elements, by (95) those for $i \geq 2$ are consequences of those for $i = 1$. Thus, by Cartan's test the p_{11}^a may be freely specified and then the

$$(99) \qquad \begin{aligned} p_{i1}^a &= p_{1j}^a \\ p_{\rho\sigma}^a &= p_{\sigma\rho}^a \end{aligned}$$

are determined. From (96) we have

$$p_{\rho i}^a = C_{\rho b}^a p_{1i}^b.$$

Thus

$$(100) \qquad p_{\rho 1}^a = C_{\rho b}^a p_{11}^b$$

and

$$(101) \qquad p_{\rho\sigma}^a = C_{\rho b}^a p_{1\sigma}^b.$$

Using the first equation in (99) and (100) in (101) gives

$$p^a_{\rho\sigma} = C^a_{\rho b} C^b_{\sigma c} p^c_{11},$$

and then the second equation in (99) gives

$$(C^a_{\rho b} C^b_{\sigma c} - C^a_{\sigma b} C^b_{\rho c}) p^c_{11} = 0.$$

Since the p^c_{11} may be freely specified, we conclude (98). Reversing the argument gives our assertion. $\qquad\square$

A slightly more general case arises when we assume that

$$(102) \qquad s'_1 = \cdots = s'_l = s_0, \qquad s'_{l+1} = \cdots = s'_n = 0.$$

With the ranges of indices

$$1 \le \lambda \le l, \qquad l + 1 \le \rho \le n$$

the complete symbol relations are

$$\pi^a_\rho \equiv C^{\lambda a}_{\rho b} \pi^b_\lambda.$$

For any $\xi = (\xi_1, \ldots, \xi_l)$ we define

$$C_\rho(\xi) = \|C^{\lambda a}_{\rho b} \xi_\lambda\|,$$

and the the same proof gives:

(103)   *A tableau satisfying* (102) *is involutive if, and only if, for all ξ, the commutation relations*

$$(104) \qquad [C_\rho(\xi), C_\sigma(\xi)] = 0,$$

       are satisfied.

Although we shall not completely write out the Guillemin normal form here, we will refine the normal form (90) and indicate how this in fact leads to a generalization of Guillemin's result.

We first claim that:

(∗)   *In the involutive case, we may choose $\theta^1, \ldots, \theta^{s_0}$ so that*

$$\bar{\pi}^a_i = 0 \ \text{for} \ a > s'_1, \quad 1 \le i \le n.$$

Proof. We may assume that the torsion is absorbed, i.e. that (83) holds, and then the 1-forms

$$\theta^a, \ \omega^i, \ \pi^a_i \ \text{where} \ a \le s'_i$$

are linearly independent. In the tableau matrix $\|\pi_i^a\|$ without reducing modulo J, even though this is not intrinsic, we have

$$s_1' = \text{number of linearly independent 1-forms } \pi_1^a.$$

Moreover, if we then set all

$$\pi_1^a = 0$$

and denote by $\tilde\varphi$ the restriction of any 1-form φ to this space, the tableau with matrix

$$\|\tilde\pi_\rho^a\|, \quad 2 \le \rho \le n \quad \text{and} \quad 1 \le a \le s_0$$

is again involutive.

We choose $\theta^1, \ldots, \theta^{s_0}$ so that

$$\pi_1^a = 0, \quad a > s_1',$$

and will show that, as a consequence of involutivity, all the remaining

$$\pi_\rho^a = 0, \quad a > s_1' \quad \text{and} \quad 2 \le \rho \le n.$$

For this we recall from the argument for Cartan's test given below Definition 5.12 that integral elements are defined by equations

(i) $$\pi_i^a - p_{ij}^a \omega^j = 0, \ p_{ij}^a = p_{ji}^a,$$

and that the p_{11}^a, $a \le s_1'$, may be freely specified. We now use the additional range of indices $1 \le \lambda \le s_1'$. Then we have equations

$$\pi_\rho^a = B_{\rho\lambda}^a \pi_1^\lambda, \quad 2 \le \rho \le n.$$

Using (i) these give equations

$$p_{\rho j}^a \omega^j = B_{\rho\lambda}^a p_{1j}^\lambda \omega^j$$

$$\Rightarrow p_{\rho 1}^a = B_{\rho\lambda}^a p_{11}^\lambda.$$

But $\pi_1^a = 0$ for $a > s_1'$ gives

$$0 = p_{1\rho}^a = p_{\rho 1}^a = B_{\rho\lambda}^a p_{11}^\lambda,$$

and the only way the p_{11}^λ can be freely specified is if all $B_{\rho\lambda}^a = 0$. □

We now may repeat (i) for the involutive tableau $\|\tilde\pi_\rho^a\|$ to conclude that, with a suitable choice of $\theta^1, \ldots, \theta^{s_1'}$, all $\tilde\pi_\rho^a = 0$ for $s_2' < a \le s_1'$. This is equivalent to

$$\pi_\rho^a \equiv 0 \mod \{\pi_1^b\}, \quad s_2' < a \le s_1'.$$

Continuing in this way we may assume that the tableau matrix has the following form

(ii)
$$
\left\|
\begin{array}{cccc}
\pi_1^1 & \pi_2^1 & \pi_3^1 & \cdots \\[1ex]
\vdots & \vdots & \vdots & \\[1ex]
\vdots & \vdots & \pi_3^{s_3'} & \\[1ex]
\vdots & \pi_2^{s_2'} & & \Psi_2 \\[1ex]
\pi_1^{s_1'} & & \Psi_1 & \\[1ex]
& \Psi_0 & &
\end{array}
\right\|
$$

where

$$
\left\{
\begin{array}{l}
\Psi_0 = 0 \\[1ex]
\Psi_1 \equiv 0 \mod \{\pi_1^1, \ldots, \pi_{s_1'}^1\} \\[1ex]
\Psi_2 \equiv 0 \mod \{\pi_1^1, \ldots, \pi_{s_1'}^1, \pi_1^2, \ldots, \pi_{s_2'}^2\} \\[1ex]
\vdots
\end{array}
\right.
$$

The Guillemin normal form arises by writing

(iii)
$$
\left\{
\begin{array}{l}
\Psi_1 = C_{11}\pi_1 \\[1ex]
\Psi_2 = C_{21}\pi_1 + C_{22}\pi_2 \\[1ex]
\vdots
\end{array}
\right.
$$

where $\pi_1, \pi_2, \ldots$ are the columns in the tableau matrix. We may then repeat the argument that gave (98) to deduce a set of quadratic conditions on the symbol relations (iii) that are necessary and sufficient in order that the tableau be involutive.[7] We shall not pursue this further here.

§6. Prolongation.

Let $\mathcal{I}$ be an exterior differential system on a manifold M. The *first prolongation* will be a linear Pfaffian system $(\mathcal{I}^{(1)}, \Omega)$ with independence

---

[7] Actually, (ii) is a slight refinement of the Guillemin normal form, which more closely corresponds to the normal form (90). The way to understand (ii) is as follows: Given a generic flag $V_1 \subset V_2 \subset \cdots \subset V_{n-1} \subset V$, $\dim V_i = i$, this determines a flag $W \supset W_1 \supset W_2 \supset \cdots \supset W_{n-1}$, $\dim W_i = s_i'$ together with a set of matrices C_{ij}, $i \geq j$, where C_{ij} is an $(s_i' - s_{i+1}') \times (s_j' - s_{j+1}')$ matrix and where a set of quadratic relations is imposed on the C_{ij}. Counting the number of independent equations would allow one to compute the dimension of the space of involutive tableaux with fixed characters, which to our knowledge has never been done.

condition on a manifold $M^{(1)}$. Roughly speaking, *the first prolongation is obtained by imposing the first order integrability conditions on the original system.* The prolongation $(\mathcal{I}^{(1)}, \Omega)$ of an exterior differential system $(\mathcal{I}, \Omega)$ with independence condition will also be defined, and then the higher prolongations are defined inductively by $\mathcal{I}^{(0)} = \mathcal{I}$ and

$$(\mathcal{I}^{(q+1)}, \Omega) = 1^{st}\text{-prolongation of } (\mathcal{I}^{(q)}, \Omega).$$

The three basic properties of prolongation may be informally and imprecisely stated as follows:

(105)   *The integral manifolds of $(\mathcal{I}, \Omega)$ and $(\mathcal{I}^{(1)}, \Omega)$ are locally in one-to-one correspondence.*

(106)   *If $(\mathcal{I}, \Omega)$ is involutive, then so is $(\mathcal{I}^{(1)}, \Omega)$.*

(107)   *There exists a q_0 such that, for $q \geq q_0$,*

         $(\mathcal{I}^{(q)}, \Omega)$ is involutive.

This last property includes the possibility that the manifolds $M^{(q)}$ are empty—this is the case when there are no integral manifolds of $(\mathcal{I}, \Omega)$. As a consequence of (105) and (107) we may (again imprecisely) say that *every integral manifold of a differential system is an integral manifold of an involutive exterior differential system.*

    We will now give the definition of $(\mathcal{I}^{(1)}, \Omega)$ in a special case; the general definition will be taken up in Chapter VI. For this we assume that the variety $G_n(\mathcal{I})$ of n-dimensional integral elements is a smooth submanifold of $G_n(TM)$ whose defining equations are derived from $\mathcal{I}$ as explained in Chapter III (cf. Proposition 1.4 and Definition 1.7 there). Thus

(108) $G_n(\mathcal{I}) = \{E \in G_n(TM) : \varphi|_E = 0 \text{ for all } \varphi \in \mathcal{I}\}$

should be a regularly defined submanifold of $G_n(TM)$.

Definition 6.1. The *first prolongation* $(\mathcal{I}^{(1)}, \Omega)$ is defined to be the restriction to $G_n(\mathcal{I})$ of the canonical system $(\mathcal{L}, \Phi)$ on $G_n(TM)$.

    To see what prolongation looks like in coordinates, we suppose that $(x^1, \ldots, x^n, y^1, \ldots, y^s)$ is a coordinate system on M and we consider the open set $U \subset G_n(TM)$ given by tangent planes E such that $dx^1 \wedge \cdots \wedge dx^n|_E \neq 0$. These tangent planes are defined by equations (see the discussion in example 3 above)

(109) $\theta^\sigma = dy^\sigma - p_i^\sigma dx^i = 0,$

and then $(x^i, y^\sigma, p_i^\sigma)$ forms a local coordinate system on $G_n(TM)$. In this open set, the canonical system $\mathcal{L}$ is generated by the 1-forms θ^σ together with their exterior derivatives and the independence condition is given by $\Phi = dx^1 \wedge \cdots \wedge dx^n$. When written out in this coordinate system, the equations (108) become a system of equations

$$(110) \qquad\qquad F_\varphi(x, y, p) = 0, \qquad \varphi \in \mathcal{I},$$

where $\varphi|_E = F_\varphi(x, y, p)\Phi$ for E given by (109). Our assumption is that these equations regularly define a submanifold in (x, y, p) space, and then $(\mathcal{I}^{(1)}, \Omega)$ is the restriction to this submanifold of the canonical system. It is clear that

$$(111) \qquad \textit{The first prolongation is a linear Pfaffian differential}$$
$$\textit{system with independence condition.}$$

Moreover, property (105) above is also clear from the fact (cf. example 1.5) that the integral manifolds of $(\mathcal{L}, \Phi)$ are locally the canonical liftings of smooth mappings $f : N \to M$. Properties (106) and (107) are more subtle and will be taken up later.

If we begin with a differential system $(\mathcal{I}, \Omega)$ with independence condition, then assuming as above that $G(\mathcal{I}, \Omega)$ is a smooth submanifold, $(\mathcal{I}^{(1)}, \Omega)$ is defined to be the restriction to $G(\mathcal{I}, \Omega)$ of the canonical system on $G_n(TM)$.

Example 6.2. We will work out the structure equations for the first prolongation of a linear Pfaffian system for which the torsion vanishes. First we make a general observation.

Using the above notation, suppose that

$$\varphi = f_\sigma(x, y)dy^\sigma - g_i(x, y)dx^i$$

is a 1-form in $\mathcal{I}$ where some $f_\sigma \neq 0$. The condition that φ vanish on the integral element (109) is

$$(112) \qquad\qquad f_\sigma(x, y)p_i^\sigma = g_i(x, y).$$

We let $\pi : M^{(1)} \to M$ be the projection, and on $M^{(1)}$ we consider the 1-form

$$\pi^*\varphi = f_\sigma(x, y)dy^\sigma - g_i(x, y)dx^i$$
$$= f_\sigma(x, y)(dy^\sigma - p_i^\sigma dx^i)$$

by (112)

$$= f_\sigma \theta^\sigma$$

by (109). In summary:

(113)   *Let $\pi : M^{(1)} \to M$ be the canonical projection, and
let $\mathcal{I}_1$ denote the differential ideal generated by
the 1-forms in $\mathcal{I}$. Then*

$$\pi^* \mathcal{I}_1 \subseteq \mathcal{I}^{(1)}.$$

In particular, if $\mathcal{I}$ is a Pfaffian system then

$$\pi^* \mathcal{I} \subset \mathcal{I}^{(1)}.$$

From now on we will usually omit the π^*'s.

Returning to our example, we suppose $\mathcal{I}$ to locally have the structure
equations (83) above. Then $M^{(1)} \to M$ is a bundle of affine linear spaces
whose associated vector bundle has fibre $A_x^{(1)}$ over $x \in M$. Integral elements
of $(\mathcal{I}, \Omega)$ are defined by

(114)
$$\begin{cases} \theta^a = 0 \\ \theta_i^a = \pi_i^a - p_{ij}^a \omega^j = 0 \end{cases}$$

where the equations

(115)
$$\begin{cases} p_{ij}^a = p_{ji}^a \\ B_a^{\lambda i}(x) p_{ij}^a = 0 \end{cases}$$

are satisfied. From (113) it follows that $\mathcal{I}^{(1)}$ is generated as a differential
ideal by the 1-forms θ^a, θ_i^a (recall that we are omitting the π^*'s). For the
structure equations of $\mathcal{I}^{(1)}$ we let $I^{(1)} \subset T^* M^{(1)}$ be the sub-bundle whose
sections are the 1-forms in $\mathcal{I}^{(1)}$, so that locally $I^{(1)} = \text{span}\{\theta^a, \theta_i^a\}$. We
shall prove that:

(116)
$$d\theta^a \equiv 0 \quad \text{mod } \{I^{(1)}\}$$

(117)
$$d\theta_i^a \equiv \pi_{ij}^a \wedge \omega^j \quad \text{mod } \{I^{(1)}\}$$

where

(118)
$$\begin{cases} \pi_{ij}^a \equiv \pi_{ji}^a \quad \text{mod } \{I^{(1)}\} \\ B_a^{\lambda i} \pi_{ij}^a \equiv C_{jk}^\lambda \omega^k \quad \text{mod } \{I^{(1)}\}. \end{cases}$$

Proof of (116). From (83) and (114) above

$$d\theta^a \equiv \pi_i^a \wedge \omega^i \mod \{I\}$$

(119)
$$\equiv \theta_i^a \wedge \omega^i \mod \{I\}$$

$$\equiv 0 \mod \{I^{(1)}\}.$$

Proof of (117). We shall use the following variant of the Cartan lemma to be proved in Chapter VIII (cf. Proposition 2.1 and its corollaries in Chapter VIII): Let $\tilde{\omega}^i$ be linearly independent vectors in a vector space U and let $\eta_i \in \Lambda^2 U$ satisfy the exterior equation

$$(120) \qquad \eta_i \wedge \tilde{\omega}^i = 0.$$

Then it follows that

$$(121) \qquad \begin{cases} \eta_i = \eta_{ij} \wedge \tilde{\omega}^j \text{ where} \\ \eta_{ij} = \eta_{ji} \in U. \end{cases}$$

To apply this, we take the exterior derivative of (119) and use (116) and (114) to obtain

$$0 \equiv d\pi_i^a \wedge \omega^i - \pi_j^a \wedge d\omega^j \mod \{I^{(1)}\}$$

$$\equiv (d\pi_i^a - p_{ij}^a d\omega^j) \wedge \omega^i \mod \{I^{(1)}\}.$$

We take U to be a typical fibre of $T^* M^{(1)}/I^{(1)}$ and $\tilde{\omega}^i = \omega^i \mod \{I^{(1)}\}$. Then from (120) and (121) we conclude that

$$(122) \qquad d\pi_i^a - p_{ij}^a d\omega^j \equiv \eta_{ij}^a \wedge \omega^j \mod \{I^{(1)}\}$$

where

$$\eta_{ij}^a \equiv \eta_{ij}^a \mod \{I^{(1)}\}.$$

Now we have from the definition (114) and (122),

$$d\theta_i^a = d\pi_i^a - p_{ij}^a d\omega^j - dp_{ij}^a \wedge \omega^j$$

$$\equiv \pi_{ij}^a \wedge \omega^j \mod \{I^{(1)}\}$$

where

$$(123) \qquad \begin{cases} \pi_{ij}^a = -dp_{ij}^a + \eta_{ij}^a \\ \pi_{ij}^a \equiv \pi_{ji}^a \mod \{I^{(1)}\}. \end{cases}$$

This gives (117) where the 1st equation in (118) is satisfied. To verify the second equation in (118) we let $J^{(1)} \subset T^*M^{(1)}$ be the sub-bundle generated by $I^{(1)}$ and the values of the $\omega^i(x)$. Then exterior differentiation of the second equation in (115) gives

$$B_a^{\lambda i}(x)dp_{ij}^a \equiv 0 \mod \{J^{(1)}\}$$

since $dB_a^{\lambda i}(x) \in T_x^*M$ and the $\theta^a(x)$, $\pi_i^a(x)$, $\omega^i(x)$ span T_x^*M. From this and the 1st equation in (123) we have (dropping reference to x)

$$B_a^{\lambda i}\pi_{ij}^a \equiv 0 \mod \{J^{(1)}\}$$

since $\eta_{ij}^a \equiv 0 \mod \{J^{(1)}\}$. This implies the 2nd-equation in (118). $\qquad\square$

Using (116)–(118) the structure equations of $(\mathcal{I}^{(1)}, \Omega)$ may be summarized as follows:

$$
(124) \qquad
\begin{aligned}
&\text{(i)} \quad \theta^a = 0 \\
&\text{(ii)} \quad \theta_i^a = 0 \\
&\text{(iii)} \quad d\theta^a \equiv 0 \mod \{I^{(1)}\} \\
&\text{(iv)} \quad d\theta_i^a \equiv \pi_{ij}^a \wedge \omega^j \mod \{I^{(1)}\} \\
&\text{(v)} \quad \pi_{ij}^a = \pi_{ji}^a \\
&\text{(vi)} \quad B_a^{\lambda i}\pi_{ij}^a \equiv C_{ij}^\lambda \mod \{I^{(1)}\}
\end{aligned}
$$

with the independence condition $\Omega = \omega^1 \wedge \cdots \wedge \omega^n \neq 0$. From this we conclude that

(125)   *$(\mathcal{I}^{(1)}, \Omega)$ is a linear Pfaffian system whose tableau*

                     over a point $y \in M^{(1)}$ is the 1st prolongation

                     $A_x^{(1)}$ of the tableau of $(\mathcal{I}, \Omega)$ over $x = \pi(y)$.

In other words, the *tableau of the prolongation is the prolongation of the tableau*.

It follows also that the integral elements of $(\mathcal{I}^{(1)}, \Omega)$ over a point $y \in M^{(1)}$ form an affine linear space whose associated vector space is the 2nd prolongation $A_x^{(2)}$ where $x = \pi(y)$. In fact, referring to (124) there integral elements are given by

$$\pi_{ij}^a - p_{ijk}^a\omega^k = 0$$

where

$$p_{ijk}^a = p_{jik}^a = p_{ikj}^a \quad \text{(by (iv) and (v))}$$
$$B_a^{\lambda i}p_{ijk}^a = C_{jk}^\lambda(y) \quad \text{(by (vi))}.$$

The homogeneous linear equations associated to the second of these equations define $A_x^{(2)} \subset W \otimes S^3 V^*$. We will return to these matters in Chapters V and VI.

Finally, for use in Chapter VIII we want to prove the relation

$$(126) \qquad C_{jk}^\lambda \omega^j \wedge \omega^k \equiv 0 \quad \mathrm{mod}\ \{\mathcal{I}^{(1)}\}.$$

Proof. From the equations

$$\begin{cases} B_a^{\lambda i} \pi_i^a \equiv 0 \ \ \mathrm{mod}\ \mathcal{I}_1 \\ B_a^{\lambda i} p_{ij}^a = 0 \\ \theta_i^a = \pi_i^a - p_{ij}^a \omega^j \end{cases}$$

together with (113) above, we infer that

$$B_a^{\lambda i} \theta_i^a \equiv 0 \ \ \mathrm{mod}\ \mathcal{I}_1$$

$$B_a^{\lambda i} d\theta_i^a \equiv 0 \ \ \mathrm{mod}\ \mathcal{I}^{(1)}.$$

Plugging this into equation (iv) in (124) above and using (vi) there gives (126). $\qquad\square$

Example 6.3. We consider a 1st order P.D.E. system

$$(127) \qquad F^\lambda(x^i, z^a, \partial x^a / \partial x^i) = \mathrm{constant}$$

(we do not specify what the constant is). On $M = J^1(\mathbb{R}^n, \mathbb{R}^{s_0})$ with coordinates (x^i, z^a, p_i^a) this P.D.E. system corresponds to the exterior differential system $(\mathcal{I}, \Omega)$ generated by the 1-forms

$$(128) \qquad \begin{cases} \text{(i)} \ \ dF^\lambda(x, z, p) \\ \text{(ii)} \ \ \theta^a = dz^a - p_i^a dx^i \end{cases}$$

and with independence condition $\Omega = dx^1 \wedge \cdots \wedge dx^n$. We want to see what, if any, P.D.E. system corresponds to the first prolongation $(\mathcal{I}^{(1)}, \Omega)$.

In fact, the *prolongation of the* P.D.E. *system* (127) is usually defined by introducing new variables p_i^a for the derivatives $\partial z^a / \partial x^i$ and differentiating (127). Explicitly, it is the 1st order P.D.E. system for unknown functions z^a, p_i^a

$$(127^1) \qquad \begin{cases} p_i^a = \partial z^a / \partial x^i \\ \dfrac{\partial F^\lambda}{\partial x^i} + \dfrac{\partial F^\lambda}{\partial z^a} p_i^a + \dfrac{\partial F^\lambda}{\partial p_j^a} \dfrac{\partial}{\partial x^i}(p_j^a) = 0, \quad i = 1, \ldots, n \end{cases}$$

where $F^\lambda = F^\lambda(x, z, p)$. Clearly the solutions of (127) and (127^1) are in one-to-one correspondence (this is the reason for the constant in (127)), and we shall check that:

$$(129) \qquad (\mathcal{I}^{(1)}, \Omega) \text{ is the exterior differential system}$$

$$\text{corresponding to the P.D.E. system (127}^1\text{).}$$

Proof. By definition, the exterior differential system corresponding to (127^1) occurs on a submanifold $\tilde{M}$ of a jet space with coordinates $(x^i, z^a, p_i^a, q_i^a, p_{ij}^a)$, where $\tilde{M}$ has defining equations

$$(130) \qquad \begin{cases} p_i^a = q_i^a \\ \dfrac{\partial F^\lambda}{\partial x^i} + \dfrac{\partial F^\lambda}{\partial z^a} q_i^a + \dfrac{\partial F^\lambda}{\partial p_j^a} p_{ji}^a = 0, \end{cases}$$

where $F^\lambda = F^\lambda(x, z, p)$. The differential ideal $\tilde{\mathcal{I}}$ is generated by the restrictions to $\tilde{M}$ of the 1-forms

$$\begin{cases} \theta^a = dz^a - q_i^a dx^i \\ \theta_i^a = dp_i^a - p_{ij}^a dx^j. \end{cases}$$

Imposing the first equations in (130), we may think of $\tilde{M}$ as being defined in $(x^i, z^a, p_{ij}^a = p_{ji}^a)$ space by the equations

$$(131) \qquad \dfrac{\partial F^\lambda}{\partial x^i} + \dfrac{\partial F^\lambda}{\partial z^a} p_i^a + \dfrac{\partial F^\lambda}{\partial p_j^a} p_{ij}^a = 0$$

and the ideal $\tilde{\mathcal{I}}$ is generated by the restrictions to $\tilde{M}$ of the 1-forms

$$(132) \qquad \begin{cases} \theta^a = dz^a - p_i^a dx^i \\ \theta_i^a = dp_i^a - p_{ij}^a dx^j. \end{cases}$$

On the other hand, integral elements to (128) are defined by

$$dp_i^a - p_{ij}^a dx^j = 0$$

subject to the conditions (from (ii) in (128))

$$-dp_i^a \wedge dx^i = -p_{ij}^a dx^j \wedge dx^i = 0,$$

which implies that $p_{ij}^a = p_{ji}^a$, and (from (i) in (128))

$$\dfrac{\partial F^\lambda}{\partial x^i} dx^i + \dfrac{\partial F^\lambda}{\partial z^a} p_i^a dx^i + \dfrac{\partial F^\lambda}{\partial p_j^a} p_{ij}^a dx^i = 0.$$

Comparing with (131) we see that we may identify $\tilde{M}$ with $M^{(1)}$ and that, when this is done, $\tilde{\mathcal{I}} = \mathcal{I}^{(1)}$. $\qquad\qquad\square$

§7. Examples.

We will give some examples of differential systems in involution.

Example 7.1 (Cauchy–Riemann equations). Let $w = u + iv$ be a holomorphic function in the n complex variables

$$z^i = x^i + \sqrt{-1}\, y^i, \qquad 1 \le i \le n.$$

The Cauchy–Riemann equations can be written as the differential system

$$\theta^1 = du - (p_i dx^i + q_i dy^i) = 0$$

$$\theta^2 = dv - (-q_i dx^i + p_i dy^i) = 0$$

in the space $(x^i, y^i, u, v, p_i, q_i)$ of $4n + 2$ dimensions, the independence condition being

$$\Lambda_i dx^i \wedge dy^i \ne 0.$$

We have

$$-d\theta^1 = dp_i \wedge dx^i + dq_i \wedge dy^i,$$

$$-d\theta^2 = -dq_i \wedge dx^i + dp_i \wedge dy^i.$$

To prove that the system is involutive we can proceed in one of the following two ways:

1) We search for a regular integral flag

$$E^1 \subset E^2 \subset \cdots \subset E^n \subset E^{n+1} \subset \cdots \subset E^{2n}$$

such that E^{2n} is defined by

$$dp_i = (h_{ij} dx^j + k_{ij} dy^j)$$

$$dq_i = (l_{ij} dx^j + m_{ij} dy^j), \qquad 1 \le i, j \le n,$$

and E^j, E^{n+j} respectively by the further equations

$$dx^{j+1} = \cdots = dx^n = 0, \; dy^1 = \cdots = dy^n = 0,$$
$$dy^{j+1} = \cdots = dy^n = 0.$$

The conditions for E^j to be integral are

$$h_{ij} = h_{ji}, \; l_{ij} = l_{ji}, \; i \le j.$$

The conditions for E^{n+j} to be integral are

$$-k_{ij} + l_{ji} = 0, \quad m_{ij} + h_{ji} = 0,$$
$$k_{ij} = k_{ji}, \quad m_{ij} = m_{ji}.$$

In both cases the equations are compatible, taking account of earlier equations expressing respectively, the conditions that $E^1 \subset \cdots \subset E^{j-1}$ and $E^1 \subset \cdots \subset E^{j-1+n}$ are integral flags. Hence the system is in involution.

2) We apply Cartan's test. The tableau matrix given by (89) in the previous section is (we omit the bars over the π_i^a)

$$\Pi = \begin{pmatrix} dp_1 & \cdots & dp_n & dq_1 & \cdots & dq_n \\ -dq_1 & & -dq_n & dp_1 & \cdots & dp_n \end{pmatrix}.$$

It is easily checked that

$$s_1' = \cdots = s_n' = 2, \quad s_{n+1}' = \cdots = s_{2n-1}' = r_{2n}' = 0.$$

Both sides of the inequality (87) are equal to $n(n+1)$, and the system is in involution.

Example 7.2 (Partial differential equations of the second order). As discussed above, the basis of second-order P.D.E.'s is the differential system

$$\theta = dz - p_i dx^i = 0$$

$$\theta_i = dp_i - p_{ij} dx^j = 0; \quad p_{ij} = p_{ji}, \quad 1 \le i, j \le n$$

in the space (x^i, z, p_i, p_{ij}) of dimension

$$2n + 1 + \frac{1}{2}n(n+1).$$

We have

$$-d\theta = dp_i \wedge dx^i \equiv 0, \qquad \mathrm{mod}\ \theta_i,$$

$$-d\theta_i = dp_{ij} \wedge dx^j.$$

To illustrate the scope of our concept of involutiveness, we wish to remark that the system is involutive, if there is no relation between the variables. In fact, define the admissible integral elements by

$$dp_{ij} = p_{ijk} dx^k, \qquad 1 \le i, j, k, l \le n.$$

Then we have $p_{ijk} = p_{ikj}$, and therefore p_{ijk} is symmetric in any two of its indices. Thus $\dim G_x(\mathcal{I}, \Omega) = n(n+1)(n+2)/6$. On the other hand, consider the tableau matrix (where we again omit the bars)

$$\pi = \begin{pmatrix} dp_{11} & \cdots & dp_{1n} \\ & \cdots & \\ dp_{n1} & \cdots & dp_{nn} \end{pmatrix}.$$

We find

$$s_1' = n, s_2' = n-1, \ldots, s_{n-1}' = 2, \; r_n' = s_n' = 1.$$

The involutiveness follows from the identity

$$\sum_{1 \leq i \leq n} i(n-i+1) = \frac{1}{6}n(n+1)(n+2).$$

Suppose now there is one equation

$$F(x^i, z, p_i, p_{ij}) = 0.$$

Then its differential gives

$$\sum \frac{\partial F}{\partial p_{ij}} dp_{ij} + \cdots = 0,$$

and the dp_{ij} are linearly dependent. An advantage in using differential forms is that we can choose a basis to write this equation in a simple form. We put

$$\omega^i = dx^i,$$

and apply the substitution

$$\tilde{\omega}^i = u_j^i \omega^j, \; \omega^j = v_i^j \tilde{\omega}^i,$$

$$\tilde{\theta}_i = v_i^j \theta_j, \; \theta_j = u_j^i \tilde{\theta}_i,$$

where (u_i^j), (v_i^j) are inverse matrices to each other so that

$$\theta_i \wedge \omega^i = \tilde{\theta}_i \wedge \tilde{\omega}^i.$$

Then we have

$$d\tilde{\theta}_i \equiv \tilde{\pi}_{il} \wedge \tilde{\omega}^l \quad \mod \{\theta, \theta_j\}$$

where

$$\tilde{\pi}_{il} = \tilde{\pi}_{li} = -\sum dp_{jk} v_i^j v_l^k,$$

or

$$-dp_{jk} = \sum u_j^i u_k^l \tilde{\pi}_{il}.$$

Suppose we make the non-degeneracy assumption $\det(\partial F/\partial p_{jk}) \neq 0$. Then we can choose u_j^i so that

$$\sum \frac{\partial F}{\partial p_{jk}} u_j^i u_k^l = \begin{cases} \varepsilon_i = \pm 1, & i = l, \\ 0, & i \neq l. \end{cases}$$

This gives

$$\sum \varepsilon_i \tilde{\pi}_{ii} + \sum C_k \tilde{\omega}^k \equiv 0 \quad \text{mod } \{\theta, \theta_j\}$$

we put

$$\tilde{\tilde{\pi}}_{ii} = \tilde{\pi}_{ii} + \varepsilon_i C_i \tilde{\omega}^k, \ \tilde{\tilde{\pi}}_{ij} = \tilde{\pi}_{ij}, \ i \neq j,$$

and we absorb θ, θ_j into $\tilde{\tilde{\pi}}_{ij}$. By dropping the tildes, we arrive at the normal form

$$(133) \qquad \begin{aligned} d\theta_i &\equiv \pi_{ij} \wedge \omega^j \quad \text{mod } \{\theta, \theta_j\}, \\ \sum \varepsilon_i \pi_{ii} &= 0. \end{aligned}$$

From this the involutiveness of the system follows immediately.

In fact, we put

$$\pi = \begin{pmatrix} \pi_{11} \ \cdots \ \pi_{1n} \\ \cdots \\ \pi_{n1} \ \cdots \ \pi_{nn} \end{pmatrix}$$

where

$$\pi_{ij} = \pi_{ji}, \ \sum \varepsilon_i \pi_{ii} = 0.$$

We find

$$s_1' = n, \ s_2' = n - 1, \ldots, s_{n-1}' = 2, \ s_n' = 0$$

so that

$$s_1' + 2s_2' + \cdots + (n-1)s_{n-1}' + ns_n' = \frac{1}{6}n(n+1)(n+2) - n.$$

On the other hand, the admissible n-dimensional integral elements on which $\theta = \theta_j = 0$ are given by

$$(134) \qquad \pi_{ij} = l_{ijk}\omega^k$$

where

$$l_{ijk} = l_{jik} = l_{ikj}, \ \sum \varepsilon_i l_{iik} = 0.$$

Its space has the dimension $\frac{1}{6}n(n+1)(n+2) - n$. Hence the system is in involution.

However, such a result, that the system arising from a non-degenerate second-order P.D.E. is in involution, does not seem to be exciting. To get an idea of the meaning of involutiveness, we will study a system of q equations

$$(135) \qquad F^\lambda(x^i, z, p_i, p_{ij}) = 0, \qquad 1 \le \lambda, \mu \le q.$$

For $q > 1$, the system is "over-determined", and we should expect strong conditions for it to be involutive.

First the integrability conditions have to be satisfied. These can be expressed in terms of the functions F^λ. We suppose this to be the case and proceed to study the conditions for involutivity in terms of the tableau. By the structure equation (83) in the previous section there exist π_{ij} such that

$$d\theta_i \equiv \pi_{ij} \wedge \omega^j \mod \{\theta_j\},$$

$$(136) \qquad \pi_{ij} = \pi_{ji},$$

$$\sum B_{ij}^\lambda \pi_{ij} = 0 \mod \{\theta_j\}, \quad B_{ij}^\lambda = B_{ji}^\lambda.$$

In the matrix π we have

$$s_1' \le n, s_2' \le n - 1, \ldots, s_{n-2}' \le 3, \ s_1' + \cdots + s_{n-1}' \le \tfrac{1}{2}n(n+1) - q,$$

$$s_1' + \cdots + s_{n-1}' + s_n' = \tfrac{1}{2}n(n+1) - q.$$

For the system to be in involution we must have

$$\dim G_x(\mathcal{I}, \Omega) = s_1' + 2s_2' + \cdots + (n-1)s_{n-1}' + nr_n'$$

$$= n\left\{\frac{1}{2}n(n+1) - q\right\} - (n-2)s_1' - \cdots - s_{n-2}' - s_1' - \cdots - s_{n-1}'$$

$$\ge (n-1)\left\{\frac{1}{2}n(n+1) - q\right\} - (n-2)n - \cdots - 1 \cdot 3$$

$$= \frac{1}{6}(n-1)(n^2 + 4n + 6) - (n-1)q.$$

On the other hand, for (134) to be an integral element we have

$$(137) \qquad \begin{aligned} l_{ijk} &= l_{jik} = l_{ikj}, \\ \sum B_{ij}^\lambda l_{ijk} &= 0. \end{aligned}$$

The l_{ijk}, being symmetric, has $\tfrac{1}{6}n(n+1)(n+2)$ components. Hence, in the involutive case, the number of linearly independent equations in the system (137) is

$$\frac{1}{6}n(n+1)(n+2) - \dim G_x(\mathcal{I}, \Omega) \le (n-1)q + 1.$$

This condition can be reformulated as follows: We introduce the quadratic forms

$$B^\lambda = B^\lambda_{ij}\xi^i\xi^j.$$

The nq cubic forms

$$\xi^k B^\lambda$$

satisfy a linear relation

$$\sum m_{k\lambda}\xi^k B^\lambda = 0,$$

if and only if

$$(138)\qquad\qquad \sum_\lambda (B^\lambda_{ij}m_{k\lambda} + B^\lambda_{jk}m_{i\lambda} + B^\lambda_{ki}m_{j\lambda}) = 0.$$

The quantities defined by

$$B^\lambda_{ijk,l} = B^\lambda_{ij}\delta_{kl} + B^\lambda_{jk}\delta_{il} + B^\lambda_{ki}\delta_{jl}$$

are the elements of an $\frac{1}{6}n(n+1)(n+2)\times nq$ matrix and are the coefficients of each of the systems (137) and (138). Hence the two systems have the same rank. It follows that *the cubic forms $\xi^k B^\lambda$ satisfy at least $q-1$ independent linear relations.*[8]
Consider the case $q = 2$. The above conclusion implies that

$$l_1(\xi)B^1 + l_2(\xi)B^2 = 0,$$

where $l_1(\xi)$ and $l_2(\xi)$ are linear forms. Hence B^1 and B^2 have a common linear factor. By a change of coordinates we have two cases:

$$\text{a)}\quad B^1 = \xi^1\xi^2,\ B^2 = \xi^1\xi^3;$$
$$\text{b)}\quad B^1 = (\xi^1)^2,\ B^2 = \xi^1\xi^2.$$

For the corresponding differential systems we have the normal forms:

$$\text{a)}\quad \pi_{12} = \pi_{13} = 0;$$
$$\text{b)}\quad \pi_{11} = \pi_{12} = 0.$$

We state our results as a proposition following from Theorem 5.16:

---

[8]In §7 of Chapter VIII this will be generalized to P.D.E. systems of any order—cf. (146)–(152) there.

Proposition 7.3. *For a system of two second-order P.D.E.'s to be in involution it is necessary and sufficient that:*

1) The integrability conditions be satisfied;

2) The symbols, as quadratic forms B^1 and B^2, have a common linear factor.

Proof. The sufficiency follows from verifying the involutiveness conditions when the system is in the normal forms a) or b). We leave this to the reader.

Example 7.4. Consider the following equations which play a role in the old theory of matter and gravitation (see Cartan and Einstein [1979], p. 33):

$$\frac{\partial X_i}{\partial x^j} - \frac{\partial X_j}{\partial x^i} = 0, \ \sum \frac{\partial X_i}{\partial x^i} = -4\pi\rho$$

$$\frac{\partial \rho}{\partial t} + \sum \frac{\partial(\rho u_i)}{\partial x^i} = 0$$

$$\frac{\partial u_i}{\partial t} + \sum u_j \frac{\partial u_i}{\partial x^j} = X_i, \qquad 1 \le i,j \le 3.$$

Here x^i are the space coordinates, t is time, and u_i and X_i are components respectively of the velocity and acceleration vectors, while ρ is the density of matter. This is a system of 8 equations in 4 independent variables x^i, t, and 7 dependent variables u_i, X_i, ρ. It is thus an overdetermined system. This system is involutive.

To make the ideas more clear we will consider the simpler system

$$(139) \qquad \frac{\partial X_i}{\partial x^j} - \frac{\partial X_j}{\partial x^i} = 0, \ \sum_i \frac{\partial X_i}{\partial x^i} = -4\pi\rho, \ 1 \le i,j,k \le 3,$$

where X_i, ρ are functions of x^1, x^2, x^3, and ρ is given. We shall show that the system (139) is involutive. For this purpose we write it as a Pfaffian system

$$(140) \qquad \begin{aligned} dX_i &= \sum X_{ij} dx^j \\ X_{ij} &= X_{ji}, \ \sum X_{ii} = -4\pi\rho, \end{aligned}$$

with the six X_{ij} as new variables. The exterior derivatives of these equations give

$$\sum dX_{ij} \wedge dx^j = 0$$
$$\sum dX_{ii} = -4\pi d\rho.$$

Consider admissible integral elements E^3 defined by

$$dX_{ij} = \sum X_{ijk} dx^k, \ X_{ijk} = X_{jik}.$$

To show the involutivity of the system it suffices to find in E^3 a regular integral flag $E^1 \subset E^2 \subset E^3$, such that E^1 and E^2 are defined respectively by

$$dx^2 = dx^3 = 0 \text{ and } dx^3 = 0.$$

The condition for E^1 to be integral is

$$(141a) \qquad \sum X_{ii1} = -4\pi \frac{\partial \rho}{\partial x^1}.$$

The condition for E^2 to be integral are, in addition to $(141a)$,

$$(141b) \qquad X_{i12} = X_{i21}, \ \sum X_{ii2} = -4\pi \frac{\partial \rho}{\partial x^2}, \ 1 \le i \le 3.$$

These equations can be solved in terms of X_{ij2}, so that E^1 is regular.

To see whether E^2 is regular we consider the conditions for E^3 to be integral, which are

$$(141c) \qquad X_{i13} = X_{i31}, \ X_{i23} = X_{i32}, \ \sum X_{ii3} = -4\pi \frac{\partial \rho}{\partial x^3}.$$

The first two equations imply

$$X_{231} = X_{132}.$$

But this is the first equation of $(141b)$, with $i = 3$. Hence it is satisfied, and we see that $(141c)$ are compatible as linear equations in X_{ij3}. Thus E^2 is regular, and so is the integral flag $E^1 \subset E^2 \subset E^3$. This proves the involutivity of the system (139) or (140).

The proof of the involutivity of the original system is exactly the same. It is only suggested that one take as the starting one-dimensional integral element the one defined by

$$dx^1 = dx^2 = dx^3 = 0, \qquad dt \ne 0.$$

In [1953] Cartan proved that Einstein's field equations for a unified field theory based on distant parallelism are involutive.[9] They are a highly overdetermined system.

---

[9]The specific references are Part III, Vol. 2, p. 1167–1185 and Part II, Vol. 2, p. 1199–1229.

§8. Families of Isometric Surfaces in Euclidean Space.

In Chapter III we gave a proof of the Cartan–Janet isometric imbedding theorem. For two dimensions it says that an analytic Riemannian manifold of two dimensions can be locally isometrically imbedded in the 3-dimensional space E^3. By the discussion in that chapter, the imbedding is not unique. The data needed to specify it uniquely will be discussed in the next chapter. In any case, we can say that it is not rigid, meaning that there is a surface isometric without being congruent to it.

We therefore try to impose further conditions. The two natural conditions are:

A) preservation of the lines of curvature;

B) preservation of the principal curvatures.

In each case we are led to an over-determined system, where the number of equations exceeds the number of unknown functions. Prolongation leads to new conditions and, in these two cases of geometrical significance, to very remarkable conditions. We shall state the two main results in this section. They are local results dealing with non-trivial families of isometric surfaces containing no umbilics, where a non-trivial family means a family which is not obtained from a given surface by a family of rigid motions.

Theorem 8.1. *A non-trivial family of isometric surfaces of non-zero Gaussian curvature, preserving the lines of curvature, is a family of cylindrical molding surfaces.*

The cylindrical molding surfaces can be kinematically described as follows: Take a cylinder Z and a curve C on one of its tangent planes. A *cylindrical molding surface* is the locus described by C as the tangent plane rolls about Z.

Theorem 8.2. *A non-trivial family of isometric surfaces preserving the principal curvatures is one of the following:*

α) (the general case) a family of surfaces of constant mean curvature;

β) (The exceptional case) a family of surfaces of non-constant mean curvature. They depend on six arbitrary constants and have the properties:

β₁) they are W-surfaces;

β₂) the metric

$$(142) \qquad d\hat{s}^2 = (\operatorname{grad} H)^2 dx^2/(H^2 - K),$$

where ds^2 is the metric of the surface and H and K are its mean curvature and Gaussian curvature respectively, has Gaussian curvature equal to -1.

Our discussions in Chapter III contain the essence of a surface theory in E^3, and we will summarize it in a form convenient for the present discussion as follows:

We begin with the diagram

$$(143) \qquad \begin{array}{c} P_0 \\ F \nearrow \downarrow \pi \\ M \xrightarrow{f} E^3 \end{array} ,$$

where P_0 is the space of all orthonormal frames $xe_1e_2e_3$ in E^3, $\pi(xe_1e_2e_3) = x \in E^3$ is the projection, f is the imbedding where we identify both the original and image point as x, $F(x) = xe_1e_2e_3$ is a "lifting" of f satisfying the condition that e_3 is the oriented unit normal at x, M being supposed to be oriented.

The lifting F defines a family of frames over M which satisfy the equations

$$(144) \qquad \begin{aligned} dx &= \omega_1 e_1 + \omega_2 e_2, \\[4pt] de_1 &= \omega_{12} e_2 + \omega_{13} e_3 \\[4pt] de_2 &= -\omega_{12} e_1 + \omega_{23} e_3 \\[4pt] de_3 &= -\omega_{13} e_1 - \omega_{23} e_2. \end{aligned}$$

Denoting by $(\ ,\ )$ the inner product in E^3, the first and second fundamental forms of the surface M are

$$(145) \qquad \begin{aligned} I &= ds^2 = (dx, dx) = \omega_1^2 + \omega_2^2, \\[4pt] II &= -(dx, de_3) = \omega_1 \omega_{13} + \omega_2 \omega_{23} \\[4pt] &= l_{11}\omega_1^2 + 2l_{12}\omega_1\omega_2 + l_{22}\omega_2^2. \end{aligned}$$

These depend only on the imbedding f, and are independent of the lifting F. The mean curvature and Gaussian curvature of M are respectively

$$(146) \qquad H = \frac{1}{2}(l_{11} + l_{22}), \ \ K = l_{11}l_{22} - l_{12}^2.$$

The eigenvalues of II with respect to I are the principal curvatures, and the eigen-directions, which are perpendicular, are the principal directions. If the lifting F is such that e_1, e_2 are in the principal directions, II is diagonalized, i.e. $l_{12} = 0$. The point x is umbilic if the principal curvatures are equal; i.e., if $H^2 - K = 0$.

We restrict ourselves to a neighborhood of M without umbilics, and choose e_1, e_2 to be tangent vectors along the principal directions. Then the ω's in (144) are all linear combinations of ω_1, ω_2 and we set

$$(147) \qquad \begin{aligned} \omega_{13} &= a\omega_1, \ \ \omega_{23} = c\omega_2 \\[4pt] \omega_{12} &= h\omega_1 + k\omega_2 \end{aligned}$$

Here a and c are the two principal curvatures; the mean curvature and the Gaussian curvature are now given by

$$(148) \qquad H = \frac{1}{2}(a + c), \;\; K = ac,$$

and the absence of umbilics is expressed by the condition $a \neq c$.

Exterior differentiation of (144) gives the structure equations

$$(149) \qquad \begin{aligned} &d\omega_1 = \omega_{12} \wedge \omega_2, \;\; d\omega_2 = \omega_1 \wedge \omega_{12} \\ &d\omega_{ij} = \sum_k \omega_{ik} \wedge \omega_{kj}, \;\; 1 \leq i, j, k \leq 3. \end{aligned}$$

The last equation, when written explicitly gives

$$(150) \qquad d\omega_{13} = \omega_{12} \wedge \omega_{23}, \;\; d\omega_{23} = \omega_{13} \wedge \omega_{12}$$

$$(151) \qquad d\omega_{12} = -K\omega_1 \wedge \omega_2.$$

Equations (150) are called the *Codazzi equations* and equation (151) the *Gaussian equation*.

We will use ω_1, ω_2 to express the differential of any function on M, thus

$$(152) \qquad df = f_1\omega_1 + f_2\omega_2,$$

so that f_1, f_2 are the "directional derivatives" of f. Using this notation and substituting the expressions in the first equations of (147) into (150), we get

$$(153) \qquad \begin{aligned} a_2 &= (a - c)h, \\ c_1 &= (a - c)k. \end{aligned}$$

We will use this formalism to study our isometry problems.

To study problem A, let M^* be a surface isometric to M such that the isometry preserves the lines of curvature. Using asterisks to denote the quantities pertaining to M^*, we have

$$(154) \qquad \begin{aligned} &\omega_1^* = \omega_1, \;\; \omega_2^* = \omega_2, \;\; \omega_3 = \omega_3^* = 0, \;\; \omega_{12}^* = \omega_{12}, \\ &\omega_{13}^* = ta\omega_1, \;\; \omega_{23}^* = \frac{c}{t}\omega_2. \end{aligned}$$

The last two equations follow from the fact that M^* has the same Gaussian curvature as M at corresponding points. Equation (153) gives, when applied to M^*,

$$(155) \qquad \begin{aligned} (ta)_2 &= (ta - \frac{c}{t})h, \\ (\frac{1}{t}c)_1 &= (ta - \frac{c}{t})k. \end{aligned}$$

Comparison of (153) and (155) gives

$$(156) \qquad \begin{aligned} t_1 &= t(1 - t^2)ac^{-1}k, \\ t_2 &= -t^{-1}(1 - t^2)a^{-1}ch \end{aligned}$$

or

$$(156a) \qquad \frac{tdt}{1 - t^2} = t^2 ac^{-1}k\omega_1 - a^{-1}ch\omega_2.$$

In fact, from now on we suppose $t^2 \neq 1$, discarding the trivial case that M^* is congruent or symmetric to M. We set

$$(157) \qquad m = a^{-1}h, \quad n = c^{-1}k,$$

so that

$$(158) \qquad \omega_{12} = m\omega_{13} + n\omega_{23}$$

and define

$$(159) \qquad \pi_1 = n\omega_{13}, \quad \pi_2 = m\omega_{23}.$$

Then (156a) can be written

$$(156b) \qquad \frac{tdt}{1 - t^2} = t^2\pi_1 - \pi_2.$$

Its exterior differentiation gives

$$(160) \qquad t^2(d\pi_1 - 2\pi_1 \wedge \pi_2) = d\pi_2 - 2\pi_1 \wedge \pi_2.$$

This equation, if not satisfied identically, completely determines t^2. On substituting into (156), we get conditions on the surfaces M, to which there exist isometric but not congruent or symmetric surfaces preserving the lines of curvature. The later are uniquely determined up to position in space.

The most interesting case is when the equation (160) is identically satisfied, i.e.,

$$(161) \qquad d\pi_1 = d\pi_2 = 2\pi_1 \wedge \pi_2.$$

This leads to a non-trivial family of isometric surfaces preserving the lines of curvature.

In fact, substitution of (159) into (161) gives

(162)
$$(dm - mn\omega_{13}) \wedge \omega_{23} = 0$$
$$(dn + mn\omega_{23}) \wedge \omega_{13} = 0.$$

We shall show that *these equations imply $mn = 0$ or equivalently $hk = 0$.*
Equations (162) allow us to set

(163)
$$dm = mn\omega_{13} + q\omega_{23},$$
$$dn = p\omega_{13} - mn\omega_{23}.$$

by (150) and (158) we have

(164)
$$d\omega_{13} = \omega_{12} \wedge \omega_{23} = m\omega_{13} \wedge \omega_{23},$$
$$d\omega_{23} = -\omega_{12} \wedge \omega_{13} = n\omega_{13} \wedge \omega_{23}.$$

Taking the exterior derivative of (158) and using (163), (164), we get

(165)
$$p - q + 1 + m^2 + n^2 = 0.$$

If m and n are considered as unknown functions, equations (163) and (165) give three relations between their derivatives. This primitive counting shows that the differential system is over-determined. To study our problem there is no other way but to examine the integrability conditions through differentiation of (163), (165). In this case the integrability conditions give a very simple conclusion.

Exterior differentiation of (163) gives

(166)
$$(dq + 2m^2 n\omega_{13}) \wedge \omega_{23} = 0,$$
$$(dp + 2mn^2\omega_{23}) \wedge \omega_{13} = 0,$$

which allow us to set

(167)
$$dp = r\omega_{13} - 2mn^2\omega_{23},$$
$$dq = -2m^2 n\omega_{13} + s\omega_{23}.$$

Differentiation of (165) then gives

(168)
$$r = 2n(-2m^2 - p),$$
$$s = 2m(-2n^2 + q).$$

As a result the prolongation "stabilizes" with

$$(169) \qquad \begin{aligned} dp &= 2n(-2m^2 - p)\omega_{13} - 2mn^2\omega_{23}, \\ dq &= -2m^2 n\omega_{13} + 2m(-2n^2 + q)\omega_{23}. \end{aligned}$$

Exterior differentiation of this equation and use of (163), (164), (169) gives

$$mn(p - q + m^2 + n^2) = 0.$$

Comparing with (165), we get $mn = 0$, or equivalently $hk = 0$, as claimed above.

We wish to describe these surfaces geometrically. Suppose $k = 0$. Then, by (163), (165),

$$p = 0, \ q = 1 + m^2.$$

It follows that the surfaces in question satisfy the equations

$$(170) \qquad \begin{aligned} \omega_3 &= 0, \ \omega_{13} = a\omega_1, \ \omega_{23} = c\omega_2, \ \omega_{12} = h\omega_1, \\ d\left(\frac{h}{a}\right) &= c\left(1 + \frac{h^2}{a^2}\right)\omega_2, \\ \omega_1 \wedge da &- h(a - c)\omega_1 \wedge \omega_2 = 0, \\ \omega_2 &\wedge dc = 0. \end{aligned}$$

The last three equations are obtained by exterior differentiation of the three equations before them. Hence the differential system (170) is closed.

To describe these surfaces observe that

$$\omega_1 = 0, \quad (\text{resp. } \omega_2 = 0)$$

defines a family of lines of curvature, to be denoted by Γ_2 (resp. Γ_1). Along a curve of Γ_2, we have $\omega_{12} = 0$, so that these curves are geodesics. Writing $\omega_2 = ds$, we have, along a curve of Γ_2,

$$\frac{dx}{ds} = e_2, \ \frac{de_2}{ds} = ce_3, \ \frac{de_3}{ds} = -ce_2, \ \frac{de_1}{ds} = 0.$$

Hence it is a plane curve with curvature c, the plane having the normal e_1. The last equation of (170) says that dc is a multiple of ω_2, which means that all the curves of Γ_2 have the same Frenet equations and hence are congruent to each other.

Since

$$de_1 = \omega_{12}e_2 + \omega_{13}e_3 = (he_2 + ae_3)\omega_1,$$

the intersection of two neighboring planes of the curves of Γ_2 is a line in the direction

$$e_1 \times (he_2 + ae_3) = -ae_2 + he_3.$$

By (144) and (170), we have

$$d(-e_2 + \frac{h}{a}e_3) = +\frac{h}{a}\omega_{23}(-e_2 + \frac{h}{a}e_3).$$

Hence this direction is fixed. It follows that the planes of the lines of curvature in Γ_2 are the tangent planes of a cylinder Z.

The curves of Γ_1, being tangent to e_1, are the orthogonal trajectories of the tangent planes of Z. Each line of curvature of Γ_1 is thus the locus of a point in a tangent plane of Z as the latter rolls about Z. The curves of Γ_2 are the orthogonal trajectories of those of Γ_1. Each of them is therefore the position taken by a fixed curve on a tangent plane through the rolling.

The surfaces defined by (170) can be kinematically described as follows: Take a cylinder Z and a curve C on one of its tangent planes. The surface M is the locus described by C as the tangent plane rolls about Z. Such a surface is called a *cylindrical molding surface*. It depends on two arbitrary functions in one variable, one defining the base curve of Z and the other the plane curve C.

On our molding surface the equation (156b) is completely integrable and has a solution t which depends on an arbitrary constant. We get in this way all non-trivial families of isometric surfaces preserving the lines of curvature.

We observe that among the molding surfaces are the surfaces of revolution.

The above discussion can be summarized as follows:

In the three-dimensional euclidean space E^3 consider two pieces of surfaces M, M^*, such that: (a) their Gaussian curvature $\neq 0$ and they have no umbilics; (b) they are connected by an isometry $f : M \to M^*$ preserving the lines of curvature. Then M and M^* are in general congruent or symmetric. There are surfaces M, for which the corresponding M^* is distinct relative to rigid motions. The cylindrical molding surfaces, and only these, are such surfaces belonging to a continuous family of distinct surfaces, which are connected by isometries preserving the lines of curvature.

In particular, we have proved Theorem 8.1.

This is an example of an elaborate nature of a non-involutive differential system whose solutions are studied through successive prolongations. If the surface is considered as a map $M \to E^3$, then the first and second fundamental forms involve respectively the first and second order jets, h, k, m, n those of the third order, and p, q those of the fourth order. Hence the surface M must be of class C^5 for our proof to be valid.

Problem B also leads to an over-determined differential system. Its treatment is more involved. We refer to Chern [1985] for details, and for a proof of Theorem 8.2.

CHAPTER V

THE CHARACTERISTIC VARIETY

In this chapter we will define the *characteristic variety* Ξ associated to a differential ideal (satisfying one non-essential restriction). This variety plays at least as important a role in the theory of differential systems as that played by the usual characteristic variety in classical P.D.E. theory. We shall give a number of examples of characteristic varieties, discuss some of their elementary properties, and shall state a number of remarkable theorems concerning characteristic varieties of involutive differential systems. The proofs of most of the results rely on certain commutative algebra properties of involutive tableaux and will be given in Chapter VIII.

In Section 1 we define and give examples of the characteristic variety of a differential ideal having no Cauchy characteristics. Roughly speaking, it is given by all hyperplanes in n-dimensional integral elements whose extension fails to be unique. This is an infinitesimal analogue of saying that an initial value problem fails to have a unique solution, and as such is parallel to the classical meaning of characteristic.

In Section 2 we define the characteristic variety of a linear Pfaffian differential system. This definition, which is the one we shall use throughout the remainder of the book, is modelled on the P.D.E. definition using the symbol. After showing that this symbol definition coincides with the previous one in the absence of Cauchy characteristics, we go ahead and explain in general the characteristics and the characteristic variety. Then we give a number of examples, including a local existence theorem for determined elliptic Pfaffian systems and the local isometric embedding of surfaces in E^3. This latter example illustrates in a non-trivial manner essentially all the basic concepts—prolongation, involution, torsion, Cauchy characteristics, characteristic variety—in the theory of exterior differential systems. It will be carried as a running example in this chapter.

In Section 3 we give some properties of the characteristic variety. The first few of these, such as the relation between the characteristic variety of a Pfaffian system and that of its prolongation, are elementary. Following this we turn to deeper properties, all of which require the use of the *complex* characteristic variety and require that the system be *involutive*. The first of these, Theorem 3.6, tells us "how many" local integral manifolds there are in terms of the dimension and degree of the complex characteristic variety. The second of these, Theorem 3.15, deals with the overdetermined case and relates the characteristic hyperplanes to the singular integral elements in dimension l where l is *the* character of the system (Definition 3.4). The third of these, Theorem 3.20, which is a differential system analogue of the theorem of Guillemin, Quillen and Sternberg [1970], states that the characteristic variety induces an involutive system in the cotangent bundle of integral manifolds. Here we shall only prove the result when the characteristic variety is smooth and consists only of isolated points, a case already found in Cartan (see Subsection (vi)), and a case that is simpler from a technical point of view. A number of examples illustrating this result will be given in Chapter VII.

In this chapter, $\mathcal{I}$ will denote a differential ideal, i.e., a homogeneous ideal in $\Omega^*(M)$ that is closed under exterior differentiation. Sometimes we shall refer to $\mathcal{I}$ as a differential system. We will denote by $I \subset \Omega^1(M)$ the degree one piece of $\mathcal{I}$, and we shall assume that I is given by the sections of a sub-bundle of T^*M that we also denote by I. When there is an independence condition it will be denoted by Ω, where Ω is a decomposable n-form defined up to non-zero scalar factors and up to adding n-forms from $\mathcal{I}$; as in Chapter IV, Ω is given by a non-zero section of $\Lambda^n(J/I)$ where J is a sub-bundle of T^*M with $I \subset J \subset T^*M$. For a submanifold $N \subset M$ and differential form $\alpha \in \Omega^*(M)$, we will denote the restriction $\alpha|_N$ by α_N. Thus a n-dimensional submanifold $N \subset M$ is an integral manifold of the differential system with independence condition $(\mathcal{I}, \Omega)$ if

$$\begin{cases} \alpha_N = 0 & \text{for all } \alpha \in \mathcal{I} \\ \Omega_N \neq 0 \end{cases}.$$

Finally, we will use the summation convention.

§1. Definition of the Characteristic Variety of a Differential System.

Let M be a manifold and $G_n(TM) \to M$ the Grassmann bundle of n-planes in the tangent spaces to M. Points of $G_n(TM)$ will usually be denoted by (x, E) where $x \in M$ and $E \subset T_xM$ is an n-plane.[1] Over $G_n(TM)$ we have the universal n-plane bundle

$$U \to G_n(TM)$$

whose fibre over (x, E) is just E. We shall consider the projectivization $\mathbb{P}U^*$ of the dual bundle U^*. A point in the fibre $\mathbb{P}U_E^*$ of $\mathbb{P}U^*$ over $E \in G_n(TM)$ will be written as $[\xi]$ where $\xi \in E^* \backslash \{0\}$ is a non-zero vector and $[\xi] \subset E^*$ is the corresponding line (the brackets are supposed to suggest homogeneous

---

[1]We shall sometimes use the abbreviated notation E when the base point $x \in M$ is unimportant.

coordinates). By projective duality $[\xi]$ determines a hyperplane $[\xi]^\perp$ in E, and geometrically we may think of $\mathbb{P}U_E^*$ as being the set of hyperplanes in $E \subset T_x M$.

Let $\mathcal{I}$ be a differential ideal and assume that $\mathcal{I}$ has no Cauchy characteristics, i.e., we assume that there are no vector fields $v \neq 0$ satisfying

$$v \lrcorner \mathcal{I} \subset \mathcal{I}.$$

This non-essential assumption is put here for convenience of exposition; it will be eliminated below. Let $G_n(\mathcal{I}) \subset G_n(TM)$ be the set of n-dimensional integral elements of $\mathcal{I}$. Associated to each hyperplane $[\xi]^\perp \subset E$ is the *polar space*

$$H(\xi) = \{v \in T_x M : \mathrm{span}\{v, [\xi]^\perp\} \text{ is an integral element}\},$$

that we may think of as all ways of enlarging $[\xi]^\perp$ to an n-dimensional integral element.

Definition 1.1. The *characteristic variety* is the subset Ξ of $\mathbb{P}U^*$ defined by

$$\Xi = \cup_{E \in G_n(\mathcal{I})} \Xi_E$$

where $\Xi_E = \Xi \cap \mathbb{P}U_E^*$ and

$$\Xi_E = \{[\xi] : H(\xi) \supsetneq E\}.$$

Thus, Ξ *consists of all hyperplanes in n-dimensional integral elements whose extension to an n-dimensional integral element fails to be unique.* To the extent that we think of integral elements as infinitesimal solutions to a differential system, the characteristic variety corresponds to non-uniqueness of an initial value problem, in close analogy to the classical notion. There is a commutative diagram of mappings

$$
\begin{array}{ccc}
\Xi & \subset & \mathbb{P}U^* \\
\downarrow & & \downarrow \\
G_n(\mathcal{I}) & \subset & G_n(TM)
\end{array}
$$

and we shall denote by Ξ_E the fibre of $\Xi \to G_n(\mathcal{I})$ lying over E.

The condition that $[\xi]$ be characteristic is

$$\dim H(\xi) > n.$$

Therefore, it does not depend on the particular E with $[\xi]^\perp \subset E$ (so long as there is at least one such). Thus, we may give the following

Definition 1.2. An $(n-1)$-plane $E^{n-1} \in G_{n-1}(\mathcal{I})$ is *non-characteristic* in case $\dim H(E^{n-1}) = n$.

From the proof of the Cartan–Kähler theorem we have the result: *Let $\mathcal{I}$ be a real-analytic differential system and $N \subset M$ an $(n-1)$-dimensional real-analytic integral manifold whose tangent planes are K-regular and non-characteristic. Then there is locally a unique extension of N to an n-dimensional integral manifold of $\mathcal{I}$.*

It is easy to see that the fibre Ξ_E of the projection

$$\Xi \to G_n(\mathcal{I})$$

is an algebraic subvariety of $\mathbb{P}E^*$, i.e., it is defined by polynomial equations. This is because the polar equations are linear in vectors $v \in T_x M$, and Ξ_E consists of hyperplanes $[\xi]$ for which the ranks of these equations jump suitably; this condition is expressed by homogeneous polynomial equations in ξ. Of importance will be the *complex characteristic variety* $\Xi_{\mathbb{C}}$, defined as the complex solutions to these same polynomial equations. Equivalently, for a complex integral element E we may consider complex hyperplanes $[\xi]^\perp$ in the complex vector space E, and then

$$\Xi_{\mathbb{C},E} = \{ [\xi] \in \mathbb{P}E^* : H(\xi) \supsetneqq E \}$$

where the polar space $H(\xi) = \{ v \in T_{\mathbb{C},x}M : \operatorname{span}\{v, [\xi]^\perp\}$ is a complex integral element$\}$.[2] Of course, it may well happen that Ξ is empty but $\Xi_{\mathbb{C}}$ is not.

The reason we assumed no Cauchy characteristics is that $v \in H(\xi)$ for any Cauchy characteristic vector v. Thus, the characteristic variety should only be defined for integral elements that contain all Cauchy characteristic vectors. Equivalently, we may consider the differential system obtained by "foliating out" the Cauchy characteristics and define the characteristic variety on this reduced system. For linear Pfaffian systems this annoyance will be circumvented.

Before turning to more substantive examples, we mention these two:

i) A *Frobenius system* when $\mathcal{I}$ is generated by $dy^1, \ldots, dy^s$ in $\mathbb{R}^{n+s}$ with coordinates $(x^1, \ldots, x^n, y^1, \ldots, y^s)$. Then $\Xi_{\mathbb{C}} = \emptyset$. A converse of this for involutive systems will be discussed below.

ii) A *Darboux system* when $\mathcal{I}$ is generated by $\Theta = \sum_i dx^i \wedge dy^i$ in $\mathbb{R}^{2n}$ with coordinates $(x^1, \ldots, x^n, y^1, \ldots, y^n)$. In this case, $\Xi_E = \mathbb{P}E^*$ is everything.

Example 1.3 (Triply orthogonal systems, cf. DeTurck and Yang [1984]). Let X be a 3-dimensional Riemannian manifold and consider the following

---

[2] An n-plane $E \subset T_{x,\mathbb{C}}M$ is a complex integral element if $\alpha_E = 0$ for all $\alpha \in \mathcal{I}$; E need not be the complexification of a real integral element.

Problem. *Determine the triples of foliations in X that intersect pairwise orthogonally.*

This problem was discussed in n-dimensions and in the real analytic case in Chapter III. Here we shall restrict to 3-dimensions and shall discuss the characteristic variety, which is the first step towards a C^∞ result. To set the problem up we let $M^6 = \mathcal{F}(X)$ be the bundle of orthonormal frames, on which we have the canonical parallelism given by the 1-forms $\omega_1, \omega_2, \omega_3, \omega_{12}, \omega_{13}, \omega_{23}$ satisfying the usual structure equations (here $1 \leq i, j \leq 3$)

$$(1) \qquad d\omega_i = \sum_j \omega_j \wedge \omega_{ji}, \qquad \omega_{ij} + \omega_{ji} = 0,$$

that uniquely determine the connection matrix $\|\omega_{ij}\|$. On M we let $\mathcal{I}$ be the differential ideal generated by the 3-forms

$$\Theta_i = \omega_i \wedge d\omega_i$$

As explained in Chapter III, the solutions to our problem are given by sections $s : X \to \mathcal{F}(X) = M$ satisfying $s^*\Theta_i = 0$. Locally then we look for integral manifolds $N^3 \subset M$ of $\mathcal{I}$ such that

$$\Omega_N \neq 0$$

where $\Omega = \omega_1 \wedge \omega_2 \wedge \omega_3$.

We shall denote p-planes in TM by E^p (i.e., we don't worry about the *foot* of E^p, which is the point $x \in M$ such that $E^p \subset T_x M$). When $p = n$ we shall generally just write E. In case E^p is an integral element of $\mathcal{I}$ we set

$$r(E^p) = \dim H(E^p) - p - 1.$$

Thus, $r(E^p) = 0$ is the condition that E^p extend to a unique integral E^{p+1}.

Using (1), it is a nice little exercise to show that $\mathcal{I}$ is algebraically generated by the forms

$$\omega_1 \wedge \omega_2 \wedge \omega_{12}$$
$$(2) \qquad \omega_1 \wedge \omega_3 \wedge \omega_{13}$$
$$\omega_2 \wedge \omega_3 \wedge \omega_{23}.$$

An integral E^3 on which $\Omega \neq 0$ is therefore given by linear equations

$$(3) \qquad \begin{pmatrix} \omega_{23} \\ \omega_{13} \\ \omega_{12} \end{pmatrix} = \begin{pmatrix} 0 & a_{12} & a_{13} \\ a_{21} & 0 & a_{23} \\ a_{31} & a_{32} & 0 \end{pmatrix} \begin{pmatrix} \omega_1 \\ \omega_2 \\ \omega_3 \end{pmatrix} = 0$$

where the a_{ij} are arbitrary. Thus, over each point of M the integral 3-planes on which $\Omega \neq 0$ form an $\mathbb{R}^6$. For E given by (3) with basis for E^* given by the restrictions $(\omega_1)_E, (\omega_2)_E, (\omega_3)_E$, a point $\xi = [\xi_1, \xi_2, \xi_3] \in \mathbb{P}E^*$ corresponds to the hyperplane $[\xi]^\perp \subset E$ defined by the additional equation

$$(4) \qquad \xi_1 \omega_1 + \xi_2 \omega_2 + \xi_3 \omega_3 = 0.$$

Lemma 1.4. *Setting $r(\xi) = r([\xi]^{\perp})$, we have*

$$r(\xi) = \begin{cases} 0 & \text{if all } \xi_i \neq 0 \\ 1 & \text{if one } \xi_i = 0 \\ 2 & \text{if two } \xi_i = 0. \end{cases}$$

Proof. We shall check the results when all the $a_{ij} = 0$; the general case is similar. By symmetry we may assume that $\xi_1 \neq 0$ in (4); multiplying by a constant we may then assume that (4) is

$$(5) \qquad\qquad \omega_1 - \alpha\omega_2 - \beta\omega_3 = 0.$$

Letting $\{e_i, e_{ij}\}$ be the dual frame to $\{\omega_i, \omega_{ij}\}$, the integral E^2 given by $\omega_{ij} = 0$ and (5) has basis

$$v_1 = \alpha e_1 + e_2$$
$$v_2 = \beta e_1 + e_3.$$

For $v = \sum \lambda_i e_i + \sum_{i<j} \lambda_{ij} e_{ij}$ we want to count the number of solutions of

$$(6) \qquad\qquad \langle \Theta_i, v_1 \wedge v_2 \wedge v \rangle = 0$$

with the transversality condition $\langle \Omega, v_1 \wedge v_2 \wedge v \rangle \neq 0$. By subtracting from v a linear combination of v_1, v_2 and multiplying by a constant we may assume that

$$v = e_3 + \sum_{i<j} \lambda_{ij} e_{ij}.$$

Then

$$v_1 \wedge v_2 \wedge v = e_1 \wedge e_2 \wedge e_3 + e_1 \wedge e_2 \wedge \left(\sum_{i<j} \lambda_{ij} e_{ij}\right)$$
$$-\alpha e_2 \wedge e_3 \wedge \left(\sum_{i<j} \lambda_{ij} e_{ij}\right) + \beta e_1 \wedge e_3 \wedge \left(\sum_{i<j} \lambda_{ij} e_{ij}\right).$$

Using (2) the equations (6) are

$$\lambda_{12} = 0$$
$$\alpha \lambda_{23} = 0$$
$$\beta \lambda_{13} = 0,$$

from which the lemma (in the case $a_{ij} = 0$) is clear. $\qquad\square$

Since a hyperplane is characteristic exactly when $r(\xi) > 0$, we see that Ξ_E consists of the union of the three coordinate lines in $\mathbb{P}E^* \cong \mathbb{P}^2$; i.e., it is the usual coordinate triangle encountered in plane projective geometry. The singular points of Ξ_E are clearly just the vertices of this coordinate triangle.

If $N \subset M$ is a local integral manifold, then N may be identified with X together with a framing, and a surface $N^2 \subset N$ is characteristic if one leg of the framing is tangent to N^2; it is doubly characteristic (i.e., has tangent planes given by an intersection point of two branches in Ξ) if two legs of the framing are tangent to N^2, which is equivalent to one leg being normal to N^2.

Example 1.5 (Linear Weingarten surfaces). Let X be an oriented Riemannian 3-manifold and set

$$M^5 = \{(x, e_3) \in TX : \|e_3\| = 1\}.$$

Thinking of $e_3 \in T_x X$ as corresponding to the oriented 2-plane $e_3^\perp \subset T_x X$ we may picture M as the manifold of oriented contact elements lying over X. We shall use the fibering picture

$$
\begin{array}{ccc}
\pi & \mathcal{F}(X) & \\
\swarrow & & \\
M & \downarrow & \pi(x, e_1, e_2, e_3) = (x, e_3), \\
\searrow & & \\
& X &
\end{array}
$$

and for computational purposes shall pull all forms back to $\mathcal{F}(X)$. Among the forms that are well-defined on M are

$$\omega_3 = e_3 \cdot dx$$

$$d\omega_3 = \omega_1 \wedge \omega_{13} + \omega_2 \wedge \omega_{23}$$

$$(7) \qquad \Omega_0 = \omega_1 \wedge \omega_2$$

$$\Omega_1 = \omega_1 \wedge \omega_{23} - \omega_2 \wedge \omega_{13}$$

$$\Omega_2 = \omega_{13} \wedge \omega_{23}.$$

If $N^2 \subset X$ is any oriented surface we have its canonical lift (Gauss map)

$$\gamma : N \to M$$

where $\gamma(y) =$ unit normal to N at the point y. The pull-backs of the forms (7) are

$$\gamma^*(\omega_3) = 0 = \gamma^*(d\omega_3)$$

$$\gamma^*(\Omega_0) = \text{ induced area form } dA$$

$$\gamma^*(\Omega_1) = (\text{Trace II})dA$$

$$\gamma^*(\Omega_2) = (\det \text{II})dA$$

where II is the 2nd fundamental form of $N \subset X$. Conversely, a smooth surface $\gamma : N^2 \to M$ is the canonical lift of an immersed oriented surface in X if

$$\text{(i)} \quad \gamma^*(\omega_3) = 0$$
$$\text{(ii)} \quad \gamma^*(\Omega_0) \neq 0.$$

If only (i) is satisfied, then we shall speak of $\gamma : N \to M$ as being a generalized surface; these are special cases of Legendre manifolds (see the following remark). Note that

$$\omega_3 \wedge (d\omega_3)^2 \neq 0$$

on M, so that the differential ideal generated by ω_3 has the Pfaff–Darboux local normal form.

Remark. Given a differential ideal on a $(2n+1)$-dimensional manifold generated by a 1-form satisfying $\theta \wedge (d\theta)^n \neq 0$, the Pfaff–Darboux local normal form shows that the maximal integral manifolds N, called *Legendre manifolds*, are of dimension n and are given by one arbitrary function of n-variables. Here we may think of $\theta = dz - \sum_i y_i dx_i$ and N given by $(x_i, y_i = \frac{\partial z(x)}{\partial x_i})$ where $z = z(x)$ is an arbitrary function. In particular, the generalized surfaces are described locally by 1 arbitrary function of 2 variables.

Definition 1.6. Let A, B, C be constants not all zero. The differential ideal

$$\mathcal{I} = \{\omega_3, \Theta = A\Omega_2 + B\Omega_1 + C\Omega_0\}$$

generated by ω_3 and Θ will be called a *linear Weingarten system*.

Remark that by the structure equations (1) together with $d\omega_{ij} + \omega_{ik} \wedge \omega_{kj} = \frac{1}{2} R_{ijkl}\omega^k \wedge \omega^l$, it is easy to see that $\mathcal{I}$ is generated *algebraically* by

$$\omega_3, d\omega_3, \Theta.$$

The study of the integral elements thus leads us to the following linear algebra data:

$$T \text{ is 5-dimensional vector space} \qquad (= T_x M)$$
$$\omega \in T^* \text{ is a 1-form} \qquad (= \omega_3 \in T_x^* M)$$
$$\Theta_1, \Theta_2 \in \Lambda^2 T^* \text{ are 2-forms} \qquad (= d\omega_3, \Theta \in \Lambda^2 T_x^* M).$$

The integral 1-planes are $E^1 = \mathbb{R}v_1$ where $\langle \omega, v_1 \rangle = 0$. Given E^1 its polar equations are

$$\text{(i)} \quad \langle \omega, v \rangle = 0$$
$$\text{(ii)} \quad \langle v_1 \lrcorner \Theta_1, v \rangle = 0$$
$$\text{(iii)} \quad \langle v_1 \lrcorner \Theta_2, v \rangle = 0.$$

In general the rank of these equations is 3, i.e.,

$$r(E^1) = 0.$$

Geometrically, we expect (locally and in the real-analytic case) to be able to find a unique linear Weingarten surface of type (A, B, C) passing through a general curve $\Gamma \subset X$.

To compute the characteristics, we observe that these occur when equations (i)–(iii) become dependent. More precisely, for an integral 2-plane $E \cong \mathbb{R}^2$ we have $\mathbb{P}E^* \cong \mathbb{P}^1$, and each $[v_1] \in \mathbb{P}E^*$ such that the equations (i)–(iii) are dependent is characteristic. Since Θ_1 and Θ_2 do not contain ω, this is equivalent to equations (ii) and (iii) becoming dependent, i.e., when

$$(8) \qquad v_1 \lrcorner\, (\lambda_1 \Theta_1 + \lambda_2 \Theta_2) = 0$$

for some $(\lambda_1, \lambda_2) \neq (0, 0)$. Note that if we restrict to integral elements E^2 on which $\Omega_0 \neq 0$ we must have $\lambda_2 \neq 0$. Now it is easy to see that (8) is equivalent to

$$(\lambda_1 \Theta_1 + \lambda_2 \Theta_2)^2 = 0.$$

Setting $\lambda = \lambda_1/\lambda_2$ it follows from (7) that this is the same as

$$(9) \qquad \lambda^2 + B^2 - AC = 0.$$

Let $\lambda_\pm$ be the two roots of this equation (we allow complex values). Since $d\omega_3$ and Θ are linearly independent, the two roots of (9) are the values for which the 2-form

$$\lambda d\omega_3 + \Theta \neq 0$$

becomes decomposable in the space of complex valued differential forms. We assume that $\lambda_+ \neq \lambda_-$ and write

$$\lambda_+ d\omega_3 + \Theta = \alpha_+ \wedge \beta_+$$
$$\lambda_- d\omega_3 + \Theta = \alpha_- \wedge \beta_-.$$

Then it follows that

$$\omega_3, \alpha_+, \beta_+, \alpha_-, \beta_-$$

are pointwise linearly independent over $\mathbb{C}$ and that $\mathcal{I}$ is generated algebraically over $\mathbb{C}$ by

$$\omega_3, \alpha_+ \wedge \beta_+, \alpha_- \wedge \beta_-.$$

Proof. Since $\lambda_+ \neq \lambda_-$, we have

$$0 \neq \omega_3 \wedge (d\omega_3)^2 = \frac{1}{\lambda_+ - \lambda_-} (\omega_3 \wedge \alpha_+ \wedge \beta_+ \wedge \alpha_- \wedge \beta_-).$$

Thus, over the complex numbers

$$\mathrm{span}\{\alpha_+ \wedge \beta_+,\ \alpha_- \wedge \beta_-\} = \mathrm{span}\{d\omega_3, \Theta\},$$

which proves our claim that $\mathcal{I} = \{\omega_3, \alpha_+ \wedge \beta_+, \alpha_- \wedge \beta_-\}$.

Now, for any integral 2-plane E

$$\alpha_+ \wedge \beta_+|_E = 0 = \alpha_- \wedge \beta_-|_E.$$

Moreover, we do not have $\alpha_+|_E = \beta_+|_E = 0$ nor $\alpha_-|_E = \beta_-|_E = 0$, since for example if the former holds then E is defined by

$$\omega_3 = \alpha_+ = \beta_+ = 0$$

and consequently $\alpha_- \wedge \beta_-|_E \neq 0$. Thus the restrictions to E of each of α_+, β_+ and α_-, β_- spans a line in E^*, and these two lines are the characteristics. That is

$$\Xi_E = [v_+] \cup [v_-]$$

where $v_\pm \neq 0$ and $\alpha_\pm(v_\pm) = \beta_\pm(v_\pm) = 0$.

§2. The Characteristic Variety for Linear Pfaffian Systems; Examples.

In Chapter IV we have defined the class of linear Pfaffian systems. Associated to such a system are its tableau and symbol, and we shall now show how to compute the characteristic variety in terms of the symbol by a process that is formally analogous to that for P.D.E.'s. We begin by recalling some notation.

From Chapter IV we recall that linear Pfaffian systems are given by sub-bundles

$$I \subset J \subset T^*M$$

satisfying

$$(10) \qquad\qquad dI \subset \{J\}$$

where $\{J\} \subset \Omega^*(M)$ is the algebraic ideal generated by the sections of J.[3] We set $L = J/I$, so that the exterior derivative induces a bundle mapping

$$(11) \qquad\qquad \bar{\delta} : I \to (T^*M/J) \otimes L$$

---

[3]As usual, we shall use I and J to denote both a sub-bundle of T^*M and the C^∞ sections of that bundle—the context will make clear which use is intended.

given in terms of bases by equation (56) in Chapter IV. Dualizing and using that $(T^*M/J)^* \cong J^\perp \subset TM$, giving $\bar{\delta}$ is equivalent to giving the *tableau mapping* (cf. equation (60) in Chapter IV):

$$(12) \qquad \pi : J^\perp \to I^* \otimes L.$$

The relations on the image of π are given by setting

$$Q = I^* \otimes L/\text{image } \pi$$

and defining the *symbol mapping* σ to be the quotient mapping

$$\sigma : I^* \otimes L \to Q.$$

Then image $\pi = $ kernel σ.

We now define the characteristic variety

$$\Xi \subset \mathbb{P}L$$

of a linear Pfaffian system. For $0 \neq \xi \subset L_x$ we let $[\xi] \subset \mathbb{P}L_x$ be the corresponding line and define

$$\sigma_\xi : I_x^* \to Q_x$$

by

$$\sigma_\xi(w) = \sigma(w \otimes \xi).$$

Definition 2.1. The *characteristic variety* $\Xi \subset \mathbb{P}L$ is given by $\Xi = \cup_{x \in M} \Xi_x$ where

$$\Xi_x = \{[\xi] \in \mathbb{P}L_x : \sigma_\xi \text{ fails to be injective}\}.$$

Equivalently,

$$\Xi_x = \{[\xi] \in \mathbb{P}L_x : \text{ there exists } 0 \neq w \in I_x^* \text{ such that } w \otimes \xi \in A_x\}$$

where A_x is the tableau of $(\mathcal{I}, \Omega)$ at x.

We will now see how this definition looks in coordinates. Following the notations in §5 of Chapter IV, locally we choose 1-forms θ^a, ω^i so that

$$I = \text{span}\{\theta^a\}$$
$$J = \text{span}\{\theta^a, \omega^i\}$$

and then

$$L \cong \text{span}\{\omega^i\}.$$

Here "span" means all linear combinations with smooth functions as coefficients. The condition (10) that the Pfaffian system be linear is

$$d\theta^a \equiv 0 \mod \{J\}$$

where $\{\theta^a, \omega^i\} = \{J\}$ is the algebraic ideal generated by the θ^a and ω^i. This means that we have

$$d\theta^a \equiv \pi^a_i \wedge \omega^i \mod \{I\}$$

where $\{I\} = \{\theta^a\}$ is the algebraic ideal generated by the θ^a's, and the π^a_i are then 1-forms well-defined modulo J, and they give the tableau mapping (12) in coordinates. We may thus think of π as given by the *tableau matrix*

$$\pi = \begin{bmatrix} \pi^1_1 & \cdots & \pi^1_n \\ \vdots & & \vdots \\ \pi^{s_0}_1 & \cdots & \pi^{s_0}_n \end{bmatrix} \quad \mod J.$$

A basis for the relations on the image of π will be written as

$$B^{\lambda i}_a \pi^a_i \equiv 0 \mod J.$$

Summarizing, the structure equations of a linear Pfaffian system are

$$
\begin{array}{ll}
\text{(i)} & \theta^a = 0 \\
\text{(ii)} & d\theta^a \equiv \pi^a_i \wedge \omega^i \mod \{I\} \\
\text{(iii)} & B^{\lambda i}_a \pi^a_i \equiv 0 \mod J \\
\text{(iv)} & \Omega = \omega^1 \wedge \cdots \wedge \omega^n \neq 0.
\end{array}
$$
(13)

The symbol mapping σ is given in coordinates by the $B^{\lambda i}_a$. More precisely, for $\xi = \xi_i \omega^i(x) \in L_x$ as above, σ_ξ is given by the matrix

$$\sigma_\xi = \|B^{\lambda i}_a(x)\xi_i\| \in \mathrm{Hom}(I^*_x, Q_x).$$

Then

$$\Xi_x = \{[\xi] : B^{\lambda i}_a(x)\xi_i w^a = 0 \text{ for some } w \neq 0\}$$

$$= \{[\xi] : \mathrm{rank}\, \|B^{\lambda i}_a(x)\xi_i\| < s_0\}$$

where $\mathrm{rank}\, I = s_0$. It is clear that Ξ_x is defined by homogeneous polynomial equations in the ξ_i, so that $\Xi = \cup_{x \in M}\Xi_x$ is a family of algebraic varieties parameterized by M.

The way to remember this definition is as follows: Associated to a P.D.E. system

$$F^\lambda(y^i, z^a, \partial z^a / \partial y^i) = 0$$

is the linear Pfaffian differential system

$$\theta^a = dz^a - p_i^a \, dy^i = 0$$

$$d\theta^a \equiv -dp_i^a \wedge dy^i \mod \{I\}$$

$$\Omega = dy^1 \wedge \cdots \wedge dy^n \neq 0$$

on the manifold

$$M = \{(y^i, z^a, p_i^a) : F^\lambda(y^i, z^a, p_i^a) = 0\}.$$

Differentiation of the defining equations of M gives the symbol relations

$$\frac{\partial F^\lambda}{\partial p_i^a} dp_i^a \equiv 0 \mod J$$

where $J = \mathrm{span}\{\theta^a, dy^i\}$. Comparing with equation (iii) in (13), we find that our definition of symbol relations for a linear Pfaffian differential system is a natural extension of the usual definition for a P.D.E. system. Correspondingly, our definition of the characteristic variety is the natural extension of the usual one for a P.D.E. system.

We now want to compare the more general definition in the preceeding section with this one. Recall that the Cauchy characteristics are the vector fields v satisfying

$$v \lrcorner \, \theta^a = 0$$

$$v \lrcorner \, d\theta^a = 0 \mod \mathcal{I}.$$

If the 1-forms θ^a, ω^i, π_i^a fail to span $T_x^* M$ for some x, then by our constant rank assumptions this will be true in a neighborhood and we can find a vector field v satisfying

$$v \lrcorner \, \theta^a = v \lrcorner \, \omega^i = v \lrcorner \, \pi_i^a = 0.$$

By the structure equations (13) this vector field will be a Cauchy characteristic. Thus, under the assumption of no Cauchy characteristics we have

$$(14) \qquad\qquad \mathrm{span}\{\theta^a(x), \omega^i(x), \pi_i^a(x)\} = T_x^* M.$$

For any integral element (x, E) of $(\mathcal{I}, \Omega)$, the independence condition

$$\Omega_E \neq 0$$

implies that the restriction mapping

$$L_x \to E^*$$

is an isomorphism. We will show that:

Under this isomorphism, $\Xi_x \subset \mathbb{P}L_x$ corresponds to $\Xi_E \subset \mathbb{P}E^*$.

In particular, in this case the characteristic variety Ξ_E depends only on the point x and not on the particular E lying over x.

Proof. We omit reference to the point x. By (14), the integral element E will be given by linear equations

$$\pi_i^a - p_{ij}^a \omega^j = 0,$$

where

$$p_{ij}^a = p_{ji}^a$$

by (ii) in (13). By a substitution $\pi_i^a \to \pi_i^a - p_{ij}^a \omega^j$ we may assume that E is given by

$$(15) \qquad \begin{cases} \theta^a = 0 \\ \pi_i^a = 0. \end{cases}$$

Additionally, by (iii) in Proposition 5.15 in Chapter IV we may assume that the symbol relations are given by

$$(16) \qquad B_a^{\lambda i} \pi_i^a \equiv 0 \quad \mod I.$$

Finally, we may assume that

$$\xi = \omega^n.$$

The polar equations of $[\xi]^\perp$ are then easily seen to be

$$(17) \qquad \begin{cases} \theta^a = 0 \\ \pi_1^a = \cdots = \pi_{n-1}^a = 0. \end{cases}$$

It follows that $E \neq H(\xi)$ if, and only if, the equations (15) are not consequences of the equations (17), equivalently,

$$(18) \qquad \textit{some } \pi_n^b \textit{ is not a linear combination of the 1-forms}$$
$$\{\theta^a, \pi_1^a, \ldots, \pi_{n-1}^a\}.$$

We choose a vector v such that some $v \lrcorner \pi_n^b \neq 0$ but where $v \lrcorner \theta^a = 0$ and $v \lrcorner \pi_i^a = 0$ for $i = 1, \ldots, n-1$ and all a. Contracting the symbol relations with v gives

$$B_a^{\lambda n}(v \lrcorner \pi_n^a) = 0,$$

so that $\|B_a^{\lambda n}\|$ is not injective.

Conversely, suppose that $\|B_a^{\lambda n}\|$ is not injective. We may assume that the kernel contains the vector $(1, 0, \ldots, 0)$, i.e., that all $B_1^{\lambda n} = 0$. But then π_n^1 does not appear in any of the symbol relations (13), (iii); in particular, it is not a linear combination of the $\{\theta^a, \pi_1^a, \ldots, \pi_{n-1}^a\}$. Thus, the condition that $\xi = \omega^n$ be characteristic according to either Definition 1.1 or Definition 2.1 is equivalent to (18). $\qquad\qquad\square$

There is a very simple relation between the Cauchy characteristics and characteristic variety of a linear Pfaffian differential systems. Let

$$A(\mathcal{I}) \subset TM$$

denote the Cauchy characteristic sub-bundle. Since $A(\mathcal{I}) \subset I^\perp$ the mapping

$$A(\mathcal{I}) \to L^*$$

given in coordinates by

$$v \to (\omega^1(v), \ldots, \omega^n(v))$$

is well-defined, and we denote its image by $S \subset L^*$. Then

$$S^\perp \subset L$$

and we shall show that:

$$(19) \qquad\qquad \begin{array}{c} \textit{The characteristic variety} \\ \Xi \subset \mathbb{P}S^\perp. \end{array}$$

In particular, if $S \neq 0$ then it follows that the fibres Ξ_x of the characteristic variety lie in the proper linear subspaces $\mathbb{P}S_x^\perp \subset \mathbb{P}L_x$.

Clearly, (19) also remains valid when we complexify. In the involutive case, there is a converse to the complex version of (19).

Proof. Choose a basis $\omega^1, \ldots, \omega^n$ for L so that $\omega^1, \ldots, \omega^k$ is a basis for $S^\perp$. Then for $k + 1 \leq \rho \leq n$ there is a Cauchy characteristic vector field v_ρ satisfying

$$v_\rho \lrcorner \omega^\sigma = \delta_\rho^\sigma, \qquad k + 1 \leq \rho, \sigma \leq n.$$

From

$$v_\rho \lrcorner d\theta^a \equiv 0 \quad \mathrm{mod}\ \mathcal{I}$$

we infer that

$$\pi^a_\rho \equiv 0 \quad \mathrm{mod}\ J.$$

Thus the tableau matrix looks like

$$(20) \qquad \begin{bmatrix} \pi^1_1 & \cdots & \pi^1_k & 0 & \cdots & 0 \\ \vdots & & & \vdots & & \vdots \\ \pi^{s_0}_1 & \cdots & \pi^{s_0}_k & 0 & \cdots & 0 \end{bmatrix} \quad \mathrm{mod}\ J.$$

In particular, among the symbol relations we have

$$\pi^a_\rho \equiv 0 \quad \mathrm{mod}\ J, \qquad k+1 \le \rho \le n.$$

From this it is easy to see that a characteristic vector $\xi = \xi_i \omega^i$ must have $\xi_{k+1} = \cdots = \xi_n = 0.$ $\qquad\square$

We remark that whenever we may choose bases so that the tableau matrix has the block form (20), then the zeroes correspond to the image $S \subset L^*$ of Cauchy characteristics as described above.

We also remark that the mapping

$$A(\mathcal{I}) \to S$$

may not be injective; using (ii) in (13) it is easy to see that this is the case exactly when the $\theta^a, \omega^i, \pi^a_i$ fail to span T^*M. Examples of this arise by adding extra variables to any Pfaffian differential system.

Example 2.2. We shall set up a linear Pfaffian differential system whose integral manifolds are the *Darboux framings* of immersed surfaces $S \subset E^3$. For this we denote by $\mathcal{F}(E^3) \to E^3$ to the bundle of orthonormal frames (x, e_1, e_2, e_3) in Euclidean 3-space. On $\mathcal{F}(E^3)$ we have the equations of a moving frame (here $1 \le i, j \le 3$)

$$(21) \qquad \begin{aligned} &\text{(i)} \quad dx = \sum_i \omega_i e_i \\ &\text{(ii)} \quad de_i = \sum_j \omega_{ij} e_j, \quad \omega_{ij} + \omega_{ji} = 0 \end{aligned}$$

and structure equations

$$(22) \qquad \begin{aligned} &\text{(i)} \quad d\omega_i = \sum_j \omega_j \wedge \omega_{ji} \\ &\text{(ii)} \quad d\omega_{ij} = \sum_k \omega_{ik} \wedge \omega_{kj}. \end{aligned}$$

We consider the Pfaffian differential system on $M = \mathcal{F}(E^3)$

(23)
$$\text{(i)} \quad \omega_3 = 0$$
$$\text{(ii)} \quad \omega_1 \wedge \omega_2 \wedge \omega_{12} \neq 0.$$

An integral manifold of this system is given by an immersion f in the diagram

(24)
$$\begin{array}{ccc} N & \xrightarrow{f} & \mathcal{F}(E^3) \\ {\scriptstyle x_f} \searrow & & \downarrow {\scriptstyle x} \\ & E^3 & \end{array}$$

where $\dim N = 3$, $f^*(\omega_3) = 0$, and $f^*(\omega_1 \wedge \omega_2 \wedge \omega_{12}) \neq 0$. From $f^*(\omega_3) = 0$ we have

$$dx_f = f^*(\omega_1)e_1 + f^*(\omega_2)e_2,$$

and it follows first that the image $x_f(N) = S$ is an immersed surface in E^3, and secondly that $f(N)$ consists of all Darboux frames $(x_f(y), e_1, e_2, e_3)$ to S; here $y \in N$ and $e_1, e_2 \in T_{x_f(y)}(S)$.

The structure equations of the Pfaffian differential system (23) are

(25)
$$d\omega_3 = -\omega_{13} \wedge \omega_1 - \omega_{23} \wedge \omega_2$$

from which it follows that (23) is a linear Pfaffian differential system.

The tableau matrix is
$$\|\omega_{13} \;\; \omega_{23} \;\; 0\|.$$

To compute the Cauchy characteristic system of the Pfaffian differential system (23) we denote by $\partial/\partial\omega_1, \partial/\partial\omega_2, \ldots$ the tangent frame dual to the coframe $\omega_1, \omega_2, \ldots$ on $\mathcal{F}(E^3)$. Then using (25) it is easy to verify that: *the Cauchy characteristic system of* (23) *is spanned by* $\partial/\partial\omega_{12}$. This is an example of the block form (20).

We now claim that $\partial/\partial\omega_{12}$ is tangent to any integral manifold of (23). For this it will suffice to show that $\partial/\partial\omega_{12}$ lies in any integral element of the system. Recall that an integral element is given by a 3-plane $E \subset T\mathcal{F}(E^3)$ satisfying the conditions

$$(\omega_3)_E = 0, \quad (d\omega_3)_E = 0$$
$$(\omega_1 \wedge \omega_2 \wedge \omega_{12})_E \neq 0.$$

It follows after a short computation that E is given by linear equations

$$\omega_{13} - a\omega_1 - b\omega_2 = 0$$
$$\omega_{23} - b\omega_1 - c\omega_2 = 0,$$

and therefore $\partial/\partial\omega_{12} \in E$ as claimed.

The conclusion concerning integral manifolds (24) that we may draw is this: *The fibering $x_f : N \to S$ is given by the integral curves of the Cauchy characteristic foliation. Moreover, the equality $\Xi = \mathbb{P}S^\perp$ holds in* (19) *above.* Geometrically, the Cauchy characteristic curves correspond to spinning the tangent frame of S. The reason that Ξ equals all of $\mathbb{P}S^\perp$ is that no conditions are being put on the immersed surfaces $x_f(N) \subset E^3$; thus we do not expect to be able to uniquely determine integral manifolds by data along a curve.

We note that the pullbacks to N of the 1st and 2nd fundamental forms of S are given by the quadratic differential forms

$$I = (\omega_1)^2 + (\omega_2)^2$$

$$II = a(\omega_1)^2 + b(\omega_1\omega_2) + c(\omega_2)^2.$$

Remark finally that this discussion generalizes to Darboux framings associated to submanifolds $Y^n \subset X^N$ where X is any Riemannian manifold and n, N are arbitrary. The Cauchy characteristics give the spinning of the tangential and normal frames to Y.

We shall now give two further types of examples of characteristic varieties. For the first, following standard terminology we give the following

Definition 2.3. The linear Pfaffian differential system $(\mathcal{I}, \Omega)$ is said to be *elliptic* in case its real characteristic variety is empty

$$\Xi = \emptyset.$$

Example 2.4. In $\mathbb{R}^{2m} = \mathbb{C}^m$ we consider the Cauchy–Riemann system

$$
(26) \qquad
\begin{aligned}
\frac{\partial u}{\partial x^i} - \frac{\partial v}{\partial y^i} &= 0 \\
\frac{\partial u}{\partial y^i} + \frac{\partial v}{\partial x^i} &= 0
\end{aligned}
\qquad i = 1,\ldots,m.
$$

As previously remarked, the symbol matrix of the Pfaffian differential system corresponding to (26) is given by its symbol matrix as a partial differential equation system, i.e., by the matrix

$$
(27) \qquad
\begin{pmatrix}
\xi_1 & \xi_2 & \xi_3 & \xi_4 & \cdots & \xi_{2m-1} & \xi_{2m} \\
-\xi_2 & \xi_1 & -\xi_4 & \xi_3 & \cdots & -\xi_{2m} & \xi_{2m-1}
\end{pmatrix}.
$$

The real characteristic variety is, as expected, empty. However, for each $z \in \mathbb{C}^m$ the complex characteristic variety $\Xi_{\mathbb{C},z}$ is given by the vanishing of all 2×2 minors of (27). It is easy to then verify that

$$(28) \qquad \Xi_{\mathbb{C},z} = \mathbb{C}P_+^{m-1} \cup \mathbb{C}P_-^{m-1} \subset \mathbb{C}P^{2m-1}$$

where
$$\mathbb{C}P_{\pm}^{m-1} = \{\xi_2 = \pm\sqrt{-1}\,\xi_1, \ldots, \xi_{2m} = \pm\sqrt{-1}\,\xi_{2m-1}\}.$$

For example, when $m = 2$ we may picture $\Xi_{\mathbb{C},z}$ as two purely imaginary and conjugate skew lines in $\mathbb{C}P^3$.

Example 2.5. We now consider a linear Pfaffian differential system $(\mathcal{I}, \Omega)$ with square symbol matrices (in this case we say that *the system is determined*). Following the notation in (23), (iii) above we write the symbol relations of a determined system as

$$B_a^{bi}\pi_i^a \equiv 0 \quad \mathrm{mod}\ J,^4$$

so that the symbol matrix is

$$\sigma_\xi = \|B_a^{bi}\xi_i\|.$$

We shall show that the system is involutive if for some ξ

$$\det \sigma_\xi \not\equiv 0;$$

i.e., *if the complex characteristic variety is not everything, then the system is involutive.*

Proof. We may assume that $\det\|B_a^{bn}\| \neq 0$, and then by a basis change in the space of relations that
$$B_a^{bn} = -\delta_a^b.$$

When this is done the symbol relations are

$$\pi_n^a \equiv B_b^{a\rho}\pi_\rho^b \quad \mathrm{mod}\ J$$

where $1 \leq \rho, \sigma \leq n - 1$. It follows that the characters of the tableau matrix are given by
$$s_1' = s_0, \ldots, s_{n-1}' = s_0, s_n' = 0.$$

Now write the symbol relations out as

$$\pi_n^a \equiv B_b^{a\rho}\pi_\rho^b + B_i^a\omega^i \quad \mathrm{mod}\ I.$$

Integral elements are given by linear equations

$$\theta^a = 0$$
$$\pi_i^a - p_{ij}^a\omega^j = 0$$

---

[4] This notation is slightly misleading, since the symbol matrices $\sigma_\xi = \|B_a^{bi}\xi_i\|$ are elements of $\mathrm{Hom}(W_1, W_2)$ where W_1, W_2 are *different* vector spaces of the same dimension. Thus we may pre- and post-multiply σ_ξ by different invertible matrices.

where

$$\text{(i)} \quad p^a_{ij} = p^a_{ji}$$

$$\text{(ii)} \quad p^a_{nj} = B^{a\rho}_b p^b_{\rho j} + B^a_j.$$

Choose $p^a_{\rho\sigma} = p^a_{\sigma\rho}$ arbitrarily and use (ii) for $1 \leq j \leq n-1$ to determine the $p^a_{n\sigma} = p^a_{\sigma n}$. Then use (ii) when $j = n$ to determine p^a_{nn}. It follows that the $s_0 n(n-1)/2$ quantities $p^a_{\rho\sigma}$ may be freely specified and that each set of these quantities uniquely determines an integral element; thus the space of integral elements is non-empty (the integrability conditions are satisfied) and has dimension equal to

$$s_0 n(n-1)/2 = s'_1 + 2s'_2 + \cdots + ns'_n.$$

By Cartan's test the system is involutive.   □

We will say that the Pfaffian differential system $(\mathcal{I}, \Omega)$ is *locally embeddable* in case it is locally induced from the canonical system on $J^1(\mathbb{R}^n, \mathbb{R}^{s_0})$. It is easy to show that this is equivalent to $J = \text{span}\{\theta^a, \omega^i\}$ being a Frobenius system (cf. Proposition 5.10 in Chapter IV). Under this circumstance $(\mathcal{I}, \Omega)$ is locally equivalent to a determined P.D.E. system

$$(29) \qquad F^b(y^i, z^a, \partial z^a / \partial y^i) = 0,$$

and known results from P.D.E. theory may be applied to construct integral manifolds.

We shall now prove that

(30)   *If the linear system $(\mathcal{I}, \Omega)$ is determined and elliptic,*

and if $n = \text{rank}(L) \geq 4$, then it is locally embeddable.

As a corollary of (30) we have the following result:

(31)   *Under the conditions of* (30) *there exist local integral*

manifolds of the Pfaffian system $(\mathcal{I}, \Omega)$.

Proof of (31). We may locally realize $(\mathcal{I}, \Omega)$ as the Pfaffian differential system associated to a determined 1st order elliptic P.D.E. system (29). Appealing to a standard result in elliptic equation theory (see Nirenberg [1973]), we infer that (29) has local solutions.   □

Proof of (30). As noted above, the system $(\mathcal{I}, \Omega)$ is involutive, and by the proof of that result we may write it as

$$\begin{aligned}
&\text{(i)} \quad \theta^a = 0\\
(32) \quad &\text{(ii)} \quad d\theta^a = \pi^a_i \wedge \omega^i + \eta^a_b \wedge \theta^b\\
&\text{(iii)} \quad \pi^a_n = B^{a\rho}_b \pi^b_\rho \ \text{mod} \ J
\end{aligned}$$

where $1 \leq \rho,\, \sigma \leq n - 1$ and where

(33)   the 1-forms π_ρ^a are linearly independent   mod $J = \mathrm{span}\{\theta^b, \omega^i\}$.

The exterior derivative of (ii) in (32) gives

$$(34) \qquad\qquad 0 \equiv \pi_i^a \wedge d\omega^i \quad \mathrm{mod}\ J.$$

We must show that, if $n \geq 3$, this implies conversely that

$$(35) \qquad\qquad d\omega^i \equiv 0 \quad \mathrm{mod}\ J.$$

For this we set

$$\varphi_1^a = \pi_1^a, \ldots, \varphi_{n-1}^a = \pi_{n-1}^a,\ \varphi_n^a = B_b^{a\rho}\pi_\rho^b,$$

and using (iii) in (32) write (34) as

$$(36) \qquad\qquad \varphi_i^a \wedge d\omega^i \equiv 0 \quad \mathrm{mod}\ J$$

for each a.

Lemma 2.6. *Assume that the system $(\mathcal{I}, \Omega)$ is elliptic. Then (i) for each a the 1-forms $\varphi_1^a, \ldots, \varphi_n^a$ are linearly independent   mod J; (ii) if $n \geq 3$, then for each $a \neq b$ the 1-forms $\{\varphi_i^a\}$, $\{\varphi_i^b\}$ are linearly independent mod J; and (iii) in general, for $a_1 < \cdots < a_m$ and $m \leq n - 1$ the 1-forms $\varphi_1^{a_a}, \ldots, \varphi_n^{a_1}, \ldots, \varphi_1^{a_m}, \ldots, \varphi_n^{a_m}$ are linearly independent   mod J.*

Proof. (i) Suppose there is a linear relation

$$\zeta^\rho \pi_\rho^a + \zeta B_b^{a\sigma} \pi_\sigma^b \equiv 0 \quad \mathrm{mod}\ J.$$

Then $\zeta \neq 0$ by (33); by homogeneity we may take $\zeta = -1$ and then

$$(\delta_b^a \zeta^\rho - B_b^{a\rho})\pi_\rho^b \equiv 0 \quad \mathrm{mod}\ J.$$

By (33) this implies

$$(37) \qquad\qquad \delta_b^a \zeta^\rho - B_b^{a\rho} = 0 \text{ for all } \rho, b.$$

Taking $0 \neq \xi = (\xi_1, \ldots, \xi_n)$ with $\xi_n = \xi_\rho \zeta^\rho$, (37) gives

$$\delta_b^a \xi_n - B_b^{a\rho}\xi_\rho = 0.$$

But this says that the ath row of the symbol matrix σ_ξ is zero, which contradicts ellipticity.

(ii) Given a linear relation

$$(38) \qquad (\delta_c^a \zeta^\rho + \zeta B_c^{a\rho})\pi_\rho^c + (\delta_c^b \eta^\rho + \eta B_c^{b\rho})\pi_\rho^c \equiv 0 \quad \mathrm{mod}\ J$$

we choose $\xi \neq 0$ with $\zeta^\rho \xi_\rho = \xi_n \zeta$ and $\eta^\rho \xi_\rho = \xi_n \eta$. This is possible since $n \geq 3$, and (33) now gives as before

$$\zeta(\delta_c^a \xi_n + B_c^{a\rho}\xi_\rho) + \eta(\delta_c^b \xi_n + B_c^{b\rho}\xi_\rho) = 0.$$

But this implies that the ath and b^{th} rows of the symbol matrix are dependent (it is easy to see that we don't have $\zeta = \eta = 0$); again this contradicts ellipticity.

(iii) The proof just given applies to the general case, provided we can find a vector $\xi \neq 0$ that is orthogonal to m given vectors; under the conditions of the lemma, this is always possible. $\qquad \square$

By (36), the above lemma and the usual Cartan lemma we have *for each* a

$$(39) \qquad d\omega^i \equiv A(a)^{ij} \wedge \varphi_j^a \quad \mathrm{mod}\ J$$

where the $A(a)^{ij} = A(a)^{ji}$ are 1-forms. By ellipticity, the symbol gives a linear mapping

$$\mathbb{R}^n \to s_0 \times s_0 \text{ matrices,}$$

denoted by $\xi \to \sigma_\xi$, such that

$$\xi \neq 0 \Rightarrow \det \sigma_\xi \neq 0.$$

Thus s_0 must be even, and since $n \geq 4$ we cannot have $s_0 = 2$, i.e.,

$$(40) \qquad s_0 \geq 4.$$

We take $a \neq b$ and use (39) for a and b to obtain

$$(41) \qquad A(a)^{ij} \wedge \varphi_j^a - A(b)^{ij} \wedge \varphi_j^b \equiv 0 \quad \mathrm{mod}\ J.$$

Again by Cartan's lemma this implies that

$$A(a)^{ij} \in \mathrm{span}\{\varphi_k^a, \varphi_k^b\} \quad \mathrm{mod}\ J.$$

Taking $c \neq a, b$ and applying this also for a, c we infer that

$$A(a)^{ij} \in \mathrm{span}\{\varphi_k^a, \varphi_k^c\} \quad \mathrm{mod}\ J.$$

But then (40) and the lemma together imply that

$$A(a)^{ij} \wedge \varphi_j^a \equiv 0 \quad \mod J$$

which by (39) is our desired statement (35).

Example 2.7. We shall study in some detail the isometric embedding system for an abstract Riemannian surface $\overline{S}$ mapping isometrically to E^3. The notation $\overline{S}$ will denote the abstract surface, and $\overline{S} \to S \subset E^3$ will denote the isometric embedding.

Although the study of this example for general dimensions and codimensions has been initiated in Chapter III and will be resumed in Chapter VII, we shall set it up somewhat differently. One motivation is that the characteristic variety will appear in a simple manner. Moreover, although we restrict here to the case of surfaces, all aspects of the general theory already appear in this special case and the extension to higher dimensions basically involves more elaborate algebra. Finally, this example illustrates in a substantial way most of the aspects of the general theory of the characteristic variety.

The general theory will be discussed further in Chapter VII below. Additional references are the original sources, Berger, Bryant and Griffiths [1983] and Bryant, Griffiths and Yang [1983], and the detailed exposition given in Griffiths and Jensen [1987] of these papers.

In discussing this example, we shall make use of certain concepts such as prolongation, that was introduced in Chapter IV and will be discussed in Chapter VI below, and elementary results such as the relationship between the characteristic variety of a Pfaffian system and the characteristic variety of its 1st derived system, that also will be discussed below. These concepts and elementary results should be pretty much self-evident in our example.

We begin by setting up the system and computing its structure equations and 1st prolongation. Geometrically the idea is to map the principal frame bundle of $\overline{S}$ to the Darboux frames of the image surface. In this regard, it may be helpful to keep in mind Example 2.2 above.

We denote by $\overline{\pi} : \overline{P} \to \overline{S}$ the principal frame bundle whose points are (y, e_1, e_2) with $y \in \overline{S}$ and e_1, e_2 an orthonormal basis of $T_y(\overline{S})$ and where $\overline{\pi}(y, e_1, e_2) = y$. On $\overline{P}$ there is the canonical parallelism given by 1-forms $\overline{\omega}_1, \overline{\omega}_2, \overline{\omega}_{12}$ satisfying the structure equations

$$\begin{aligned} d\overline{\omega}_1 &= -\overline{\omega}_2 \wedge \overline{\omega}_{12} \\[2mm] d\overline{\omega}_2 &= \overline{\omega}_1 \wedge \overline{\omega}_{12} \\[2mm] d\overline{\omega}_{12} &= K\overline{\omega}_1 \wedge \overline{\omega}_2 \end{aligned}$$

(42)

where K is the Gaussian curvature of $\overline{S}$.

Now we set $M = \overline{P} \times \mathcal{F}(E^3)$ and on M consider the Pfaffian differential system given by

$$
\begin{aligned}
&\text{(i)} \quad \theta_i = \omega_i - \overline{\omega}_i = 0, \quad i = 1, 2, \\
\text{(43)} \qquad &\text{(ii)} \quad \theta_3 = \omega_3 = 0 \\
&\text{(iii)} \quad \overline{\omega}_1 \wedge \overline{\omega}_2 \wedge \overline{\omega}_{12} \neq 0.
\end{aligned}
$$

Throughout this example we shall use the index range $i, j = 1, 2$ and $a = 1, 2, 3$. We shall see below that the integrals of this system are locally graphs of maps $f : \overline{P} \to \mathcal{F}(E^3)$ where

(i) $f(\overline{P})$ is the set $\mathcal{F}(S)$ of Darboux frames associated to an immersed surface $S \subset E^3$ and

(ii) there is a commutative diagram

$$
\text{(44)} \qquad
\begin{array}{ccc}
\overline{P} & \xrightarrow{\ f\ } & \mathcal{F}(S) \\
{\scriptstyle \overline{\pi}} \downarrow & & \downarrow {\scriptstyle \pi} \\
\overline{S} & \xrightarrow{\ x_f\ } & S
\end{array}
$$

where x_f is an isometric immersion. One difference between this method and that in Chapter III is that here we make no a priori choice of frame field on $\overline{S}$. Using the equations (21) and (22) above of a moving frame and (42), the structure equations of (43) are

$$
\begin{aligned}
&\text{(i)} \quad d\theta_1 \equiv (\omega_{12} - \overline{\omega}_{12}) \wedge \omega_2 \ \ \bmod \{\theta_a\} \\
\text{(45)} \qquad &\text{(ii)} \quad d\theta_2 \equiv -(\omega_{12} - \overline{\omega}_{12}) \wedge \omega_1 \ \ \bmod \{\theta_a\} \\
&\text{(iii)} \quad d\theta_3 \equiv -\omega_{13} \wedge \omega_1 - \omega_{23} \wedge \omega_2 \ \ \bmod \{\theta_a\}.
\end{aligned}
$$

It follows that (43) is a linear Pfaffian differential system.

Any 3-plane $E \subset T(M)$ on which all the 1-forms $\theta_a = 0$ and $\overline{\omega}_1 \wedge \overline{\omega}_2 \wedge \overline{\omega}_{12} \neq 0$ is given by linear equations

$$
\begin{aligned}
\text{(46)} \qquad
\omega_{12} &= p_{121}\overline{\omega}_1 + p_{122}\overline{\omega}_2 + p_{123}\overline{\omega}_{12} \\
\omega_{13} &= p_{11}\overline{\omega}_1 + p_{12}\overline{\omega}_2 + p_1\overline{\omega}_{12} \\
\omega_{23} &= p_{21}\overline{\omega}_1 + p_{22}\overline{\omega}_2 + p_2\overline{\omega}_{12}.
\end{aligned}
$$

Using (i) in (43) we may replace $\overline{\omega}_i$ by ω_i in these equations. The conditions that (46) be an integral element is that all 2-forms $d\theta_a$ restrict to zero on E. By (45) this is

$$
\begin{aligned}
p_{121} &= p_{122} = 0, \ p_{123} = 1 \\
p_1 &= p_2 - 0 \\
p_{12} &= p_{21}.
\end{aligned}
$$

Setting $p_{11} = a$, $p_{12} = p_{21} = b$, $p_{22} = c$ it follows that the space $M^{(1)}$ of integral elements of the system (43) is $M \times \mathbb{R}^3$, where $(a, b, c) \in \mathbb{R}^3$ and where the integral elements are given by

$$\omega_{12} - \overline{\omega}_{12} = 0$$

$$\omega_{13} - a\omega_1 - b\omega_2 = 0$$

$$\omega_{23} - b\omega_1 - c\omega_2 = 0.$$

By definition, the 1st prolongation of (43) is the Pfaffian differential system on $M^{(1)}$ given by

$$(47) \qquad \begin{aligned} &\text{(i)} \quad \theta_i = \omega_i - \overline{\omega}_i = 0 \\ &\text{(ii)} \quad \theta_3 = \omega_3 = 0 \\ &\text{(iii)} \quad \theta_{12} = \omega_{12} - \overline{\omega}_{12} = 0 \\ &\text{(iv)} \quad \theta_{13} = \omega_{13} - a\omega_1 - b\omega_2 = 0 \\ &\text{(v)} \quad \theta_{23} = \omega_{23} - b\omega_1 - c\omega_2 = 0 \end{aligned}$$

together with the independence condition $\overline{\omega}_1 \wedge \overline{\omega}_2 \wedge \overline{\omega}_{12} \neq 0$.[5] We note that the 1st prolongation (47) contains two parts: the original system (i) and (ii), and the equations (iii), (iv), (v) of integral elements of the original system. As discussed in Chapter IV, this is a completely general fact. We also note that equation (iii) is defined on M but did not appear in our original system, which was thus "incomplete". More precisely, we will shortly see that this means that the original system fails to be involutive.

Next we shall compute the Cauchy characteristic system of (43). For this we note first that the 1-forms appearing in (43) and (45) span all of T^*M, so that by our remarks following the proof of (19) above the mapping $A(\mathcal{I}) \to S$ is injective. Setting $\theta_{12} = \omega_{12} - \overline{\omega}_{12}$, by (45) the tableau matrix of (43) is

$$(48) \qquad \begin{bmatrix} 0 & -\theta_{12} & 0 \\ \theta_{12} & 0 & 0 \\ \omega_{13} & \omega_{23} & 0 \end{bmatrix} \quad \mathrm{mod}\ \{\theta_a, \overline{\omega}_i, \overline{\omega}_{ij}\}.$$

Referring to (20) above we see that there is, up to non-zero multiples, one Cauchy characteristic vector field. In fact, it is

$$v = \partial/\partial\omega_{12} + \partial/\partial\overline{\omega}_{12}.$$

Geometrically it corresponds to *spinning the tangent frames to $\overline{S}$ and S at the same rate*. More precisely, as in example 3 above we may see that v lies

---

[5]Later on, we shall see that $M^{(1)}$ should be taken to be $M \times (\mathbb{R}^3 \backslash \{0\})$.

in any integral element of (43), and therefore any integral manifold of this system is fibered by the circle group action whose infinitesimal generator is v. Using this observation it is easy to see that the integral manifolds of (43) are locally graphs of maps $f : \overline{P} \to \mathcal{F}(E^3)$ for which there is a commutative diagram (44) where x_f is an isometric immersion.

Now we observe that (43) fails to be involutive, essentially due to the fact that equation (iii) in (47)

$$\omega_{12} - \overline{\omega}_{12} = 0,$$

which is implied by (i) and (ii) in (45) (uniqueness of the Levi–Civita connection), is missing. Referring to (48) the reduced characters are

$$s_1' = 2, \ s_2' = 1, \ s_3' = 0$$

$$\Rightarrow s_1' + 2s_2' = 4,$$

while as previously noted the space of integral elements is an $\mathbb{R}^3$. Thus, Cartan's test is not satisfied and (43) fails to be involutive.

It is for this reason that we went ahead and wrote down the 1st prolongation (47) of (43). The next step is to compute the integrability condition and tableau for (47), and for this some additional notation will be helpful (we want to eliminate the indices—this is especially useful in the higher dimensional case). We introduce a vector space $V \cong \mathbb{R}^2$ and consider the following vector-valued differential forms

$$\omega = {}^t(\omega_1, \omega_2) \text{ and } \overline{\omega} = {}^t(\overline{\omega}_1, \overline{\omega}_2) \text{ are } V\text{-valued 1-forms}$$

$$\psi = \begin{bmatrix} 0 & \omega_{12} \\ -\omega_{12} & 0 \end{bmatrix} \text{ and } \overline{\psi} = \begin{bmatrix} 0 & \overline{\omega}_{12} \\ -\overline{\omega}_{12} & 0 \end{bmatrix} \text{ are } V \otimes V^*\text{-valued 1-forms}$$

$$\eta = (\omega_{13}, \omega_{23}) \text{ is a } V^*\text{-valued 1-form}$$

$$\omega \wedge \omega = \begin{bmatrix} 0 & \omega_1 \wedge \omega_2 \\ -\omega_1 \wedge \omega_2 & 0 \end{bmatrix} \text{ is a } V \otimes V^*\text{-valued 2-form.}$$

$$\varphi = \omega_3 \text{ is a scalar 1-form.}$$

The structure equations of a moving frame now appear as

$$\begin{aligned}
&\text{(i)} \quad d\omega = -\psi \wedge \omega + {}^t\eta \wedge \varphi \\
(49) \quad &\text{(ii)} \quad d\psi = {}^t\eta \wedge \eta \\
&\text{(iii)} \quad d\eta + \eta \wedge \psi = 0.
\end{aligned}$$

Here, for example, $\psi \wedge \omega$ is the natural pairing

$$(V \otimes V^*\text{-valued 1-form}) \otimes (V\text{-valued 1-form}) \to V\text{-valued 2-form.}$$

The structure equations (42) on the frame bundle $\overline{P}$ of $\overline{S}$ are

$$(50) \qquad \begin{aligned} &\text{(i)} \quad d\overline{\omega} = -\overline{\psi} \wedge \overline{\omega} \\ &\text{(ii)} \quad d\overline{\psi} = K\overline{\omega} \wedge \overline{\omega} \end{aligned}$$

where $\overline{\omega} \wedge \overline{\omega}$ is the obvious analogue of $\omega \wedge \omega$ and K is the Gaussian curvature.

The Pfaffian differential system (43) is now

$$(51) \qquad \begin{aligned} &\text{(i)} \quad \theta = \omega - \overline{\omega} = 0 \\ &\text{(ii)} \quad \varphi = 0 \end{aligned}$$

together with the independence condition (iii) in (43). The structure equations (45) of (51) are

$$(52) \qquad \begin{aligned} &\text{(i)} \quad d\theta \equiv -(\overline{\psi} - \psi) \wedge \omega \quad \text{mod } \{\theta, \varphi\} \\ &\text{(ii)} \quad d\varphi \equiv -\eta \wedge \omega \quad \text{mod } \{\theta, \varphi\}. \end{aligned}$$

The integral elements (46) for this system are given by

$$\psi - \overline{\psi} = 0$$
$$\eta - B\omega = 0$$

where B is the $S^2 V^*$-valued function corresponding to the symmetric matrix $\begin{bmatrix} a & b \\ b & c \end{bmatrix}$. The 1st prolongation (47) of (51) is given on $M^{(1)} = \overline{P} \times \mathcal{F}(E^3) \times S^2 V^*$ by the Pfaffian differential system

$$(53) \qquad \left\{ \begin{aligned} &\text{(i)} \quad \theta = \omega - \overline{\omega} = 0 \\ &\text{(ii)} \quad \varphi = 0 \\ &\text{(iii)} \quad \psi - \overline{\psi} = 0 \\ &\text{(iv)} \quad \eta - B\omega = 0 \end{aligned} \right.$$

with the independence condition $\overline{\omega}_1 \wedge \overline{\omega}_2 \wedge \overline{\omega}_{12} \neq 0$. We want now to compute the integrability condition, tableau, and involutive prolongation of this system.

For this we let

$$\gamma : S^2 V^* \times S^2 V^* \to \mathbb{R}$$

be the symmetric bilinear function obtained by polarizing the quadratic function

$$\gamma(B, B) = \det B = ac - b^2.$$

Then letting $\equiv$ denote congruence modulo the algebraic ideal generated by the system (53), we obtain from (49) and (50) that

(54)
$$\begin{aligned}
&\text{(i)} \quad d\theta \equiv 0 \\
&\text{(ii)} \quad d\varphi \equiv 0 \\
&\text{(iii)} \quad d(\psi - \bar{\psi}) \equiv -\{\gamma(B,B) - K\}\omega \wedge \omega \\
&\text{(iv)} \quad d(\eta - B\omega) \equiv -DB \wedge \omega.
\end{aligned}$$

To explain equation (iv) we have

(55)
$$\begin{aligned}
d(\eta - B\omega) &= -\eta \wedge \psi - dB \wedge \omega + B\psi \wedge \omega - B^t \eta \wedge \varphi \\
&\equiv -B\omega \wedge \psi - dB \wedge \omega + B\psi \wedge \omega \text{ modulo (53)} \\
&= -DB \wedge \omega
\end{aligned}$$

where DB is the $S^2 V^*$-valued 1-form defined by the coefficient of $\wedge \omega$ in the right hand side of the middle equation in (55). From (54) we may draw two important conclusions:

(56)
$$\begin{aligned}
&\textit{The integrability conditions of (53) are given by} \\
&\qquad \tau = \gamma(B,B) - K = 0.
\end{aligned}$$

Since this equation is not identically satisfied, the system is not involutive.

(57)   *The tableau of (53) is given by the $S^2 V^*$-valued 1-form DB.*

We may picture it as the symmetric matrix

(58)
$$\pi = \begin{bmatrix} \pi_{11} & \pi_{12} \\ \pi_{21} & \pi_{22} \end{bmatrix}$$

where $\pi_{11} \equiv da \mod (53)$, $\pi_{12} \equiv db \mod (53)$, $\pi_{22} \equiv dc \mod (53)$. The symbol relations are given by

(59)
$$\pi_{21} - \pi_{12} \equiv 0 \mod (53).$$

Actually the tableau matrix of (54) should be

(60)
$$\begin{bmatrix} 0 & 0 & 0 \\ 0 & 0 & 0 \\ 0 & 0 & 0 \\ \pi_{11} & \pi_{12} & 0 \\ \pi_{21} & \pi_{22} & 0 \end{bmatrix} = \begin{bmatrix} 0 & 0 \\ \pi & 0 \end{bmatrix}.$$

The top two rows of zeros correspond to the original system (43), which as reflected by (i) and (ii) in (54) has gone into the 1st derived system of (53). (As noted in §6 of Chapter IV, this also is a general property of prolongation.) The third row of zeros corresponds to (iii) in (54), where we recall that the tableau matrix is always considered modulo $J = \operatorname{span}\{\theta_a, \overline{\omega}_i\}$ where the θ_a span the system (53). The last column of zeros reflects the Cauchy characteristic vector field. It is an easily verified fact that the involutivity of a tableau (60) is equivalent to the involutivity of the non-zero block (58). In other words, *in testing for involutivity of a tableau we may throw out the first derived system and Cauchy characteristic system.*

According to the general prolongation scheme, as explained more fully in Chapter VI below, we must set the integrability conditions equal to zero. This gives the *Gauss equations*

$$(61) \qquad\qquad \gamma(B, B) = K.$$

It is easy to see that these equations always have solutions, and that the subset of $M^{(1)}$ defined by (61) and $B \neq 0$ is a smooth manifold Y. We therefore restrict the Pfaffian differential system (53) to Y; in effect this means that we impose the condition (61) and the exterior derivative of this equation. To compute the latter we have the easy

Lemma 2.8. $d\gamma(B, B) = 2\gamma(B, DB)$ *where DB is defined by* (55).

It follows that *on Y we must add to* (59) the additional symbol relation (here $\equiv$ denotes congruence modulo the system (53))

$$(62) \qquad\qquad \gamma(B, \pi) \equiv dK$$

where π is the $S^2 V^*$-valued 1-form $2DB$.

Definition 2.9. We shall call the restriction of (53) to Y, as defined by the Gauss equation (61) and $B \neq 0$, the *involutive prolongation* of the isometric embedding system (51).

This terminology will be justified in a moment. Remark that the involutive prolongation is given by the system (53) on Y where the structure equations are now

$$(63) \qquad
\begin{aligned}
\text{(i)} \quad & d\theta \equiv 0 \\
\text{(ii)} \quad & d\varphi \equiv 0 \\
\text{(iii)} \quad & d(\psi - \overline{\psi}) \equiv 0 \\
\text{(iv)} \quad & d(\eta - B\omega) \equiv -\pi \wedge \omega
\end{aligned}$$

with

$$\pi = \begin{bmatrix} \pi_{11} & \pi_{12} \\ \pi_{21} & \pi_{22} \end{bmatrix}$$

and where the symbol relations are

$$(64) \qquad \begin{aligned} &\text{(i)} \quad \pi_{12} - \pi_{21} \equiv 0 \\ &\text{(ii)} \quad \gamma(B, \pi) \equiv dK.^6 \end{aligned}$$

Summarizing, the procedure is this:

(i) Begin with the naive isometric system $(43) = (51)$. The tableau of this system is not involutive and so we must prolong. In effect, prolonging means that we equate the connection forms and add all candidates $a\omega_1^2 + 2b\omega_1\omega_2 + c\omega_2^2$ for the 2nd fundamental form as new variables. (ii) For the prolonged system $(47) = (53)$ the integrability condition is given by the Gauss equations (61). So it also fails to be involutive, and we adjoin the Gauss equations together with their exterior derivatives. *We shall now prove that the resulting system* (63) *is involutive.*

By the remarks above we may restrict our attention to the essential piece

$$\pi = \begin{bmatrix} \pi_{11} & \pi_{12} \\ \pi_{21} & \pi_{22} \end{bmatrix}$$

of the tableau, whose symbol relations (64) are

$$(65) \qquad \begin{aligned} &\text{(i)} \quad \pi_{12} - \pi_{21} \equiv 0 \\ &\text{(ii)} \quad a\pi_{22} - b\pi_{12} - b\pi_{21} + c\pi_{11} \equiv dK \end{aligned}$$

where $B = \begin{bmatrix} a & b \\ b & c \end{bmatrix}$. Remark that

$$(66) \qquad dK \equiv 0 \quad \mod \{\omega_1, \omega_2\}.$$

Since $B \neq 0$, by choosing a general basis for V we may assume that $a \neq 0$ (this will correspond to choosing a regular flag). It follows that π_{11}, π_{21}, may be assigned arbitrarily, and then π_{12} is determined by (i) in (65) and π_{22} by (ii) in (65). Moreover, *by* (66) *we may choose* π_{22} *to annihilate the integrability condition.* The characters are

$$s_1' = 2, \quad s_2' = 0.$$

Integral elements are given by linear equations

$$\pi_{ij} - p_{ijk}\omega_k = 0$$

---

[6]We recall our notation that $\equiv$ denotes congruence modulo the algebraic ideal generated by the 1-forms in (53).

where, upon setting $dK = K_j\omega_j$,

$$(\text{i}) \quad p_{ijk} = p_{ikj}$$
$$(\text{ii}) \quad p_{12j} = p_{21j}$$
$$(\text{ii}) \quad ap_{22j} - 2bp_{12j} + cp_{11j} = K_j.$$

From (i) and (ii) it follows that p_{ijk} is symmetric in all indices, and from (iii) it follows that p_{ijk} is determined by p_{111} and p_{112}. Thus the integral elements lying over a point of Y have dimension equal to

$$2 = s_1' + 2s_2',$$

and so by Cartan's test the system is involutive.

Finally, we want to compute the characteristic variety of the involutive prolongation of the isometric embedding system. We have noted above that if the tableau matrix looks like

$$\pi = \begin{bmatrix} 0 \\ \tilde{\pi} \end{bmatrix},$$

then as is easily verified the characteristic variety for π is the same as that for $\tilde{\pi}$. Thus we must compute the characteristic variety Ξ for a tableau matrix

$$\begin{bmatrix} \pi_{11} & \pi_{12} & 0 \\ \pi_{21} & \pi_{22} & 0 \end{bmatrix}$$

with symbol relations (65) where we have absorbed dK into π_{22}. Recall from (19) above that for, $y \in Y$, the Cauchy characteristics as reflected in the last column of zeros, give

$$\Xi_y \subset \mathbb{P}^1 \subset \mathbb{P}^2.$$

The determination of $\Xi_y \subset \mathbb{P}^1$ is consequently the same as that of determining the characteristic variety of the tableau matrix

$$\begin{bmatrix} \pi_{11} & \pi_{12} \\ \pi_{21} & \pi_{22} \end{bmatrix}$$

The symbol matrix at $\xi = [\xi_1, \xi_2] \in \mathbb{P}^1$ is

$$\sigma_\xi = \begin{bmatrix} \xi_2 & -\xi_1 \\ (c\xi_1 - b\xi_2) & (a\xi_2 - b\xi_1) \end{bmatrix}.$$

Then

$$\det \sigma_\xi = a(\xi_2)^2 - 2b\xi_1\xi_2 + c(\xi_1)^2$$

is a quadratic form with discriminant $\Delta(\sigma_\xi)$ given by

$$\Delta(\sigma_\xi) = ac - b^2 = K$$

where K is the Gaussian curvature. Thus:

(67)   *If $K(y) < 0$ then $\Xi_y \subset \mathbb{P}^2$ consists of two distinct points.*

If $K(y) = 0$ then Ξ_y consists of one point counted twice.

Finally, if $K(y) > 0$ then $\Xi_y = \emptyset$ but $\Xi_{\mathbb{C},y} \subset \mathbb{C}P^1$ consists

of a pair of distinct conjugate points.

Following the usual P.D.E. terminology we may say that in the cases $K < 0$, $K = 0$, $K > 0$ the involutive prolongation of the isometric embedding system is respectively *hyperbolic, parabolic and elliptic*.

For a surface $S \subset E^3$ with $K < 0$, at each point $p \in S$ the two characteristic lines in $T_p(S)$ are the *asymptotic directions*.

A striking fact is that *the isometric embedding system for $M^n \subset E^{n(n+1)/2}$, is never elliptic when $n \geq 3$* (cf. Bryant, Griffiths and Yang [1983] and the references cited there).

§3. Properties of the Characteristic Variety.

In this section we shall state a number of properties of the characteristic variety of a linear Pfaffian differential system $(\mathcal{I}, \Omega)$. The proofs of the more substantial of these will be given in Chapter VIII.

(i) **The 1st derived system and Ξ.** We consider a Pfaffian differential system given by a filtration $I \subset J \subset T^*X$ and with 1st derived system (cf. Chapter I)

$$I_1 = \ker\{\delta : I \to \Lambda^2 T^* M \quad \mathrm{mod} \ \{I\}\}.$$

For an adapted basis $\{\theta^1, \ldots, \theta^p; \theta^{p+1}, \ldots, \theta^{s_0}\} = \{\theta^\rho, \theta^\varepsilon\}$ for $I_1 \subset I$ (here $1 \leq \rho, \sigma \leq p$ and $p + 1 \leq \varepsilon, \delta \leq s_0$) we have

$$\text{(i)} \quad d\theta^\rho \equiv 0 \ \mathrm{mod} \ \{I\}$$
$$\text{(ii)} \quad d\theta^\varepsilon \equiv \pi_i^\varepsilon \wedge \omega^i \ \mathrm{mod} \ \{I\}.$$

Here we recall that $\{I\}$ is the *algebraic* ideal generated by the sections of I. The symbol relations are of the form

$$\text{(i)} \quad \pi_i^\rho \equiv 0 \ \mathrm{mod} \ J$$
$$\text{(ii)} \quad B_\varepsilon^{\lambda i} \pi_i^\varepsilon \equiv 0 \ \mathrm{mod} \ J,$$

and the tableau matrix is

$$\begin{bmatrix} 0 \\ \pi_i^\varepsilon \end{bmatrix}.$$

To put this in an intrinsic algebraic settting, a sub-bundle $I_1 \subset I$ gives a quotient dual bundle, i.e., we have

$$0 \to (I/I_1)^* \to I^* \to I_1^* \to 0.$$

The tableau corresponding to the above matrix is given by a sub-bundle $A \subset I^* \otimes L$ with the property that A projects to zero in $I_1^* \otimes L$, i.e.,

$$A \subset (I/I_1)^* \otimes L \subset I^* \otimes L.$$

For $0 \neq \xi \in L$ the symbol mapping

$$\sigma_\xi : I^* \to Q$$

restricts to

$$\sigma_{1,\xi} : (I/I_1)^* \to Q,$$

and it follows directly from the definitions that

$$\ker \sigma_\xi = \ker \sigma_{1,\xi}.$$

Thus the characteristic variety is the same as if we consider only the bottom non-zero block in the tableau matrix, i.e., we consider only the "smaller" symbol matrices

$$\|B_\varepsilon^{\lambda i} \xi_i\|.$$

Informally, we may rephrase this as:

(68)   *In computing the characteristic variety, we may*

             ignore the 1st derived system.

This property of characteristic varieties may be viewed as a generalization of the fact that the characteristic variety of a P.D.E. system depends only on its highest order terms.

As was discussed in §6 of Chapter IV, if we prolong a Pfaffian differential system the original system appears in the 1st derived system of its prolongation. We have already encountered this phenomenon in Example 2.7 above (compare (51), (53), and (54)), where property (68) was in fact used.

(ii) 2nd **order systems and** Ξ. We begin with a Pfaffian differential system that "looks like" the system associated to a 2nd order P.D.E. system. Rather than giving an involved intrinsic formulation of this we shall use indices. Thus we assume the system to be given by

$$\begin{aligned}
&\theta^a = 0 \\
&\theta^a_i = 0 \\
&d\theta^a \equiv 0 \ \mod \{I\} \\
&d\theta^a_i \equiv \pi^a_{ij} \wedge \omega^j \ \mod \{I\}
\end{aligned}$$

(69)

where $\{I\} = \{\theta^a, \theta^a_i\}$ and the symbol relations are

(70)
$$\begin{aligned}
&\text{(i)} \ \ \pi^a_{ij} \equiv \pi^a_{ji} \ \mod J \\
&\text{(ii)} \ \ B^{\lambda ij}_a \pi^a_{ij} \equiv 0 \ \mod J
\end{aligned}$$

and where in (ii) it is understood that $B^{\lambda ij}_a = B^{\lambda ji}_a$.

Example 3.1. We consider a 2nd order P.D.E. system

$$(71) \qquad F^\lambda \left(y^i, z^a, \frac{\partial z^a}{\partial y^i}, \frac{\partial^2 z^a}{\partial y^i \partial y^j} \right) = 0.$$

To set this up as a Pfaffian differential system we use the space $J^2(\mathbb{R}^n, \mathbb{R}^{s_0})$ of 2-jets of maps from $\mathbb{R}^n$ to $\mathbb{R}^{s_0}$, and on $J^2(\mathbb{R}^n, \mathbb{R}^{s_0})$ we use coordinates $(y^i, z^a, p^a_i, p^a_{ij})$ where $p^a_{ij} = p^a_{ji}$. In $J^2(\mathbb{R}^n, \mathbb{R}^{s_0})$ we consider an open subset M of smooth points on the locus

$$F^\lambda(y^i, z^a, p^a_i, p^a_{ij}) = 0.$$

Setting

$$\begin{aligned}
&\theta^a = dz^a - p^a_i dx^i|_M \\
&\theta^a_i = dp^a_i - p^a_{ij} dx^j|_M \\
&\pi^a_{ij} = -dp^a_{ij}|_M \\
&\omega^i = dx^i|_M
\end{aligned}$$

(this is just the restriction to M of the canonical system) and using

$$d\theta^a = -dp^a_i \wedge dx^i \equiv 0 \ \mod \{\theta^a_i\},$$

we see that the Pfaffian differential system corresponding to (71) has the form (69) with symbol relations (70) where

$$B^{\lambda ij}_a = \frac{\partial F^\lambda}{\partial p^{ij}_a}.$$

The usual symbol matrix of the P.D.E. system (71) is given by

$$(72) \qquad \|B_a^{\lambda ij}\xi_i\xi_j\|,$$

and we want to extend this to the Pfaffian differential system (69). More precisely, it is well known that the characteristic variety associated to the symbol matrix (72), in which ξ appears quadratically in each term, is the same as that obtained by writing (71) as a 1st order system and computing its symbol matrix, in which ξ appears linearly in each term. It is the differential system analogue of this that we wish to establish.

We remark that the condition that (69) be locally induced from a 2nd order P.D.E. system, i.e., that there should locally be an embedding $f : X \to J^2(\mathbb{R}^n, \mathbb{R}^{s_0})$ satisfying

$$\mathrm{span}\{\theta^a, \omega^i\} = \mathrm{span}\{f^*dx^a, f^*dy^i\}$$

$$\mathrm{span}\{\theta_i^a, \theta^a, \omega^i\} = \mathrm{span}\{f^*dp_i^a, f^*dz^a, f^*dy^i\},$$

is the Frobenius condition

$$d\theta^a \equiv 0 \mod \{\theta^a, \omega^i\}$$

$$d\omega^i \equiv 0 \mod \{\theta^a, \omega^i\}$$

(these plus (69) imply that $d\theta_i^a \equiv 0 \mod \{\theta^a, \theta_i^a, \omega^i\}$).

Example 3.2. The involutive prolongation (53) of the isometric embedding system for S in E^3 has the form (69) (where $\mathrm{span}\{\theta^a\} = \mathrm{span}\{\theta, \varphi\}$ and $\mathrm{span}\{\theta^a, \theta_i^a\} = \mathrm{span}\{\theta, \varphi, \psi - \overline{\psi}\}$), so that it "looks like" a 2nd order P.D.E. system. (This is certainly natural to expect, since the curvature is involved.) But it follows from (49) and (50) that it is *not* locally equivalent to such a system.

Returning to the general discussion, we want to determine the characteristic variety of the system (69). The tableau matrix has the block form

$$\pi = \begin{bmatrix} 0 \\ \pi_{ij}^a \end{bmatrix},$$

and by (68) we may ignore the block of zeros. For $\xi = \xi_i\omega^i$, the symbol matrix applied to a vector $w = \{w_i^a\}$ is

$$(73) \qquad \sigma_\xi(w) = \begin{pmatrix} w_i^a\xi_j - w_j^a\xi_i \\ B_a^{rij}w_i^a\xi_j \end{pmatrix}.$$

The two blocks of this column vector correspond to the blocks (i) and (ii) of symbol relations (70). If $\sigma_\xi(w) = 0$, then from

$$w_i^a\xi_j = w_j^a\xi_i, \quad \xi \neq 0,$$

we conclude that

$$w_i^a = w^a \xi_i$$

is a decomposable tensor. Using this the second block in (73) gives

$$(74) \qquad B_a^{\lambda ij} w^a \xi_i \xi_j = 0.$$

In other words, the symbol matrix (72) is singular. Conversely, if (74) holds, then for $w_i^a = w^a \xi_i$ we have $\sigma_\xi(w) = 0$. In conclusion:

(75)   *The characteristic variety for the Pfaffian differential*

           system (69) and (70) is the same as the characteristic

           variety formed from the symbol matrices (72).

Informally, we may say that if the tableau matrix of a Pfaffian differential system looks like the tableau matrix of a 2nd order P.D.E. system, then the characteristic variety according to Definition 2.1 above may be computed as one ordinarily would for 2nd order P.D.E. systems. Of course, this may be generalized to higher order systems.

Example 3.3. Referring to (64), the characteristic variety of the involutive prolongation of the isometric embedding system is given immediately by

$$a(\xi_2)^2 - 2b\xi_1\xi_2 + c(\xi_1)^2 = 0,$$

a result we arrived at there by a somewhat longer calculation following the original definition.

(iii) **Characteristic variety of the 1st prolongation.** We have briefly introduced the 1st prolongation in Chapter IV and will more fully discuss it in Chapter VI below. Here, we shall show by computation how the characteristic variety behaves under prolongation.

   We consider a linear Pfaffian differential system whose integrability conditions are assumed to be satisfied; thus there are integral elements over each point. Omitting reference to the independence condition, such a system may be assumed to be given by (13) and (16) above

$$\text{(i)} \quad \theta^a = 0$$

$$\text{(ii)} \quad d\theta^a \equiv \pi_i^a \wedge \omega^i \quad \mathrm{mod} \ \{I\}$$

$$\text{(iii)} \quad B_a^{\lambda i} \pi_i^a \equiv 0 \quad \mathrm{mod} \ \{I\}$$

where (iii) are a basis for the symbol relations. Writing these as modulo $\{I\} = \{\theta^b\}$ means that the torsion has been absorbed. Integral elements are given by linear equations

$$\pi_i^a - p_{ij}^a \omega^j = 0$$

where

$$p^a_{ij} = p^a_{ji}$$

$$B^{\lambda i}_a p^a_{ij} = 0.$$

The 1st prolongation is a differential system on the manifold $M^{(1)}$ obtained locally from M by adding the p^a_{ij}'s satisfying these conditions. On $M^{(1)}$ the 1st prolongation is the Pfaffian differential system $I^{(1)}$ generated by the equations

$$
\begin{aligned}
&\text{(i)} \quad \theta^a = 0 \\
\text{(76)} \qquad &\text{(ii)} \quad \theta^a_i = \pi^a_i - p^a_{ij}\omega^j = 0 \\
&\text{(iii)} \quad B^{\lambda i}_a p^a_{ij} = 0.
\end{aligned}
$$

The exterior derivatives of these equations give, using the original structure equations,

$$
\begin{aligned}
\text{(77)} \qquad & d\theta^a \equiv 0 \mod \{I^{(1)}\} \\
& d\theta^a_i \equiv \pi^a_{ij} \wedge \omega^j \mod \{I^{(1)}\}
\end{aligned}
$$

where locally $\{I^{(1)}\} = \{\theta^a, \theta^a_i\}$ and

$$\pi^a_{ij} = -dp^a_{ij} + (\text{horizontal forms relative to } M^{(1)} \to M).^7$$

Comparing (76) and (77) we see as before that the original Pfaffian differential system goes into the 1st derived system of its prolongation, and hence, by (68), when computing the characteristic variety of $\mathcal{I}^{(1)}$ we need only consider the tableau $\|\pi^a_{ij}\|$. Differentiating equation (iii) in (76) and using that

$$\left(\begin{array}{c} \text{horizontal forms} \\ \text{relative to } M^{(1)} \to M \end{array} \right) = \text{span}\{\theta^a, \omega^i, \pi^a_i\}^8$$

$$= \text{span}\{\theta^a, \pi^i, \theta^a_i\}$$

where span allows linear combinations with coefficients in $C^\infty(M^{(1)})$, it follows that the symbol relations on the π^a_{ij} are

$$
\begin{aligned}
\text{(78)} \qquad & \pi^a_{ij} \equiv \pi^a_{ji} \mod \{\theta^b, \theta^b_i, \omega^i\} \\
& B^{\lambda i}_a \pi^a_{ij} \equiv 0 \mod \{\theta^b, \theta^b_i, \omega^i\}.
\end{aligned}
$$

---

[7]A differential form $\varphi \in \Omega^q X$ is *horizontal* relative to a smooth mapping $f : X \to Y$ if $\varphi(x) \in \pi^* \wedge^q T^*_{f(x)} Y$ for all $x \in X$.

[8]This equality is valid if there are no Cauchy characteristics. The general case may be done by "foliating out" the Cauchy characteristics or by the more intrinsic argument used in §6 of Chapter IV to prove the same equations as (78) below—cf. equations (118) there.

Note that the 2nd set of relations is indexed by pairs (λ, j) of indices. Thus we should write these relations as

$$B_a^{(\lambda,j)ik}\pi_{ik}^a \equiv 0 \quad \mathrm{mod}\ \{\theta^b, \theta_i^b, \omega^i\}$$

where

$$B_a^{(\lambda,j)ik} = \delta_k^j B_a^{\lambda i}.$$

If $\xi = \{\xi_i\}$ is in the characteristic variety for $(\mathcal{I}^{(1)}, \Omega)$, then there exists $w = \{w_i^a\}$ satisfying

$$\text{(i)} \quad w_i^a \xi_j = w_j^a \xi_i$$

$$\text{(ii)} \quad B_a^{(\lambda,j)ik} w_i^a \xi_k = 0.$$

The reasoning here is now analogous to that used in establishing (75) just above. From (i) it follows that

$$w_i^a = w^a \xi_i,$$

and then from (ii) it follows that

$$B_a^{\lambda i} w^a \xi_i \xi_j = 0, \qquad 1 \le j \le n.$$

It follows that

$$B_a^{\lambda i} w^a \xi_i = 0,$$

which says that ξ is characteristic for $(\mathcal{I}, \Omega)$. Since the converse is obviously true, we have established that:

(79)   *Under the projection $\tilde\omega : M^{(1)} \to M$, there is a*

         natural isomorphism

$$\Xi_x^{(1)} \cong \Xi_{\tilde\omega(x)}$$

         between the fibre of the characteristic variety $\Xi^{(1)}$ for

         $(\mathcal{I}^{(1)}, \Omega)$ lying over $x \in M^{(1)}$ and the fibre of

         the characteristic variety Ξ for $(\mathcal{I}, \Omega)$ lying over

         $\tilde\omega(x) \in M$.

Informally, we may say that, in the absence of integrability conditions, the characteristic variety remains unchanged when we prolong. If there are integrability conditions, then they will contribute additional symbol relations to the prolonged system and the characteristic variety may get smaller—i.e., $\Xi_x^{(1)}$ may be a proper subvariety of $\Xi_{\tilde\omega(x)}$.

Remark. The question of whether the symbol relations (78) may be refined
to

$$\pi_{ij}^a \equiv \pi_{ji}^a \mod \{I^{(1)}\}$$

(78*bis*)

$$B_a^{\lambda i}\pi_{ij}^a \equiv 0 \mod \{I^{(1)}\},$$

or equivalently whether the absence of integrability conditions on M implies
the absence of integrability conditions on $M^{(1)}$, is an interesting one. In
general the answer is no; however, it will be proved in Chapter VI below
that
$$\left\{ \begin{array}{c} (\mathcal{I},\Omega) \text{ involutive} \\ \text{on } M \end{array} \right\} \Rightarrow \left\{ \begin{array}{c} (\mathcal{I}^{(1)},\Omega) \text{ involutive} \\ \text{on } M^{(1)} \end{array} \right\}.$$

The proof will show that the involutivity of the tableau of $(\mathcal{I},\Omega)$ implies
both that the tableau of $(\mathcal{I}^{(1)},\Omega)$ are involutive *and* that there are no
integrability conditions for $(\mathcal{I}^{(1)},\Omega)$, which is just (78*bis*). This argument
will be put in a conceptual framework in Chapter VIII when we discuss
Spencer cohomology.

(iv) Relationship between the characteristic variety and the Cartan–Kähler theorem.

The remaining properties of the characteristic variety are more substantial; they require involutivity and deal with the complex characteristic variety.

Let $(\mathcal{I},\Omega)$ be a linear Pfaffian differential system on a manifold M given
by a filtration $I \subset J \subset T^*M$. We give the structure equations of $(\mathcal{I},\Omega)$
in the form (13) above. For simplicity of exposition we assume that there
are no Cauchy characteristics, so that $\text{span}\{\theta^a, \omega^i, \pi_i^a\} = T^*M$. The statements of the results given below remain valid without this assumption.
From section 5 in Chapter IV we recall the tableau matrix given by equation (88) there

(80)
$$\pi = \left\| \begin{array}{ccc} \pi_1^1 & \cdots & \pi_n^1 \\ \vdots & & \vdots \\ \pi_1^{s_0} & \cdots & \pi_n^{s_0} \end{array} \right\| \mod J,$$

(we now omit the bars over the π_i^a's and understand that all 1-forms in π
are considered modulo J), and the reduced Cartan characters $s_1', s_2', \ldots, s_n'$
defined inductively by

(81) $$s_1' + \cdots + s_k' = \left\{ \begin{array}{l} \text{number of linearly independent} \\ \text{forms in the first } k\text{-columns of } \pi \end{array} \right\}.$$

Here we assume that the basis $\omega^1, \ldots, \omega^n$ for J/I is chosen generically. Also,
in reality (81) is defined at each point $x \in M$, and we assume that these

pointwise defined ranks are constant. We also recall that, in the absence of integrability conditions, the reduced characters are equal to the usual characters s_k—cf. equation (86) in Chapter IV.

Definition 3.4. The *character l* and *Cartan integer κ* of $(\mathcal{I}, \Omega)$ are defined by

$$\begin{cases} s'_1, \ldots, s'_l \neq 0, \quad s'_{l+1} = \cdots = s'_n = 0 \\ \kappa = s'_l. \end{cases}$$

As will be seen below, both the character and Cartan integer are invariant under prolongation so long as the system has no integrability conditions.

Now suppose that the system is involutive and real analytic. According to the Cartan–Kähler theorem we may construct local integral manifolds for $(\mathcal{I}, \Omega)$ by solving a succession of initial value or Cauchy problems. Such a succession of initial value problems corresponds to nested sequence of integral manifolds

$$\begin{cases} N^0 \subset N^1 \subset \cdots \subset N^{n-1} \subset N \\ \dim N^p = p \end{cases}$$

whose tangent spaces form a regular flag. From Chapter III we recall that:

N is uniquely determined by N_l and N_l is uniquely obtained from

N_{l-1} by prescribing "κ arbitrary functions of l variables".

Thus we may think of (l, κ) as telling us something about how many local integral manifolds there are. To an algebraic geometer, l resembles a transcendence degree and κ a field extension degree—this analogy will turn out to be precise. We will state results that express l, κ and the condition to be a regular flag in terms of algebro-geometric properties of the complex characteristic variety $\Xi_{\mathbb{C}} \subset \mathbb{P}L_{\mathbb{C}}$. In practice these theorems usually allow us to determine l, κ and regular flags without going through the sometimes laborious procedure of calculating the s'_i. This will be illustrated by examples.

Assume that $(\mathcal{I}, \Omega)$ is involutive and let

$$\sigma : I^*_{\mathbb{C}} \otimes L_{\mathbb{C}} \to Q_{\mathbb{C}}$$

be the complexified symbol map for $(\mathcal{I}, \Omega)$. For each $x \in M$ and $0 \neq \xi \in L_{\mathbb{C},x}$ we have

$$\sigma_x : I^*_{\mathbb{C},x} \otimes L_{\mathbb{C},x} \to Q_{\mathbb{C},x}$$

$$\sigma_{x,\xi} : I^*_{\mathbb{C},x} \to Q_{\mathbb{C},x}$$

where for $w \in I^*_{\mathbb{C},x}$

$$\sigma_{x,\xi}(w) = \sigma_x(w \otimes \xi).$$

By definition

$$\Xi_{\mathbb{C},x} = ([\xi] \in \mathbb{P}L_{\mathbb{C},x} : \dim \ker \sigma_{x,\xi} \geq 1\}.$$

It is clear that $\Xi_{\mathbb{C},x}$ is a complex algebraic variety, in fact, the ideal of $\Xi_{\mathbb{C},x}$ is by definition the homogeneous ideal generated by suitable minors of the symbol matrix

$$\|B_a^{\lambda i}(x)\xi_i\|.$$

Assuming first for simplicity that $\Xi_{\mathbb{C},x}$ is irreducible we set

$$d = \dim \Xi_{\mathbb{C},x}$$

$$\delta = \deg \Xi_{\mathbb{C},x}$$

$$\mu = \dim \ker \sigma_{x,\xi} \text{ where } [\xi] \in \Xi_{\mathbb{C},x} \text{ is a general point.}$$

As a consequence of the involutivity of $(\mathcal{I}, \Omega)$ we will see that d, δ, and μ are independent of $x \in M$. In general, if

$$\Xi_{\mathbb{C},x} = \bigcup_{\alpha} \Xi_{\mathbb{C},x}^{(\alpha)}$$

is the unique decomposition of $\Xi_{\mathbb{C},x}$ into irreducible components we set

$$d = \max_{\alpha} \dim \Xi_{\mathbb{C},x}^{(\alpha)}$$

$$\delta = {\sum}' \delta^{(\alpha)}(x)$$

$$\mu(x) = {\sum}' \mu^{(\alpha)}(x)$$

where $\delta^{(\alpha)}(x) = \deg \Xi_{\mathbb{C},x}^{(\alpha)}$, $\mu^{(\alpha)}(x) = \dim \ker \sigma_{x,\xi}(\alpha)$ where $[\xi^{(\alpha)}] \in \Xi_{\mathbb{C},x}^{(\alpha)}$ is a general point, and ${\sum}'$ denotes the sum over components of maximal dimension. Again, d, δ, and ${\sum}' \mu^{(\alpha)}(x)\delta^{(\alpha)}(x)$ will be independent of x.

Example 3.5. For the involutive prolongation of the isometric embedding system we have

$$\sigma_{x,\xi} = \left\|\begin{matrix} \xi_2 & -\xi_1 \\ c\xi_1 - b\xi_2 & a\xi_2 - b\xi_1 \end{matrix}\right\|$$

where

$$ac - b^2 = K(x)$$

is the Gaussian curvature. If $K(x) \neq 0$ then $\Xi_{\mathbb{C},x}$ consists of two distinct points with each having

$$\delta^{(\alpha)}(x) = 1, \quad \mu^{(\alpha)}(x) = 1.$$

If $K(x) = 0$ then $\Xi_{\mathbb{C},x}$ consists of one point with $\delta = 2$ and $\mu = 1$. Note that

$$(82) \quad \begin{cases} \mu^{(1)}(x)\delta^{(1)}(x) + \mu^{(2)}(x)\delta^{(2)}(x) = 2, & K(x) \neq 0 \\ \mu(x)\delta(x) = 2, & K(x) = 0 \end{cases}$$

in accordance with the above remark. Indeed, if $K(x) \neq 0$ then over $\mathbb{C}$ we may assume that the 2nd fundamental form is

$$\left\| \begin{matrix} a & 0 \\ 0 & c \end{matrix} \right\|, \quad ac = K \neq 0,$$

in which case

$$\sigma_{x,\xi} = \left\| \begin{matrix} \xi_2 & -\xi_1 \\ c\xi_1 & a\xi_2 \end{matrix} \right\|$$

$$\Xi_{\mathbb{C},x} = \left[i, \sqrt{\frac{c}{a}} \right] \cup \left[i, -\sqrt{\frac{c}{a}} \right].$$

For each point of $\Xi_{\mathbb{C},x}$ clearly this matrix has rank one. If $K(x) = 0$ then we may assume that the 2nd fundamental form is

$$\left\| \begin{matrix} a & 0 \\ 0 & 0 \end{matrix} \right\|, \quad a \neq 0.$$

In this case, $\Xi_{\mathbb{C},x} = [1,0]$ counted with multiplicity 2 and at this point

$$\sigma_{x,\xi} = \left\| \begin{matrix} 0 & -1 \\ 0 & 0 \end{matrix} \right\|$$

has rank one. By our assumption of involutivity the 2nd fundamental form can never be zero.

Our result is the following

Theorem 3.6. *Let $(\mathcal{I}, \Omega)$ be an involutive Pfaffian system of character l and having Cartan integer κ. Then*

$$\begin{cases} l = d + 1 \\ \kappa = \sum{}' \mu^{(\alpha)}(x)\delta^{(\alpha)}(x). \end{cases}$$

Corollary 3.7. *Suppose that all $\mu^{(\alpha)}(x) = 1$ (this is frequently the case). Then*

$$\begin{cases} l = \dim \Xi_{\mathbb{C},x} + 1 \\ \kappa = \deg \Xi_{\mathbb{C},x}. \end{cases}$$

If we omit reference to the particular point $x \in X$ we may rephrase this as:

(83)   *The integral manifolds of an involutive, real analytic Pfaffian-differential system locally depend on $\deg \Xi_{\mathbb{C}}$ arbitrary functions of $\dim \Xi_{\mathbb{C}} + 1$ variables.*

Example 3.8. Referring to (82) above, the local isometric embeddings of a real analytic surface in E^3 depend on two arbitrary functions of one variable. We want to explain this, and the result we shall find is essentially: *To locally isometrically embed $\overline{S}$ in E^3, we choose a connected curve $\overline{\gamma} \subset \overline{S}$. Then the isometric embeddings*

$$\overline{\gamma} \to \gamma \subset E^3$$

depend on two functions of one variable, and such an embedding extends essentially uniquely to a local isometric embedding

$$\overline{S} \to S \subset E^3$$

provided that γ is suitably general.

We let $(\mathcal{I}, \Omega)$ on the manifold M denote the involutive prolongation of the isometric embedding system. Then

$$M \subset \overline{P} \times \mathcal{F}(E^3) \times S^2 V^*.$$

There is an essentially unique lifting of $\overline{\gamma} \subset \overline{S}$ to $\overline{P}$ given by

$$s \to (y(s), \overline{e}_1(s), \overline{e}_2(s))$$

where $\overline{\gamma}$ is given by $s \to y(s)$ and $\overline{e}_1(s)$ is the unit tangent to $\overline{\gamma}$ (s is an arclength parameter). To be 'essentially unique' will mean that the lifting is unique once we have specified it at one point. The connection form

$$\overline{\omega}_{12} = k_g(s)ds,$$

where k_g is the geodesic curvature of $\overline{\gamma}$ in $\overline{S}$. Let $\overline{\gamma} \to \gamma$ be given by

$$s \to x(s) \in E^3$$

and set $e_1(s) = dx(s)/ds$, which is a unit vector. This embedding locally depends on two functions of one variable. We claim that there are essentially unique vectors $e_2(s), e_3(s)$ and functions $a(s), b(s)$ such that

$$\text{(i)} \quad \frac{de_1}{ds} = k_g e_2 + a e_3$$

$$\text{(ii)} \quad \frac{de_2}{ds} = -k_g e_1 + b e_3$$

$$\text{(iii)} \quad \frac{de_3}{ds} = -a e_1 - b e_2.$$

To see this we note that

$$\left| \frac{de_1}{ds} \right|^2 = k_g^2 + a^2 = \kappa^2$$

where $\kappa(s)$ is the curvature of γ in E^3. Solving this gives $a(s)$ up to ± 1. Referring to (i) above, we may find unique unit vectors $e_2(s)$ and $e_3(s)$ in $e_1(s)^\perp$ such that de_1/ds is the hypothenuse of the right triangle with sides $k_g e_2$ and $a e_3$. Having determined e_2 we may determine b up to ± 1 from the length of de_2/ds. Having determined $a(s)$ and $b(s)$, we may determine $c(s)$ by the Gauss equation

$$a(s)c(s) - b(s)^2 = K(y(s)).$$

In this way, given $\overline{\gamma} \to \gamma$ we have determined a frame $e_1(s), e_2(s), e_3(s)$ and 2nd fundamental form $a(s)\omega_1^2 + 2b(s)\omega_1\omega_2 + c(s)\omega_2^2$ along γ, i.e. we have a 1-dimensional integral manifold $N^1 \subset M$ lying over the graph of $\overline{\gamma} \to \gamma$. Since $s_2 = 0$ this integral manifold extends locally to a unique integral manifold N^2 of $(\mathcal{I}, \Omega)$.

Example 3.9. We consider a single linear P.D.E.

$$\text{(84)} \qquad\qquad P(x, D)u = 0$$

of order m with one unknown function. Here we use the standard notations

$$P(x, D) = \sum_{|\alpha| \le m} P_\alpha(x) D_x^\alpha$$

$$D_x^\alpha = (\partial/\partial x^1)^{\alpha_1} \ldots (\partial/\partial x^n)^{\alpha_n}$$

$$|\alpha| = \alpha_1 + \cdots + \alpha_n.$$

As follows from the discussion above, when we write (84) as a Pfaffian differential system and compute its characteristic variety we get the expected answer

$$\Xi_{\mathbb{C}} = \{(x, \xi) : P_m(x, \xi) = \sum_{|\alpha|=m} P_\alpha(x)\xi^\alpha = 0\}.$$

This system (84) also turns out to be involutive (this is a nice exercise using Cartan's test). Since clearly also all $\mu^{(\alpha)}(x) = 1$ (the symbol is a 1×1 matrix) it follows from (83) that, in the real analytic case, the solutions to (84) depend on m arbitrary functions of $n-1$ variables. These are just the values of u and its 1st $m-1$ normal derivatives along a non-characteristic hypersurface. In this case the result is a well known consequence of the Cauchy–Kowaleski theorem.

Example 3.10. We reconsider the Cauchy–Riemann system given by (26) above with symbol matrix (27) and complex characteristic variety (28) there. Then $\delta^{(1)} = \delta^{(2)} = \mu^{(1)} = \mu^{(2)} = 1$,

$$\begin{cases} d = m - 1 \\ \kappa = 2 \end{cases}$$

and (83) is in accordance with the well-known fact that holomorphic functions in an open set $U \subset \mathbb{C}^m$ depend on two real functions of m real variables (think of locally extending a complex-valued real analytic function from $\mathbb{R}^m$ to $\mathbb{C}^m$).

As another consequence of Theorem 3.6 we have the following:

Corollary 3.11. *Let $(\mathcal{I}, \Omega)$ be a C^∞ involutive linear Pfaffian differential system whose complex characteristic variety is empty, i.e.,*

$$\Xi_{\mathbb{C}} = \emptyset.$$

Then $\mathcal{I}$ is completely integrable. In particular, through each point of M there passes a unique integral manifold of $\mathcal{I}$.

Proof. By Theorem 3.6 we have $s_1 = \cdots = s_n = 0$, and since there are no integrability conditions the structure equations are

$$d\theta^a \equiv 0 \quad \mod \{\theta^a\}.$$

$\square$

There is another consequence of this corollary. Because of its many uses we state the result as a theorem.

Theorem 3.12. *Let $(\mathcal{I}, \Omega)$ be a C^∞ exterior differential system and assume that:*
   (i) the complex characteristic variety is empty,
   (ii) (technical assumption) the process of prolongation makes sense (i.e., at each stage we get a locally finite union of manifolds; this is automatic in the real analytic case). Then a prolongation of $(\mathcal{I}, \Omega)$ is either empty or is

a Frobenius system. In particular, for a suitable q each connected integral manifold of $(\mathcal{I}, \Omega)$ is uniquely determined by its q-jet at one point.

Informally, we may say that in case $\Xi_{\mathbb{C}} = \emptyset$ the integral manifolds of $(\mathcal{I}, \Omega)$ depend on a finite number of constants.

Proof. We will prove in Chapter VI that a suitable prolongation $(\mathcal{I}^{(q)}, \Omega)$ of $(\mathcal{I}, \Omega)$ is either empty or involutive. Since the integral manifolds of $(\mathcal{I}, \Omega)$ and $(\mathcal{I}^{(q)}, \Omega)$ are locally in one-to-one correspondence, we may restrict to the latter case. From the remarks following (79) above we infer that the complex characteristic variety of $(\mathcal{I}^{(q)}, \Omega)$ is empty. By Corollary 3.11, $(\mathcal{I}^{(q)}, \Omega)$ is a Frobenius system. Its integral manifolds are uniquely specified by constants; through each point of $M^{(q)}$ there is a unique connected integral manifold. This translates into the assertion that connected integral manifolds of $(\mathcal{I}, \Omega)$ are uniquely determined by their q-jets at one point.

Example 3.13. We shall give an example of the finiteness Theorem 3.12 for a linear P.D.E. system that arose initially in algebraic geometry. This example is for illustrative purposes and will not be referred to elsewhere in the book.

Let $E \to M$ be a vector bundle over a manifold, $\mathcal{E}$ the sheaf of C^∞ sections of E, and $\Theta \subset \mathcal{E}$ a subsheaf. We ask for another vector bundle $F \to M$ and linear differential operator

$$(85) \qquad\qquad D : \mathcal{E} \to \mathcal{F}$$

whose kernel is Θ. A first candidate for F may be obtained as follows: For each k let $J^k(E) \to M$ be the bundle of k-jets of sections of $E \to M$, and denote by $\mathcal{J}^k(E)$ the sheaf of C^∞ sections of this jet bundle. There is the *universal k^{th} order differential operator*

$$j_k : \mathcal{E} \to \mathcal{J}^k(E)$$

that sends a section s of $\mathcal{E}$ to its k-jet $j_k(s)$ (in local coordinates, $j_k(s)(x)$ is the Taylor series up through order k of s at the point x). We then consider the images $j_k(\Theta) \subset \mathcal{J}^k(E)$ of k-jets of sections of Θ, and we assume that the values $j_k(s)(x)$ ($s \in \Theta$ and $x \in M$) form a sub-bundle $J^k(\Theta)$ of $J^k(E)$. For some k for which $J^k(\Theta) \neq J^k(E)$ we set $F = J^k(E)/J^k(\Theta)$ and consider the kth order operator

$$D : \mathcal{E} \to \mathcal{F}$$

defined by

$$(Ds)(x) = \pi j_k(s)(x)$$

where $\pi : J^k(E) \to F$ is the projection. Clearly, $\Theta \subset \ker D$; in this way we obtain candidates for linear differential operators (85) whose solution sheaf is Θ. (It is also clear that *any* solution to this problem must essentially be of this form.)

Now let $G = G(k, V)$ be the Grassmann manifold of k-planes in a vector space V (which may be real or complex; it doesn't matter). Over G we have the universal sub-bundle $S \to G$, the trivial bundle $\tilde{V} = G \times V$, and the universal quotient bundle $Q \to G$, all fitting in the standard exact sequence

$$(86) \qquad\qquad 0 \to S \to \tilde{V} \to Q \to 0.$$

The *constant sections* V of $\tilde{V} \to G$ project to give a subsheaf $V \subset \mathcal{Q}$, and we ask for a linear differential operator

$$(87) \qquad\qquad D : \mathcal{Q} \to \mathcal{R}$$

whose kernel is V (here we abuse notation by identifying V with a subsheaf of $\mathcal{Q}$). More generally, for any submanifold $M \subset G$ we may restrict (86) to M and ask the same question. It is this latter situation that arose in algebraic geometry. What we shall do here is:

i) for any $M \subset G$ define a linear, 1st order operator (87) such that $V \subset \ker D$;

ii) determine geometric conditions on M such that the complex characteristic variety $\Xi_{\mathbb{C}}$ of D is empty. By Theorem 3.12 this implies that: *Over any open subset $U \subset M$, $\ker D$ is a finite-dimensional subspace of $H^0(U, \mathcal{E})$;* and

iii) show that if $M = G$, then over any open subset $U \subset G$, $\ker D = V$.

We note that since the symbol of D will not be identically zero, *the linear P.D.E. system $Ds = 0$ cannot be involutive* (this is a consequence of Corollary 3.11 above).

We begin with some notation. Given $M \subset G$ and a point $x \in M$ we denote by $S_x \subset V$ the corresponding k-plane and set $Q_x = V/S_x$. Since the question is local we may choose a frame $e_1(x), \ldots, e_N(x)$ $(N = \dim V)$ for $\tilde{V} \to N$ such that $e_1(x), \ldots, e_k(x)$ is a frame for $S \to M$. Using the range of indices

$$\begin{cases} 1 \le \alpha, \beta, \gamma \le k, \\ k+1 \le \mu, \nu \le N \end{cases}$$

we set

$$(88) \qquad \begin{cases} de_\alpha = \omega_\alpha^\beta e_\beta + \omega_\alpha^\mu e_\mu \\ de_\mu = \omega_\mu^\alpha e_\alpha + \omega_\mu^\nu e_\nu. \end{cases}$$

For each $x \in M$ the map

$$T_x M \to \mathrm{Hom}(S_x, Q_x) = S_x^* \otimes Q_x$$

given for $v \in T_x M$ by

$$v \to \langle \omega_\alpha^\mu(x), v \rangle e_\alpha^*(x) \otimes e_\mu(x)$$

is well-defined and gives an inclusion

$$(89) \qquad\qquad TM \subset \mathrm{Hom}(S, Q).$$

When M is an open set on G this inclusion is an equality and gives the well known identification $TG \cong \mathrm{Hom}(S, Q)$.

We now fix a point $x_0 \in M$ and set $T = T_{x_0} M$, $S = S_{x_0}$, $Q = Q_{x_0}$; we thus have a subspace

$$T \subset \mathrm{Hom}(S, Q).$$

For this subspace we consider the following two conditions:

$$(90) \qquad\qquad \bigcap_{\varphi \in T} \ker \varphi = (0);$$

For any hyperplane $H \subset T$

$$(91) \qquad\qquad \bigcap_{\varphi \in H} \ker \varphi = (0).$$

If the condition (90) is satisfied at each point then we shall prove that $J^1(V) \subset J^1(Q)$ *is a proper sub-bundle.* We then set $R = J^1(Q)/J^1(V)$ and define

$$(92) \qquad\qquad D : \mathcal{Q} \to \mathcal{R}$$

by the above procedure ($D = \pi D_1$ where $\pi : J^1(Q) \to R$ is the projection). This is a linear, 1st order differential operator and we shall prove that:

$$(93) \qquad \textit{If (91) is satisfied, then the complex characteristic}$$
$$\textit{variety } \Xi_{\mathbb{C}} \textit{ of } D \textit{ is empty.}$$

As noted above, this implies that locally $\ker D$ is a finite-dimensional vector space.

$$(94) \qquad \textit{If } \dim V \geq k + 2 \textit{ and } M \textit{ is an open set of } G \textit{ then (90)}$$
$$\textit{and (91) are satisfied, and in fact } \ker D = V.$$

The dimension restriction simply means that G is not a projective space.

To get some feeling for the conditions (90) and (91) we set $\dim V = N$ and assume that

$$k \leq N - k;$$

i.e., $\dim S \leq \dim Q$. Then it is easy to see that: *Condition (90) is generic if* $\dim M \geq 1$. *Condition (91) is generic if either*

$$\begin{cases} k \leq N - k - 1 \text{ and } \dim M \geq 2, \text{ or} \\ k = N - k \text{ and } \dim M \geq 3. \end{cases}$$

We are now ready to compute. If $v \in V$ gives a constant section of $\mathcal{Q}$, then writing $v = v^\alpha e_\alpha + v^\mu e_\mu$ we have

$$0 = dv \equiv (dv^\mu + v^\lambda \omega_\lambda^\mu + v^\alpha \omega_\alpha^\mu) e_\mu \quad \mathrm{mod}\ \{e_\alpha\}$$

which implies that

$$(95) \qquad\qquad dv^\mu + v^\lambda \omega_\lambda^\mu + v^\alpha \omega_\alpha^\mu = 0.$$

Now the map

$$(96) \qquad\qquad V \to J^0(Q)_x = Q_x$$

is obviously *surjective* for each $x \in M$. We shall show that if (90) is satisfied then the map

$$V \to J^1(Q)_x$$

is *injective* for each $x \in M$ (in fact, this is equivalent to (90)). For this we may assume that $v \in S_x$ (i.e., $v = 0$ in Q_x), so that all $v^\lambda(x) = 0$. Then (95) gives

$$(97) \qquad\qquad dv^\mu(x) = -v^\alpha(x) \omega_\alpha^\mu(x).$$

Now in the sequence

$$0 \to T_x^* M \otimes Q_x \to J^1(Q)_x \xrightarrow{\rho} J^0(Q)_x \to 0,$$

$\rho(v) = 0$, and so the 1-jet $j_1 v(x)$ is given by $dv^\mu(x) e_\mu(x) \in T_x^* M \otimes Q_x$. By (97) this is zero if, and only if,

$$v^\alpha(x) \omega_\alpha^\mu(x) = 0$$

for all μ. Clearly this is just a reformulation of (90).

Next we want to compute the symbol σ of the operator (92). Working over a fixed point $x_0 \in M$ and using our above notations, the symbol is a map

$$(98) \qquad \sigma : T^* \otimes Q \to R$$

where R is the fibre of $J^1(Q)/J^1(V)$ over x_0. To identify R, we note that by the surjectivity of (96) it is a quotient space of $T^* \otimes Q$. Denoting by

$$j : T \subset S^* \otimes Q$$

the inclusion (89) over x_0, we define a linear mapping $\lambda : S \to T^* \otimes Q$ by the commutative diagram

$$
\begin{array}{ccc}
S & \xrightarrow{\ \lambda\ } & T^* \otimes Q \\
\mathrm{id}_S \otimes I \searrow & & \swarrow j^* \otimes \mathrm{id}_Q \\
& S \otimes Q^* \otimes Q &
\end{array}
$$

where $I \in Q^* \otimes Q = \mathrm{Hom}(Q, Q)$ is the identity. Our main observation is the following consequence of (95):

$$R = T^* \otimes Q/\lambda(S).$$

Thus the symbol (98) is just the quotient mapping

$$T^* \otimes Q \to T^* \otimes Q/\lambda(S).$$

The characteristic variety is non-empty if, and only if, we have

$$(99) \qquad \lambda(s) = \eta \otimes q$$

for some $0 \neq s \in S$, $\eta \in T^*$, $q \in Q$. Setting

$$H = \eta^{\perp} \cap T,$$

(99) is easily seen to be equivalent to

$$s \in \bigcap_{\varphi \in H} \ker(j(\varphi)).$$

Comparing with (91) we obtain a proof of (93).
We will now prove (94). Let

$$q = v^{\mu} e_{\mu}$$

be a section of Q over an open set in G and assume that $Dq = 0$. By definition this means that there exists $s = v^\alpha e_\alpha$ such that, for each μ,

$$(100) \qquad dv^\mu + v^\lambda \omega_\lambda^\mu + v^\alpha \omega_\alpha^\mu = 0.$$

The exterior derivatives of (88) give

$$d\omega_i^j = \omega_i^k \wedge \omega_k^j, \qquad 1 \le k, j, k \le N.$$

Using this the exterior derivative of (100) gives

$$dv^\lambda \wedge \omega_\lambda^\mu + v^\lambda \omega_\lambda^\alpha \wedge \omega_\alpha^\mu + v^\lambda \omega_\lambda^\nu \wedge \omega_\nu^\mu$$
$$+ dv^\alpha \wedge \omega_\alpha^\mu + v^\alpha \omega_\alpha^\beta \wedge \omega_\beta^\mu + v^\alpha \omega_\alpha^\lambda \wedge \omega_\alpha^\mu = 0.$$

Plugging (100) into this equation several cancellations occur and it becomes

$$(dv^\alpha + v^\lambda \omega_\lambda^\alpha + v^\beta \omega_\beta^\alpha) \wedge \omega_\alpha^\mu = 0$$

for $\mu = k+1, \ldots, N$. Now all the forms ω_α^μ are linearly independent (they give a local coframe for G), and the Cartan lemma implies that

$$dv^\alpha + v^\lambda \omega_\lambda^\alpha + v^\beta \omega_\beta^\alpha \in \mathrm{span}\{\omega_1^\mu, \ldots, \omega_k^\mu\}$$

for *each* $\mu = k+1, \ldots, N$. If $N \ge k+2$ then we may choose $\mu \ne \nu$ and use

$$\mathrm{span}\{\omega_1^\mu, \ldots, \omega_k^\mu\} \cap \mathrm{span}\{\omega_1^\nu, \ldots, \omega_k^\nu\} = (0)$$

to conclude that

$$(101) \qquad dv^\alpha + v^\lambda \omega_\lambda^\alpha + v^\beta \omega_\beta^\alpha = 0.$$

For the V-valued function

$$v = v^\alpha e_\alpha + v^\mu e_\mu$$

(100) and (101) imply that $dv = 0$. $\qquad\qquad\qquad\qquad\qquad\square$

The proof shows that if $N \ge k+2$ and $M \subset G$ is a generic submanifold with

$$\dim M \ge 2k,$$

then $\ker D = V$. (Note that $\dim G = k(N - k)$ so that M must be an open set on G if $N = k+2$.) Unfortunately, for the cases that arise in algebraic geometry we have that approximately $\dim M = k$, so that prolongation is necessary to decide if $\mathrm{Ker}\, D = V$.

Returning to the general discussion, in section 2 above we have defined the sub-bundle

$$S \subset L^*$$

to be the image of the Cauchy characteristics and have proved (cf. equation (19) there) that

$$\Xi_{\mathbb{C}} \subset \mathbb{P}S_{\mathbb{C}}^{\perp}.$$

In the involutive case there is a converse. To explain it, for each subset $\Sigma \subset \mathbb{P}L_{\mathbb{C}}$ whose fibres Σ_x are algebraic subvarieties of $\mathbb{P}L_{x,\mathbb{C}} \cong \mathbb{P}^{n-1}$, we define the *span* of Σ_x to be

$$\{\Sigma_x\} = \bigcap (\text{linear spaces containing } \Sigma_x)$$

and set

$$\{\Sigma\} = \bigcup_{x \in M} \{\Sigma_x\}.$$

Theorem 3.13. *In case* $(\mathcal{I}, \Omega)$ *is involutive, we have*

$$S_{\mathbb{C}}^{\perp} = \{\Xi_{\mathbb{C}}\}.$$

In the extreme case when $\Xi_{\mathbb{C}}$ *is empty, this gives* $S_{\mathbb{C}}^{\perp} = L_{\mathbb{C}}$ *so that* $(\mathcal{I}, \Omega)$ *is a Frobenius system, which is Corollary 3.11 above.*

We shall not prove this result in this book.

(v) The characteristic variety and K-singular integral elements.

As we have defined it, the characteristic variety essentially has to do with characteristic, or singular, hyperplanes in n-dimensional integral elements. On the other hand, if $(\mathcal{I}, \Omega)$ is an involutive Pfaffian differential system of character l then the uniqueness of extensions in the Cauchy problem for n-dimensional integral manifolds occurs along l-dimensional submanifolds. Thus, we may expect that the characteristic variety should consist of singular l-dimensional integral elements. In other words, when $l < n - 1$ (i.e., roughly speaking when we are in the *overdetermined case*) there are two possible characteristic varieties, and it is obviously important to relate them.

To explain this we recall that $G_p(\mathcal{I}) \subset G_p(TM)$ denotes the set of p-dimensional integral elements of a differential ideal $\mathcal{I}$ on manifold M, and that the rank of the polar equations at $(x, E) \in G_p(\mathcal{I})$ is denoted by $\rho(E)$. Next, we recall that (i) $(x, E) \in G_p(\mathcal{I})$ is K-*regular* if near (x, E) the set $G_p(\mathcal{I})$ is a manifold with defining equations

$$\varphi|_{E'} = 0, \qquad (x', E') \in G_p(TM)$$

for $\varphi \in \mathcal{I}$, and (ii) $\rho(E')$ is constant near (x, E). If (x, E) is not K-regular then it is said to be K-*singular*.

Now suppose that $(\mathcal{I}, \Omega)$ is a linear Pfaffian differential system given by a filtration $I \subset J \subset T^*M$ and with characteristic variety

$$\Xi \subset \mathbb{P}L$$

where $L = J/I$ with rank $L = n$. For any n-dimensional integral manifold $N \subset M$, the restriction $J \to T^*(N)$ induces isomorphisms

$$L|_N \cong T^*N$$

$$\mathbb{P}L|_N \cong G_{n-1}(TN),$$

so that the complex characteristic variety $\Xi_{\mathbb{C}}$ maps to a set of hyperplanes in the tangent spaces to N that we think of as giving characteristic hyperplanes for a determined Cauchy problem posed along hypersurfaces in N. On the other hand, if $(\mathcal{I}, \Omega)$ has character l then, from the proof of the Cartan–Kähler theorem, there is a sequence of uniquely determined Cauchy problems beginning with one posed along general l-dimensional submanifolds of N. The characteristics for the first of these problems should appear in $G_l(TN)$, which leads us to look for some sort of characteristic variety in $G_l(L^*) \cong G_{n-l}(L)$. In fact, for each p with $1 \leq p \leq n-1$ we shall now define

$$\Lambda_p \subset G_p(L^*)$$

with the properties that $\Lambda_{n-1} = \Xi$ and that $\Lambda_l = \Lambda$ is the characteristic variety of primary interest.

Using the identification

$$G_p(L^*) \cong G_{n-p}(L)$$

given by sending a p-plane E in an n-dimensional vector space to its $(n-p)$-dimensional annihilator $E^\perp$ in the dual space, for each $(x, E) \in G_p(L^*)$ we choose our basis $\omega^1, \ldots, \omega^n$ for L so that $\omega^{p+1}(x), \ldots, \omega^n(x)$ gives a basis for $E^\perp$. Consider now the tableau matrix

$$\pi(x) = \begin{bmatrix} \pi_1^1(x) & \cdots & \pi_n^1(x) \\ \vdots & & \vdots \\ \pi_1^{s_0}(x) & \cdots & \pi_n^{s_0}(x) \end{bmatrix} \mod J_x,$$

and denote by $\sigma(E)$ the number of 1-forms in the first p columns of $\pi(x)$ that are linearly independent $\mod J_x$. Under a substitution

$$\tilde{\omega}^\lambda = \omega^\lambda \qquad\qquad p+1 \leq \lambda \leq n$$

$$\tilde{\omega}^\rho = \omega^\rho + A_\lambda^\rho \omega^\lambda \qquad 1 \leq \rho \leq p$$

we have

$$\tilde{\pi}^a_{\ \rho} = \pi^a_{\ \rho}$$

$$\tilde{\pi}^a_{\ \lambda} = \pi^a_{\ \lambda} + A^\rho_\lambda \pi^a_{\ \rho},$$

from which it follows that $\sigma(E)$ is well-defined.

Definition 3.14. i) We define $\Lambda_p \subset G_p(L^*)$ by

$$\Lambda_p = \{E : \sigma(E) < s'_1 + \cdots + s'_p\}$$

ii) If $(\mathcal{I}, \Omega)$ has character l, then we define the *Cartan characteristic variety* Λ to be Λ_l.

We may intuitively think of Λ_p as the set of K-singular p-dimensional integral elements. In fact, as is easily verified from the discussions in Chapter IV, the precise statement is this: If there are no integrability conditions, so that the symbol relations may be assumed to be

$$B^{\lambda i}_a \pi^a_i \equiv 0 \quad \mod I,$$

then Λ_p is the set of p-dimensional integral elements whose polar equations have smaller rank than is generically the case. In particular, in the absence of integrability conditions, Λ_{n-1} coincides with characteristic variety Ξ as given above in terms of the symbol.

According to the proof of the Cartan–Kähler theorem, the Cartan characteristic variety determines the set of integral elements that are characteristic for the last Cauchy–Kowaleski system in which there is any freedom in assigning initial data.

It is clear how to define $\Lambda_{p,\mathbb{C}}$ and $\Lambda_\mathbb{C}$. A fundamental result is given by the following

Theorem 3.15. *In case $(\mathcal{I}, \Omega)$ is involutive we have*

$$\Lambda_\mathbb{C} = \{E \in G_l(L^*_\mathbb{C}) : E \in [\xi]^\perp \text{ for some } [\xi] \in \Xi_\mathbb{C}\}$$

$$\Xi_\mathbb{C} = \{[\xi] \in \mathbb{P}L_\mathbb{C} : E \in \Lambda_\mathbb{C} \text{ for all } E \subset [\xi]^\perp\}.$$

Simple examples show that the result is false without the assumption of involutiveness. Since we may have $\Xi = \emptyset$ but $\Lambda \neq \emptyset$ (see below), the result is false over $\mathbb{R}$. What we can say is that the *real* Cartan characteristic variety Λ is given in terms of the *complex* usual characteristic variety $\Xi_\mathbb{C}$ by

$$(102) \qquad \Lambda = \{E \in G_l(L^*) : E \subset [\xi]^\perp \text{ for some } [\xi] \in \Xi_\mathbb{C}\}.$$

In particular, we may have $\Xi = \emptyset$ but $\Xi_{\mathbb{C}}$ and Λ both $\neq \emptyset$; see below.

To picture Theorem 3.15 it may help to use the incidence correspondence

$$\Sigma \subset G_l(L_{\mathbb{C}}^*) \times \mathbb{P}L_{\mathbb{C}}$$

defined by

$$\Sigma = \{(E, [\xi]) : E \subset [\xi]^\perp\}.$$

There are projections

$$\begin{array}{ccc} & \Sigma & \\ \pi_1 \swarrow & & \searrow \pi_2 \\ G_l(L_{\mathbb{C}}^*) & & \mathbb{P}L_{\mathbb{C}} \end{array}$$

and the first assertion in Theorem 3.8 and (102) are equivalent to

$$\begin{cases} \Lambda_{\mathbb{C}} = \pi_1(\pi_2^{-1}(\Xi_{\mathbb{C}})) \\ \Lambda = \Lambda_{\mathbb{C}} \cap G_l(L^*). \end{cases}$$

Example 3.16. On $\mathbb{R}^{2m} \cong \mathbb{C}^m$ with coordinates $z^i = x^i + \sqrt{-1}\,y^i$ and complex structure $J : \mathbb{R}^{2m} \to \mathbb{R}^{2m}$ given by

$$\begin{cases} J(\partial/\partial x^i) = \partial/\partial y^i \\ J(\partial/\partial y^i) = -\partial/\partial x^i \end{cases}$$

we consider the Cauchy–Riemann system (cf. Example 2.4 above). Since it is elliptic, the real characteristic variety $\Xi = \emptyset$.

To describe the Cartan characteristic variety, since the system is translation invariant it will suffice to describe the fibre Λ_0 of Λ over the origin, and using (102) this is given by

$$\Lambda_0 = \{E \in G_m(\mathbb{R}^{2m}) : E \cap J(E) \neq (0)\}.$$

In other words, Λ_0 consists of real m-planes $E \subset \mathbb{R}^{2m}$ that contain at least one *complex* line (the latter being a *real* 2-plane $F \subset E$ with $J(F) = F$).

It is, of course, well known that real m-dimensional submanifolds $Y^m \subset \mathbb{C}^m$ such that $T_y(Y) \cap JT_y(Y) = (0)$ for every $y \in Y$ are locally determining sets for holomorphic functions.

In general we have

$$\Lambda_{p,0} = \{E \in G_p(\mathbb{R}^{2m}) : \dim E \cap J(E) \geq \max(1, 2(p-m)+1)\}.$$

For instance, $\Lambda_{2m-1,0} = G_{2m-1}(\mathbb{R}^{2m})$ contains no information.

The second assertion in Theorem 3.15 gives $\Xi_{\mathbb{C}}$ in terms of $\Lambda_{\mathbb{C}}$ as follows

$$\Xi_{\mathbb{C}} = \{[\xi] : \pi_2^{-1}([\xi]) \subset \pi_1^{-1}(\Lambda_{\mathbb{C}})\}.$$

In other words, an $(n-1)$-plane is characteristic only if *every* l-plane contained in it is Cartan characteristic.

(vi) **Integrability of the characteristic variety.**

(a) Let N be a manifold and $\Sigma \subset \mathbb{P}T^*N$ a subset. There is an associated *eikonal equation* E_Σ defined as follows:

> *A function $\varphi(y)$ on N is a solution of E_Σ if*
>
> $[d\varphi(y)] \in \Sigma_y$ *whenever* $d\varphi(y) \neq 0$.

More precisely, we let $\tilde{\Sigma} \subset T^*N$ be defined by

$$\tilde{\Sigma} = \pi^{-1}\Sigma \cup \{0\}$$

where $\pi : T^*N\backslash\{0\} \to \mathbb{P}T^*N$ is the projection and $\{0\} \subset T^*N$ is the zero section. Then $\tilde{\Sigma}$ is a *conical subvariety* of T^*N, i.e., it is invariant under the natural $\mathbb{R}^*$ action on T^*N. Moreover, any conical subvariety is of this form. If $y^1, \ldots, y^n = (y^i)$ are local coordinates on N with induced coordinates (y^i, ξ_i) on T^*N, then we shall always assume that Σ is a subset with the property that $\tilde{\Sigma}$ is defined by equations

$$(103) \qquad F^\lambda(y^i, \xi_i) = 0 \qquad \lambda = 1, \ldots, R$$

where the F^λ are either C^∞ or real analytic functions depending on the category in which we are working.

Definition 3.17. The *eikonal equation* is

$$(E_\Sigma) \qquad F^\lambda\left(y^i, \frac{\partial\varphi(y)}{\partial y^i}\right) = 0 \qquad \lambda = 1, \ldots, R.$$

Remarks. The functions F^λ need only be defined *microlocally*, i.e., in open sets $U \subset T^*N$ invariant under the natural $\mathbb{R}^*$ action. In the cases of interest to us the fibres Σ_y of $\Sigma \to N$ will be algebraic varieties, so that the $F^\lambda(y^i, \xi_i)$ may be chosen to be homogeneous polynomials in the ξ_i whose coefficients are functions of the y^i. In this case, the complexifications

$$\begin{cases} \Sigma_\mathbb{C} \subset \mathbb{P}T^*_\mathbb{C}N \\ \tilde{\Sigma}_\mathbb{C} \subset T^*_\mathbb{C}N \end{cases}$$

are naturally defined, and so the complex eikonal equation makes sense by allowing the function φ to have complex values and requiring that $d\varphi \in \tilde{\Sigma}_\mathbb{C}$. We remark that we may have $\Sigma = \emptyset$ but $\Sigma_\mathbb{C} \neq \emptyset$. *From now on we assume that the $F^\lambda(y^i, \xi_i)$ may be chosen to be homogeneous polynomials in the ξ_i.*

Definition 3.18. The subset $\Sigma_\mathbb{C} \subset \mathbb{P}T^*_\mathbb{C}N$ is *involutive* in case the eikonal equation $E_{\Sigma_\mathbb{C}}$ is involutive.

To state the main result, we let $\mathcal{I}$ be a differential system on a manifold M and with complex characteristic variety

$$\Xi_{\mathbb{C}} \subset \mathbb{P}U_{\mathbb{C}}^*$$

(cf. Definitions 1.1, 1.2 and the subsequent discussion). For any integral manifold

$$(104) \qquad\qquad f : N \to M$$

of $\mathcal{I}$ there is the *induced characteristic variety*

$$\Xi_{\mathbb{C},N} \subset \mathbb{P}T_{\mathbb{C}}^*N$$

defined as follows: Given (104) we have a diagram

$$(105)$$

$$
\begin{array}{ccc}
 & & \mathbb{P}U_{\mathbb{C}}^* \\
 & \nearrow & \downarrow \tilde{\omega} \\
\mathbb{P}T_{\mathbb{C}}^*N & \xrightarrow{\ \hat{f}_*\ } & G_n(TM) \\
\downarrow & & \downarrow \\
N & \xrightarrow{\ f\ } & M
\end{array}
$$

where

$$
\begin{cases}
\hat{f}_*(y) = f_*(T_yN) \\
\tilde{\omega}^{-1}(x, E) = \mathbb{P}E_{\mathbb{C}}^*.
\end{cases}
$$

The condition that (104) be an integral manifold of $\mathcal{I}$ is that

$$\hat{f}_*(N) \subset G_n(\mathcal{I}),$$

and the dotted arrow in (105) means that there is a natural mapping

$$\tilde{\omega}^{-1}(\hat{f}_*(N)) \to \mathbb{P}T_{\mathbb{C}}^*N.$$

By definition, $\Xi_{\mathbb{C},N}$ is the image of $\Xi_{\mathbb{C}} \cap \tilde{\omega}^{-1}(\hat{f}_*(N))$ under this mapping. Informally, we may say that $\Xi_{\mathbb{C},N}$ *is induced from the characteristic variety* $\Xi_{\mathbb{C}}$ *in each of the integral elements* $f_*(T_yN) \in G_n(TM)$.

Definition 3.19. We shall say that $\Xi_{\mathbb{C}}$ is *involutive* in case the eikonal equation $E_{\Xi_{\mathbb{C},N}}$ is involutive for any integral manifold (104) of $\mathcal{I}$.

The main result is essentially the following: *If $\mathcal{I}$ is involutive, then its characteristic variety is involutive.* More precisely, the result is

Theorem 3.20. *Let $E \in G_n(\mathcal{I})$ be an ordinary integral element. Then $\Xi_{\mathbb{C}}$ is involutive in a (possibly smaller) neighborhood V of E.*

What this means is that $\Xi_{\mathbb{C},N}$ is involutive for all integral manifolds (104) satisfying $\hat{f}_*(N) \subset V$.

We will prove this result only in the case when the characteristic varieties $\Xi_{\mathbb{C},E} = \Xi_{\mathbb{C}} \cap \mathbb{P}E_{\mathbb{C}}$ are points. By Theorem 3.6 this corresponds to the case where the Cartan characters are given by

$$(*) \qquad\qquad s_1' = s_0, \ s_2' = \cdots = s_n' = 0.$$

Referring to (97) in §5 of Chapter IV we see that the symbol relations given by (96) in that section are given by commuting matrices C_ρ. We shall make the additional assumption that

$$(**) \qquad\qquad \textit{the } C_\rho \textit{ are simultaneously diagonalizable.}$$

It is interesting to note that Cartan stated the above theorem under the assumption $(*)$ (cf. Cartan [1953][9]). He also proved the result under our additional assumption $(**)$, and by the computation of several examples he showed that the result is much more subtle in case the C_ρ may have non-diagonal Jordan normal forms.

More recently in Guillemin, Quillen and Sternberg [1970] Theorem 3.20 is stated and proved for involutive P.D.E. systems. There they also make a technical assumption analogous to but weaker than our assumption $(**)$, but there is no restriction on the dimension of the characteristic variety. Subsequently the general result was proved in Gabber [1981], where additional references may be found.

Since not every exterior differential system is derived from a P.D.E. system, the Guillemin, Quillen and Sternberg result does not immediately imply Theorem 3.20. However, the result is not really in doubt; even the stronger theorem corresponding to the result proved by Gabber is certainly true. What is important, in our opinion, is that we do not know a proof of Theorem 3.20 using moving frames in the spirit of the one we shall give below in our special case. That argument will show that the theorem falls out by exterior differentiation of the structure equations of a involutive system, where the assumption $(**)$ is used to put these equations in a particular form. We think it is a very worthwhile problem to give a similar proof of the full result.

It will simplify our notations if we now work only with the real characteristic variety and observe that the arguments remain valid in the complex case. Unless mentioned to the contrary this will now be done.

---

[9]The result is announced on page 1127 in Part II of the 1984 edition of the collected works. Numerous special cases were worked out in the paper preceding the announcement.

(b) We will now derive conditions for the involutivity of a conical submanifold $\tilde{\Sigma}$ of T^*N. For this we assume that we may micro-locally choose functions $F^\lambda(y^i, \xi_i)$ such that (103) gives a regular set of defining functions for $\tilde{\Sigma}$. We denote by $\{f, g\}$ the *Poisson bracket* of functions locally defined on T^*N, and recall that by definition

$$(106) \qquad \{f, g\} = \sum_i \frac{\partial f}{\partial \xi_i} \frac{\partial g}{\partial y^i} - \frac{\partial f}{\partial y^i} \frac{\partial g}{\partial \xi_i}.$$

More intrinsically, if

$$(107) \qquad \eta = \xi_i dy^i$$

is the tautological 1-form on T^*N with associated symplectic form

$$(108) \qquad \Theta = d\eta = d\xi_i \wedge dy^i,$$

then

$$\{f, g\}\Theta^n = n\, df \wedge dg \wedge \Theta^{n-1}.$$

It is well known that the condition for the involutivity of E_Σ is

$$(109) \qquad \{F^\lambda, F^\mu\} = 0 \text{ on } \tilde{\Sigma}, \qquad 1 \le \lambda, \mu \le R.$$

We shall derive this result from differential system point of view. This computation will also allow us to reformulate the condition for integrability in a way that leads naturally to the proof of Theorem 3.20.

The result is local (even micro-local), and so we work in the above local coordinates and consider $T^*N \times \mathbb{R}$ as $J^1(\mathbb{R}^n, \mathbb{R})$ with coordinates $(y^1, \ldots, y^n, \psi, \xi_1, \ldots, \xi_n)$. In $J^1(\mathbb{R}^n, \mathbb{R})$ we consider the submanifold $P \cong \tilde{\Sigma} \times \mathbb{R}$ defined by

$$F^\lambda(y^i, \xi_i) = 0 \qquad \lambda = 1, \ldots, R.$$

On P we have the contact system $(\mathcal{J}, \Phi)$ generated by the equations

$$(110) \qquad \begin{aligned} \theta &= d\psi - \xi_i dy^i = 0 \\ \Phi &= dy^1 \wedge \cdots \wedge dy^n \ne 0. \end{aligned}$$

Lemma 3.21. *Locally on P we may choose a coframe $\varphi^1, \ldots, \varphi^n, \theta, \pi_{R+1}, \ldots, \pi_n$ such that*

$$(111) \qquad \begin{aligned} \Phi &\equiv \varphi^1 \wedge \cdots \wedge \varphi^n \mod \{\theta\} \text{ and} \\ d\theta &\equiv \pi_\varepsilon \wedge \varphi^\varepsilon + \frac{1}{2} c_{ij} \varphi^i \wedge \varphi^j \mod \{\theta\} \end{aligned}$$

where $\varepsilon = R+1, \ldots, n$ and $c_{ij} + c_{ji} = 0$.

Proof. The structure equation of (110) is

$$(112) \qquad d\theta \equiv -d\xi_i \wedge dy^i \mod \{\theta\},$$

and the symbol relations on the $d\xi_i$ are

$$(113) \qquad \frac{\partial F^\lambda(y^i, \xi_i)}{d\xi_i} d\xi_i \equiv 0 \mod \{\theta, dy^i\}.$$

By our assumption, the matrix $\|\partial F^\lambda/\partial \xi_i\|$ has everywhere rank R, and so we may find invertible matrices $\|A_\mu^\lambda\|$ and $\|B_j^i\|$ such that

$$A_\mu^\lambda \partial F^\mu/\partial \xi_j B_j^i = \delta_i^\lambda.$$

Setting $d\xi_i = -B_i^j \pi_j$ the symbol relations (113) imply that

$$(114) \qquad \pi_\lambda \equiv 0 \mod \{\theta, dy^i\}, \qquad \lambda = 1, \ldots, R.$$

Then for $\varphi^i = B_j^i dy^j$ we have

$$-d\xi_i \wedge dy^i = B_i^j \pi_j \wedge dy^i$$
$$= \pi_j \wedge \varphi^j,$$

so that by (112) and (114)

$$d\theta \equiv \pi_\varepsilon \wedge \varphi^\varepsilon \mod \{\theta, \varphi^i\}.$$

Writing this out modulo $\{\theta\}$ gives (114) for suitable functions $c_{ij} = -c_{ji}$.
$\square$

To establish (109), we consider a differential system $(\mathcal{J}, \Phi)$ with structure equations on an open set $U \subset \mathbb{R}^{2n-R+1}$

$$
\begin{aligned}
&\text{(i)} \quad \theta = 0 \\
(115) \qquad &\text{(ii)} \quad d\theta \equiv \pi_\varepsilon \wedge \varphi^\varepsilon + \tfrac{1}{2} c_{ij} \varphi^i \wedge \varphi^j \mod \{\theta\}, \qquad c_{ij} + c_{ji} = 0 \\
&\text{(iii)} \quad \Phi = \varphi^1 \wedge \cdots \wedge \varphi^n \neq 0
\end{aligned}
$$

where $\varphi^1, \ldots, \varphi^n, \theta, \pi_{R+1}, \ldots, \pi_n$ are a local coframe. We may make a substitution

$$(116) \qquad \pi_\varepsilon \to \pi_\varepsilon - p_{\varepsilon i} \varphi^i, \qquad p_{\varepsilon\delta} + p_{\delta\varepsilon} = 0,$$

without effecting the form of the structure equations (115). When this is done we may eliminate the terms

$$c_{\varepsilon\delta}, c_{\varepsilon\mu} \quad 1 \le \varepsilon, \delta \le R \text{ and } R + 1 \le \mu \le n$$

in $d\theta$, and then

(117)
$$d\theta \equiv \pi_\varepsilon \wedge \varphi^\varepsilon + \frac{1}{2} c_{\lambda\mu} \varphi^\lambda \wedge \varphi^\mu \quad \text{mod } \{\theta\}.$$

It is easy to see that the 2-form

$$T = \frac{1}{2} c_{\lambda\mu} \varphi^\lambda \wedge \varphi^\mu$$

is, up to a conformal factor, invariantly associated to $(\mathcal{J}, \Phi)$. Indeed, the only substitution (116) that leaves the form (117) invariant is when all $p_{\varepsilon i} = 0$, and consequently the condition that the integrability conditions be satisfied for $(\mathcal{J}, \Phi)$ is

(118)
$$c_{\lambda\mu} = 0.$$

In this case $(\mathcal{J}, \Phi)$ is involutive.

In fact, assuming (118) we may "integrate $(\mathcal{J}, \Phi)$ by O.D.E.'s" as follows: Denoting by $\{\partial/\partial\varphi^i, \partial/\partial\theta, \partial/\partial\pi_\varepsilon\}$ the dual frame to $\{\varphi^i, \theta, \pi_\varepsilon\}$, the vector fields $\partial/\partial\varphi^\lambda$ satisfy

$$\partial/\partial\varphi^\lambda \lrcorner \theta = 0$$

$$\partial/\partial\varphi^\lambda \lrcorner d\theta \equiv c_{\lambda\mu} \varphi^\mu \quad \text{mod } \{\theta\}.$$

Thus, assuming (118) the vector fields $\partial/\partial\varphi^\lambda$ are Cauchy characteristics, and in fact it is easy to see that, in the notation of Chapter I,

$$A(\Phi) = \text{span}\{\partial/\partial\varphi^1, \dots, \partial/\partial\varphi^R\}.$$

Local integral manifolds for $(\mathcal{J}, \Phi)$ may be found by prescribing arbitrary values of $\psi(x)$ along an $\mathbb{R}^{n-R} \cap U$ in a general linear coordinate system, and then flowing this initial data out along the R-dimensional integrable distribution $C(\Phi)$, just as is done in the case $R = 1$ as explained in Chapter II.

It remains to mutually identify the conditions (109) and (118). Both are invariantly attached to the submanifold $\tilde{\Sigma} \subset T^*N$ and do not depend on the local defining equations or choice of coframing. It will therefore suffice to identify these two sets of conditions at a point $(\underline{y}^i, \underline{\xi}_i) \in \tilde{\Sigma}$.

For this we will show that (109) expresses the condition that there be integral elements of $(\mathcal{J}, \Phi)$ at each point of P. We may choose the $F^\lambda(y^i, \xi_i)$ such that

$$
(119) \qquad \frac{\partial F^\lambda}{\partial \xi_j}(\underline{y}^i, \underline{\xi}_i) = \delta_j^\lambda.
$$

We view integral elements of $(\mathcal{J}, \Phi)$ as n-planes E^n in $T_{(\underline{y}^i, \underline{\xi}_i)}(T^*N \times \mathbb{R})$, and by (110) and (119) any E^n on which $\Phi \neq 0$ has equations

$$
(120) \qquad
\begin{cases}
d\xi_\lambda + \displaystyle\sum_j \frac{\partial F^\lambda}{\partial y^j}(\underline{y}^i, \underline{\xi}_i) dy^j = 0 & \lambda = 1, \ldots, R \\[2ex]
d\xi_\rho + \displaystyle\sum_j p_{\rho j} dy^j = 0 & \rho = R+1, \ldots, n \\[2ex]
d\psi = \displaystyle\sum_i \underline{\xi}_i dy^i
\end{cases}
$$

where $p_{\rho j}$ are to be chosen to annihilate $d\theta$. By (112) the restriction of $d\theta$ to the n-plane (120) is

$$
(121) \qquad \sum_{\lambda, j} \frac{\partial F^\lambda}{\partial y^j} dy^j \wedge dy^\lambda + \sum_{\rho, j} p_{\rho j} dy^j \wedge dy^\rho,
$$

where the $\partial F^\lambda / \partial y^j$ are to be evaluated at $(\underline{y}^i, \underline{\xi}_i)$. We may set $p_{\rho \sigma} = 0$ and $p_{\rho \lambda} = -\partial F^\lambda / \partial y^\rho$ so that (121) reduces to

$$
\sum_{\lambda, \mu} \partial F^\lambda / \partial y^\mu dy^\mu \wedge dy^\lambda.
$$

Thus the vanishing of all

$$
(122) \qquad (\partial F^\lambda / \partial y^\mu)(\underline{y}^i, \underline{\xi}_i) = \{F^\mu, F^\lambda\}(\underline{y}^i, \underline{\xi}_i)
$$

is equivalent to the existence of an integral element of $(\mathcal{J}, \Phi)$ at $(\underline{y}^i, \underline{\xi}_i)$ and this establishes the equivalence of (109) and (118).

(c) In preparation for the proof of Theorem 3.7 we want to express the involutivity conditions (109) in more geometric form. For this we recall that $\tilde{\Sigma} \subset T^*N$ has dimension given by

$$
\dim \tilde{\Sigma} = 2n - R,
$$

and we denote by Θ the symplectic form (108) on T^*N.

Proposition 3.22. *The involutivity condition (109) is equivalent to*

$$(123) \qquad\qquad \operatorname{rank} \Theta = 2(n - R)$$

on $\tilde{\Sigma}$.

Proof. It will suffice to verify the equivalence of (109) and (123) at a point of $\tilde{\Sigma}$ where (119) is satisfied. Then, as in (121),

$$\begin{aligned}
\Theta &= -\sum_{\lambda,j} \frac{\partial F^\lambda}{\partial y^j} dy^j \wedge dy^\lambda + \sum_\rho d\xi_\rho \wedge dy^\rho \\
&= \sum_\rho (d\xi_\rho + \sum_\lambda \frac{\partial F^\lambda}{\partial y^\rho} dy^\lambda) \wedge dy^\rho + \sum_{\lambda,\mu} \frac{\partial F^\lambda}{\partial y^\mu} dy^\lambda \wedge dy^\mu \\
&= \sum_\rho \gamma_\rho \wedge dy^\rho + \sum_{\lambda,\mu} \frac{\partial F^\lambda}{\partial y^\mu} dy^\lambda \wedge dy^\mu
\end{aligned}$$

where the forms $\gamma_\rho = d\xi_\rho + \sum_\lambda \frac{\partial F^\lambda}{\partial y^\rho} dy^\lambda, dy^\rho, dy^\lambda$ give a coframe for $T^*\tilde{\Sigma}$ at the point in question. It follows from this last expression and (122) that (123) is equivalent to (109). $\qquad\qquad\square$

A noteworthy special case arises when $R = n - 1$, which is the case corresponding to our assumption $(*)$. Then $\Sigma \subset \mathbb{P}T^*N$ consists of points, say d of these, in each fibre Σ_y lying over $y \in N$. Geometrically, each of these d points gives a hyperplane $H_\alpha(y)$ in T_yN, and we claim that

$$(124) \qquad \textit{The involutivity of } \tilde{\Sigma} \textit{ is equivalent to the integrability}$$
$$\textit{of each of the distributions } H_\alpha.$$

Proof. It will suffice to locally treat the case $d = 1$. Then Σ_y is given by $[\eta(y)]$ where

$$\eta(y) = \sum_i \xi_i(y) dy^i$$

is a non-zero section of T^*N, and

$$\tilde{\Sigma} = \{\lambda\eta(y) : \lambda \in \mathbb{R} \text{ and } y \in N\}.$$

On $\tilde{\Sigma}$

$$\begin{aligned}
\Theta &= d(\lambda\eta(y)) \\
&= d\lambda \wedge \eta(y) + \lambda d\eta(y).
\end{aligned}$$

If the integrability condition

$$d\eta(y) = \gamma \wedge \eta$$

is satisfied, then

$$\Theta = (d\lambda + \lambda\gamma) \wedge \eta$$

has rank 2. Conversely, if we locally choose forms $\beta^1, \ldots, \beta^{n-1}$ such that

$$\mathrm{span}\{\eta, \beta^1, \ldots, \beta^{n-1}\} = \mathrm{span}\{dy^1, \ldots, dy^n\},$$

then writing

$$d\eta = \gamma \wedge \eta + \beta$$

where β involves only the β^i we have

$$\Theta = (d\lambda + \lambda\gamma) \wedge \eta + \beta.$$

From this it is apparent that Θ has rank 2 only if $\beta = 0$. $\qquad\square$

Proof of Theorem 3.20. (under the further hypotheses $(*)$ and $(**)$)

We may prolong $\mathcal{I}$ to obtain an involutive linear Pfaffian system with the same induced characteristic variety on integral manifolds. Thus it will suffice to prove the result in this case.

Let $I \subset \Omega^1(M)$ be an involutive linear Pfaffian system and denote by $\{I\} \subset \Omega^*(M)$ the algebraic ideal generated by I. Its structure equations may be written in the form

$$d\theta^\alpha \equiv 0 \mod \{I\}$$

$$d\theta^a \equiv \pi_i^a \wedge \omega^i \mod \{I\}$$

where the θ^α span the 1st derived system, the forms θ^a with $1 \le a \le s$ span the remainder of I, and $\Omega = \omega^1 \wedge \cdots \wedge \omega^n$ is the independence condition.

Our assumption $(*)$ means that, if we set $\pi_1^a = \pi^a$ and denote by $\pi = (\pi^a)$ the first column of the tableau matrix and by $\pi_\rho = (\pi_\rho^a)$ the ρth column for $2 \le \rho \le n$, then the 1-forms π^a are linearly independent modulo $\{\theta^\alpha, \theta^a, \omega^i\}$ and the symbol relations are

$$\pi_\rho \equiv B_\rho \pi \mod J$$

for $s \times s$ matrices B_ρ. The integrability conditions mean that we may assume this equation holds modulo I. The assumption of involutivity is then equivalent to commutation relations

$$[B_\rho, B_\sigma] = 0.$$

Our assumption $(**)$ implies that we may make a linear change among the π^a's so that the B_ρ are all diagonal. We note that the π_i^a may now be complex valued. Thus we have

$$\pi_\rho^a \equiv \lambda_\rho^a \pi^a \quad \bmod I$$

(no summation), and if we set $\lambda_1^a = 1$ and

$$\hat{\omega}^a = \lambda_i^a \omega^i,$$

then the characteristic variety is given by the s points $[\hat{\omega}^a] \in \mathbb{P}L_{\mathbb{C}}$. The structure equations of I are

$$(125) \qquad \begin{aligned} d\theta^\alpha &\equiv 0 && \bmod \{I\} \\ d\theta^a &\equiv \pi^a \wedge \hat{\omega}^a && \bmod \{I\} \end{aligned}$$

(no summation), and by (124) we must show that

$$(126) \qquad \textit{On any integral manifold } N \textit{ we have}$$

$$d\hat{\omega}_N^a \equiv 0 \ \bmod \hat{\omega}_N^a.$$

Here, we recall our notation $\psi_N = \psi|_N$ for any submanifold $N \subset M$.

The idea is to differentiate the equations (125), and in so doing we shall make use of a trick that was employed by Cartan. Namely, fixing N we may make a change $\pi_i^a \to \pi_i^a - p_{ij}^a \omega^j$ so that

$$(127) \qquad (\pi_i^a)_N = 0.$$

This does *not* mean that we adjoin the equations (127) to I; it means only that on the particular integral manifold N we may assume that (127) holds.

We now write out the second equation in (125) as

$$d\theta^a = \pi^a \wedge \hat{\omega}^a + \varphi_\beta^a \wedge \theta^\beta + \varphi_b^a \wedge \theta^b.$$

Exterior differentiation and use of (125) gives for each fixed a

$$(128) \qquad \pi^a \wedge d\hat{\omega}^a + \varphi_b^a \wedge \pi^b \wedge \hat{\omega}^b \equiv 0 \quad \bmod \{I, \hat{\omega}^a\}.$$

In any case we have expansions of the form

$$(129) \qquad \begin{aligned} \varphi_b^a &\equiv F_{bc}^a \pi^c + F_{bi}^a \omega^i \ \bmod I \\ d\hat{\omega}^a &\equiv \frac{1}{2} C_{ij}^a \omega^i \wedge \omega^j + B_{bi}^a \pi^b \wedge \omega^i + \frac{1}{2} A_{bc}^a \pi^b \wedge \pi^c \ \bmod \{I\}. \end{aligned}$$

Using the trick (127) we must show that the $C_{ij}^a = -C_{ji}^a$ satisfy suitable conditions. The ω-quadratic terms in (128) give (no summation on a)

$$\frac{1}{2}\pi^a \wedge C_{ij}^a \omega^i \wedge \omega^j + F_{bi}^a \omega^i \wedge \pi^b \wedge \hat{\omega}^b \equiv 0 \quad \mathrm{mod}\ \{I, \hat{\omega}^a\}.$$

For each fixed $b \neq a$ this implies that

$$F_{bi}^a \omega^i \wedge \pi^b \wedge \hat{\omega}^b \equiv 0 \quad \mathrm{mod}\ \{I, \hat{\omega}^a\},$$

and the remaining equation becomes (no summation on a)

$$\frac{1}{2}\pi^a \wedge C_{ij}^a \omega^i \wedge \omega^j \equiv 0 \quad \mathrm{mod}\ \{I, \hat{\omega}^a\}.$$

This gives

$$\frac{1}{2}C_{ij}^a \omega^i \wedge \omega^j \equiv 0 \quad \mathrm{mod}\ \{I, \hat{\omega}^a\}$$

which when plugged into (129) becomes

$$d\hat{\omega}^a \equiv 0 \quad \mathrm{mod}\ \{I, \hat{\omega}^a, \pi^1, \ldots, \pi^s\}.$$

Using (127) this implies our desired result (126). $\qquad\square$

CHAPTER VI

PROLONGATION THEORY

As has been seen in earlier chapters, it often happens that a given differential system with independence condition fails to be involutive. The process of *prolongation* is designed to remedy this situation and will be discussed in this chapter. At the P.D.E. level, the process of prolongation is nothing more than introducing the partial derivatives of the unknown functions as new variables and then adjoining new P.D.E. to the original P.D.E. system which ensure that the new variables are, in fact, the partial derivatives of the original unknown functions. The objective in doing this is that it may happen that the new system of P.D.E. is involutive even though the original system is not. (For an explicit example of this, see Examples 1.1 and 1.2 in Section 1 below.) Geometrically, for an exterior differential system, prolongation is essentially the process of replacing the original exterior differential system $\mathcal{I} \subset \Omega^*(M)$ by the canonical Pfaffian system with independence condition $(\mathcal{I}^{(1)}, \Omega)$ defined on the space $V_n(\mathcal{I})$ of n-dimensional integral elements of $\mathcal{I}$. This is made precise in Section 1 under the assumption that the space $V_n(\mathcal{I})$ is sufficiently "well-behaved". More precisely, we assume that $V_n(\mathcal{I})$ has a stratification into smooth submanifolds of $G_n(TM)$, an assumption which is always satisfied in practice or when $\mathcal{I}$ is real analytic. The remainder of Section 1 is devoted to three examples which illustrate several phenomena which may arise during the process of prolongation.

In Section 2, we investigate the effect that prolongation has on a component Z of $V_n(\mathcal{I})$ which consists of ordinary integral elements. We prove the expected result that the prolongation $(\mathcal{I}^{(1)}, \Omega)$ is involutive on Z. Moreover, (see Theorem 2.1) we show that the Cartan characters of $(\mathcal{I}^{(1)}, \Omega)$ on Z can be computed by the expected formula from the Cartan characters of Z as a component of $V_n(\mathcal{I})$. This result is to be found in Cartan's work in the case that $\mathcal{I}$ is a Pfaffian system in linear form. The more general case (which does not follow from the Pfaffian system case) is due to Matsushima [1953]. For the (simpler) proof in the Pfaffian system case, the reader may want to compare the discussion at the end of Chapter VIII, Section 2, where Cartan's original argument (albeit in more modern language) is given.

There remains the question of the effect of prolongation on a non-involutive exterior differential system. It was a conjecture of Cartan (based on his having computed a large number of examples) that, for any real analytic differential system $\mathcal{I}$, a finite number of iterations of the process of prolongation applied to $\mathcal{I}$ would lead either to an involutive differential system or else to a system with no integral elements (and hence, no integral manifolds). Although Cartan made attempts to prove this important result (for example, see Cartan [1946]), he was never able to do so. It was Kuranishi [1957] who first provided a proof of Cartan's conjecture under certain technical hypotheses on $\mathcal{I}$, which, in practice, were always fulfilled. Since then, various improvements in the technical hypotheses have been made, (cf. Goldschmidt [1968a, 1968b] and Theorem 1.14, Chapter X), although the general result remains open. Since the technical hypotheses are difficult to check without computing the successive prolongations up to a certain order, the general Cartan–Kuranishi prolongation theorem has so far been of more theoretical importance than practical importance. On the other hand, the fully linear theory is better behaved and the reader should consult Chapter X for that case. In any case, the validity of the theorem (even with the technical hypotheses) serves, in practice, as motivation for the computation of the prolongations.

In the last two sections of this chapter, we discuss (a version of) the Cartan–Kuranishi prolongation theorem. In Section 3, we introduce the important notion of a *prolongation sequence*, which slightly generalizes the case of a sequence of prolongations for which the space of integral elements at each stage forms a smooth submanifold of the appropriate Grassmann bundle for which the projection to the appropriate base manifold is a submersion. In the terminology of Chapter IV, this corresponds to the case of a sequence of prolongations for which the "torsion" is always "absorbable". We prove that, in this case, after a finite number of steps, the remaining differential systems in the sequence are all involutive (see Theorem 3.2). The crucial step is the reduction of the problem to a commutative algebra statement which is related to the Hilbert syzygy theorem (an important step in all of the known proofs of the Cartan–Kuranishi theorem). This commutative algebra statement is then proved in Chapter VIII.

Finally, in Section 4, we relate Theorem 3.2 to the "classical" version of the Cartan–Kuranishi Prolongation Theorem, namely the case where the torsion is always absorbable. We then enter into a discussion of the general case and point out some of the difficulties and what is expected to be the general result. The upshot of our discussion is that, when dealing with a non-involutive differential system, the process of prolongation is an essential step in the study of the integral manifolds. Moreover, in practice, in the analytic category, the process satisfactorily answers the existence question for integrals of an exterior differential system.

§1. The Notion of Prolongation.

We begin by recalling some relevant constructions from earlier sections. Let M be a smooth manifold of dimension m and let $n \leq m$ be an integer.

We let $\pi : G_n(TM) \to M$ denote the Grassmann bundle whose fiber at $x \in M$ consists of the space of n-planes $E \subset T_x M$. Every smooth immersion $f : N^n \to M$ induces a canonical smooth map $f_1 : N^n \to G_n(TM)$ by the formula $f_1(p) = f_*(T_p N) \subset T_{f(p)} M$. This f_1 is clearly a lifting of f. The dimension of $G_n(TM)$ is $m + n(m - n)$ and it carries a canonical Pfaffian system I of rank $m - n$ which has the property that for every immersion $f : N^n \to M$ the induced lifting $f_1 : N^n \to G_n(TM)$ is a integral manifold of the exterior differential system $\mathcal{I}$ generated by I. Moreover, a smooth map $\varphi : N^n \to G_n(TM)$ is of the form $\varphi = f_1$ for some immersion $f : N^n \to M$ if and only if $\pi \circ \varphi : N^n \to M$ is an immersion and φ is an integral of the system $\mathcal{I}$. In this case, it then follows that $\varphi = (\pi \circ \varphi)_1$. Alternatively, we note that there is a canonical rank n independence condition Ω on $G_n(TM)$ with the property that the integrals $\varphi : N^n \to G_n(TM)$ of $(\mathcal{I}, \Omega)$ are precisely the canonical lifts of immersions $f : N^n \to M$. For a more explicit description of this system, we refer the reader to Chapter IV.

Let $S \subset G_n(TM)$ be a subset. We say that an immersion $f : N^n \to M$ is an *S-immersion* if $f_1(N) \subset S$. In most cases of interest, S will be a submanifold, so let us assume this for the moment. It is clear that a map $\varphi : N^n \to S$ will be the canonical lift of an S-immersion if and only if it is also an integral of the system $(\mathcal{I}, \Omega)$. Thus, the study of S-immersions is equivalent to the study of the integrals of the system $(\mathcal{I}|_S, \Omega|_S)$ on S.

A special case will be of great importance. If $\mathcal{I} \subset \Omega_*(M)$ is a closed differential ideal and $S = V_n(\mathcal{I})$, then the system $(\mathcal{I}|_S, \Omega|_S)$ is what we would like to call the (first) *prolongation* of $\mathcal{I}$. The difficulty with this as a definition is that the space $V_n(\mathcal{I})$ often fails to be a smooth manifold. In practice, however, we can usually write

$$\tag{1} V_n(\mathcal{I}) = \bigcup_{\beta \in B} S_\beta$$

where $\{ S_\beta \mid \beta \in B \}$ is a stratification of $V_n(\mathcal{I})$ into irreducible smooth components. (This is always possible when $\mathcal{I}$ is real analytic.) Then we may consider the exterior differential system

$$\tag{2} (\mathcal{I}^{(1)}, \Omega^{(1)}) = \bigcup_{\beta \in B} (\mathcal{I}|_{S_\beta}, \Omega|_{S_\beta}).$$

defined on the *disjoint* union of the strata S_β as the (first) *prolongation* of $\mathcal{I}$. Note that, on each component S_β, the prolongation of $\mathcal{I}$ is always a Pfaffian system with independence condition.

For a Pfaffian system with independence condition $(\mathcal{I}, \Omega)$ the first prolongation is defined to consist of the canonical system with independence condition on $V_n(M, \Omega)$ restricted to the components of $V_n(\mathcal{I}, \Omega)$.

In practice, we are usually interested in the integrals of $\mathcal{I}$ (or $(\mathcal{I}, \Omega)$) which satisfy some additional conditions. This often has the effect of restricting our attention to a particular smooth stratum in $V_n(\mathcal{I}, \Omega)$ anyway.

Before going on to the general theory of prolongation, we will discuss several examples. These examples will be used to motivate the development of the theory in the later sections of this chapter.

Example 1.1. Consider the differential system which describes the simultaneous solutions of the pair of differential equations

$$(3) \qquad \frac{\partial^2 u}{\partial x^2} - \frac{\partial^2 u}{\partial y^2} = \frac{\partial^2 u}{\partial x^2} - \frac{\partial^2 u}{\partial z^2} = 0.$$

These equations describe a submanifold M of $J^2(\mathbb{R}^3, \mathbb{R})$. With the usual coordinates $(x^i, u, p_i, p_{ij} = p_{ji})$ on $J^2(\mathbb{R}^3, \mathbb{R})$, M is given by the equations $p_{22} = p_{33} = p_{11}$. Comparing this system with the example in Chapter IV about pairs of second order equations, we note that the symbol quadrics are spanned by $\{(\xi_1)^2 - (\xi_2)^2, (\xi_1)^2 - (\xi_3)^2\}$ and, since these have no common divisor, the system (3) of P.D.E. is not involutive.

The differential system $\mathcal{I}$ is the restriction of the contact system on $J^2(\mathbb{R}^3, \mathbb{R})$ to M and thus is generated by the Pfaffian system I spanned by the four 1-forms

$$
\begin{aligned}
\vartheta &= du - p_1 dx^1 - p_2 dx^2 - p_3 dx^3 \\[4pt]
\vartheta_1 &= dp_1 - p_{11} dx^1 - p_{12} dx^2 - p_{13} dx^3 \\[4pt]
\vartheta_2 &= dp_2 - p_{12} dx^1 - p_{11} dx^2 - p_{23} dx^3 \\[4pt]
\vartheta_3 &= dp_3 - p_{13} dx^1 - p_{23} dx^2 - p_{11} dx^3.
\end{aligned}
$$

(4)

The independence condition is given by $\Omega = dx^1 \wedge dx^2 \wedge dx^3$. Setting $\pi_{ij} = dp_{ij}$ and $\omega^i = dx^i$, the structure equations become

$$(5) \qquad d\begin{bmatrix} \vartheta \\ \vartheta_1 \\ \vartheta_2 \\ \vartheta_3 \end{bmatrix} \equiv - \begin{bmatrix} 0 & 0 & 0 \\ \pi_{11} & \pi_{12} & \pi_{13} \\ \pi_{12} & \pi_{11} & \pi_{23} \\ \pi_{13} & \pi_{23} & \pi_{11} \end{bmatrix} \wedge \begin{bmatrix} \omega^1 \\ \omega^2 \\ \omega^3 \end{bmatrix} \quad \mathrm{mod}\ I.$$

It is straightforward to compute that $s'_0 = 4$, $s'_1 = 3$, $s'_2 = 1$, and $s'_3 = 0$. By Cartan's Test, the system will be involutive if the integral elements of $(\mathcal{I}, \Omega)$ at each point of M form an (affine) space of dimension $5 = s'_1 + 2s'_2 + 3s'_3$. However, calculation yields that the integral elements at each point of M are described by 4 parameters. Explicitly, for any four real numbers r_1, r_2, r_3, r_4, the 3-plane based at any point of M which is annihilated by the 1-forms $\vartheta, \vartheta_1, \vartheta_2, \vartheta_3$ and the 1-forms

$$(6) \qquad \begin{bmatrix} \vartheta_4 \\ \vartheta_5 \\ \vartheta_6 \\ \vartheta_7 \end{bmatrix} = \begin{bmatrix} \pi_{11} \\ \pi_{12} \\ \pi_{13} \\ \pi_{23} \end{bmatrix} - \begin{bmatrix} r_1 & r_2 & r_3 \\ r_2 & r_1 & r_4 \\ r_3 & r_4 & r_1 \\ r_4 & r_3 & r_2 \end{bmatrix} \begin{bmatrix} \omega^1 \\ \omega^2 \\ \omega^3 \end{bmatrix}$$

is an integral element of $(\mathcal{I}, \Omega)$ and every such integral element is of this form.

Thus, $V_3(\mathcal{I}, \Omega) = M \times \mathbb{R}^4$ and $I^{(1)}$ is generated on $M^{(1)} = M \times \mathbb{R}^4$ by the eight 1-forms $\{\vartheta, \vartheta_1, \vartheta_2, \ldots, \vartheta_7\}$. The structure equations of $I^{(1)}$ are

$$d\vartheta \equiv d\vartheta_1 \equiv d\vartheta_2 \equiv d\vartheta_3 \equiv 0 \mod I^{(1)}$$

$$(7) \qquad d\begin{bmatrix} \vartheta_4 \\ \vartheta_5 \\ \vartheta_6 \\ \vartheta_7 \end{bmatrix} = -\begin{bmatrix} \pi_1 & \pi_2 & \pi_3 \\ \pi_2 & \pi_1 & \pi_4 \\ \pi_3 & \pi_4 & \pi_1 \\ \pi_4 & \pi_3 & \pi_2 \end{bmatrix} \wedge \begin{bmatrix} \omega^1 \\ \omega^2 \\ \omega^3 \end{bmatrix} \mod I^{(1)}.$$

where we have set $\pi_i = dr_i$.

The sequence of reduced Cartan characters of $\mathcal{I}^{(1)}$ is easily computed to be $(s_0', s_1', s_2', s_3') = (8, 4, 0, 0)$. Moreover, the space of integral elements of $(\mathcal{I}^{(1)}, \Omega)$ at each point of $M^{(1)}$ can be seen to be parametrized by 4 ($= s_1' + 2s_2' + 3s_3'$) parameters t_1, t_2, t_3, t_4 in such a way that the annihilator of the corresponding integral element at any point of $M^{(1)}$ is spanned by the eight 1-forms $\{\vartheta, \vartheta_1, \vartheta_2, \ldots, \vartheta_7\}$ together with the four 1-forms

$$(8) \qquad \begin{bmatrix} \vartheta_8 \\ \vartheta_9 \\ \vartheta_{10} \\ \vartheta_{11} \end{bmatrix} = \begin{bmatrix} \pi_1 \\ \pi_2 \\ \pi_3 \\ \pi_4 \end{bmatrix} - \begin{bmatrix} t_1 & t_2 & t_3 \\ t_2 & t_1 & t_4 \\ t_3 & t_4 & t_1 \\ t_4 & t_3 & t_2 \end{bmatrix} \begin{bmatrix} \omega^1 \\ \omega^2 \\ \omega^3 \end{bmatrix}.$$

It then follows, by Cartan's Test, that the system $(\mathcal{I}^{(1)}, \Omega)$ is involutive on $M^{(1)}$ with Cartan character sequence $(s_0, s_1, s_2, s_3) = (8, 4, 0, 0)$ even though $(\mathcal{I}, \Omega)$ is not involutive on M. Since the integrals of $(\mathcal{I}^{(1)}, \Omega)$ on $M^{(1)}$ and $(\mathcal{I}, \Omega)$ on M are in one-to-one correspondence, we see that we may actually study the integrals of $(\mathcal{I}, \Omega)$ by applying the Cartan–Kähler theorem to $(\mathcal{I}^{(1)}, \Omega)$.

The fact that $s_1 = 4$ is the last non-zero Cartan character of $\mathcal{I}^{(1)}$ indicates that the "general" integral manifold of $(\mathcal{I}^{(1)}, \Omega)$ should depend on four functions of one variable. Note also that the characteristic variety in each integral element E of $(\mathcal{I}^{(1)}, \Omega)$ consists of the four points $[dx^1 \pm dx^2 \pm dx^3]$ in $\mathbb{P}E^*$. Now, in this particular example, it is possible to explicitly write down the general solution of (3). This general solution takes the form

$$(9) \qquad \begin{aligned} u(x^1, x^2, x^3) &= f_{++}(x^1 + x^2 + x^3) + f_{+-}(x^1 + x^2 - x^3) \\ &\quad + f_{-+}(x^1 - x^2 + x^3) + f_{--}(x^1 - x^2 - x^3) \end{aligned}$$

where the functions $f_{\pm\pm}$ are four arbitrary functions of 1 variable.

However, it is not usually possible to write down the "general" solution of a P.D.E. so explicitly. For example, the reader might try analyzing the system

$$(10) \qquad \frac{\partial^2 u}{\partial x^2} + \lambda_1 u = \frac{\partial^2 u}{\partial y^2} + \lambda_2 u = \frac{\partial^2 u}{\partial z^2} + \lambda_3 u$$

where the λ_i are arbitrary constants. Again, it turns out that the first prolongation on $M^{(1)}$ is involutive even though the natural system on $M \subset J^2(\mathbb{R}^3, \mathbb{R})$ is not.

The next example illustrates the fact that the prolongation process may need to be iterated several times before the resulting system becomes involutive.

Example 1.2. Consider the pair of equations for u as a function of x and y

$$(11) \qquad \frac{\partial^n u}{\partial x^n} = \frac{\partial^n u}{\partial y^n} = 0.$$

In a departure from our usual notation for coordinates on $J^n(\mathbb{R}^2, \mathbb{R})$, let us use $p_{i,j}$ to represent $\partial^{i+j} u/\partial x^i \partial y^j$. Thus, $p_{0,0} = u$ and the equations (11) correspond to the submanifold $M \subset J^n(\mathbb{R}^2, \mathbb{R})$ given by the equations $p_{0,n} = p_{n,0} = 0$. The contact Pfaffian system on $J^n(\mathbb{R}^2, \mathbb{R})$ is then generated by the 1-forms $\{\vartheta_{i,j} \mid i + j < n\}$ where

$$(12) \qquad \vartheta_{i,j} = dp_{i,j} - p_{i+1,j} dx - p_{i,j+1} dy.$$

We let I denote the restriction of this Pfaffian system to M. The structure equations of I then can be written as

$$(13a) \qquad d\vartheta_{i,j} \equiv 0 \quad \text{for} \quad 0 \le i + j < n - 1$$

$$(13b) \qquad \begin{pmatrix} d\vartheta_{n-1,0} \\ d\vartheta_{n-2,1} \\ \vdots \\ d\vartheta_{0,n-1} \end{pmatrix} \equiv - \begin{pmatrix} 0 & \pi_{n-1,1} \\ \pi_{n-1,1} & \pi_{n-2,2} \\ \vdots & \vdots \\ \pi_{1,n-1} & 0 \end{pmatrix} \wedge \begin{pmatrix} dx \\ dy \end{pmatrix}$$

where the congruences are taken $\mod I$ and $\pi_{k,l} = dp_{k,l}$ for $k + l = n$.

It is straightforward to compute that $s_1' = n - 1$ and $s_2' = 0$. Thus, in order for $(\mathcal{I}, \Omega)$ to be involutive (where $\Omega = dx \wedge dy$), it would be necessary for there to exist an $(n-1)$-parameter family of integral elements of $(\mathcal{I}, \Omega)$

at each point of M. However, it is easy to see that there is only an $(n-2)$-parameter family of integral elements of $(\mathcal{I}, \Omega)$ at each point of M. In fact, $M^{(1)} = M \times \mathbb{R}^{n-2}$ and we can introduce coordinates $\{p_{i,j} \mid 2 \leq i, j; i+j = n+1\}$ on the $\mathbb{R}^{n-2}$-factor so that the annihilator of the corresponding integral element is spanned by the 1-forms $\{\vartheta_{i,j} \mid 1 \leq i+j \leq n-1\}$ and the 1-forms

$$(14) \qquad \begin{pmatrix} \vartheta_{n-1,1} \\ \vartheta_{n-2,2} \\ \vdots \\ \vartheta_{1,n-1} \end{pmatrix} \equiv \begin{pmatrix} \pi_{n-1,1} \\ \pi_{n-2,2} \\ \vdots \\ \pi_{1,n-1} \end{pmatrix} - \begin{pmatrix} 0 & p_{n-1,2} \\ p_{n-1,2} & p_{n-2,3} \\ \vdots & \vdots \\ p_{2,n-1} & 0 \end{pmatrix} \begin{pmatrix} dx \\ dy \end{pmatrix}.$$

Note that $M^{(1)}$ can be regarded as the submanifold of $J^{n+1}(\mathbb{R}^2, \mathbb{R})$ defined by the equations

$$(15) \qquad p_{0,n} = p_{n,0} = p_{0,n+1} = p_{1,n} = p_{n,1} = p_{n+1,0} = 0.$$

Under this identification, $I^{(1)}$ becomes the restriction to $M^{(1)}$ of the contact Pfaffian system on $J^{n+1}(\mathbb{R}^2, \mathbb{R})$.

The structure equations of $I^{(1)}$ are easily seen to be

$$(16a) \qquad\qquad d\vartheta_{i,j} \equiv 0 \quad \text{for} \quad 0 \leq i+j < n$$

$$(16b) \qquad \begin{pmatrix} d\vartheta_{n-1,1} \\ d\vartheta_{n-2,2} \\ \vdots \\ d\vartheta_{1,n-1} \end{pmatrix} \equiv - \begin{pmatrix} 0 & \pi_{n-1,2} \\ \pi_{n-1,2} & \pi_{n-2,3} \\ \vdots & \vdots \\ \pi_{2,n-1} & 0 \end{pmatrix} \wedge \begin{pmatrix} dx \\ dy \end{pmatrix}$$

where the congruences are taken $\mod I^{(1)}$ and $\pi_{k,l} = dp_{k,l}$ for $k+l = n+1$.

We compute that $s_1' = n-2$ and $s_2' = 0$, but that, for $n \geq 3$, there is only an $(n-3)$-parameter family of integral elements of $(\mathcal{I}^{(1)}, \Omega)$ at each point of $M^{(1)}$. Thus, for $n \geq 3$, $(\mathcal{I}^{(1)}, \Omega)$ is not involutive.

It is easy to continue this process. If we inductively define $(\mathcal{I}^{(k+1)}, \Omega) = ((\mathcal{I}^{(k)})^{(1)}, \Omega)$, we find that, for $k \leq n-1$, the system $\mathcal{I}^{(k)}$ is diffeomorphic to the restriction of the contact Pfaffian system on $J^{n+k}(\mathbb{R}^2, \mathbb{R})$ to a submanifold $M^{(k)}$ defined by the equations

$$(17) \qquad p_{i,j} = 0 \text{ whenever } n \leq i+j \leq n+k \text{ and } \min\{i,j\} \leq k.$$

For this system, $s_1' = n-k-1$ and $s_2' = 0$. However, when $k \leq n-2$, the dimension of the space of integral elements of $(\mathcal{I}^{(k)}, \Omega)$ at any point of $M^{(k)}$ is only $n-k-2$. Thus, the system $(\mathcal{I}^{(k)}, \Omega)$ is not involutive on $M^{(k)}$ for $k \leq n-2$.

However, it is also easy to see that the Pfaffian system $I^{(n-1)}$ on $M^{(n-1)}$ is a Frobenius system and hence the system $(\mathcal{I}^{(n-1)}, \Omega)$ is involutive on $M^{(n-1)}$.

Since the rank of $I^{(n-1)}$ is n^2, it follows that there is an n^2-parameter family of local solutions of the system (11). This was expected since the solutions of (11) are clearly the polynomials in x and y whose x-degree and y-degree are less than or equal to $n - 1$.

In the next example, we treat a problem involving the geometry of Riemannian submersions in which the corresponding $V_n(\mathcal{I})$ has several components.

Example 1.3 (Riemannian submersions). Let (N^3, dx^2) be a Riemannian manifold with constant sectional curvature K. Our problem is to classify the Riemannian submersions $f : (N^3, dx^2) \to (\Sigma^2, d\sigma^2)$ where $(\Sigma^2, d\sigma^2)$ is a Riemannian surface. We do not specify the metric on Σ^2 in advance.

Given such an f, we may locally choose an orthonormal frame field $\mathbf{e} = (e_1, e_2, e_3)$ on N so that e_3 is tangent to the fibers of f. Then the hypothesis that f be a Riemannian submersion is equivalent to the condition that $f^*(d\sigma^2) = (\eta_1)^2 + (\eta_2)^2$ where $\boldsymbol{\eta} = (\eta_1, \eta_2, \eta_3)$ is the coframing dual to the frame field $\mathbf{e}$. It follows that there exists a "connection form" γ in the domain of $\mathbf{e}$ so that $d\eta_1 = \gamma \wedge \eta_2$ and $d\eta_2 = -\gamma \wedge \eta_1$.

Conversely, if $\boldsymbol{\eta} = (\eta_1, \eta_2, \eta_3)$ is any orthonormal coframing on an open set $U \subset N$ such that there exists a 1-form γ satisfying $d\eta_1 = \gamma \wedge \eta_2$ and $d\eta_2 = -\gamma \wedge \eta_1$, then it is easy to see that $(\eta_1)^2 + (\eta_2)^2$ is a well defined quadratic form on $\Sigma^2 = U/\mathcal{F}$ where $\mathcal{F}$ is the foliation of U by the integral curves of the vector field e_3 on U, i.e., the integral curves of the system $\eta_1 = \eta_2 = 0$ on U. It follows that the projection $f : U \to U/\mathcal{F}$ is a Riemannian submersion.

For any local orthonormal coframing $\boldsymbol{\eta} = (\eta_1, \eta_2, \eta_3)$ on $U \subset N$, there exist unique 1-forms (known as the Levi–Civita connection forms) $\eta_{ij} = -\eta_{ji}$ which satisfy the "symmetry" condition $d\eta_i = -\eta_{ij} \wedge \eta_j$. The following equations for γ

$$
\begin{aligned}
-\eta_{12} \wedge \eta_2 - \eta_{13} \wedge \eta_3 &= d\eta_1 = \gamma \wedge \eta_2 \\
-\eta_{21} \wedge \eta_1 - \eta_{23} \wedge \eta_3 &= d\eta_2 = -\gamma \wedge \eta_1
\end{aligned}
\tag{18}
$$

are satisfiable only if there exist functions a, b, c on U so that

$$
\begin{aligned}
\eta_{13} &= a\eta_2 + b\eta_3 \\
\eta_{23} &= -a\eta_1 + c\eta_3.
\end{aligned}
\tag{19}
$$

Conversely, the existence of such functions is sufficient to yield a solution, namely $\gamma = \eta_{21} + a\eta_3$, to (18).

Thus the search for the desired Riemannian submersions f is locally equivalent to the search for orthonormal coframings $\boldsymbol{\eta}$ which satisfy certain first order differential equations. Actually, the coframing $\boldsymbol{\eta}$ carries slightly more information since two coframings determine the same Riemannian submersion iff their corresponding e_3-foliations are the same. We shall see what becomes of this ambiguity in what follows.

Using the above discussion as our guide, we may now set up a differential system to find the desired framings as follows. Let $\mathcal{F} \to N^3$ denote the orthonormal frame bundle of (N^3, dx^2) and let $\{\omega_i, \omega_{ij} = -\omega_{ji}\}$ denote the canonical and connection 1-forms on $\mathcal{F}$. They satisfy the first and second structure equations of Cartan:

$$(20) \qquad \begin{aligned} d\omega_i &= -\omega_{ij} \wedge \omega_j \\ d\omega_{ij} &= -\omega_{ik} \wedge \omega_{kj} + K\omega_i \wedge \omega_j. \end{aligned}$$

(Here, we use the summation convention. Recall that ds^2 has constant sectional curvature K.)

A (local) framing $\mathbf{e}$ is a (local) section $\mathbf{e} : U \to \mathcal{F}$ for an open set $U \subset N$. The forms $\{\omega_i, \omega_{ij} = -\omega_{ji}\}$ have the property that $\mathbf{e}^*(\omega_i) = \eta_i$ and $\mathbf{e}^*(\omega_{ij}) = \eta_{ij}$ where $\boldsymbol{\eta} = (\eta_1, \eta_2, \eta_3)$ is the local coframing which is dual to $\mathbf{e}$. As we have already said, we are seeking (local) framings $\mathbf{e}$ for which there exists functions a, b, c so that (19) holds. This motivates us to define $M = \mathcal{F} \times \mathbb{R}^3$ (where we use a, b, and c as linear coordinates on the $\mathbb{R}^3$-factor) and let I be the Pfaffian system generated by the 1-forms

$$(21) \qquad \begin{aligned} \vartheta_1 &= \omega_{13} - a\omega_2 - b\omega_3 \\ \vartheta_2 &= \omega_{23} + a\omega_1 - c\omega_2. \end{aligned}$$

It is clear that every framing $\mathbf{e}$ which satisfies (19) gives rise to a unique integral manifold $N_\mathbf{e}$ of I in M on which the form $\Omega = \omega_1 \wedge \omega_2 \wedge \omega_3$ does not vanish. Conversely, if $\mathcal{I}$ is the differential system generated by I on M then every integral of $(\mathcal{I}, \Omega)$ in M is locally of the form $N_\mathbf{e}$ for some framing $\mathbf{e}$ which satisfies (19).

It is easy to compute that the structure equations of I are given by

$$(22) \qquad \begin{aligned} d\vartheta_1 &\equiv -\pi_3 \wedge \omega_2 - \pi_4 \wedge \omega_3 \\ d\vartheta_2 &\equiv \pi_3 \wedge \omega_1 - \pi_5 \wedge \omega_3 \end{aligned} \qquad \mathrm{mod}\ I$$

where

$$(23) \qquad \begin{aligned} \pi_3 &= da & &-2a(b\omega_1 + c\omega_2) \\ \pi_4 &= db + c\omega_{12} + (a^2 - K)\omega_1 & &-b(b\omega_1 + c\omega_2) \\ \pi_5 &= dc - b\omega_{12} + (a^2 - K)\omega_2 & &-c(b\omega_1 + c\omega_2). \end{aligned}$$

Notice that the structure equations (22) imply that I has a one-dimensional Cauchy characteristic system spanned by the vector field X on M which satisfies $\omega_{12}(X) = 1$ and $\alpha(X) = 0$ for $\alpha = \omega_1, \omega_2, \omega_3, \vartheta_1, \vartheta_2, \pi_3, \pi_4$, or π_5. It is easy to see that the flow of this Cauchy vector field generates an S^1-action on M which corresponds to the rotation of a frame $\mathbf{e}$ which fixes the e_3-component. We could get rid of this S^1-action by passing to the quotient $(M \times \mathbb{R}^3)/S^1$ and working with the corresponding Pfaffian system there, but this causes computational difficulties since the quotient manifold has no natural coframing. Instead, our approach will be to augment the independence condition to $\Omega_+ = \omega_1 \wedge \omega_2 \wedge \omega_3 \wedge \omega_{12}$ and look for integrals of the system $(\mathcal{I}, \Omega_+)$. This has the added advantage that the integrals of this system correspond essentially uniquely to the local Riemannian submersions, as the reader can easily see.

Now by the structure equations (22), it is clear that the reduced Cartan character sequence of $(\mathcal{I}, \Omega_+)$ is $(s'_0, s'_1, s'_2, s'_3, s'_4) = (2, 2, 1, 0, 0)$. In order to have involutivity, Cartan's Test thus requires that there be a 4 ($= s'_1 + 2s'_2 + 3s'_3 + 4s'_4$) parameter family of integral elements of $(\mathcal{I}, \Omega_+)$ at every point of M. However, it is easy to see that the integral elements of $(\mathcal{I}, \Omega_+)$ at a point of M are parametrized by only 3 parameters. Namely, for any real numbers p, q, r, the 4-plane at each point of M which is annihilated by the 1-forms ϑ_1, ϑ_2 and the 1-forms

$$
(24) \qquad \begin{bmatrix} \vartheta_3 \\ \vartheta_4 \\ \vartheta_5 \end{bmatrix} = \begin{bmatrix} \pi_3 \\ \pi_4 \\ \pi_5 \end{bmatrix} - \begin{bmatrix} 0 & 0 & p \\ 0 & p & q \\ -p & 0 & r \end{bmatrix} \begin{bmatrix} \omega_1 \\ \omega_2 \\ \omega_3 \end{bmatrix}
$$

is an integral element of $(\mathcal{I}, \Omega_+)$ and every such integral element is of this form.

Thus, $V_4(\mathcal{I}, \Omega_+) = M^{(1)}$ is diffeomorphic to $M \times \mathbb{R}^3$ (where we use p, q, r as coordinates on the $\mathbb{R}^3$-factor) and, on $M^{(1)}$, $I^{(1)}$ is spanned by the 1-forms $\vartheta_1, \vartheta_2, \vartheta_3, \vartheta_4$, and ϑ_5.

The structure equations of $I^{(1)}$ can now be calculated to be

$$
(25a) \qquad\qquad d\vartheta_1 \equiv d\vartheta_2 \equiv 0
$$

$$
(25b) \qquad \begin{bmatrix} d\vartheta_3 \\ d\vartheta_4 \\ d\vartheta_5 \end{bmatrix} \equiv - \begin{bmatrix} 0 & 0 & \pi_6 \\ 0 & \pi_6 & \pi_7 \\ -\pi_6 & 0 & \pi_8 \end{bmatrix} \wedge \begin{bmatrix} \omega_1 \\ \omega_2 \\ \omega_3 \end{bmatrix} + \begin{bmatrix} 6ap\,\omega_1 \wedge \omega_2 \\ 0 \\ 0 \end{bmatrix}
$$

where all of the congruences are taken   mod $I^{(1)}$ and π_6, π_7, and π_8 restrict to each fiber of $M^{(1)} \to M$ to become dp, dq, and dr respectively.

Note that (25b) implies that the locus $V_4(\mathcal{I}^{(1)}, \Omega_+) \subset V_4(M^{(1)})$ lies entirely over the locus in $M^{(1)}$ defined by the equation $ap = 0$. In fact, if we let $A \subset M^{(1)}$ denote the submanifold defined by the equation $a = 0$ and let $P \subset M^{(1)}$ denote the submanifold defined by the equation $p = 0$, then $V_4(\mathcal{I}^{(1)}, \Omega_+)$ can be written as the (non-disjoint) union $A^{(1)} \cup P^{(1)}$ where $A^{(1)} \cong A \times \mathbb{R}^3$ (respectively, $P^{(1)} \cong P \times \mathbb{R}^3$) is the submanifold of $V_4(M^{(1)})$ consisting of those elements of $V_4(\mathcal{I}^{(1)}, \Omega_+)$ whose base point lies in A (resp., P) and (using coordinates t, u, v on $\mathbb{R}^3$) such that the corresponding integral element is annihilated by the 1-forms $\vartheta_1, \vartheta_2, \vartheta_3, \vartheta_4, \vartheta_5$ and the 1-forms

$$
(26) \qquad \begin{bmatrix} \vartheta_6 \\ \vartheta_7 \\ \vartheta_8 \end{bmatrix} = \begin{bmatrix} \pi_6 \\ \pi_7 \\ \pi_8 \end{bmatrix} - \begin{bmatrix} 0 & 0 & t \\ 0 & t & u \\ -t & 0 & v \end{bmatrix} \begin{bmatrix} \omega_1 \\ \omega_2 \\ \omega_3 \end{bmatrix}.
$$

Thus, $V_4(\mathcal{I}^{(1)}, \Omega_+)$ is singular, being the union of two smooth manifolds of dimension 14 in $G_4(M^{(1)})$ which intersect transversely along a submanifold diffeomorphic to $(A \cap P) \times \mathbb{R}^3$ of dimension 13. According to the stratification procedure outlined at the beginning of this section, we should regard each of the strata $A^{(1)} \cap P^{(1)}$, $A^{(1)} \backslash (A^{(1)} \cap P^{(1)})$, and $P^{(1)} \backslash (A^{(1)} \cap P^{(1)})$ as separate manifolds on which to define the systems $\mathcal{I}^{(2)}$.

Let us begin with the stratum $(A \backslash (A \cap P)) \times \mathbb{R}^3 = A^{(1)} \backslash (A^{(1)} \cap P^{(1)})$ in $M^{(2)}$. On $A \backslash (A \cap P) \subset M^{(1)}$, we have a $a = 0$ and $p \neq 0$. Looking back at the formulas (23) and (24), we see that, restricted to $A \backslash (A \cap P)$, we have $\vartheta_3 = -p\omega_3$. In particular, $\omega_3 \equiv 0 \mod I^{(1)}$, so there cannot be any integral elements of $(\mathcal{I}^{(2)}, \Omega_+)$ at points of $A^{(1)} \backslash (A^{(1)} \cap P^{(1)}) \subset M^{(2)}$. Thus, there are no integral manifolds of $(\mathcal{I}, \Omega_+)$ whose canonical lifts lie in $A \backslash (A \cap P) \subset M^{(1)}$. It follows that we may ignore the stratum $A^{(1)} \backslash (A^{(1)} \cap P^{(1)})$ in $M^{(2)}$ for the remainder of the discussion.

The remaining two strata fit together to be the smooth submanifold $P^{(1)} \subset V_4(\mathcal{I}^{(1)}, \Omega_+)$. For simplicity, we shall therefore let $(\mathcal{I}^{(2)}, \Omega_+)$ denote the restriction of the canonical differential system on $V_4(M^{(1)})$ to $P^{(1)} \subset M^{(2)}$.

Since $p = 0$ on $P^{(1)}$, an easy calculation gives that, on $P^{(1)}$

$$
\pi_6 = 2a((ac - q)\omega_1 - (ab + r)\omega_2)
$$

and hence that

$$
\vartheta_6 = 2a(ac - q)\omega_1 - 2a(ab + r)\omega_2 - t\omega_3.
$$

Thus, away from the locus Z defined by the equations

$$
2a(ac - q) = 2a(ab + r) = t = 0,
$$

we see that no integral element of $I^{(2)}$ can satisfy the independence condition. This locus Z is the union of two smooth submanifolds: Z_1 which is the locus defined by the equations $p = t = a = 0$ and Z_2 which is the locus defined by the equations $p = t = q - ac = r + ab = 0$.

First, consider the locus $Z_1 \cong (A \cap P) \times \mathbb{R}^2$ where we use u, v as coordinates on the $\mathbb{R}^2$-factor. If we restrict the system $I^{(1)}$ to $A \cap P \subset M^{(1)}$, then we may compute that $\vartheta_3 = \pi_6 = 0$. The structure equations for $I^{(1)}$ restricted to $A \cap P$ may now be computed to be

$$d\vartheta_1 \equiv d\vartheta_2 \equiv 0$$

$$(27) \qquad d\vartheta_4 \equiv -\pi_7 \wedge \omega_3 \qquad \mathrm{mod}\ I^{(1)}$$

$$d\vartheta_5 \equiv -\pi_8 \wedge \omega_3.$$

The reduced Cartan character sequence for $(\mathcal{I}^{(1)}, \Omega_+)$ restricted to $A \cap P$ is then clearly $(s_0', s_1', s_2', s_3', s_4') = (4, 2, 0, 0, 0)$. Moreover, it is clear that there *does* exist a 2-parameter family of integral elements of $(\mathcal{I}^{(1)}, \Omega_+)$ at each point of $A \cap P$. In fact, $Z_1 = V_4(\mathcal{I}^{(1)}, \Omega_+)$ when the underlying manifold is $A \cap P$. Thus, by Cartan's Test, $(\mathcal{I}^{(1)}, \Omega_+)$ is involutive on $A \cap P$. From this it is easy to see that $(\mathcal{I}^{(2)}, \Omega_+)$ is involutive on Z_1. (A direct proof is easy in this case, but see the next section.) It is interesting to note that if we let $Z_0 \subset M$ denote the locus defined by the equation $a = 0$, then $(\mathcal{I}, \Omega_+)$ is actually involutive (with Cartan character sequence $(s_0, s_1, s_2, s_3, s_4) = (2, 2, 0, 0, 0)$) when restricted to Z_0. Moreover, when $(\mathcal{I}, \Omega_+)$ is restricted to Z_0, we get $V_4(\mathcal{I}, \Omega_+) = A \cap P$.

Next, consider the locus Z_2. It is straightforward to calculate that, restricted to Z_2, we have $\pi_6 = \pi_7 = \pi_8 = 0$. It follows from (26) that there are no integral elements of $(\mathcal{I}^{(2)}, \Omega_+)$ restricted to Z_2 except along the sublocus $Z_3 \subset Z_2$ defined by the additional conditions $u = v = 0$. When we restrict to Z_3, then the forms ϑ_6, ϑ_7, and ϑ_8 all vanish and the structure equations (26) imply that the remaining 1-forms $\vartheta_1, \vartheta_2, \vartheta_3, \vartheta_4$, and ϑ_5 in $I^{(2)}$ form a Frobenius system. Note that Z_3 is actually diffeomorphic to M via its natural projection to M.

Thus, by prolongation, we arrive at the following classification of the framings which correspond to Riemannian submersions:

There are two types of such framings.

The first type consists of the integrals of the involutive system $(\mathcal{I}, \Omega_+)$ on $\mathcal{F} \times \mathbb{R}^2$ where $\mathcal{I}$ is generated by the rank 2 Pfaffian system I which is generated by the two 1-forms

$$\vartheta_1 = \omega_{13} - b\omega_3$$

$$(28)$$

$$\vartheta_2 = \omega_{23} - c\omega_3.$$

The Cartan character sequence of $(\mathcal{I}, \Omega_+)$ is $(s_0, s_1, s_2, s_3, s_4) = (2, 2, 0, 0, 0)$. Thus, the general solutions depend on two functions of one variable.

In fact, it is quite easy to describe these solutions geometrically using the Cauchy characteristic foliation of this system. The reader may want to verify the following description of the corresponding Riemannian submersions: Note that the totally geodesic surfaces in N^3 depend on 3 parameters. Let Γ^3 denote this space. Any curve γ in Γ which is in "general position" represents a 1-parameter family of totally geodesic surfaces in N^3 which foliates an open set $U \subset N$. Let $\mathcal{F}_\gamma$ denote this foliation on U. Let $\mathcal{L}_\gamma$ denote the orthogonal foliation of U by curves. Then the projection $U \to U/\mathcal{L}_\gamma$ is a Riemannian submersion where the quotient metric is such that each of the leaves of $\mathcal{F}_\gamma$ projects isometrically onto $U/\mathcal{L}_\gamma$. Of course, in this local description, we are ignoring all the difficulties caused by the (possible) non-Hausdorf nature of the quotient.

The second type corresponds to the integrals of the Frobenius system $\mathcal{I}$ on $\mathcal{F} \times \mathbb{R}^3$ generated by the 1-forms

$$
\begin{aligned}
&\vartheta_1 = \omega_{13} && -a\omega_2 && -b\omega_3 \\
&\vartheta_2 = \omega_{23} && +a\omega_1 && -c\omega_3 \\
(29) \quad &\vartheta_3 = da - 2a(b\omega_1 + c\omega_2) && && \\
&\vartheta_4 = db + c\omega_{12} && +(a^2 - K)\omega_1 \ -b(b\omega_1 + c\omega_2) \ -ac\omega_3 \\
&\vartheta_5 = dc - b\omega_{12} && +(a^2 - K)\omega_2 \ -c(b\omega_1 + c\omega_2) \ +ab\omega_3.
\end{aligned}
$$

We leave the geometric analysis of these integrals and the corresponding Riemannian submersions as an interesting exercise for the reader.

§2. Ordinary Prolongation.

In this section, we examine the effect that prolongation has when applied to a component $Z \subset V_n(\mathcal{I})$ consisting of ordinary integral elements of a differential system $\mathcal{I}$. For convenience, we shall assume that $\mathcal{I}$ is generated in positive degree. The following result is due to Matsushima [1953].

Theorem 2.1. *Let $\mathcal{I} \subset \Omega^*(M)$ be a differential ideal which is generated in positive degree (i.e., $\mathcal{I}$ contains no non-zero functions). Let $Z \subset V_n(\mathcal{I})$ be a connected component of the space of ordinary integral elements of $\mathcal{I}$. Let $(s_0, \ldots, s_n)$ be the sequence of Cartan characters of Z. Let $(\mathcal{I}^{(1)}, \Omega)$ be the restriction to Z of the canonical Pfaffian differential system with independence condition on $G_n(TM)$. Then $(\mathcal{I}^{(1)}, \Omega)$ is linear and is involutive on Z. Moreover, the sequence of Cartan characters of $V_n(\mathcal{I}^{(1)}, \Omega)$ is given by $s_p^{(1)} = s_p + s_{p+1} + \cdots + s_n$ for all $0 \leq p \leq n$.*

Proof. Without loss of generality, we may assume that $\mathcal{I}$ contains all forms on M of degree $n + 1$ or greater. (If not, enlarging $\mathcal{I}$ by adjoining all such forms will not affect our hypotheses on Z nor will it affect the sequence of

Cartan characters of Z.) Thus, $\mathcal{I}$ has no integral elements of dimension larger than n and $H(E) = E$ for $E \in Z$.

Let $s = \dim M - n$. To avoid trivialities, we shall assume that s is positive. As usual, we let $c_p = s_0 + \cdots + s_p$ for $0 \leq p \leq n$, and set $c_{-1} = 0$ for convenience. Note that $c_n = s_0 + \cdots + s_n = s$ is the rank of the polar equations of any integral element $E \in Z$. Recall from Chapter III that Z is a smooth submanifold of $G_n(TM)$ of codimension $c_0 + c_2 + \cdots + c_{n-1} = ns_0 + (n-1)s_1 + \cdots + s_{n-1} = ns - (s_0 + 2s_2 + \cdots + ns_n)$.

The conclusions of the theorem are local statements about the system $\mathcal{I}^{(1)}$, so it suffices to examine the differential ideal $\mathcal{I}^{(1)}$ in a neighborhood of an arbitrary element of Z. Let E be an element of Z and let $z \in M$ be its base point. Then there exists a local coordinate system $x^1, \ldots, x^n, y^1, \ldots, y^s$ centered at z on a z-neighborhood $U \subset M$ so that E is spanned by the vectors $\{\partial/\partial x^i \mid 1 \leq i \leq n\}$ at z. Moreover, we may assume that, for all $p \leq n$, the subspace $E_p \subset E$ spanned by the vectors $\{\partial/\partial x^i \mid 1 \leq i \leq p\}$ at z is a regular p-dimensional element of $\mathcal{I}$ and that the polar space $H(E_p)$ is spanned by the vectors $\{\partial/\partial x^i \mid 1 \leq i \leq n\}$ at z together with the vectors $\{\partial/\partial y^a \mid a > c_p\}$ at z. In particular, the polar equations $\mathcal{E}(E_p)$ are spanned by the 1-forms $\{dy^a \mid a \leq c_p\}$ at z.

Let $\Omega = dx^1 \wedge dx^2 \wedge \cdots \wedge dx^n$. As is our usual convention, let $G_n(TU, \Omega)$ denote the space of n-planes in $G_n(TU)$ on which Ω restricts to be non-zero. We define the functions p_i^a on $G_n(TU, \Omega)$ as usual so that $\tilde{E} \in G_n(TU, \Omega)$ is annihilated by the 1-forms $dy^a - p_i^a(\tilde{E})dx^i$. The functions (x, y, p) then form a coordinate system on $G_n(TU, \Omega)$ centered at E. Moreover, the 1-forms $\vartheta^a = dy^a - p_i^a dx^i$ span the canonical Pfaffian system on $G_n(TU, \Omega) \subset G_n(TM)$. Also, in accordance with our earlier notation, for each $\tilde{E} \in G_n(TU, \Omega)$ which is based at $w \in U$, we let

$$(30) \qquad X_i(\tilde{E}) = (\partial/\partial x^i + p_i^a(\tilde{E})\partial/\partial y^a)|_w.$$

denote the basis of $\tilde{E}$ dual to the 1-forms $dx^1, dx^2, \ldots, dx^n$.

Let $\pi : G_n(TU, \Omega) \to U$ be the base-point projection. Then for every exterior form φ on U which is of degree $p + 1 \leq n$, the corresponding $(p+1)$-form $\pi^*(\varphi)$ has a unique expansion on $G_n(TU, \Omega)$ of the form

$$(31) \qquad \pi^*(\varphi) = 1/(p+1)! \sum F_K dx^K + 1/p! \sum f_{bJ}\vartheta^b \wedge dx^J + Q.$$

In (31), the summation in the first term is over all (skew-symmetric) multi-indices K from the range $1, \ldots, n$ and of degree $p + 1$, the summation in the second term is over (skew-symmetric) multi-indices J from the same range but of degree p and over all b in the range $1, \ldots, s$, while the last term Q is a form of degree $p + 1$ which is at least quadratic in the terms $\{\vartheta^b \mid 1 \leq b \leq s\}$.

It is elementary that the functions F_K on $G_n(TU, \Omega)$ satisfy

$$(32) \qquad F_K(\tilde{E}) = \varphi(X_{k_0}(\tilde{E}), X_{k_1}(\tilde{E}), \ldots, X_{k_p}(\tilde{E}))$$

for each multi-index $K = (k_0, k_1, \ldots, k_p)$. To get corresponding formulae for the functions f_{bJ}, we compute the exterior derivative of both sides of (31) and reduce modulo the ideal generated by the contact forms $\{\vartheta^b\}$. This gives the formula

$$(33) \qquad dF_K \equiv 1/p! \sum_{b,iJ=K} f_{bJ}dp_i^b \quad \mathrm{mod}\ \{\vartheta, dx\}.$$

Now, recall, from Chapter III, our convention which defined the *level*, $\lambda(a)$, of an integer a in the range $1 \le a \le s$ to be the integer k so that $c_{k-1} < a \le c_k$. We let $\mathcal{P} = \{(i, a) \mid 1 \le i \le \lambda(a)\}$ denote the set of *principal pairs* of indices. Since there are s_k integers a in the range $1, \ldots, s$ satisfying $\lambda(a) = k$, it follows that $\mathcal{P}$ contains $s_1 + 2s_2 + \cdots + ns_n$ pairs of indices. Any pair (j, a) satisfying $\lambda(a) < j \le n$ will be referred to as *non-principal*.

Let $\varphi^1, \ldots, \varphi^s$ be a *polar sequence* (see Chapter III) for the integral flag $(0) \subset E_1 \subset E_2 \subset \cdots \subset E_n = E$. Thus, φ^a is a form in $\mathcal{I}$ of degree $\lambda(a) + 1$. By choosing our polar sequence appropriately, we may even suppose that, for all a with $\lambda(a) = 0$, we have $\varphi^a|_z = dy^a$ while, for all a with $\lambda(a) > 0$, we have

$$(34) \qquad \varphi^a(v, \partial/\partial x_1, \ldots, \partial/\partial x^{\lambda(a)}) = dy^a(v)$$

for all $v \in T_z M$.

Let

$$(35) \qquad \pi^*(\varphi^a) = 1/(p+1)! \sum F_K^a dx^K + 1/p! \sum f_{bJ}^a \vartheta^b \wedge dx^J + Q^a$$

be the expansion of $\pi^*(\varphi^a)$ as in (31). Then by (32), the functions F_K^a vanish identically on $Z \cap G_n(TU, \Omega)$ for all a and all multi-indices K of degree $\lambda(a) + 1$.

It is easy to show that the relations (34) imply that $f_{b12\ldots\lambda(a)}^a(E) = \delta_b^a$ (Kronecker δ) for all a and b. It then follows from (33) that, at E, for each non-principal pair (j, a) we have

$$(36) \qquad dF_{j12\ldots\lambda(a)}^a \equiv dp_j^a \quad \mathrm{mod}\ \{\vartheta, dx, \{dp_i^b\}_{i \le \lambda(a)}\}.$$

Thus, the collection of functions $\mathcal{F} = \{F_{j12\ldots\lambda(a)}^a \mid (j, a)\ \text{non-principal}\}$ has linearly independent differentials on a neighborhood of E in $G_n(TU, \Omega)$.

Since the locus of common zeroes of $\mathcal{F}$ contains $Z \cap G_n(TU, \Omega)$ by construction and since the codimension of Z in $G_n(TU, \Omega)$ is equal to $ns - (s_1 + 2s_2 + \cdots + ns_n) =$ the number of non-principal pairs, there exists an open neighborhood W of $E \in G_n(TU, \Omega)$ so that the functions in $\mathcal{F}$ have linearly independent differentials on W and so that the set $Z \cap W$ is the common set of zeroes of these functions in W.

Moreover, on $Z \cap W$ the equations $F^a_{j12\ldots\lambda(a)} = 0$ imply

$$
\begin{aligned}
0 = dF^a_{j12\ldots\lambda(a)} &\equiv 1/p! \sum_{b,iJ=j12\ldots\lambda(a)} f^a_{bJ} dp^b_i \mod \{\vartheta, dx\} \\
&\equiv \sum_b f^a_{b12\ldots\lambda(a)} dp^b_j \mod \{\vartheta, dx, \{dp^b_i\}_{i \leq \lambda(a)}\}.
\end{aligned}
$$
(37)

Combining this with the fact that $f^a_{b12\ldots\lambda(a)}(E) = \delta^a_b$, it follows that, by shrinking W if necessary, we may suppose that the $ns - (s_1 + 2s_2 + \cdots + ns_n)$ relations (37) may be expressed in the form

$$
(38) \qquad dp^a_j \equiv \sum_{i \leq \min(\lambda(b), j)} B^{ai}_{jb} dp^b_i \mod \{\vartheta, dx\}
$$

where (j, a) is non-principal.

In particular, on $Z \cap W$, for each non-principal pair (j, a), the function p^a_j can be expressed as a function of the variables x, y, and $\{p^b_i \mid i \leq \min(\lambda(b), j)\}$. Thus, the functions x, y, $\{p^a_j \mid j \leq \lambda(a)\}$ form a coordinate system on $Z \cap W$ centered at E and the $n + s + s_1 + 2s_2 + \cdots + ns_n$ 1-forms $\{dx^i \mid 1 \leq i \leq n\}$, $\{\vartheta^a \mid 1 \leq a \leq s\}$, and $\{dp^a_i \mid i \leq \lambda(a)\}$ are a coframing on $Z \cap W$.

Finally, we may suppose, by shrinking W if necessary, that, for any $\tilde{E} \in Z \cap W$, the integral flag $(0) \subset \tilde{E}_1 \subset \tilde{E}_2 \subset \cdots \subset \tilde{E}_n = \tilde{E}$, defined by letting $\tilde{E}_p$ be the span of the vectors $X_1(\tilde{E}), \ldots, X_p(\tilde{E})$, is a regular flag.

Let $I^{(1)}$ be the Pfaffian system on $Z \cap W$ generated by the 1-forms $\{\vartheta^b \mid 1 \leq b \leq s\}$. Then $I^{(1)}$ generates the differential ideal $\mathcal{I}^{(1)}$ restricted to $Z \cap W$. In order to prove involutivity of $(\mathcal{I}^{(1)}, \Omega)$, we need to compute the expressions $\{d\vartheta^b \mid 1 \leq b \leq s\}$ modulo $I^{(1)}$.

Now, on $Z \cap W$, we have

$$
(39) \qquad d\vartheta^a = -dp^a_i \wedge dx^i.
$$

The equations (38, 39) constitute the structure equations of the system $I^{(1)}$. Using the reduced flag determined by the sequence $(dx^1, dx^2, \ldots, dx^n)$, we see by (38) that the reduced characters of $I^{(1)}$ are given for $0 \leq p \leq n$ by

$$
(40) \qquad \tilde{s}'_p = s_p + s_{p+1} + \cdots + s_n.
$$

Now, let $\bar{\pi}_i^a = dp_i^a$ for all principal pairs (i, a) and define

$$(41) \qquad \bar{\pi}_j^a = - \sum_{i \leq \min(\lambda(b), j)} B_{jb}^{ai} \bar{\pi}_i^b$$

for every non-principal pair (j, a). Then by (38) we have structure equations for $I^{(1)}$ of the form

$$(42) \qquad d\vartheta^a \equiv -\bar{\pi}_i^a \wedge dx^i + \frac{1}{2} T_{ij}^a dx^i \wedge dx^j \mod I^{(1)}$$

$$(43) \qquad \bar{\pi}_j^a \equiv - \sum_{i \leq \min(\lambda(b), j)} B_{jb}^{ai} \bar{\pi}_i^b \mod I^{(1)}.$$

In order to prove that the torsion of this system vanishes, we need to prove the existence of functions L_{ij}^a on $Z \cap W$ so that the equations

$$(44) \qquad \begin{aligned} L_{ij}^a - L_{ji}^a &= T_{ij}^a \\ L_{jk}^a + \sum_{i \leq \min(\lambda(b), j)} B_{jb}^{ai} L_{ik}^b &= 0 \quad (\lambda(a) < j) \end{aligned}$$

hold. In order to prove that the symbol relations of $I^{(1)}$ are involutive with Cartan characters given by (40), we need to show that the space of solutions of the homogeneous equations associated to (44) is of dimension $\tilde{s}_1' + 2\tilde{s}_2' + \cdots + n\tilde{s}_n' = s_1 + 3s_2 + \cdots + \frac{1}{2}p(p+1)s_p + \cdots + \frac{1}{2}n(n+1)s_n$ at each point of $Z \cap W$. (Note that Cartan's inequality already tells us that the homogeneous solution space cannot have dimension larger than this number.) We will now prove both of these assertions together.

Let $\beta : Z \cap W \to U$ be the restriction of the base-point projection π to $Z \cap W$. Note that for any $\varphi \in \mathcal{I}$, the functions F_K in the expansion (35) of $\beta^*(\varphi)$ must all be zero. Thus, $\beta^*(\varphi) \equiv 0 \mod I^{(1)}$ for all $\varphi \in \mathcal{I}$. In particular, the expansion (35) simplifies now to

$$(45) \qquad \beta^*(\varphi^a) = 1/p! \sum f_{bJ}^a \vartheta^b \wedge dx^J + Q^a$$

where the forms Q^a are of degree at least 2 in the terms $\{\vartheta^b \mid 1 \leq b \leq s\}$ and the $\{f_{bJ}^a \mid \deg J = \lambda(a)\}$ are functions defined on $Z \cap W$.

Note that on $Z \cap W$ we also have

$$(46) \qquad 0 = dF_K^a \equiv 1/p! \sum_{b, iJ=K} f_{bJ}^a dp_i^b \mod \{\vartheta, dx\},$$

which implies for all a and K that

$$(47) \qquad 0 = \sum_{b, iJ=K} f_{bJ}^a \bar{\pi}_i^b.$$

Of course, these relations are simply linear combinations of the relations
(41). Indeed, the relations (41) are also linear combinations of these relations.

Now, since $\mathcal{I}$ is a differential ideal, we have $d\varphi^a \in \mathcal{I}$, so it follows that
$\beta^*(d\varphi^a) \equiv 0 \mod I^{(1)}$. Since Q^a is at least quadratic in the generators
of $I^{(1)}$, it follows that $dQ^a \equiv 0 \mod I^{(1)}$. Thus, computing the exterior
derivative of (45) and reducing $\mod I^{(1)}$, we get the formula

$$(48) \qquad \sum f^a_{bJ} d\vartheta^b \wedge dx^J \equiv 0 \mod I^{(1)}.$$

Substituting the equation (42) into (48) and making use of the equation
(47), we get

$$(49) \qquad \sum f^a_{bJ} T^b_{ij} dx^i \wedge dx^j \wedge dx^J \equiv 0 \mod I^{(1)}.$$

Now, for each $\tilde{E} \in Z \cap W$, let $\mathcal{W}(\tilde{E}) \subset \mathbb{R}^s \otimes \Lambda^2(\mathbb{R}^n)$ denote the vector
space which consists of the solutions $\tau = (\tau^a_{ij})$ of the linear equations $\tau^a_{ij} =
-\tau^a_{ji}$ and

$$(50) \qquad \sum f^a_{bJ}(\tilde{E}) \tau^b_{ij} dx^i \wedge dx^j \wedge dx^J = 0.$$

(This sum extends over all b, J (of degree $\lambda(a)$), i, and j. Of course, a is
fixed.)

We claim that the dimension of $\mathcal{W}(E)$ is at most

$$D = \sum_{p=0}^{n} s_p \left(\binom{n}{2} - \binom{n-p}{2} \right).$$

To see this, note that when we set $\tilde{E} = E$, then for any triple (i, j, a) with
$\lambda(a) < i < j$, the coefficient of $dx^{ij12\ldots\lambda(a)}$ in the a'th equation of (49)
is given by $2\tau^a_{ij} +$ (terms involving τ^b_{kl} where $\min(k,l) \leq \lambda(b)$). It follows
at once that there are at least as many linearly independent equations in
(49) as there are triples (i, j, a) with $\lambda(a) < i < j$. This verifies our upper
estimate for the dimension of $\mathcal{W}(E)$.

Of course, by shrinking W if necessary, we may assume that D is also
an upper bound for the dimension of $\mathcal{W}(\tilde{E})$ for all $\tilde{E} \in Z \cap W$.

Now, for each $\tilde{E} \in Z \cap W$, let $\mathcal{L}(\tilde{E}) \subset \mathbb{R}^s \otimes \mathbb{R}^n \otimes \mathbb{R}^n$ denote the vector
space which consists of the solutions $l = (l^a_{ij})$ of the linear equations

$$(51) \qquad \sum_{b, iJ=K} f^a_{bJ}(\tilde{E}) l^b_{ij} = 0.$$

Because of our remarks following (47) above, we know that these equations are equivalent to the equations

$$(52) \qquad l^a_{jk} + \sum_{i \le \min(\lambda(b),j)} B^{ai}_{jb}(\tilde{E}) l^b_{ik} = 0 \quad (\lambda(a) < j).$$

It follows that the dimension of $\mathcal{L}(\tilde{E})$ is $n(s_1 + 2s_2 + \cdots + ns_n)$.

Finally, for each $l \in \mathcal{L}(\tilde{E})$, let $\delta(l) \in \mathbb{R}^s \otimes \Lambda^2(\mathbb{R}^n)$ be given by the formula

$$(53) \qquad \delta(l)^a_{ij} = l^a_{ij} - l^a_{ji}.$$

It is a consequence of our definitions, that $\delta(\mathcal{L}(\tilde{E})) \subset \mathcal{W}(\tilde{E})$. Moreover, as we have already noted after (44), the kernel of δ cannot have dimension greater than $s_1 + 3s_2 + \cdots + \frac{1}{2}p(p+1)s_p + \cdots + \frac{1}{2}n(n+1)s_n$. It follows that the dimension of $\delta(\mathcal{L}(\tilde{E}))$ must be at least

$$\sum_{p=0}^{n} s_p \left(np - \binom{p+1}{2} \right) = \sum_{p=0}^{n} s_p \left(\binom{n}{2} - \binom{n-p}{2} \right) = D.$$

Since $\delta(\mathcal{L}(\tilde{E})) \subset \mathcal{W}(\tilde{E})$ and $\dim \mathcal{W}(\tilde{E}) \le D$, it follows that we must have both

$$(54) \qquad \delta(\mathcal{L}(\tilde{E})) = \mathcal{W}(\tilde{E})$$

and

$$(55) \qquad \dim \ker \delta = \sum_{p=0}^{n} \binom{p+1}{2} s_p$$

for all $\tilde{E} \in Z \cap W$. Thus, $\mathcal{L}$ and $\mathcal{W}$ are smooth vector bundles over $Z \cap W$ and the map $\delta : \mathcal{L} \to \mathcal{W}$ is a smooth surjection. Since by (49), T is a smooth section of $\mathcal{W}$ where $T(\tilde{E}) = (T^a_{ij}(\tilde{E}))$, it follows that there exists a smooth section $L = (L^a_{ij})$ of $\mathcal{L}$ so that $T = \delta(L)$. This completes the verification of the existence of a solution to the equation (44) and the verification of the required dimension of the space of solutions to the homogeneous equations (which is the rank of the bundle $\ker \delta$.)

By Cartan's Test, it follows that $(\mathcal{I}^{(1)}, \Omega)$ is involutive. $\qquad\qquad \square$

§3. The Prolongation Theorem.

In this section, we prove a version of the Cartan–Kuranishi prolongation theorem. The aim of this theorem is to reduce the problem of finding the integrals of a differential system to that of finding the integrals of a differential system which is in involution. While the actual theorem we prove is not quite strong enough to do this, it suffices for the analysis of most differential systems which arise in practice.

We begin with the following fundamental definition.

Definition 3.1. Given an exterior differential ideal $\mathcal{I} \subset \Omega^*(M)$ and an integer n, a *prolongation sequence* for $\mathcal{I}$ is a sequence of manifolds $\{M_k \mid k \geq 0\}$ (where $M_0 = M$) together with immersions $\iota_k : M_k \to G_n(TM_{k-1})$ for $k > 0$ with the following properties:

(i) The map $\bar{\iota}_k : M_k \to M_{k-1}$ is a submersion for all $k > 0$. Here, $\bar{\iota}_k$ is the composition $M_k \to G_n(TM_{k-1}) \to M_{k-1}$.

(ii) $\iota_1(M_1) \subset G_n(\mathcal{I})$ and for all $k \geq 1$, $\iota_{k+1}(M_{k+1}) \subset G_n(\mathcal{I}^{(k)}, \Omega^{(k)})$ where $(\mathcal{I}^{(k)}, \Omega^{(k)})$ is the pull-back to M_k of the canonical differential system with independence condition on $G_n(TM_{k-1})$.

To define a prolongation sequence for a pair $(\mathcal{I}, \Omega)$, we modify (ii) to require that $\iota_1(M_1) \subset G_n(\mathcal{I}, \Omega)$.

Note that one particular example of a prolongation sequence when $\mathcal{I} = 0$ is to let $M_1 = G_n(TM)$, let $(\mathcal{I}^{(1)}, \Omega^{(1)})$ be the canonical exterior differential system with independence condition on $G_n(TM)$, and then, by induction, define $M_{k+1} = G_n(\mathcal{I}^{(k)}, \Omega^{(k)})$ and let $(\mathcal{I}^{(k+1)}, \Omega^{(k+1)})$ be the restriction to M_{k+1} of the canonical exterior differential system with independence condition on $G_n(TM_k)$ for all $k > 0$. The resulting sequence is called the *tautological prolongation sequence.*

We can now state our main theorem concerning prolongation sequences.

Theorem 3.2. *If* $\mathbf{S} = \{(M_k, \iota_k) \mid k > 0\}$ *is a prolongation sequence for* $\mathcal{I}$ *over* $M = M_0$*, then there exists an integer* k_0 *such that, for* $k \geq k_0$*, each of the systems* $(\mathcal{I}^{(k)}, \Omega^{(k)})$ *is involutive and moreover,* $\iota_{k+1}(M_{k+1})$ *is an open subset of* $G_n(\mathcal{I}^{(k)}, \Omega^{(k)})$.

Before we begin the proof of Theorem 3.2, we shall establish a piece of notation concerning prolongation sequences that will be useful in the sequel. Fix a prolongation sequence $\mathbf{S} = \{(M_k, \iota_k) \mid k > 0\}$ over a base manifold $M = M_0$.

A sequence of elements $\mathbf{y} = (y_0, y_1, y_2, \dots)$ with $y_k \in M_k$ which satisfies the condition $\bar{\iota}_k(y_k) = y_{k-1}$ for all $k > 0$ will be called a *coherent sequence.* Note that $y_1 \subset T_{y_0}M$ is an n-plane by definition. Let us define $Q_{\mathbf{y}} = T_{y_0}M/y_1$ and $E_{\mathbf{y}} = y_1$.

Also, we remind the reader of the following algebraic notation from Chapters IV and V (also, see Chapter VIII) which will be used extensively in the proof. If V is a (real) vector space of dimension n, then there is a natural pairing $S^k(V^*) \otimes V \to S^{k-1}(V^*)$ which, in the interpretation of $S^k(V^*)$ as the space of polynomial functions on V of degree k, corresponds to partial differentiation. The extension of this mapping, by tensoring with a space W, $W \otimes S^k(V^*) \otimes V \to W \otimes S^{k-1}(V^*)$ is the obvious one. Moreover, if α and β are differential forms on a manifold M which have values in the spaces $W \otimes S^k(V^*)$ and V respectively, then $\alpha \wedge \beta$ (or simply $\alpha\beta$ if α is of degree 0) will denote the $W \otimes S^{k-1}(V^*)$-valued differential form obtained by using exterior form multiplication and the above pairing. One algebraic lemma which we shall use

rather frequently is the following consequence of the polynomial version of Poincaré's lemma:

If β is a V-valued 1-form on M whose components are linearly independent and $\mathbf{a}$ is a function on M with values in $W \otimes S^k(V^*) \otimes V^*$ with the property that $(\mathbf{a}\beta) \wedge \beta = 0$, then $\mathbf{a}$ actually has values in $W \otimes S^{k+1}(V^*) \subset W \otimes S^k(V^*) \otimes V^*$.

Proof of Theorem 3.2. Fix a coherent sequence $\mathbf{y}$ in $\mathbf{S}$. Let $\dim M = n + s$. To avoid trivialities, we assume that n and s are both positive. Let U_0 be an open neighborhood of y_0 on which there exist local coordinates $(x^1, \ldots, x^n, u^1, \ldots, u^s) = (x, u)$ centered on y_0 so that $\Omega^{(0)} = dx^1 \wedge dx^2 \wedge \cdots \wedge dx^n$ does not vanish on y_1. Clearly, $du|_{y_0}$ induces an isomorphism of $Q_{\mathbf{y}}$ with $\mathbb{R}^s$, and $dx|_{y_0}$ induces an isomorphism of $E_{\mathbf{y}}$ with $\mathbb{R}^n$. We shall identify these spaces from now on and speak of du as having values in $Q_{\mathbf{y}}$ and dx as having values in $E_{\mathbf{y}}$.

Let $U_1 \subset \bar{\iota}_1^{-1}(U_0)$ be an open neighborhood of y_1 with the property that $\Omega^{(0)}$ does not vanish on any of the n-planes in $\iota_1(U_1) \subset G_n(TU_0)$. There exists a unique function $p_1 : U_1 \to \mathrm{Hom}(E_{\mathbf{y}}, Q_{\mathbf{y}}) \cong Q_{\mathbf{y}} \otimes E_{\mathbf{y}}^*$ with the property that, for all $z_1 \in U_1$, $\iota_1(z_1)$ is the null space of the $Q_{\mathbf{y}}$-valued 1-form $du - p_1(z_1)dx$ at $z_0 = \bar{\iota}_1(z_1)$. By our hypothesis that ι_1 be an immersion, it follows that $(\bar{\iota}_1, p_1) : U_1 \to U_0 \times Q_{\mathbf{y}} \otimes E_{\mathbf{y}}^*$ is an immersion. In particular, p_1 is an immersion when restricted to any fiber of $\bar{\iota}_1$. It follows that, for each $z_1 \in U_1$, dp_1 induces an injection $\ker(d\bar{\iota}_1)|_{z_1} \to Q_{\mathbf{y}} \otimes E_{\mathbf{y}}^*$. We let $A^{(1)}(z_1) \subset Q_{\mathbf{y}} \otimes E_{\mathbf{y}}^*$ denote $dp_1(\ker(d\bar{\iota}_1)|_{z_1})$. Then $A^{(1)}$ is a smooth sub-bundle of the trivial bundle over U_1 whose fiber is $Q_{\mathbf{y}} \otimes E_{\mathbf{y}}^*$. By definition, dp_1 induces an isomorphism of the sub-bundle $\ker(d\bar{\iota}_1) \subset TU_1$ with the bundle $A^{(1)}$. For convenience of notation, we define $A^{(0)}$ to be the trivial bundle over U_0 whose typical fiber is $Q_{\mathbf{y}}$. By pull-back, we regard $A^{(0)}$ as being well-defined over U_1 as well.

To keep our notation as simple as possible, we shall write x and u for the functions $\bar{\iota}_1^*(x)$ and $\bar{\iota}_1^*(u)$ on U_1. Then the components of the $Q_{\mathbf{y}}$-valued 1-form $\vartheta_0 = du - p_1 dx$ span the pull-back of the canonical Pfaffian system $I^{(1)}$ on $G_n(TU_0)$ to U_1. Moreover, the canonical independence condition may be taken to be $\Omega^{(1)} = dx^1 \wedge dx^2 \wedge \cdots \wedge dx^n$. We now want to derive the structure equations of $(\mathcal{I}^{(1)}, \Omega^{(1)})$.

Let $\sigma_1 : U_1 \to M_2$ be a section of the submersion $\bar{\iota}_2$ which satisfies $\sigma_1(y_1) = y_2$. (We may have to shrink U_1 to do this.) Let $P(z_1) = \iota_2(\sigma_1(z_1))$. Then $P \subset TU_1$ is a rank n sub-bundle whose fiber at each point of U_1 is an integral element of $(\mathcal{I}^{(1)}, \Omega^{(1)})$. It follows easily that there exists a unique $Q_{\mathbf{y}} \otimes E_{\mathbf{y}}^*$-valued 1-form π_1 on U_1 with the following properties:

(1) $dp_1 = \pi_1 + B\vartheta_0 + Tdx$ on U_1 for some B and T.

(2) At each $z_1 \in U_1$, π_1 takes values in $A^{(1)}(z_1)$.

(3) At each $z_1 \in U_1$, $P(z_1)$ is the kernel of π_1 and ϑ_1.

Note that on $\ker(d\bar{\iota}_1)$, we have $dp_1 = \pi_1$. Also note that the functions B and T have values in $Q_{\mathbf{y}} \otimes E_{\mathbf{y}}^* \otimes Q_{\mathbf{y}}^*$ and $Q_{\mathbf{y}} \otimes E_{\mathbf{y}}^* \otimes E_{\mathbf{y}}^*$, respectively.

We now have the structure equations

$$(56) \qquad d\vartheta_0 = -dp_1 \wedge dx \equiv -(\pi_1 + Tdx) \wedge dx \quad \mathrm{mod}\ I^{(1)}.$$

Moreover, since the distribution P (which is annihilated by π_1 and ϑ_0) is a distribution of integral elements of $(\mathcal{I}^{(1)}, \Omega^{(1)})$, it follows that $(Tdx) \wedge dx = 0$ on U_1. Thus, we get the structure equation

$$(57) \qquad d\vartheta_0 \equiv -\pi_1 \wedge dx \quad \mathrm{mod}\ I^{(1)}.$$

Note that the $E_{\mathbf{y}} \oplus Q_{\mathbf{y}} \oplus Q_{\mathbf{y}} \otimes E_{\mathbf{y}}^*$-valued 1-form (dx, ϑ_0, π_1) has constant rank and induces an isomorphism of TU_1 with the bundle $E_{\mathbf{y}} \oplus A^{(0)} \oplus A^{(1)}$.

At this point, we may continue our construction by induction. Suppose that for each integer k in the range $1 \leq k < q$, we have constructed an open neighborhood U_k of y_k so that $U_k \subset \bar{\iota}_k^{-1}(U_{k-1})$. Suppose also that for each k in this range, we have constructed a vector bundle $A^{(k)}$ over U_k which is a sub-bundle of the trivial bundle with typical fiber $Q_{\mathbf{y}} \otimes S^k(E_{\mathbf{y}}^*)$. By the obvious pull-back, we regard forms and bundles defined over U_j for $j < k$ as being well-defined over U_k. We suppose that these bundles satisfy $A^{(k)} \subset (A^{(k-1)})^{(1)}$ for all $1 \leq k < q$. (Note that we do not assume that $(A^{(k-1)})^{(1)}$ is a vector bundle, indeed, it may not have constant rank.) Finally, we suppose that we have constructed a sequence of 1-forms $(dx, \vartheta_0, \vartheta_2, \ldots, \vartheta_{q-2}, \pi_{q-1})$ with the following properties:

(1) For $k < q - 1$, the $Q_{\mathbf{y}} \otimes S^k(E_{\mathbf{y}}^*)$-valued 1-form ϑ_k is well defined on U_{k+1} and takes values in the sub-bundle $A^{(k)}$. The $Q_{\mathbf{y}} \otimes S^{q-1}(E_{\mathbf{y}}^*)$-valued 1-form π_{q-1} is well defined on U_{q-1} and takes the values in the sub-bundle $A^{(q-1)}$.

(2) At each point of U_{q-1}, the 1-form $(dx, \vartheta_0, \vartheta_1, \ldots, \vartheta_{q-2}, \pi_{q-1})$ induces an isomorphism of TU_{q-1} with $E_{\mathbf{y}} \oplus A^{(0)} \oplus A^{(1)} \cdots \oplus A^{(q-1)}$.

(3) For each $k < q$, the components of the forms $\{\vartheta_0, \vartheta_1, \ldots, \vartheta_{k-1}\}$ span the Pfaffian system $I^{(k)}$ restricted to U_k. Furthermore, they satisfy the structure equations

$$(58) \qquad d\vartheta_j \equiv -\vartheta_{j+1} \wedge dx \quad \mathrm{mod}\ \{\vartheta_0, \vartheta_1, \ldots, \vartheta_j\}$$

for $j < q - 2$ and

$$(59) \qquad d\vartheta_{q-2} \equiv -\pi_{q-1} \wedge dx \quad \mathrm{mod} \ \{\vartheta_1, \vartheta_1, \ldots, \vartheta_{q-2}\}.$$

Now, let $U_q \subset \bar{\iota}_q^{-1}(U_{q-1})$ be an open neighborhood of y_q. Then there exists a unique function $p_q : U_q \to Q_\mathbf{y} \otimes S^{q-1}(E_\mathbf{y}^*) \otimes E_\mathbf{y}^*$ with the property that if $z_{q-1} = \bar{\iota}_q(z_q)$ where $z_q \in U_q$, then $p_q(z_q) \in A^{(q-1)}(z_{q-1}) \otimes E_\mathbf{y}^*$ and so that the $A^{(q-1)}(z_{q-1})$-valued 1-form $\pi_{q-1} - p_q(z_q)dx$ annihilates the integral element $\iota_q(z_q)$ of $(\mathcal{I}^{(q-1)}, \Omega^{(q-1)})$. Since $\pi_{q-1} \wedge dx \in \mathcal{I}^{(q-1)}$ must vanish on $\iota_q(z_q)$, we must have $(p_q(z_q)dx) \wedge dx = 0$. By the lemma, this implies that $p_q(z_q) \in Q_\mathbf{y} \otimes S^q(E_\mathbf{y}^*)$. Thus, we must have

$$(60) \quad p_q(z_q) \in (A^{(q-1)}(z_{q-1}) \otimes E_\mathbf{y}^*) \cap Q_\mathbf{y} \otimes S^q(E_\mathbf{y}^*) = (A^{(q-1)}(z_{q-1}))^{(1)}.$$

Just as in the above discussion of the case $q = 1$, the assumption that ι_q is an immersion implies p_q is an immersion when restricted to the fibers of $\bar{\iota}_q$. We define $A^{(q)}(z_q) \subset Q_\mathbf{y} \otimes S^q(E_\mathbf{y}^*)$ to be the vector space $dp_q(\ker(d\bar{\iota}_q)|_{z_q})$. It then follows that $A^{(q)}$ is a smooth sub-bundle of the trivial bundle over U_q whose typical fiber is $Q_\mathbf{y} \otimes S^q(E_\mathbf{y}^*)$. Since, by (60), we have $p_q(\bar{\iota}_q^{-1}(z_{q-1})) \subset (A^{(q-1)}(z_{q-1}))^{(1)}$, it follows that $A^{(q)}(z_q) \subset (A^{(q-1)}(z_{q-1}))^{(1)}$ for all $z_q \in \bar{\iota}_q^{-1}(z_{q-1})$. Thus $A^{(q)} \subset (A^{(q-1)})^{(1)}$.

Now define $\vartheta_{q-1} = \pi_{q-1} - p_q dx$. Then the equation (58) holds for all $j < q - 1$. Moreover, it is clear that the components of the forms $\{\vartheta_0, \vartheta_1, \ldots, \vartheta_{q-2}, \vartheta_{q-1}\}$ generate the Pfaffian system $I^{(q)}$ on U_q. It remains to construct the appropriate form π_q with values in $A^{(q)}$ and verify the analogue of (59) to complete the induction step.

First, we note that since π_{q-1} is well-defined on U_{q-1}, it follows that $d\pi_{q-1}$ can be expressed in terms of the forms $\{dx, \vartheta_0, \vartheta_1, \ldots, \vartheta_{q-2}, \pi_{q-1}\}$. Using the fact that $\pi_{q-1} \equiv p_q dx \ \mathrm{mod} \ I^{(q)}$, we then obtain a formula of the form

$$(61) \qquad d\pi_{q-1} \equiv T_{q-1}(dx \wedge dx) \quad \mathrm{mod} \ I^{(q)}$$

where T_{q-1} is some function on U_q with values in $A^{(q-1)} \otimes \Lambda^2(E_\mathbf{y}^*)$.

It follows that we have the structure equation

$$(62) \qquad d\vartheta_{q-1} \equiv -dp_q \wedge dx + T_{q-1}(dx \wedge dx) \quad \mathrm{mod} \ I^{(q)}.$$

Next let $\sigma_q : U_q \to M_{q+1}$ be a section of the submersion $\bar{\iota}_{q+1}$ which satisfies $\sigma_q(y_q) = y_{q+1}$. (We may have to shrink U_q to do this.) Let $P(z_q) = \iota_{q+1}(\sigma_q(z_q))$. Then $P \subset TU_q$ is a rank n sub-bundle whose value at each point of U_q is an integral element of $(\mathcal{I}^{(q)}, \Omega^{(q)})$. It follows easily that there exists a unique $Q_\mathbf{y} \otimes S^q(E_\mathbf{y}^*)$-valued 1-form π_q on U_q with the following properties:

(1) $dp_q \equiv \pi_1 + R_q dx \mod I^{(q)}$ on U_q for some function R_q which has values in $Q_{\mathbf{y}} \otimes S^q(E_{\mathbf{y}}^*) \otimes E_{\mathbf{y}}^*$.

(2) At each $z_q \in U_q$, π_q takes values in $A^{(q)}(z_q)$.

(3) At each $z_q \in U_q$, $P(z_q)$ is in the kernel of π_q.

Note that on $\ker(d\bar{\iota}_q)$, we have $dp_q = \pi_q$. It follows that the 1-form $(dx, \vartheta_0, \vartheta_1, \ldots, \vartheta_{q-1}, \pi_q)$ induces an isomorphism of TU_q with $E_{\mathbf{y}} \oplus A^{(0)} \oplus A^{(1)} \cdots \oplus A^{(q)}$. Also, equation (7) becomes

$$(63) \qquad d\vartheta_{q-1} \equiv -\pi_q \wedge dx - (R_q dx) \wedge dx + T_{q-1}(dx \wedge dx) \mod I^{(q)}.$$

However, since the distribution P is a distribution of integral elements of $\mathcal{I}^{(q)}$ and since π_q vanishes on P, it follows that $(R_q dx) \wedge dx - T_{q-1}(dx \wedge dx) = 0$ on U_q. Thus, (63) simplifies to

$$(64) \qquad\qquad d\vartheta_{q-1} \equiv -\pi_q \wedge dx \mod I^{(q)}.$$

Setting $\Omega^{(q)} = dx^1 \wedge dx^2 \wedge \cdots \wedge dx^n$, this completes the induction step.

Now, for every coherent sequence $\mathbf{z} = (z_0, z_1, z_2, \ldots)$ with $z_k \in U_k$, we have the sequence of vector spaces $A^{(k)}(\mathbf{z}) = A^{(k)}(z_k) \subset Q_{\mathbf{y}} \otimes S^k(E_{\mathbf{y}}^*)$ which have the property that $A^{(k)}(\mathbf{z}) \subset (A^{(k-1)}(\mathbf{z}))^{(1)}$ for all $k > 0$ and the property that $\dim A^{(k)}(\mathbf{z})$ is independent of the coherent sequence $\mathbf{z}$. By Proposition 3.10 of Chapter VIII, it follows that there exists an integer $k_0 >> 0$ so that, for all $k \geq k_0$ and all coherent sequences $\mathbf{z}$, $A^{(k)}(\mathbf{z})$ is involutive and moreover $A^{(k+1)}(\mathbf{z}) = A^{(k)}(\mathbf{z})$. Moreover, k_0 can be bounded above by a constant which depends only on the sequence of integers $d_k = \dim A^{(k)}(\mathbf{z})$.

Now assume that $k \geq k_0$ is fixed. The structure equations of $(\mathcal{I}^{(k)}, \Omega^{(k)})$ on U_k are then given by

$$(65) \qquad \begin{aligned} d\vartheta_j &\equiv 0 \mod I^{(k)} \quad 0 \leq j < k-1 \\[2mm] d\vartheta_{k-1} &\equiv -\pi_k \wedge dx \mod I^{(k)}. \end{aligned}$$

Since π_k is a 1-form which maps $\ker(d\bar{\iota}_k)$ surjectively onto $A^{(k)} \subset Q_{\mathbf{y}} \otimes S^k(E_{\mathbf{y}}^*)$, and since, by Proposition 3.10 of Chapter VIII, $A^{(k)}(z_k)$ is involutive as a tableau in $(A^{(k-1)}(z_k)) \otimes E_{\mathbf{y}}^*$, it follows immediately from the above structure equations that $(\mathcal{I}^{(k)}, \Omega^{(k)})$ is involutive on U_k. Since $A^{(k+1)}(z_{k+1}) = A^{(k)}(z_k)$ for all $z_{k+1} \in \bar{\iota}_{k+1}^{-1}(z_k)$, it follows for dimension reasons that $\iota_{k+1}(U_{k+1})$ is an open subset of $G_n(\mathcal{I}^{(k)}, \Omega^{(k)})$.

Since this construction was undertaken with respect to any coherent sequence $\mathbf{y}$, it follows that there is a k_0 sufficiently large so that $(\mathcal{I}^{(k)}, \Omega^{(k)})$ is involutive on M_k for all $k \geq k_0$ and that, for dimension reasons, $\iota_{k+1}(M_{k+1})$ is an open subset of $G_n(\mathcal{I}^{(k)}, \Omega^{(k)})$ for all such k. $\qquad\square$

§4. The Process of Prolongation.

The reader may well wonder about the relevance of Theorem 3.2 for the computation of examples. The aim of prolongation, of course, is to reduce the study of the integral manifolds of an arbitrary differential system to the case of an *involutive* differential system, the case to which the vast majority of the theory applies. In this last section of the chapter, we will discuss just how successful this program is. Let us begin with the simplest case.

Theorem 4.1. *Let $\mathcal{I} \subset \Omega^*(M)$ be a differential ideal, and let*

$$\{(M^{(k)}, \mathcal{I}^{(k)}, \Omega^{(k)}) \mid k > 0\}$$

be the sequence of its prolongations. Suppose that, for each $k > 0$, the space $V_n(\mathcal{I}^{(k-1)}, \Omega^{(k-1)}) = M^{(k)}$ is a smooth submanifold of $G_n(TM^{(k-1)})$ and that the projection $M^{(k)} \to M^{(k-1)}$ is a surjective submersion. Then there exists an integer $k_0 \geq 0$ such that $(\mathcal{I}^{(k)}, \Omega^{(k)})$ is involutive on $M^{(k)}$ for all $k \geq k_0$.

Proof. This follows immediately from Theorem 3.2 since $\{(M^{(k)}, \mathcal{I}^{(k)}, \Omega^{(k)}) \mid k > 0\}$ is clearly a prolongation sequence. $\qquad\square$

While Theorem 4.1 is somewhat satisfying, it is of limited use in practice for the following reason. In order to verify the hypotheses of Theorem 4.1 for a given differential system $\mathcal{I}$, one must be able to compute the entire prolongation sequence $\{(M^{(k)}, \mathcal{I}^{(k)}, \Omega^{(k)}) \mid k > 0\}$. In the process of doing this computation, of course, one usually checks whether $(\mathcal{I}^{(k)}, \Omega^{(k)})$ is involutive while one is computing $M^{(k+1)}$. Thus, in practice, before one can apply Theorem 4.1 (assuming that it does, indeed, apply), one finds an involutive prolongation of $\mathcal{I}$ anyway, and then Theorem 2.1 takes over to ensure that all higher prolongations are involutive.

Nevertheless, there are cases where Theorem 4.1 is useful. Although the terminology is not explained until Chapter VIII, we give one such example here because of its close association with Theorem 4.1. The reader may also want to compare Theorem 2.16 of Chapter IX, where the following result is interpreted in the language of jet bundles.

Theorem 4.2. *Let $I \subset J \subset T^* = T^*(M)$ be a pair of sub-bundles defining a linear Pfaffian system on a manifold M. Set $L = J/I$ and let $A \subset I^* \otimes L$ be the tableau bundle of (I, J). Suppose that*

   (i)   *There is an integral element of (I, J) at every point of M,*

  (ii) *A is 2-acyclic at each point of M, i.e., $H^{p,2}(A_x) = 0$ for all $x \in M$ and $p > 0$, and*

 (iii) *The subspaces $A^{(k)} \subset I^* \otimes S^{(k+1)}(L)$ have constant rank on M for all $k \geq 0$.*

Then the hypotheses of Theorem 4.1 are fulfilled for the prolongation sequence

$$\{(M^{(k)}, \mathcal{I}^{(k)}, \Omega^{(k)}) \mid k > 0\}.$$

In particular, $(\mathcal{I}^{(k)}, \Omega^{(k)})$ is involutive for all k sufficiently large.

The proof of Theorem 4.2 will only be indicated here. The hypotheses (i) and (iii) (for $k = 0$) guarantee that $V_n(\mathcal{I}, \Omega) = M^{(1)}$ is a smooth manifold which submerses onto M. The hypothesis (ii) then guarantees that the Pfaffian system $(\mathcal{I}^{(1)}, \Omega^{(1)})$ has all of its torsion absorbable, i.e., that the set $V_n(\mathcal{I}^{(1)}, \Omega^{(1)}) = M^{(2)}$ surjects onto $M^{(1)}$. Then (iii) (for $k = 1$) guarantees that $M^{(2)}$ is a smooth submanifold. This then continues indefinitely. The important point is that the hypotheses (ii) and (iii) ensure that $M^{(k)}$ is a smooth submanifold submersing onto $M^{(k-1)}$ for all $k > 1$. For more details, see §2 of Chapter VIII.

It may seem that Theorem 4.2 would only be marginally more useful than Theorem 4.1. However, for an interesting application, the reader may consult Gasqui [1979b]. The essential point is that the hypotheses of Theorem 4.2 are *algebraic* pointwise conditions on the structure equations of (I, J) and hence are checkable without having to compute the prolongations (which may depend on high derivatives of the original system).

Let us say that a linear Pfaffian system (I, J) which satisfies the hypotheses of Theorem 4.2 is *2-acyclic*. We then have the following easy corollary of Theorem 4.2, which may be regarded as a generalization of the Cartan–Kähler theorem for linear Pfaffian systems.

Corollary 4.3. *If (I, J) is a real analytic, 2-acyclic linear Pfaffian system, then there exist real analytic integral manifolds of (I, J).*

Proof. By Theorem 4.2, some finite prolongation of (I, J) is real analytic and involutive. Now apply the Cartan–Kähler theorem. $\square$

Looking over the examples from §1, we see that the first two examples had the property that, at each stage, the prolongation $M^{(k)}$ was a smooth submanifold for which the basepoint projection $M^{(k)} \to M^{(k-1)}$ was a surjective submersion. (Actually, we only checked this until we reached a value of k for which the system $(\mathcal{I}^{(k)}, \Omega^{(k)})$ was involutive on $M^{(k)}$, for then Theorem 2.1 implies that all higher prolongations will have this property.)

In practice, however, examples such as Example 1.3 are often encountered. In that example, the reader will recall, that when $M^{(1)} = V_4(\mathcal{I}, \Omega_+)$ we had a submersion $M^{(1)} \to M$, but that the basepoint projection $V_4(\mathcal{I}^{(1)}, \Omega_+) \to M^{(1)}$ was neither surjective nor submersive. Nevertheless, we *were* able to reduce the analysis of Example 1.3 to the involutive case by applying a sequence of prolongations. It is natural to ask if this can be done for any differential system $\mathcal{I}$.

Practically nothing can be said about the prolongations of a general *smooth* differential system without making various constant rank assumptions which quickly become too cumbersome to be useful. Therefore, for the remainder of this section we shall assume that the exterior differential system $\mathcal{I} \subset \Omega^*(M)$ is real analytic with respect to some fixed real analytic structure on M. Then the set $V_n(\mathcal{I})$ is a real analytic subset of $G_n(M)$ and, as such, has a canonical coarsest real-analytic stratification

$$V_n(\mathcal{I}) = \bigcup_{\beta \in B} S_\beta,$$

for which each stratum S_β is a smooth, analytically irreducible submanifold of $G_n(M)$. Just as in §1, we define the (first) *prolongation* of $\mathcal{I}$ to be the exterior differential system

$$\left(\mathcal{I}^{(1)}, \Omega^{(1)}\right) = \bigcup_{\beta \in B} \left(\mathcal{P}|_{S_\beta}, \Psi|_{S_\beta}\right),$$

where the underlying manifold $M^{(1)}$ is defined to be the disjoint union of the strata S_β and $(\mathcal{P}, \Psi)$ is the canonical differential system with independence condition on $G_n(M)$. (In the case that an independence condition Ω has been specified, we restrict our attention to $G_n(M, \Omega) \subset G_n(M)$.) The higher prolongations are then defined inductively: $M^{(k)}$ is the disjoint union of the strata of the canonical stratification of $V_n(\mathcal{I}^{(k-1)})$ and $(\mathcal{I}^{(k)}, \Omega^{(k)})$ is the differential system with independence condition got by restricting the canonical differential system with independence condition on $G_n(M^{(k-1)}, \Omega^{(k-1)})$ to each stratum of $V_n(\mathcal{I}^{(k-1)})$. Let us denote the inclusion mapping of $M^{(k)}$ into $G_n(M^{(k-1)}, \Omega^{(k-1)})$ by ι^k and, for $0 \le j < k$, denote the natural projection mapping from $M^{(k)}$ to $M^{(j)}$ by π_j^k.

Theorem 4.4. *If $\mathcal{I} \subset \Omega^*(M)$ is a real analytic differential system on M and $M^{(k)}$ is empty for some $k > 0$, then there are no n-dimensional real analytic integral manifolds of $\mathcal{I}$.*

Proof. Suppose that $f : N^n \hookrightarrow M$ were an n-dimensional, irreducible, real analytic integral manifold of $\mathcal{I}$. Then $V_n(\mathcal{I})$ is non-empty and N has a natural lifting $f^{(1)} : N \hookrightarrow V_n(\mathcal{I})$. Since N is irreducible, it follows that there is a unique stratum of $M^{(1)}$ which intersects $f^{(1)}(N)$ in an analytic manifold of dimension n. Let $N_1 \subset N$ denote the inverse image of this stratum under $f^{(1)}$. Then $f^{(1)} : N_1 \hookrightarrow M^{(1)}$ is an n-dimensional, irreducible, real analytic integral manifold of $(\mathcal{I}^{(1)}, \Omega^{(1)})$. This process can clearly be continued inductively to produce a non-trivial n-dimensional, irreducible, real analytic integral manifold of $(\mathcal{I}^{(k)}, \Omega^{(k)})$, denoted by $f^{(k)} : N_k \hookrightarrow M^{(k)}$, for all $k > 0$. In particular, it follows that $M^{(k)}$ is non-empty for all k. $\qquad \square$

It is natural to ask whether the contrapositive converse of Theorem 4.4 is true, namely, whether or not the non-emptiness of $M^{(k)}$ for all $k > 0$ is sufficient to imply that there are non-empty n-dimensional real analytic integral manifolds of $\mathcal{I}$. Unfortunately, a definitive answer to this question does not seem to be available, though, for a related result due to Malgrange and phrased in the language of jets, the reader may consult Theorem 2.2 of Chapter IX. One statement which would imply this converse is the following:

Prolongation Conjecture. *If $\mathcal{I} \subset \Omega^*(M)$ is a real analytic differential system on M and $M^{(k)}$ is non-empty for all $k > 0$, then there exists a $k_0 \geq 0$ so that for all $k \geq k_0$ there exists an analytic subvariety $\mathcal{S}^{(k)} \subset M^{(k)}$ which intersects each component of $M^{(k)}$ in a (possibly empty) proper analytic sub-variety so that $(\mathcal{I}^{(k)}, \Omega^{(k)})$ is involutive on $M^{(k)} \setminus \mathcal{S}^{(k)}$. Moreover, for every real analytic integral manifold $f : N^n \hookrightarrow M$ of $\mathcal{I}$ there exists an open submanifold $N_k \subset N$ together with an immersion $f^{(k)} : N_k \hookrightarrow M^{(k)} \setminus \mathcal{S}^{(k)}$ which is an ordinary integral manifold of $(\mathcal{I}^{(k)}, \Omega^{(k)})$ and which satisfies $f = \pi_0^k \circ f^{(k)}$ on $N_k \subset N$.*

The reader may be surprised by the appearance of the "singular subvariety" $\mathcal{S}^{(k)}$ in the above statement. However, it is easy to see that such a singular subvariety can occur in such a way that it will not be removed by any finite prolongation. Such examples are furnished by the theory of non-regular singular points of ordinary differential equations. Thus, consider the differential system $(\mathcal{I}, \Omega)$ on $M = \mathbb{R}^2$ generated by the single 1-form $\theta = x^2\, du - u\, dx$ and let $\Omega = dx$. The curve $u = 0$ is clearly an integral manifold of $(\mathcal{I}, \Omega)$, and it is easy to see that $M^{(k)} = \mathbb{R}^2$ for all $k > 0$. However, the integral elements which lie over the locus $x = 0$ are not ordinary for any $k > 0$. The problem is, of course, caused by the fact that the differentials of the maps π_j^k drop rank along the locus $x = 0$.

If the Prolongation Conjecture were known to be true, then we could conclude that all of the real analytic integrals of a differential system could be constructed by suitable applications of the Cartan–Kähler theorem, a highly desirable situation. However, at present, we can only say that the evidence for the Prolongation Conjecture is rather strong: Although many examples of prolongation have been computed, no counterexamples to the conjecture have ever been found. Moreover, under appropriate non-degeneracy hypotheses, the Prolongation Conjecture *has* been proved. One version, due to Kuranishi [1957], is known as the Cartan–Kuranishi prolongation theorem. Unfortunately, the non-degeneracy hypotheses in Kuranishi's theorem are rather difficult to make explicit (and often difficult to check in practice), so we shall refer the reader to Kuranishi's paper for the precise statement of his result. The reader may also consult Kuranishi [1967] and Chapter IX for versions of this theorem which apply directly to P.D.E.

In practical calculations, all of the maps π_j^k are submersions away from proper analytic sub-varieties for j sufficiently large, and, in that case, the conjecture follows from Theorem 3.2. Nevertheless, a proof of the full Prolongation Conjecture remains an interesting problem.

An alternative approach, which avoids the difficulty caused by the fact that the maps π_j^k need not have constant rank, is to consider another definition of prolongation (which, for clarity's sake, we shall call *fine prolongation*). Let $\mathcal{I} \subset \Omega^*(M)$ be a real analytic differential system. Then the set $V_n(\mathcal{I})$ has a coarsest real analytic stratification

$$V_n(\mathcal{I}) = \bigcup_{\beta \in B'} S'_\beta,$$

(which may be finer than the original stratification defined above) with the property that each stratum S'_β is connected and smooth and moreover that the basepoint projection $\pi_0^1 : V_n(\mathcal{I}) \to M$ has constant rank when restricted to each stratum S'_β. Let us define $M^{\langle 1 \rangle}$ to be the disjoint union of the strata S'_β, and let $(\mathcal{I}^{\langle 1 \rangle}, \Omega^{\langle 1 \rangle})$ denote the differential system with independence condition induced on $M^{\langle 1 \rangle}$ by its canonical inclusion into $G_n(M)$. We then continue the construction inductively, except that we require that $V_n(\mathcal{I}^{\langle k-1 \rangle}, \Omega^{\langle k-1 \rangle})$ be given the coarsest smooth stratification with connected strata for which *all* of the mappings $\{\pi_j^k \mid 0 \leq j < k\}$ have constant rank when restricted to any stratum. We then define $M^{\langle k \rangle}$ to be the disjoint union of these strata and define $(\mathcal{I}^{\langle k \rangle}, \Omega^{\langle k \rangle})$ to be the differential system with independence condition induced on $M^{\langle k \rangle}$ by its inclusion into $G_n(M^{\langle k-1 \rangle}, \Omega^{\langle k-1 \rangle})$. Note that according to this definition, fine prolongation is not purely inductive, i.e., we do not necessarily have that $M^{\langle 2 \rangle}$ is equal to $(M^{\langle 1 \rangle})^{\langle 1 \rangle}$. We continue to denote the inclusion mapping of $M^{\langle k \rangle}$ into $G_n(M^{\langle k-1 \rangle}, \Omega^{\langle k-1 \rangle})$ by ι^k and, for $0 \leq j < k$, we continue to denote the natural projection mapping from $M^{\langle k \rangle}$ to $M^{\langle j \rangle}$ by π_j^k. We shall call the sequence

$$\mathcal{S}(\mathcal{I}) = \{(M^{\langle k \rangle}, \mathcal{I}^{\langle k \rangle}, \Omega^{\langle k \rangle}) \mid k > 0\}$$

the *fine prolongation sequence* of $\mathcal{I}$, with similar terminology for a differential system with independence condition $(\mathcal{I}, \Omega)$. As in §3, we shall call a sequence $\mathbf{y} = (y_0, y_1, y_2, \dots)$ with $y_0 \in M$ and $y_k \in M^{\langle k \rangle}$ which satisfies $\pi_{k-1}^k(y_k) = y_{k-1}$ for all $k > 0$ a *fine coherent sequence* for $\mathcal{I}$.

The proof of Theorem 4.4 now goes over with only slight modifications to establish the following result.

Theorem 4.5. *If $\mathcal{I} \subset \Omega^*(M)$ is a real analytic differential system on M and there exists an n-dimensional real analytic integral manifold of $\mathcal{I}$, then there exist fine coherent sequences $\mathbf{y}$ for $\mathcal{I}$. In particular, $M^{\langle k \rangle}$ is non-empty for all $k > 0$.*

Proof. As in the proof of Theorem 4.4, given an irreducible real analytic n-dimensional integral manifold of $\mathcal{I}$, $f : N^n \hookrightarrow M$, we can construct a decreasing sequence of submanifolds $N_k \subset N$ with the property that each N_k is equal to N minus a proper analytic subvariety and so that there exists a lifting $f^{\langle k \rangle} : N_k \hookrightarrow M^{\langle k \rangle}$ of f restricted to N_k which is an integral of $(\mathcal{I}^{\langle k \rangle}, \Omega^{\langle k \rangle})$. The intersection of all of these submanifolds, $N_\infty \subset N$ is equal to N minus a countable number of proper real-analytic submanifolds and hence is non-empty. Clearly, if we let $y \in N_\infty$ be fixed, then the sequence $\mathbf{y} = (f(y), f^{\langle 1 \rangle}(y), f^{\langle 2 \rangle}(y), \dots)$ is a fine coherent sequence for $\mathcal{I}$.

$\square$

It is certainly reasonable to conjecture that the Prolongation Conjecture is true when "prolongation sequence" is replaced by "fine prolongation sequence," and it seems that, in the "fine" case, one can even dispense with the singular locus in the statement. However, this version of the Prolongation Conjecture, which might be called the "Fine Prolongation Conjecture," has not yet been proved either.

Finally, let us indicate our reasons for believing a weaker conjecture which is a sort of converse to Theorem 4.5.

Conjecture. *If $\mathcal{I} \subset \Omega^*(M)$ is a real analytic differential system on M and there exists a fine coherent sequence for $\mathcal{I}$, then there exists an n-dimensional real analytic integral manifold of $\mathcal{I}$.*

The outline of an argument for this conjecture is: Let $\mathcal{S}(\mathcal{I})$ be the fine prolongation sequence of $\mathcal{I}$, and let $\mathbf{y}$ be a fine coherent sequence for $\mathcal{I}$. For each $k > 0$, consider the set of linear maps $(d\pi_k^j)_{y_j} : T_{y_j} M^{\langle j \rangle} \to T_{y_k} M^{\langle k \rangle}$ for all $j > k$. Because of the identity $\pi_j^i \circ \pi_k^j = \pi_k^i$ for all $i > j > k$, there exists an integer $i_k > k$ so that $(d\pi_k^j)_{y_j}(T_{y_j} M^{\langle j \rangle}) = (d\pi_k^{i_k})_{y_{i_k}}(T_{y_{i_k}} M^{\langle i_k \rangle})$ for all $j \geq i_k$. By the local constancy of the ranks of differentials of the mappings π_k^j, it follows that there exists a unique, smooth, real analytic submanifold $M_{\mathbf{y}}^{\langle k \rangle} \subset M^{\langle k \rangle}$ in a neighborhood of y_k with the property that, for all $j > k$, there is a neighborhood of $y_j \in M^{\langle j \rangle}$ so that π_k^j restricted to this maps into $M_{\mathbf{y}}^{\langle k \rangle}$ and is a submersion when regarded as a mapping into $M_{\mathbf{y}}^{\langle k \rangle}$. Although the argument is tedious, it seems to be true that the natural inclusion of $M_{\mathbf{y}}^{\langle k+1 \rangle}$ into $G_n(M^{\langle k \rangle}, \Omega^{\langle k \rangle})$ actually has image in $G_n(M_{\mathbf{y}}^{\langle k \rangle}, \Omega^{\langle k \rangle})$, at least on a neighborhood of y_{k+1}. Assuming this, it then would then follow that the sequence $\{M_{\mathbf{y}}^{\langle k \rangle} \mid k > 1\}$ is a prolongation sequence, as defined in §3. By Theorem 3.2, it would follow that the system $(M_{\mathbf{y}}^{\langle k \rangle}, \mathcal{I}^{\langle k \rangle}, \Omega^{\langle k \rangle})$ is involutive for k sufficiently large. In particular, there would exist integral manifolds of such a system, and hence of the original system $\mathcal{I}$.

Of course, the Fine Prolongation Conjecture would be a considerably stronger statement.

CHAPTER VII

EXAMPLES

This chapter is a collection of examples designed to illustrate the various phenomena which occur in the application of differential systems to problems arising in differential geometry and, more generally, in partial differential equations. We have chosen these examples partly on the basis of their intrinsic interest but mainly in the hopes that the reader can use them as a guide to developing facility in computation.

§1. First Order Equations for Two Functions of Two Variables.

In this example, we shall make a fairly thorough study of the exterior differential systems which arise in the study of systems of first order partial differential equations for two functions of two variables. These cases have the advantage of displaying many of the features of differential systems in general (characteristic variety, torsion, prolongation, etc.) while they remain sufficiently simple that an essentially complete treatment can be undertaken. In the interests of simplicity, we will make constant rank and genericity assumptions whenever they are convenient.

If two variables, say z and w, are regarded as functions of two other variables, say x and y, then the general system of r first order partial differential equations for z and w as functions of x and y can be written in the form

$$(1) \qquad F^\rho(x, y, z, w, z_x, z_y, w_x, w_y) = 0 \qquad 1 \le \rho \le r.$$

Here the functions F^ρ are assumed to be smooth functions of their arguments and, for the sake of simplicity, we assume that at each common zero of the functions F^ρ in $(x, y, z, w, z_x, z_y, w_x, w_y)$-space, the equations (1) implicitly define some set of r of the functions z_x, z_y, w_x, w_y as smooth functions of the remaining variables. Note that this assumption implies that the number of equations r is at most 4.

Examples of such systems of partial differential equations arising in geometry are the volume preserving equation,

$$(2) \qquad z_x w_y - z_y w_x = 1,$$

the Cauchy–Riemann equations,

$$(3) \qquad z_x - w_y = z_y + w_x = 0,$$

and the equations which assert that the pair of functions $z(x,y)$ and $w(x,y)$ induce an isometry between the metrics

$$h_2 = E(z,w)dz^2 + 2F(z,w)dz \circ dw + G(z,w)dw^2$$

and

$$h_1 = e(x,y)dx^2 + 2f(x,y)dx \circ dy + g(x,y)dy^2,$$

$$E(z,w)z_x^2 + 2F(z,w)z_x w_x + G(z,w)w_x^2 = e(x,y)$$

$$(4) \qquad E(z,w)z_x z_y + F(z,w)(z_x w_y + z_y w_x) + G(z,w)w_x w_y = f(x,y)$$

$$E(z,w)z_y^2 + 2F(z,w)z_y w_y + G(z,w)w_y^2 = g(x,y).$$

We want to see how the methods of exterior differential systems apply in such cases. For the sake of uniformity and simplicity of notation, we shall introduce the following change of notation. We rename the variables $x, y, z, w, z_x, z_y, w_x, w_y$ as $x^1, x^2, z^1, z^2, p_1^1, p_2^1, p_1^2, p_2^2$ respectively. Of course, the reader will recognize these as the standard coordinates on the jet space $J^1(\mathbb{R}^2, \mathbb{R}^2)$. The system of equations (1) then become

$$(5) \qquad F^\rho(x^1, x^2, z^1, z^2, p_1^1, p_2^1, p_1^2, p_2^2) = 0 \qquad 1 \leq \rho \leq r.$$

By our hypothesis, these equations define a submanifold $M \subset J^1(\mathbb{R}^2, \mathbb{R}^2)$ of codimension r such that the source-target projection $M \to \mathbb{R}^2 \times \mathbb{R}^2$ is a submersion.

The contact system on $J^1(\mathbb{R}^2, \mathbb{R}^2)$ is generated by the 1-forms

$$(6) \qquad \underline{\vartheta}^a = dz^a - \sum p_i^a dx^i \qquad 1 \leq a \leq 2$$

and the canonical independence condition is given by the 2-form $\Omega = dx^1 \wedge dx^2$. Clearly, the 1-forms $\{\underline{\vartheta}^1, \underline{\vartheta}^2, dx^1, dx^2\}$ remain independent when restricted to M. We let $I \subset \Omega^1(M)$ denote the Pfaffian system generated by the pair $\{\underline{\vartheta}^1, \underline{\vartheta}^2\}$ after restriction to M and let J denote the Pfaffian system generated by the forms $\{\underline{\vartheta}^1, \underline{\vartheta}^2, dx^1, dx^2\}$ after restriction to M. The 1-forms $\{dp_i^a \mid 1 \leq i, a \leq 2\}$ are clearly not linearly independent modulo J after they have been restricted to M. In fact we must have r relations of the form

$$(7) \qquad \sum \underline{b}_a^{\rho i} dp_i^a \equiv 0 \mod J \qquad 1 \leq \rho \leq r$$

where $\underline{b}_a^{\rho i} = \partial F^\rho / \partial p_i^a$. These relations are obtained by expanding the identities on M given by $dF^\rho = 0$.

In studying the structure relations (7), it is often helpful to adapt the given bases of I and J to the problem at hand. Thus, on an open set $U \subset M$, let $\{\vartheta^1, \vartheta^2, \omega^1, \omega^2\}$ denote any set of 1-forms which have the property that $\{\vartheta^1, \vartheta^2\}$ is a basis for the sections of I restricted to U and $\{\vartheta^1, \vartheta^2, \omega^1, \omega^2\}$ is a basis for the sections of J restricted to U. Then there exist functions A^a_b and B^i_j on U so that the following relations hold:

$$
\begin{aligned}
\vartheta^a &= A^a_b \underline{\vartheta}^b \\[2mm]
dx^i &\equiv B^i_j \omega^j \quad \text{mod } I
\end{aligned}
$$
(8)

If we now choose 1-forms π^a_i subject to the condition that

$$
\pi^a_i = \sum A^a_b B^j_i dp^b_j + \sum S^a_{ij} \omega^j \quad \text{mod } I
$$
(9)

where $S^a_{ij} = S^a_{ji}$, then we have the familiar structure equations

$$
d\vartheta^a \equiv - \sum \pi^a_i \wedge \omega^i \quad \text{mod } I
$$
(10)

together with symbol relations of the form

$$
\sum b^{\rho i}_a \pi^a_i \equiv \sum C^\rho_{ij} \omega^j \quad \text{mod } I \qquad 1 \leq \rho \leq r
$$
(11)

where each 2×2 matrix $b^\rho = (b^{\rho i}_a)$ is obtained from the corresponding matrix $\underline{b}^\rho = (\underline{b}^{\rho i}_a)$ by the formula $b^\rho = B^{-1} \underline{b}^\rho A^{-1}$.

For example, in the volume preserving example, the submanifold M is defined by the equation

$$
p^1_1 p^2_2 - p^1_2 p^2_1 = 1.
$$
(12)

The symbol relation corresponding to (7) is the equation

$$
p^2_2 dp^1_1 - p^2_1 dp^1_2 - p^1_2 dp^2_1 + p^1_1 dp^2_2 = 0.
$$
(13)

If we set $\vartheta^a = \underline{\vartheta}^a$ and set

$$
\begin{pmatrix} dx^1 \\ dx^2 \end{pmatrix} = \begin{pmatrix} p^2_2 & -p^2_1 \\ -p^1_2 & p^1_1 \end{pmatrix} \begin{pmatrix} \omega^1 \\ \omega^2 \end{pmatrix}
$$
(14)

then, taking $S^a_{ij} = 0$, the relation corresponding to (11) becomes simply

$$
\pi^1_1 + \pi^2_2 \equiv 0 \quad \text{mod } I.
$$
(15)

For the purpose of computing Cartan characters and characteristics, relation (15) is much easier to deal with than (13).

As another example, consider the isometry problem of the two metrics h_1 and h_2 mentioned above. Let ω^1, ω^2 be an orthonormal coframing for the metric h_1 and let η^1, η^2 be an orthonormal coframing for the metric h_2. Define the matrices A and B so that $\eta^a = A^a_b dz^b$ and $dx^i = B^i_j \omega^j$. (Note that A is a function of the z variables and B is a function of the x variables.) If $P = (p^a_i)$ is the matrix of solutions to the equations given in (4), then one easily sees that the matrix APB is an orthogonal matrix. Conversely, if we set $P = A^{-1} g B^{-1}$ where g is any 2×2 orthogonal matrix, then the matrix P satisfies the equations given in (4). Thus, in this case, M is diffeomorphic to $\mathbb{R}^2 \times \mathbb{R}^2 \times O(2)$. Regarding $\underline{\vartheta} = (\vartheta^a)$, $dz = (dz^a)$, and $dx = (dx^i)$, etc. as columns, we may write on M,

$$(16) \qquad \vartheta = g^{-1} A \underline{\vartheta} = g^{-1} A (dz - P dx) = g^{-1} \eta - \omega.$$

Using the fact that $d\eta = -\varphi \wedge \eta$ and $d\omega = -\psi \wedge \omega$ where φ and ψ are skew-symmetric matrices, we may compute that

$$
\begin{aligned}
d\vartheta &= -g^{-1} \varphi \wedge \eta + dg^{-1} \wedge \eta + \psi \wedge \omega \\
(17) \qquad &\equiv -(g^{-1} dg + g^{-1} \varphi g - \psi) \wedge \omega \mod I \\
&\equiv -\pi \wedge \omega \mod I
\end{aligned}
$$

where π is a skew-symmetric matrix by virtue of the fact that g is an orthogonal matrix. Thus, in this basis, the symbol relations of the Pfaffian system I are simply

$$(18) \qquad \pi^1_1 = \pi^2_2 = \pi^1_2 + \pi^2_1 = 0.$$

The reader should compare these with the symbol relations one gets from (4) by naively differentiating. We will return to this example below.

In the general case, we want to take advantage of changes of basis of the form (10)–(11) to reduce the symbol relations to as simple a form as possible. It is clear that we want to normalize the linear span of the matrices $\{b^\rho \mid 1 \le \rho \le r\}$ in the vector space of all 2×2 matrices under the obvious action of the group $GL(2, \mathbb{R}) \times GL(2, \mathbb{R})$. We will now treat the four possible values of r as separate cases.

Case 1: $r = 1$.

In this case there is a single symbol matrix which we may as well denote by b instead of b^1. The admissible substitutions (10)–(11) allow us to pre- and post-multiply the 2×2 matrix b by arbitrary invertible matrices. It follows that the only invariant of the matrix b is its rank, which must be 1 or 2. We will assume that this rank is constant on M. It follows that there are two subcases.

Subcase 1.1: b has rank 1.

We may now choose our bases so that the single symbol relation is of the form

$$(19) \qquad \pi_2^2 \equiv C_1\omega^1 + C_2\omega^2 \quad \mod I.$$

Replacing π_2^2 by $\pi_2^2 - C_1\omega^1 - C_2\omega^2$ and π_1^2 by $\pi_1^2 - C_1\omega^2$, we may assume that $C_1 = C_2 = 0$. Of course, we still have the structure equations

$$(20) \qquad d\vartheta^a \equiv -\pi_i^a \wedge \omega^i \quad \mod I.$$

It follows that the torsion of the system vanishes identically. By inspection, we have $s_1' = 2$ and $s_2' = 1$. Moreover, the integral elements at a point depend on $s_1' + 2s_2' = 4$ parameters, namely $\vartheta^a = 0$ and

$$(21) \qquad \begin{aligned} \pi_1^1 &= \lambda_1\omega^1 + \lambda_2\omega^2 \\ \pi_2^1 &= \lambda_2\omega^1 + \lambda_3\omega^2 \\ \pi_1^2 &= \lambda_4\omega^1 \end{aligned}$$

and of course $\pi_2^2 = 0$. Thus, the system is involutive.

The $r \times s$ symbol matrix σ_ξ at the covector $\xi = \xi_i\omega^i$ is, in this case, the 1×2 matrix $\sigma_\xi = (0, \xi_2)$. Thus, the symbol matrix has rank 1 except when $\xi = \xi_1\omega^1$ (when it has rank 0). Thus, the characteristic variety is $\Xi_x = \mathbb{P}(J_x/I_x) \cong \mathbb{P}^1$ for all $x \in M$. However, the characteristic *sheaf* consists of Ξ_x plus the "embedded component" $[\omega^1] \in \mathbb{P}^1$.

Examples of this type of equation are given by $w_y = 0$ and $w_y = z$. It is easily shown that the Pfaffian systems I associated to these two equations are not diffeomorphic. (Consider the form ϑ^2 in an adapted coframing satisfying $\pi_2^2 = 0$ for each of the above systems. This form is canonically defined up to a multiple and yet it is of different Pfaff type for each of the two systems.) Nevertheless, there exists an O.D.E. method for constructing the integral manifolds of such I, thereby avoiding the use of the Cartan–Kähler theorem and proving a smooth existence theorem for the solutions of the given equation. The method is as follows: Construct an adapted coframing ϑ^1, ϑ^2 of I with the property that $\pi_2^2 = 0$ and that ϑ^1 is of Pfaff rank 5. (This can always be done.) Placing ϑ^1 in Pfaff normal form (which requires only O.D.E.), we may specify integral manifolds of ϑ^1 in terms of a single function of 2 variables. These integral manifolds are of dimension 4 in M. If R is such an integral, then the structure equations show that ϑ^2 restricts to be of Pfaff rank 3 on R and hence its integrals can be specified by a single function of 1 variable and have codimension 2 in R. These resulting 2-dimensional integrals are the desired integrals of I. We leave further details to the reader.

Subcase 1.2: b has rank 2.

This is, in some sense, the generic case for single equations. We may now choose our bases so that the single relation has the form

$$(22) \qquad \pi_2^1 - \pi_1^2 = C_1\omega^1 + C_2\omega^2 \quad \mathrm{mod}\ I.$$

Replacing π_1^1 by $\pi_1^1 - C_2\omega^2$ and π_2^2 by $\pi_2^2 + C_1\omega^1$, we may assume that $C_1 = C_2 = 0$. Thus, the torsion always vanishes.

We now observe that the system is involutive. By inspection, we see that $s_1' = 2$ and $s_2' = 1$. Moreover, the integral elements at a point of M depend on $s_1' + 2s_2' = 4$ parameters, namely $\vartheta^a = 0$ and

$$(23) \qquad \begin{aligned} \pi_1^1 &= \lambda_1\omega^1 + \lambda_2\omega^2 \\ \pi_2^1 = \pi_1^2 &= \lambda_2\omega^1 + \lambda_3\omega^2 \\ \pi_2^2 &= \lambda_2\omega^1 + \lambda_4\omega^2. \end{aligned}$$

The symbol matrix at a covector $\xi = \xi_1\omega^1 + \xi_2\omega^2$ is the 1×2 matrix $\sigma_\xi = (-\xi_2, \xi_1)$. Thus, the symbol matrix always has rank 1. In particular, it is never injective. It follows that the characteristic variety satisfies $\Xi_x = \mathbb{P}(J_x/I_x) \cong \mathbb{P}^1$ for all $x \in M$. All of this is in accordance with the general theory of characteristic varieties developed in Chapter V.

Note that the Cartan–Kähler theory predicts that the general solution of such a system will depend on 1 function of 2 variables. The standard example of an equation which falls into this subcase is the equation $z_y = w_x$. The general solution of this equation is given by the formula

$$(24) \qquad z = f_x \text{ and } w = f_y$$

where f is an arbitrary function of x and y. A more interesting equation whose "general solution" can be found explicitly is the volume preserving equation (2). The solutions where $z_x \neq 0$ can be described locally in parametric form by letting h be an arbitrary function of two auxiliary variables s and t which satisfies $h_{st} \neq 0$ and setting

$$(25) \qquad \begin{aligned} x &= h_t(s,t) & y &= t \\ z &= h_s(s,t) & w &= s. \end{aligned}$$

A similar formula holds for those solutions where $z_y \neq 0$. Of course, for the generic single equation for two functions of two variables, the general solution cannot be written down so explicitly.

Case 2: $r = 2$.

This is, in many ways, the most interesting of the cases. The symbol matrices $\{b^1, b^2\}$ span a two-dimensional subspace of the space of 2×2 matrices. We begin by classifying the possible two-dimensional subspaces under the equivalence generated by pre- and post-multiplication by invertible matrices. It turns out that there are exactly 5 equivalence classes. This is proved by noting that the determinant function on the space of 2×2 matrices is a conformally invariant quadratic form under the natural action of $GL(2, \mathbb{R}) \times GL(2, \mathbb{R})$. To see this, note that if R is a 2×2 matrix and $(A, B) \in GL(2, \mathbb{R}) \times GL(2, \mathbb{R})$, then $\det(ARB^{-1}) = (\det(A)/\det(B))\det(R)$. Thus, a natural invariant of a two-dimensional subspace of the space of 2×2 matrices is the type of the quadratic form det after it has been restricted to the given subspace. This crude classification can be refined slightly to give the following list of representatives of the 5 equivalence classes:

$$(26.1) \qquad B = \left\{ \begin{pmatrix} x^1 & x^2 \\ 0 & 0 \end{pmatrix} \middle| x^i \in \mathbb{R} \right\}$$

$$(26.2) \qquad B = \left\{ \begin{pmatrix} x^1 & 0 \\ x^2 & 0 \end{pmatrix} \middle| x^i \in \mathbb{R} \right\}$$

$$(26.3) \qquad B = \left\{ \begin{pmatrix} 0 & -x^1 \\ x^1 & x^2 \end{pmatrix} \middle| x^i \in \mathbb{R} \right\}$$

$$(26.4) \qquad B = \left\{ \begin{pmatrix} 0 & x^1 \\ x^2 & 0 \end{pmatrix} \middle| x^i \in \mathbb{R} \right\}$$

$$(26.5) \qquad B = \left\{ \begin{pmatrix} x^1 & x^2 \\ x^2 & -x^1 \end{pmatrix} \middle| x^i \in \mathbb{R} \right\}$$

For the sake of simplicity, we shall assume that the symbol relations of our system of two equations have constant type in the above classification. We shall now proceed to analyse each of these subcases separately.

Subcase 2.1: The symbol relations are of type (26.1).

In this case, we may change bases so that the symbol relations take the form

$$(27) \qquad \left. \begin{aligned} \pi^1_1 &\equiv C_{11}\omega^1 + C_{12}\omega^2 \\ \pi^1_2 &\equiv C_{21}\omega^1 + C_{22}\omega^2 \end{aligned} \right\} \quad \mathrm{mod}\ I$$

Since we may modify the forms π^1_j by a symmetric linear combination of the ω^j, we see that the torsion of the system vanishes if and only if

$C_{12} \equiv C_{21}$. If the torsion of the system does not vanish identically, then we may restrict to the locus $C_{12} - C_{21} = 0$. In the generic case, this gives an extra equation which, adjoined to the given two, gives a system of 3 equations. (We will treat this case below in Case 3.) On the other hand, if the identity $C_{12} \equiv C_{21}$ holds, then the structure equations of the system reduce to the form

$$(28) \qquad \left. \begin{array}{c} d\vartheta^1 \equiv 0 \\ d\vartheta^2 \equiv -\pi_1^2 \wedge \omega^1 - \pi_2^2 \wedge \omega^2 \end{array} \right\} \quad \mathrm{mod}\ I.$$

By inspection, we have $s_1' = s_2' = 1$. The space of integral elements at a point of M clearly depends on 3 parameters. Thus, the system is in involution and the general solution (at least in the analytic case) depends on one function of two variables.

The symbol matrix σ_ξ at a non-zero covector $\xi = \xi_i \omega^i$ is the 2×2 matrix

$$\begin{pmatrix} \xi_1 & 0 \\ \xi_2 & 0 \end{pmatrix}$$

Since σ_ξ always has rank 1, every covector is characteristic. Of course, this implies that a 2-dimensional integral of I cannot be determined by knowledge of any its 1-dimensional subintegrals.

Finally, we remark that, in fact, every involutive system of this type is locally equivalent to the standard example

$$(29) \qquad\qquad z_x = z_y = 0.$$

In particular, the analytic assumption is not needed. This equivalence follows by examining the structure equations (28) closely and showing that they actually "uncouple" into the the equations

$$(30) \qquad \begin{array}{l} d\vartheta^1 \equiv 0 \ \ \mathrm{mod}\ \vartheta^1 \\[2mm] d\vartheta^2 \equiv -\pi_1^2 \wedge \omega^1 - \pi_2^2 \wedge \omega^2 \ \ \mathrm{mod}\ \vartheta^2. \end{array}$$

The result then follows by applying the Frobenius theorem and the Pfaff–Darboux theorem. (See Chapter II.)

Subcase 2.2: The symbol relations are of type (26.2).

In this case, we may, by admissible basis change, assume that the symbol relations are of the form

$$(31) \qquad \left. \begin{array}{l} \pi_2^1 \equiv C_{11}\omega^1 + C_{12}\omega^2 \\ \pi_2^2 \equiv C_{21}\omega^1 + C_{22}\omega^2 \end{array} \right\} \quad \mathrm{mod}\ I$$

Replacing π_2^j by $\pi_2^j - C_{j1}\omega^1 - C_{j2}\omega^2$ and π_1^j by $\pi_1^j - C_{j1}\omega^2$ for $j = 1$ and 2, we see that we may assume that $C_{ij} = 0$. Thus, the torsion always vanishes for systems of this type. The structure equations of the system are now of the form

$$(32) \qquad \left. \begin{aligned} d\vartheta^1 &\equiv -\pi_1^1 \wedge \omega^1 \\ d\vartheta^2 &\equiv -\pi_1^2 \wedge \omega^1 \end{aligned} \right\} \quad \mathrm{mod}\ I.$$

Inspection now shows that $s_1' = 2$ and $s_2' = 0$. Moreover, the space of integral elements at each point of M is clearly of dimension 2. Thus, the system is involutive. Thus, by the Cartan–Kähler theorem, the integral manifolds (in the real-analytic category) depend on two functions of two variables.

The symbol matrix σ_ξ at a non-zero covector $\xi = \xi_1\omega^1 + \xi_2\omega^2$ is the 2×2 matrix

$$\begin{pmatrix} \xi_2 & 0 \\ 0 & \xi_2 \end{pmatrix}.$$

It follows that the characteristic variety at each point x of M is the point $[\omega^1] \in \mathbb{P}(J_x/I_x)$. Note that it should be counted with multiplicity 2.

An example of this type of P.D.E. is given by the pair of equations

$$(33) \qquad \begin{aligned} zz_x + wz_y &= f(x, y, z, w) \\[2mm] zw_x + ww_y &= g(x, y, z, w) \end{aligned}$$

where, in order to avoid singularities, we assume that the functions f and g do not simultaneously vanish.

Actually, using the techniques of Chapter II, an alternative method of describing the integral manifolds of systems in this subcase is available. By the structure equations (32), it follows that the Cartan system of the Pfaffian system I is the Pfaffian system $C(I)$ generated by the 1-forms $\vartheta^1, \vartheta^2, \omega^1, \pi_1^1, \pi_1^2$. Since M is of dimension 6, it follows that the Cauchy leaves of I are curves. In fact, any two dimensional integral manifold $N^2 \subset M^6$ of $(\mathcal{I}, \Omega)$ is a union of integral curves of $C(I)$. To see this, note that on any such integral N, the 2-forms $\pi_1^j \wedge \omega^1$ must vanish. This implies that, on N, there must be relations of the form $\pi_1^j = \lambda^j \omega^1$ for some functions λ^j. It follows that all of the forms in $C(I)$ vanish when restricted to the integral curves of ω^1 on N. Thus, these curves are integral curves of $C(I)$. Conversely, by the general theory of Cauchy characteristics developed in Chapter II, if P^1 is any integral curve of the system $(\mathcal{I}, \omega^1)$, then P is transverse to the leaves of $C(I)$. Hence the union of the leaves of $C(I)$ which pass through P is a smooth surface which is an integral of $(\mathcal{I}, \Omega)$. Thus, the construction of integrals of $(\mathcal{I}, \Omega)$ reduces to the two O.D.E. problems of constructing the integral curves of $C(I)$ and constructing integrals of $(\mathcal{I}, \omega^1)$. In particular, one does not need to apply the Cartan–Kähler theorem. Thus, our description of the integrals of $(\mathcal{I}, \Omega)$ does not depend on the assumption of real analyticity.

In fact, even more is true. Since the system generated by ϑ^1, ϑ^2, and ω^1 is the restriction to M of the Frobenius system generated by dz^1, dz^2, dx^1, and dx^2, it follows that the system generated by ϑ^1, ϑ^2, and ω^1 is itself Frobenius. It easily follows that every point of M has local coordinates $(a^1, a^2, b^1, b^2, x, y)$ so that the system I has generators of the form $\tilde\vartheta^1 = da^1 - b^1 dx$ and $\tilde\vartheta^2 = da^2 - b^2 dx$. (These coordinates can be found using O.D.E. methods alone.) Then the general integral of $(\mathcal{I}, \Omega)$ in this neighborhood can be described implicitly by the 4 equations $a^j = f^j(x)$ and $b^j = df^j/dx$ $(j = 1, 2)$, where f^1 and f^2 are arbitrary functions of x. In particular, all of the differential systems in this subcase are locally equivalent. It is an interesting exercise to reduce the system (33) to this standard form.

Subcase 2.3: The symbol relations are of type (26.3).

In this case, we may change bases so that the symbol relations take the form

$$(34) \qquad \left. \begin{aligned} \pi_2^1 - \pi_1^2 &\equiv C_{11}\omega^1 + C_{12}\omega^2 \\ \pi_2^2 &\equiv C_{21}\omega^1 + C_{22}\omega^2 \end{aligned} \right\} \quad \mathrm{mod}\ I$$

Just as before, by adding appropriate linear combinations of the forms ω^1 and ω^2 to the forms π_j^i, we may assume that the C_{ij} are all zero. Thus, the torsion vanishes. The structure equations now take the form

$$(35) \qquad \left. \begin{aligned} d\vartheta^1 &\equiv -\pi_1^1 \wedge \omega^1 - \pi_2^1 \wedge \omega^2 \\ d\vartheta^2 &\equiv -\pi_2^1 \wedge \omega^1 \end{aligned} \right\} \quad \mathrm{mod}\ I.$$

Inspection shows that $s_1' = 2$ and $s_2' = 0$. Moreover, the integral elements of $(\mathcal{I}, \Omega)$ at a point x of M depend on 2 parameters, namely

$$(36) \qquad \begin{aligned} \pi_1^1 &= \lambda_1\omega^1 + \lambda_2\omega^2 \\[4pt] \pi_2^1 &= \pi_1^2 = \lambda_2\omega^1 \\[4pt] \pi_2^2 &= 0 \end{aligned}$$

and of course $\vartheta^a = 0$. Thus, the system is involution and the general solution, in the analytic category, depends on two functions of one variable.

The symbol matrix σ_ξ at a covector $\xi = \xi_1\omega^1 + \xi_2\omega^2$ is the 2×2 matrix

$$\begin{pmatrix} \xi_2 & \xi_1 \\ 0 & \xi_2 \end{pmatrix}.$$

It follows that σ_ξ has rank 2 except when $\xi_2 = 0$, in which case, it has rank 1. It follows that the characteristic variety at each point of M is of the form $\Xi_x = [\omega^1] \in \mathbb{P}(J_x/I_x)$. Of course, the characteristic sheaf will count this point with multiplicity 2.

An example of this type of system is the pair of equations

$$(37) \qquad z_y - w_x = w_y - z = 0.$$

Note that if we solve the first equation by introducing a potential function u so that $z = u_x$ and $w = u_y$, then the remaining equation becomes the familiar parabolic equation $u_x = u_{yy}$. Thus, we shall say that systems which fall into this subcase are *parabolic*.

Note that the system I has no Cauchy characteristics. However, the integrals of I are still foliated by the "characteristic curves" $\omega^1 = 0$. Moreover, on any integral of $(\mathcal{I}, \Omega)$, we have $\pi_2^1 \wedge \omega^1 = 0$, so $\pi_2^1 = \lambda\omega^1$ for some smooth function λ. With this in mind, we define the Pfaffian system $\mathcal{M}$ on M to be the system spanned by the 1-forms $\vartheta^1, \vartheta^2, \omega^1, \pi_2^1$. It is easy to show that this span is well-defined independent of the choice of bases of J and I which put the structure equations of I in the form (35). Moreover, every integral of (I, Ω) is foliated by integral curves of $\mathcal{M}$ and these curves are clearly the characteristic curves. These are, of course, the characteristics in the classical sense of Monge (as well as in the modern sense).

The case where the system $\mathcal{M}$ is of Frobenius type (so that the characteristics obey four "conservation laws") is important because one can show that the system $\mathcal{M}$ is of Frobenius type if and only if the system I is locally diffeomorphic to the system given by the equations

$$(38) \qquad z_y - w_x = w_y = 0.$$

This is one of those cases which Cartan described as "having characteristics depending on constants," meaning that the maximal integrals of $\mathcal{M}$ depend only on constants (in this case, four constants). The fact that one can write down the general integral of such a system I by means of O.D.E. is no accident, but follows from a general procedure due to Darboux and Cartan whenever the system has characteristics depending on constants. Of course, one does not expect the general parabolic system to be solvable by O.D.E., and indeed one need only consider the system (37) whose solutions are equivalent to the solutions of the one-dimensional heat equation to see that the general solution cannot be written as an explicit function of two functions of one variable and a finite number of their derivatives. In this case, $\mathcal{M}$ fails to be a Frobenius system. Roughly speaking, the non-integrability of $\mathcal{M}$ corresponds to the original system of P.D.E. having parabolic "diffusion effects."

Subcase 2.4: The symbol relations are of type (26.4).

In this case, by admissible basis change, we may assume that the symbol relations are of the form

$$(39) \qquad \left. \begin{array}{l} \pi_2^1 \equiv C_{11}\omega^1 + C_{12}\omega^2 \\ \pi_1^2 \equiv C_{21}\omega^1 + C_{22}\omega^2 \end{array} \right\} \quad \mathrm{mod}\ I.$$

Again, by adding appropriate multiples of ω^1 and ω^2 to the 1-forms π_j^i, we may assume that the functions C_{ij} are zero. Thus, the torsion vanishes and the structure equations now take the form

$$(40) \qquad \left. \begin{array}{l} d\vartheta^1 \equiv -\pi_1^1 \wedge \omega^1 \\ d\vartheta^2 \equiv -\pi_2^2 \wedge \omega^2 \end{array} \right\} \quad \mathrm{mod}\ I.$$

Inspection shows that $s_1' = 2$ and $s_2' = 0$. Moreover, the integral elements at a point of M clearly depend on 2 parameters, namely $\vartheta^a = 0$ and $\pi_i^i = \lambda^i \omega^i$ and $\pi_2^1 = \pi_1^2 = 0$. Thus, the system is involutive and, in the analytic category, the general integral of $(\mathcal{I}, \Omega)$ depends on two functions of one variable.

The symbol matrix σ_ξ at a non-zero covector $\xi = \xi_1 \omega^1 + \xi_2 \omega^2$ is the 2×2 matrix

$$\begin{pmatrix} \xi_2 & 0 \\ 0 & \xi_1 \end{pmatrix}.$$

It follows that the characteristic variety Ξ_x at each point x of M is the pair of points $\{[\omega^1], [\omega^2]\} \in \mathbb{P}(J_x/I_x)$.

A simple example of this type of system is the pair of equations

$$(41) \qquad z_y = w_x = 0.$$

The general solution is, of course, $z = f(x)$ and $w = g(y)$ where f and g are arbitrary functions of one variable. Note that this system is hyperbolic in the classical sense. In fact, it is easy to see that the general pair of equations whose symbol relations can be taken to be of the form (26.4) is hyperbolic in the classical sense. Thus, we shall call the equations which fit into this subcase *hyperbolic*.

Again, there are no Cauchy characteristics. Nevertheless, the two characteristic foliations of an integral given by $\omega^1 = 0$ and $\omega^2 = 0$ define important geometric features of the integral manifolds. We shall not enter into a discussion of the classification of hyperbolic systems up to diffeomorphism. This would require a discussion of the equivalence problem of Élie Cartan which is too lengthy to enter into here. However, we can indicate some basic invariants of the system I which can be used to determine whether a given system is equivalent to the "flat" system (41). Suppose, for the sake of convenience, that M is oriented. Then any local basis $\{\vartheta^1, \vartheta^2\}$ of I which satisfies the structure equations (40) and the condition $\vartheta^1 \wedge d\vartheta^1 \wedge \vartheta^2 \wedge d\vartheta^2 > 0$ is unique up to a change of basis of the form $\tilde{\vartheta}^a = \lambda^a \vartheta^a$ where $\lambda^1, \lambda^2 \neq 0$, as is easily verified. Using this, we can define two rank 3 Pfaffian systems $\mathcal{M}^1$ and $\mathcal{M}^2$ on M as follows. Write

$$
(42.1) \qquad \left.
\begin{aligned}
d\vartheta^1 &\equiv 0 \\
d\vartheta^2 &\equiv -\pi_2^2 \wedge \omega^2 \\
d\pi_1^1 &\equiv a_1 \pi_2^2 \wedge \omega^2 \\
d\omega^1 &\equiv a_2 \pi_2^2 \wedge \omega^2
\end{aligned}
\right\} \quad \mod \{\vartheta^1, \vartheta^2, \omega^1, \pi_1^1\}
$$

$$
(42.2) \qquad \left.
\begin{aligned}
d\vartheta^2 &\equiv 0 \\
d\vartheta^1 &\equiv -\pi_1^1 \wedge \omega^1 \\
d\pi_2^2 &\equiv a_3 \pi_1^1 \wedge \omega^1 \\
d\omega^2 &\equiv a_4 \pi_1^1 \wedge \omega^1
\end{aligned}
\right\} \quad \mod \{\vartheta^1, \vartheta^2, \omega^2, \pi_2^2\}
$$

where the functions a_i are some smooth functions locally defined on M. Replacing $\pi_1^1, \omega^1, \pi_2^2,$ and ω^2 respectively by $\pi_1^1 + a_1 \vartheta^2$, $\omega^1 + a_2 \vartheta^2$, $\pi_2^2 + a_3 \vartheta^1$, and $\omega^2 + a_4 \vartheta^1$, we may assume that all of the a_i are zero. We now have, in addition to (40), the refined structure equations

$$
(43.1) \qquad d\vartheta^1 \equiv d\omega^1 \equiv d\pi_1^1 \equiv 0 \quad \mod \{\vartheta^1, \vartheta^2, \omega^1, \pi_1^1\}
$$

$$
(43.2) \qquad d\vartheta^2 \equiv d\omega^2 \equiv d\pi_2^2 \equiv 0 \quad \mod \{\vartheta^1, \vartheta^2, \omega^2, \pi_2^2\}.
$$

It is now an elementary matter to see that the two Pfaffian systems

$$
(44.1) \qquad \mathcal{M}^1 = \mathrm{span}\{\vartheta^1, \omega^1, \pi_1^1\}
$$

$$
(44.2) \qquad \mathcal{M}^2 = \mathrm{span}\{\vartheta^2, \omega^2, \pi_2^2\}
$$

are well-defined globally on M, that is, they depend only on the (hyperbolic) Pfaffian system I and the orientation of M (reversing the orientation of M switches the two systems). On each two dimensional integral N^2 of $(\mathcal{I}, \Omega)$, each of the systems $\mathcal{M}^a$ restrict to have rank 1. Thus, each such N is foliated by integral curves of $\mathcal{M}^a$. This pair of foliations is exactly the pair of foliations by characteristic curves in the classical sense.

As in the parabolic case, we say that the characteristics of $(\mathcal{I}, \Omega)$ depend on constants if both of the $\mathcal{M}^a$ are Frobenius systems. In this case, the integrals of $(\mathcal{I}, \Omega)$ can locally be written down explicitly using only O.D.E. as follows: Supposing that each of the $\mathcal{M}^a$ is completely integrable, we know that M can be covered by open sets of the form $U = Y^1 \times Y^2$ where each Y^a is a three dimensional ball and the leaves of the Frobenius system $\mathcal{M}^a$ are given by fixing a point in the Y^a-factor. It easily follows that ϑ^a is a non-zero multiple of a contact form, say $\tilde{\vartheta}^a$, well defined on Y^a. (The explicit construction of the factors Y^a and the forms $\tilde{\vartheta}^a$ requires O.D.E. techniques.) It is now immediate that the integrals of $(\mathcal{I}, \Omega)$ which lie in U are simply products of the form $P^1 \times P^2$ where each P^a is a contact curve in the contact manifold $(Y^a, \tilde{\vartheta}^a)$. Clearly, such a system is equivalent to the "uncoupled" system (41).

Actually, for the reduction of the initial value problem for integrals of $(\mathcal{I}, \Omega)$ to an O.D.E. problem, it suffices that at least one of the systems $\mathcal{M}^a$ be Frobenius. Of course, for the general hyperbolic system, we do not expect either of the systems $\mathcal{M}^a$ to be Frobenius. In Darboux [1870], a far-reaching generalization of the above construction is presented which takes into account any possible "higher" conservation laws for characteristics.

Subcase 2.5: The symbol relations are of the form (26.5).

In this case, we may, by admissible basis change, assume that the symbol relations are of the form

$$(45) \qquad \left.\begin{array}{l} \pi_1^1 - \pi_2^2 \equiv C_{11}\omega^1 + C_{12}\omega^2 \\ \pi_2^1 + \pi_1^2 \equiv C_{21}\omega^1 + C_{22}\omega^2 \end{array}\right\} \quad \bmod I.$$

Again, by adding appropriate multiples of ω^1 and ω^2 to the 1-forms π_j^i, we may assume that the functions C_{ij} are zero. Thus, the torsion vanishes and the structure equations now take the form

$$(46) \qquad \left.\begin{array}{l} d\vartheta^1 \equiv -\pi_1^1 \wedge \omega^1 + \pi_1^2 \wedge \omega^2 \\ d\vartheta^2 \equiv -\pi_1^2 \wedge \omega^1 - \pi_1^1 \wedge \omega^2 \end{array}\right\} \quad \bmod I.$$

Inspection shows that $s_1' = 2$ and $s_2' = 0$. Moreover, the integral elements of $(\mathcal{I}, \Omega)$ at a point of M depend on 2 parameters, namely $\vartheta^1 = 0$ and

$$(47) \qquad \begin{array}{l} \pi_1^1 = \pi_2^2 = \lambda_1\omega^1 + \lambda_2\omega^2 \\[2mm] \pi_2^1 = -\pi_1^2 = \lambda_2\omega^1 - \lambda_1\omega^2 \end{array}$$

It follows that the system $(\mathcal{I}, \Omega)$ is involution. In the analytic category, the general integral depends on two functions of one variable.

The symbol matrix σ_ξ at a non-zero covector $\xi = \xi_1\omega^1 + \xi_2\omega^2$ is the 2×2 matrix

$$\begin{pmatrix} \xi_1 & -\xi_2 \\ \xi_2 & \xi_1 \end{pmatrix}.$$

It follows that the characteristic variety Ξ_x at each point x of M is empty. On the other hand, the *complex* characteristic variety is not empty. In fact, since $\det(\sigma_\xi) = (\xi_1)^2 + (\xi_2)^2$, it follows that $\Xi_x^{\mathbb{C}}$ consists of the two points $[\omega^1 \pm \sqrt{-1}\,\omega^2]$. Again, this is in agreement with the predictions of the general theory for the dimension and degree of $\Xi_x^{\mathbb{C}}$.

A simple example of a system whose symbol relations are of this type are the Cauchy–Riemann equations:

$$(48) \qquad z_x - w_y = z_y + w_x = 0.$$

In fact, it is easy to see that a pair of equations is elliptic in the classical sense if and only if its symbol relations are of the type given by the subspace (26.5). Thus, we shall refer to this type as *elliptic*.

The "complexity" of the characteristic variety suggests that we study the system in a complex basis. Let us temporarily use the notation

$$\vartheta = \vartheta^1 + \sqrt{-1}\,\vartheta^2$$

$$(49) \qquad \pi = \pi_1^1 + \sqrt{-1}\,\pi_1^2$$

$$\omega = \omega^1 + \sqrt{-1}\,\omega^2.$$

Then the structure equations (46) become

$$(50) \qquad \left.\begin{array}{l} d\vartheta \equiv -\pi \wedge \omega \\ d\bar\vartheta \equiv -\bar\pi \wedge \bar\omega \end{array}\right\} \quad \mathrm{mod}\ I.$$

In fact, if we choose an orientation for M, then these equations plus the condition $\sqrt{-1}\,\vartheta \wedge d\vartheta \wedge \bar\vartheta \wedge d\bar\vartheta > 0$ uniquely specify $\vartheta \in I^{\mathbb{C}}$ up to a complex multiple. Moreover, we have

$$(51) \qquad \left.\begin{array}{l} d\vartheta \equiv 0 \\ d\bar\vartheta \equiv -\bar\pi \wedge \bar\omega \\ d\pi \equiv a_1\bar\pi \wedge \bar\omega \\ d\omega \equiv a_2\bar\pi \wedge \bar\omega \end{array}\right\} \quad \mathrm{mod}\ \{\vartheta, \bar\vartheta, \omega, \pi\}.$$

Replacing ω by $\omega + a_2\bar\vartheta$ and π by $\pi + a_1\bar\vartheta$, we obtain, in addition to (49), the refined structure equations

$$(52) \qquad d\vartheta \equiv d\omega \equiv d\pi \equiv 0 \quad \mathrm{mod}\ \{\vartheta, \bar\vartheta, \omega, \pi\}.$$

(Note the analogy with the hyperbolic case.) It is now an elementary matter to see that the complex Pfaffian system $\mathcal{M}$ spanned by the 1-forms $\{\vartheta, \pi, \omega\}$ is well defined on M, depending only on the elliptic Pfaffian system I and the given orientation. (If one reverses the orientation, then the system $\mathcal{M}$ will be replaced by $\bar{\mathcal{M}}$.) Note that $\mathcal{M}$ defines a unique almost complex structure on M for which the system $\mathcal{M}$ is the space of forms of type $(1,0)$. By construction, the integrals of $(\mathcal{I}, \Omega)$ are, in the terminology of Gromov [1985], "pseudo-holomorphic curves" for the given almost complex structure. In fact, the integrals of $(\mathcal{I}, \Omega)$ are precisely the $\mathcal{M}$-pseudo-holomorphic curves in M which are also integrals of the complex Pfaffian form ϑ (well-defined up to a complex multiple).

The integrability of the almost complex structure $\mathcal{M}$ is the elliptic analogue of the notion of "characteristics depending on constants" in the hyperbolic case. By an application of the Newlander–Nirenberg theorem, it can be shown that the necessary and sufficient condition that the system I be locally equivalent to the one derived from the Cauchy–Riemann equations is that the system $\mathcal{M}$ be Frobenius (in the complex sense). This is a geometric indication of the special place the Cauchy–Riemann equations occupy in the study of elliptic systems for two functions of two variables.

Case 3: $r = 3$.

In the case where there are 3 symbol relations, we may choose bases of I and J as in (8)–(10) and the resulting 1-forms π_i^a will satisfy 3 linear relations modulo J. It follows that, modulo J, we may assume that all of the 1-forms π_i^a are multiples of a single 1-form π. Thus, we may write

$$(53) \qquad \pi_i^a \equiv R_i^a \pi \quad \mathrm{mod}\ J.$$

The 2×2 matrix R may be changed by pre- and post-multiplication by invertible 2×2 matrices when we change to another adapted basis of I and J. Thus, the only invariant of R is its rank, which must be either 1 or 2. For the sake of simplicity, we shall assume that the rank of R is constant on M. This allows us to divide Case 3 into two subcases.

Subcase 3.1: The rank of R is 1.

In this case, we may choose our bases of I and J so that the symbol relations become

$$(54) \qquad \pi_2^1 \equiv \pi_1^2 \equiv \pi_2^2 \equiv 0 \quad \mathrm{mod}\ J.$$

(Note that we are reducing modulo J, not I, at this point). It follows that the structure equations for I can be written in the form

$$(55) \qquad \left.\begin{aligned} d\vartheta^1 &\equiv -\pi_1^1 \wedge \omega^1 + C^1 \omega^1 \wedge \omega^2 \\ d\vartheta^2 &\equiv C^2 \omega^1 \wedge \omega^2 \end{aligned}\right\} \quad \mathrm{mod}\ I$$

where C^1 and C^2 are smooth functions on M. Replacing π_1^1 by $\pi_1^1 + C^1\omega^2$, we see that we may assume that $C^1 = 0$. On the other hand, we clearly cannot get rid of the term involving C^2, which represents the unabsorbable torsion. In fact, it is clear that there are no integrals of $(\mathcal{I}, \Omega)$ outside of the locus where $C^2 = 0$.

If $C^2 \neq 0$, then in the generic case, C^2 will vanish along a hypersurface in H in M. This locus may be regarded as a set of 4 equations for the two unknowns functions. We will return to this case below. We now pass on to the case where C^2 vanishes identically. In this case, the structure equations (55) simplify to

$$(56) \qquad \left. \begin{array}{l} d\vartheta^1 \equiv -\pi_1^1 \wedge \omega^1 \\ d\vartheta^2 \equiv 0 \end{array} \right\} \quad \mathrm{mod}\ I.$$

Inspection shows that $s_1' = 1$ and $s_2' = 0$. Moreover, the integral elements at a general point of M depend on 1 parameter, namely $\vartheta^a = 0$ and

$$(57) \qquad \pi_1^1 = \lambda\omega^1.$$

It follows that the system $(\mathcal{I}, \Omega)$ is involution. In the real analytic case, the general integral depends on one function of one variable.

The symbol matrix σ_ξ at a non-zero covector $\xi = \xi_1\omega^1 + \xi_2\omega^2$ has the form

$$\begin{pmatrix} \xi_2 & 0 \\ 0 & \xi_1 \\ 0 & \xi_2 \end{pmatrix}$$

Consequently, σ_ξ is injective if and only if $\xi_2 \neq 0$. Thus, at each point of M, Ξ_x consists of the point $[\omega^1]$.

As a point of interest, although we shall not prove it here, we remark that there are locally only two systems of this kind up to diffeomorphisms which preserve the Pfaffian system I. The first is described by the equations

$$(58) \qquad z_y = w_x = w_y = 0$$

and its general solution is given by $z = f(x)$ and $w = c$ where f is an arbitrary function of x and c is a constant. The second is described by

$$(59) \qquad z_y = w_x - z = w_y = 0$$

with general solution $z = f'(x)$ and $w = f(x)$ where f is an arbitrary function of x. The difference in the two systems is that, for the former, the first derived system $I^{(1)}$ is a Frobenius system while for the latter, $I^{(1)}$ is not a Frobenius system.

Subcase 3.2: The rank of R is 2.

This is the generic case for systems of 3 equations for two functions of two variables. In this case we may choose bases of I and J so that the matrix R becomes the identity matrix. Thus, the symbol relations take the form

$$(56) \qquad \pi_1^1 - \pi_2^2 \equiv \pi_2^1 \equiv \pi_1^2 \equiv 0 \mod J.$$

Thus, writing τ for π_1^1, the structure equations may be written in the form

$$(61) \qquad \left. \begin{aligned} d\vartheta^1 &\equiv -\tau \wedge \omega^1 + C^1 \omega^1 \wedge \omega^2 \\ d\vartheta^2 &\equiv -\tau \wedge \omega^2 + C^2 \omega^1 \wedge \omega^2 \end{aligned} \right\} \mod I.$$

Replacing τ by $\tau + C^1 \omega^2 - C^2 \omega^1$, we see that we may assume that the functions C^a vanish identically. Thus, the torsion of this system is identically zero. The structure equations now simplify to

$$(62) \qquad \left. \begin{aligned} d\vartheta^1 &\equiv -\tau \wedge \omega^1 \\ d\vartheta^2 &\equiv -\tau \wedge \omega^2 \end{aligned} \right\} \mod I.$$

Inspection shows that $s_1' = 1$ and $s_2' = 0$. However, there is a unique integral element of $(\mathcal{I}, \Omega)$ at each point of M, given by $\vartheta^a = \tau = 0$. Thus, the system is *not* involutive.

The symbol matrix σ_ξ at a non-zero covector ξ is the 3×2 matrix

$$\begin{pmatrix} \xi_1 & \xi_2 \\ \xi_2 & 0 \\ 0 & \xi_1 \end{pmatrix}.$$

Consequently, σ_ξ is always injective. Thus, the complex characteristic variety is empty.

Since there is a unique integral element at each point of M, the prolongation of $(\mathcal{I}, \Omega)$ is particularly easy to compute. We may identify the space of integral elements of $(\mathcal{I}, \Omega)$ with M itself and the differential system $\mathcal{I}^{(1)}$ is just the Pfaffian system I_+ of rank 3 which is generated by the 1-forms ϑ^1, ϑ^2, and τ. Since I_+ is a rank 3 Pfaffian system on a manifold of dimension 5, we know that I_+ is involutive if and only if it is a Frobenius system. By the structure equations (62), we already know that $d\vartheta^a \equiv 0 \mod I_+$. It is clear that there exists a function C on M so that $d\tau \equiv C\omega^1 \wedge \omega^2 \mod I_+$. This function C vanishes if and only if the system I_+ is differentially closed. In this case, i.e., $C \equiv 0$, the integrals of the system $(\mathcal{I}, \Omega)$ clearly depend on 3 constants. In the generic case, however, not only will C not be identically zero, the locus where $C = 0$ will define a hypersurface in M which represents yet a fourth first order equation which must be adjoined to the given three. We will consider this case below.

As an example of this type of system of P.D.E., let us consider the problem introduced at the beginning of this section of determining the isometries between two metrics on regions of the plane (see the discussion in the paragraph containing equations (16) through (18). As equation (18) shows, this system falls into Subcase 3.2. In order to investigate the closure of the related system I_+, we shall explicitly parametrize the group $O(2)$. Recall that $O(2)$ has two components, each of which is diffeomorphic to a circle. Thus, the general element of $O(2)$ can be written in the form

$$(63) \qquad g = \begin{pmatrix} \cos t & \sin t \\ \mp \sin t & \pm \cos t \end{pmatrix}.$$

It then follows from equation (17) that the matrix π is skew-symmetric with upper right-hand entry $\pi_2^1 = dt \pm \varphi_2^1 - \psi_2^1$. (The $\pm$-sign is to be taken in agreement with the same sign in (61).) If we set

$$\tau = dt \pm \varphi_2^1 - \psi_2^1,$$

then the differential system I_+ is generated by ϑ^1, ϑ^2, and τ.

We have the well-known formulae

$$d\varphi_2^1 = K(z,w)\eta^1 \wedge \eta^2$$

$$d\psi_2^1 = k(x,y)\omega^1 \wedge \omega^2,$$

where K is the Gauss curvature of the metric h_2 and k is the Gauss curvature of the metric h_1. Using the easily computed identity $\eta^1 \wedge \eta^2 \equiv \pm \omega^1 \wedge \omega^2$ mod I, we obtain

$$d\tau \equiv (K - k)\omega^1 \wedge \omega^2 \quad \mod I_+.$$

Thus, the function $C = K - k$ plays the role of the torsion of I_+. The function C can vanish identically only if each of the functions K and k are equal to the same constant. Thus, we recover the well-known fact that there exists a three-parameter family of local isometries between two metrics in the plane if and only if they each have the same constant Gauss curvature. More generally, since any integral manifold of $(\mathcal{I}, \Omega)$ must lie in the locus $C = 0$, we see that any isometry between two such metrics must preserve their Gauss curvatures. It would be possible to pursue the study of this system further and arrive at a complete answer to the problem of determining when two metrics on the plane are locally isometric, but since our interest in this problem is only in the fact that it provides an example of a system in Subcase 3.2, we shall not go further into its analysis. Note, however, that this problem is not generic in the above sense because the locus $C = 0$ actually represents an equation of order zero relating the unknown functions z and w to the independent variables x and y.

Case 4: $r = 4$.

In this case there are 4 independent relations of the form (11). This means that we may write these relations in the form

$$(64) \qquad \pi_i^a \equiv \sum C_{ij}^a \omega^j \quad \mathrm{mod}\ I.$$

It follows that the structure equations of I are of the form

$$(65) \qquad d\vartheta^a \equiv (C_{12}^a - C_{21}^a)\omega^1 \wedge \omega^2 \quad \mathrm{mod}\ I.$$

Thus, it follows that the torsion of the system is represented by the two functions $C^a = C_{12}^a - C_{21}^a$. These functions vanish identically if and only if the system I is a Frobenius system. In particular, I is Frobenius if and only if it is involutive.

At the other extreme, in the generic case, the equations $C^1 = C^2 = 0$ implicitly define the functions z and w as functions, say f and g, of x and y. If these functions f and g do not solve the given equations, then there is no solution.

§2. Finiteness of the Web Rank.

The theory of webs had its origin in algebraic geometry and the theory of abelian integrals. For a more detailed introduction to the theory than we provide here, the reader should consult Chern and Griffiths [1978]. For our purposes, the following description will suffice. Let N^n be a smooth manifold of dimension n. A *d-web of codimension r* on N is a d-tuple of foliations $\mathcal{W} = (\mathcal{F}_1, \mathcal{F}_2, \ldots, \mathcal{F}_d)$ on N, each of codimension r, which satisfies the condition that the foliations $\mathcal{F}_a$ are pairwise transverse. (Actually, in algebraic geometry, this transversality condition is only supposed to hold on a dense open set in N.) Associated to each foliation $\mathcal{F}_a$ is the Pfaffian system I_a of rank r consisting of those 1-forms which vanish on the leaves of $\mathcal{F}_a$. Of course, since $\mathcal{F}_a$ is a foliation, the Pfaffian system I_a is a Frobenius system. For each non-negative integer p which is less than or equal to r, we let $\Omega^p(I_a)$ denote the space of p-forms on N which may be expressed locally as sums of products of elements of I_a. Thus, if $x_a^1, x_a^2, \ldots, x_a^r$ is a set of local functions whose differentials span I_a on the domain of their definition, then every $\Phi \in \Omega^p(I_a)$ can be expressed uniquely on this domain in the form

$$(66) \qquad \Phi = \sum_{|J|=p} f_J dx_a^J$$

for some functions f_J. Given a d-web of codimension r, $\mathcal{W}$, an abelian p-equation for $\mathcal{W}$ is a d-tuple $(\Phi_1, \Phi_2, \ldots, \Phi_d)$ where each $\Phi_a \in \Omega^p(I_a)$ is a closed form and the d-tuple satisfies

$$(67) \qquad \Phi_1 + \Phi_2 + \cdots + \Phi_d = 0.$$

The space of all such abelian p-equations for $\mathcal{W}$ obviously forms a vector space which we shall denote by $\mathcal{A}^p(\mathcal{W})$. For the webs which occur in algebraic geometry, the dimension of this vector space is finite for global reasons. This dimension is usually called the *abelian p-rank* of $\mathcal{W}$, and is an important invariant of $\mathcal{W}$.

It turns out that the finiteness of the abelian p-rank of $\mathcal{W}$ is a consequence of our general theorems about the characteristic variety of a differential system. In particular, the finiteness result we shall prove does not depend on any global conditions. For simplicity, we shall restrict ourselves to a discussion of the 1-rank. However, we will generalize the definition of web somewhat. Let us define a *d-pseudoweb* $\mathcal{W}$ on N to be a d-tuple $(I_1, I_2, \ldots, I_d)$ of (non-singular) Pfaffian systems on N which are everywhere transverse: $(I_a)_x \cap (I_b)_x = 0$ for all $x \in N$ and all $a \neq b$. Note that we do not assume that the Pfaffian systems I_a are Frobenius and we do not assume that they have the same rank. Just as above, we define the vector space of abelian 1-equations $\mathcal{A}^1(N, \mathcal{W})$ to be the space of d-tuples $(\eta_1, \eta_2, \ldots, \eta_d)$ of closed 1-forms on N with $\eta_a \in I_a$ and satisfying

$$(68) \qquad \eta_1 + \eta_2 + \cdots + \eta_d = 0.$$

We then have the following proposition.

Proposition 2.1. *If N is connected, the dimension of $\mathcal{A}^1(N, \mathcal{W})$ is finite.*

Proof. Let V be a real vector space of dimension n and let U be an open neighborhood in N on which there exists a V-valued coordinate system $y : U \to V$. Let r_a be the rank of the Pfaffian system I_a and let S_a be a real vector space of dimension r_a. Let $A_a(y) : V \to S_a$ be a surjective linear map (depending on y) so that $\omega_a = A_a(y)dy$ is an S_a-valued 1-form satisfying the condition that the components of ω_a span $I_a|_U$.

Let $X = \mathbb{R}^d \times (S_1)^* \times (S_2)^* \times \ldots \times (S_d)^* \times U$ and let $z_1, z_2, \ldots, z_d$ be coordinates on the $\mathbb{R}^d$-factor while $p_a : X \to (S_a)^*$ is the projection onto the $(S_a)^*$ factor. Let $M \subset X$ be the sub-locus defined by the n equations

$$(69) \qquad \sum_a p_a A_a = 0.$$

(Note that the left hand side of (4) is a function on X with values in V^*.) Let I denote the Pfaffian system on M which is generated by the d 1-forms ϑ_a where

$$(70) \qquad \vartheta_a = dz_a - p_a \omega_a = dz_a - (p_a A_a)dy.$$

We let $\mathcal{I}$ denote the differential ideal generated by I and we let Ω be the independence condition $\Omega = dy^1 \wedge dy^2 \wedge \cdots \wedge dy^n$. Then it is clear that $(\mathcal{I}, \Omega)$ is in linear form. An integral of $(\mathcal{I}, \Omega)$ is a submanifold of M of the form $z_a = f_a(y)$ and $p_a = g_a(y)$ on which the forms ϑ_a vanish. On such an integral, the forms $\eta_a = g_a(y)A_a(y)dy$ clearly satisfy $(\eta_1, \eta_2, \ldots, \eta_d) \in \mathcal{A}(U, \mathcal{W})$. Conversely, given $(\eta_1, \eta_2, \ldots, \eta_d) \in \mathcal{A}(U, \mathcal{W})$, there exist unique functions $g_a(y)$ with values in $(S_a)^*$ so that $\eta_a = g_a(y)A_a(y)dy$. Also, since each of the forms η_a is closed, there exist functions f_a, unique up to additive constants, so that $\eta_a = df_a$. It follows that the map given by $(f_a, g_a) \mapsto (g_a(y)A_a(y)dy)$, from the integrals of $(\mathcal{I}, \Omega)$ with domain N to $\mathcal{A}(U, \mathcal{W})$ is a surjective vector space mapping whose kernel consists of the vector space of dimension d given by setting all of the f_a equal to constants and the g_a equal to zero. Thus, in order to show that $\mathcal{A}(U, \mathcal{W})$ has finite dimension, it suffices to show that the space of integrals of $(\mathcal{I}, \Omega)$ with domain N is a finite dimensional space. In turn, in order to show this, by Theorem 3.12 of Chapter V, it will suffice to show that the associated complex characteristic variety is empty.

To prove that the complex characteristic variety is empty, we examine the tableau of $(\mathcal{I}, \Omega)$. By the above description, if we let $W = \mathbb{R}^d$, then the tableau at a point $m = (y, z, p) \in M$ is the space

$$(71) \qquad A_m = \left\{ (t_a A_a(y)) \in W \otimes V^* \,\middle|\, \sum_a t_a A_a = 0 \right\}$$

(here the variable t_a runs over the vector space $(S_a)^*$. If $\xi \in V^*$ were a non-zero covector such that $[\xi]$ were characteristic, then there would be a non-zero $w \in W$ so that $w \otimes \xi \in A_m$. Writing $w = (w_a)$, this becomes $w_a \xi = t_a A_a$ for some set of $t_a \in (S_a)^*$ where the sum $w_1 + \cdots + w_d = 0$. By this latter condition, it follows that at least two of the w_a are non-zero. However, if $w_a \neq 0$, then we clearly have $\xi \in A_a^*((S_a)^*) \subset V^*$. Since the assumption of transversality of the Pfaffian systems I_a clearly implies that $A_a^*((S_a)^*) \cap A_b^*((S_b)^*) = 0$ for all $a \neq b$, we are lead to a contradiction. Thus, there cannot be any (complex) characteristic covectors. Thus, $\Xi_m^{\mathbb{C}}$ is empty. $\qquad\qquad\square$

We remark that the problem of determining good upper bounds for the dimension of $\mathcal{A}^1(N, \mathcal{W})$ for a general d-web of codimension r is rather subtle, being connected with Castelnuovo's bound and the theory of special divisors in algebraic geometry. See Chern and Griffiths [1978].

§3. Orthogonal Coordinates.

This example is a continuation of Example 3.2 of Chapter III and of Example 1.3 of Chapter V. We will follow the notation developed there. Recall that we are given a Riemannian metric g on a manifold N of dimension n and we wish to know when there exist local coordinates (called orthogonal coordinates) $x^1, x^2, \ldots, x^n$ on N so that the metric takes the diagonal form

$$(72) \qquad g = g_{11}(dx^1)^2 + g_{22}(dx^2)^2 + \cdots + g_{nn}(dx^n)^2.$$

As in Example 3.2 of Chapter III, we let $\mathcal{F} \to N$ denote the orthonormal frame bundle of the metric g and we let $\omega_i, \omega_{ij} = -\omega_{ji}$ denote the canonical 1-forms on $\mathcal{F}$. These forms satisfy the structure equations

$$(73) \qquad \begin{aligned} d\omega_i &= -\sum_j \omega_{ij} \wedge \omega_j \\ d\omega_{ij} &= -\sum_k \omega_{ik} \wedge \omega_{kj} + \tfrac{1}{2} \sum_{k,l} R_{ijkl}\omega_k \wedge \omega_l. \end{aligned}$$

We saw, in the aforementioned example, that our problem was equivalent to finding integrals of the differential system $\mathcal{I}$ generated by the 3-forms $\Phi_i = \omega_i \wedge d\omega_i$ subject to the independence condition $\Omega = \omega_1 \wedge \omega_2 \wedge \cdots \wedge \omega_n$. Moreover, we saw that the space of integral elements of $(\mathcal{I}, \Omega)$ at a point of $\mathcal{F}$ was naturally an affine space of dimension $n^2 - n$. Namely, if $\{p_{ij}\}_{i \neq j}$ is any collection of $n^2 - n$ real numbers, and $f \in \mathcal{F}$ is fixed, then the n)plane in $T_f\mathcal{F}$ annihilated by the $\frac{1}{2}(n^2 - n)$ 1-forms $\vartheta_{ij} = \omega_{ij} + p_{ij}\omega_i - p_{ji}\omega_j$ is an integral element of $(\mathcal{I}, \Omega)$ and conversely, every integral element of $(\mathcal{I}, \Omega)$ based at f is of this form for some unique collection of real numbers $\{p_{ij}\}_{i \neq j}$.

It follows that the first prolongation of $(\mathcal{I}, \Omega)$ may be described in the following simple manner: Let $\mathcal{F}^{(1)} = \mathcal{F} \times \mathbb{R}^{n(n-1)}$ and let $\{p_{ij}\}_{i \neq j}$ be a set of (linear) coordinates on the second factor. Then $\mathcal{I}^{(1)}$ is the differential system generated by the $\frac{1}{2}(n^2 - n)$ 1-forms $\vartheta_{ij} = \omega_{ij} + p_{ij}\omega_i - p_{ji}\omega_j$. We may now compute the exterior derivatives of these 1-forms as follows: First, set $\Omega_{ij} = d\omega_{ij} + \sum \omega_{ik} \wedge \omega_{kj}$. Then, we have

$$\begin{aligned} d\vartheta_{ij} &= dp_{ij} \wedge \omega_i - dp_{ji} \wedge \omega_j + p_{ij} \wedge d\omega_i - p_{ji} \wedge d\omega_j \\ &\quad + \Omega_{ij} - \sum_k \omega_{ik} \wedge \omega_{kj} \\ (74) \qquad &\equiv dp_{ij} \wedge \omega_i - dp_{ji} \wedge \omega_j + \sum_k [p_{ij}p_{ik}\omega_i \wedge \omega_k - p_{ji}p_{jk}\omega_j \wedge \omega_k] \\ &\quad + \Omega_{ij} + \sum_k (p_{ik}\omega_i - p_{ki}\omega_k) \wedge (p_{jk}\omega_j - p_{kj}\omega_k) \mod I^{(1)} \\ &\equiv \tilde{\pi}_{ij} \wedge \omega_i - \tilde{\pi}_{ji} \wedge \omega_j + \Omega_{ij} \mod I^{(1)} \end{aligned}$$

where we have set, for all $i \neq j$,

$$(75) \qquad \tilde{\pi}_{ij} = dp_{ij} - \sum_k [(p_{ij} - p_{kj})p_{ik}\omega_k - \tfrac{1}{2}p_{ik}p_{jk}\omega_j].$$

It follows from (74) that, on an integral manifold of $\mathcal{I}^{(1)}$, the 4-form $\Omega_{ij} \wedge \omega_i \wedge \omega_j$ must vanish for all $i \neq j$. Recalling from the Cartan structure equations that

$$(76) \qquad \Omega_{ij} = \tfrac{1}{2} \sum_{k,l} R_{ijkl}\omega_k \wedge \omega_l,$$

we see that, *at any coframe $f \in \mathcal{F}$ which is part of an integrable orthonormal coframe, we must have $R_{ijkl}(f) = 0$ whenever all of i, j, k, and l are distinct.*

Of course, if $n = 3$, then this last condition is trivially fulfilled at all coframes f. Moreover, we have already seen in §3 of Chapter III that when $n = 3$ the system $(\mathcal{I}, \Omega)$ is involutive on $\mathcal{F}$. By the prolongation theorem of §2 of Chapter VI, it follows that $(\mathcal{I}^{(1)}, \Omega)$ is involutive in this dimension. Moreover, as computed in Example 1.3 of Chapter V, the characteristic variety at each integral element consists of 3 lines in general position.

On the other hand, if $n \geq 4$, then the structure equations of the Pfaffian system $I^{(1)}$ on $\mathcal{F}^{(1)}$ are written in the form

$$(77) \qquad d\vartheta_{ij} \equiv \tilde{\pi}_{ij} \wedge \omega_i - \tilde{\pi}_{ji} \wedge \omega_j + \tfrac{1}{2} \sum_{k,l} R_{ijkl}\omega_k \wedge \omega_l \quad \mathrm{mod}\ I^{(1)}.$$

This is clearly a system in linear form. By the above remark, the torsion of this system does not vanish at any point $(f, p) \in \mathcal{F}^{(1)}$ where $R_{ijkl}(f) \neq 0$ for some quadruple of distinct indices (i, j, k, l).

Rather than try for an exhaustive treatment, let us just consider the case where g is a metric on M with the property that R_{ijkl} vanishes identically as a function on $\mathcal{F}$ whenever i, j, k, and l are distinct. Because this is a linear, constant coefficient system of equations on the Riemann curvature tensor which must hold in all orthonormal coframes, it follows that this condition is equivalent to assuming the vanishing of a certain number of the irreducible components of the Riemann curvature tensor under the action of the orthogonal group. Now it is well-known (see Besse [1986]) that for $n \geq 4$, $\mathcal{K}_n$, the space of Riemann curvature tensors in dimension n, decomposes into 3 irreducible subspaces under the action of $O(n)$. These subspaces correspond to the scalar curvature, the traceless Ricci curvature, and the Weyl curvature. This gives an $O(n)$-invariant decomposition of a general Riemann curvature tensor into the form

$$(78) \quad R_{ijkl} = R(\delta_{ik}\delta_{jl} - \delta_{il}\delta_{jk}) + (S_{ik}\delta_{jl} - S_{il}\delta_{jk} + S_{jl}\delta_{ik} - S_{jk}\delta_{il}) + W_{ijkl},$$

where R is a scalar, $S_{ij} = S_{ji}$ and satisfies $\sum_i S_{ii} = 0$, and W_{ijkl} has the same symmetries as the Riemann curvature tensor but in addition satisfies the trace condition $\sum_i W_{ijik} = 0$. It is clear that if W (the Weyl curvature) vanishes, then R_{ijkl} vanishes whenever all of the indices i, j, k, and l are distinct. Moreover, it is not difficult to exhibit a Weyl curvature tensor which satisfies $W_{1234} \neq 0$. It follows that the necessary and sufficient condition that R_{ijkl} vanish identically in all frames whenever i, j, k, l are distinct is that the Weyl component of the curvature be zero. Thus, we shall assume $W \equiv 0$ from now on. It is not difficult to show that when $n \geq 4$ this condition is equivalent to the condition that the metric g be conformally flat (see Besse [1986]).

Since we are assuming that the Weyl curvature is zero, we have the formula

$$(79) \qquad R_{ijkl} = H_{ik}\delta_{jl} - H_{il}\delta_{jk} + H_{jl}\delta_{ik} - H_{jk}\delta_{il}$$

where, for simplicity, we have set $H_{ij} = S_{ij} + \frac{1}{2}R\delta_{ij}$. This gives the following simple formula for the curvature form Ω_{ij}:

$$(80) \qquad \Omega_{ij} = \sum_k [H_{ik}\omega_k \wedge \omega_j - H_{jk}\omega_k \wedge \omega_i].$$

It follows that by writing π_{ij} for $\tilde{\pi}_{ij} - \sum_k H_{jk}\omega_k$ for $i \neq j$, the structure equations for $I^{(1)}$ simplify to the equations

$$(81) \qquad d\vartheta_{ij} \equiv \pi_{ij} \wedge \omega_i - \pi_{ji} \wedge \omega_j \quad \mod I^{(1)}.$$

Thus, *the torsion of $I^{(1)}$ is zero whenever the Weyl tensor vanishes identically.*

We are now going to show that the system $(\mathcal{I}^{(1)}, \Omega)$ is involutive. We begin by calculating the space of integral elements of $(\mathcal{I}^{(1)}, \Omega)$. On an integral element of $(\mathcal{I}^{(1)}, \Omega)$ at a given point of $\mathcal{F}^{(1)}$, the 2-forms $\pi_{ij} \wedge \omega_i - \pi_{ji} \wedge \omega_j$ must vanish even though the 1-forms ω_i must remain linearly independent. It follows, in particular, that $\pi_{ij} \wedge \omega_i \wedge \omega_j$ must vanish also whenever $i \neq j$. Thus, on any integral element E, π_{ij} must be a linear combination of the two 1-forms ω_i and ω_j. Let us write

$$(82) \qquad \pi_{ij} = A_{ij}\omega_i + B_{ij}\omega_j$$

for this linear relation which holds on E. Substituting (82) into the 2-form $\pi_{ij} \wedge \omega_i - \pi_{ji} \wedge \omega_j$ yields the 2-form $(B_{ij} + B_{ji})\omega_i \wedge \omega_j$. Thus, it follows that the matrix B must be skew-symmetric. Conversely, if A_{ij} and $B_{ij} = -B_{ji}$ are any collection of $\frac{3}{2}(n^2 - n)$ numbers (remember that we always have $i \neq j$), then the relations (82) and the conditions $\vartheta_{ij} = 0$ clearly suffice to define an integral element of $(\mathcal{I}^{(1)}, \Omega)$ at every point of $\mathcal{F}^{(1)}$. Thus, the space of integral elements of $(\mathcal{I}^{(1)}, \Omega)$ at each point of $\mathcal{F}^{(1)}$ is an affine space of dimension $\frac{3}{2}(n^2 - n)$.

To complete the proof of involutivity, it will suffice to show that we have $s_1' = s_2' = \frac{1}{2}(n^2 - n)$ while $s_k' = 0$ for all $k > 2$. We will do this by explicitly computing the polar equations for a pair of vectors v and w lying in the integral element E. Let $e_1, e_2, \ldots, e_n$ be the basis of E which is dual to $\omega_1, \omega_2, \ldots, \omega_n$. Let $v = a_1 e_1 + \cdots + a_n e_n$ and $w = b_1 e_1 + \cdots + b_n e_n$ where a_k, b_k are (at the moment) an arbitrary set of $2n$ numbers. The polar equations of the zero-dimensional subspace $E_0 \subset E$ are clearly the 1-forms ϑ_{ij}. Thus, we have $s_0' = \frac{1}{2}(n^2 - n)$. The polar equations of the 1-dimensional subspace $E_1 \subset E$ which is spanned by v consists of the forms ϑ_{ij} and forms α_{ij} which satisfy $\alpha_{ij} \equiv a_i \pi_{ij} - a_j \pi_{ji} \mod \omega$. As long as none of the numbers a_i are zero, the rank of the polar equations of E_1 is clearly $(n^2 - n)$. It follows that we must have $s_1' = \frac{1}{2}(n^2 - n)$. Now let $E_2 \subset E$ be the vector space spanned by v and w. (In order for this space to be of dimension 2, we must have $a_i b_j - a_j b_i \neq 0$ for at least one pair $i \neq j$). The polar equations of E_2 are spanned by the forms ϑ_{ij}, the forms α_{ij}, and 1-forms β_{ij} which satisfy $\beta_{ij} \equiv b_i \pi_{ij} - b_j \pi_{ji} \mod \omega$. It follows that if $a_i b_j - a_j b_i \neq 0$ for *all* pairs $i \neq j$, then the rank of the polar equations of E_2 will be $3 \cdot \frac{1}{2} \cdot (n^2 - n)$. Of course, this implies that $s_2' = \frac{1}{2} \cdot (n^2 - n)$ and that $s_k' = 0$ for all $k > 2$ (since $3 \cdot \frac{1}{2} \cdot (n^2 - n)$ is the codimension of E itself). Thus, $(\mathcal{I}^{(1)}, \Omega)$ is involutive.

It is interesting to compute the characteristic variety of this differential system. Appealing to Theorem 3.2 of Chapter V, we see that since $s_2' = \frac{1}{2} \cdot (n^2 - n)$ and $s_k' = 0$ for all $k > 2$, it follows that the characteristic variety is a curve of degree $\frac{1}{2} \cdot (n^2 - n)$ in $\mathbb{RP}^{n-1}$. From the above computation of the characters, it follows that a 2-plane $E_2 \subset E$ is regular if and only if it has a basis v and w as above satisfying the condition $a_i b_j - a_j b_i \neq 0$ for *all* pairs $i \neq j$. In other words, E_2 is regular if and only if the 2-forms $\omega_i \wedge \omega_j$ are non-zero for all $i \neq j$. It follows immediately that a covector $\xi = \xi_1 \omega_1 + \cdots + \xi_n \omega_n$ is characteristic if and only if it is of the form $\xi = \xi_i \omega_i + \xi_j \omega_j$ for some pair (i, j). Thus, the characteristic variety consists of the $\frac{1}{2} \cdot (n^2 - n)$ lines joining the n points $[\omega_i] \in \mathbb{P}(E^*)$.

In closing, we note that since every metric in dimensions greater than 3 for which the Weyl tensor vanishes is conformally flat, and since a conformal change of metric does not change the status of orthogonal coordinates, we see that in the case where the Weyl curvature vanishes, we may assume that the metric g is actually flat. In particular, we may assume that the metric is real analytic. Then the Cartan–Kähler theorem may be applied to show that the space of local orthogonal coordinates on flat space depends on $\frac{1}{2} \cdot (n^2 - n)$ functions of two variables. Note also that, as the theory predicts, the characteristic variety restricted to any solution is integrable. In fact, if $x^1, x^2, \ldots, x^n$ are local orthogonal coordinates, then the character-istic hypersurfaces are those on which some pair of functions x^i, x^j become functionally dependent.

§4. Isometric Embedding.

In §3 of Chapter III, we applied differential systems to prove the Cartan–Janet embedding theorem. This theorem asserts that if g is a real analytic metric on a manifold N of dimension n, then g can be locally induced by lo-cal embeddings into the Euclidean space $\mathbb{E}^{n+r}$ for any $r \geq \frac{1}{2} \cdot n \cdot (n-1)$. In this section, we develop the application of differential systems to the study of the isometric embedding problem in the "overde-termined" case $r < \frac{1}{2} \cdot n \cdot (n-1)$. Here, prolongation and the characteristic variety play an important role.

We shall adopt a slightly different approach to isometric embedding than that in Chapter III. There, we chose specific framings of various manifolds in order to avoid the complications which would have been induced into the calculations by the variable framings. Here, in order to simplify calculation of the characteristic variety and other geometric quantities, we will employ a more invariant formulation. We begin by reviewing the structure equations of a Riemannian metric. We shall adopt the index ranges

$$1 \leq i, j, k \leq n$$

$$n + 1 \leq a, b, c \leq n + r$$

$$1 \leq A, B, C \leq n + r.$$

Let g be a Riemannian metric on a manifold N of dimension n. Let $x : \mathcal{F} \to N$ be the orthonormal frame bundle on N which consists of $(n + 1)$-tuples $f = (x; e_1, e_2, \ldots, e_n)$ where $x \in N$ and $e_1, e_2, \ldots, e_n$ is an orthonormal basis of $T_x N$. Of course, the group $O(n)$ acts on $\mathcal{F}$ on the right in the usual way and makes $\mathcal{F}$ into a principal right $O(n)$-bundle over N. The tautological 1-forms ω_i on $\mathcal{F}$ are defined by setting $\omega_i(v) = g(x_*(v), e_i)$ for all $v \in T_f \mathcal{F}$ with $f = (x; e_1, e_2, \ldots, e_n)$. The Levi–Civita connection forms on $\mathcal{F}$ are the unique 1-forms $\omega_{ij} = -\omega_{ji}$ which satisfy the first structure equation of Cartan

$$(83) \qquad d\omega_i = -\sum_j \omega_{ij} \wedge \omega_j.$$

These forms also satisfy the second structure equations of Cartan

$$(84) \qquad d\omega_{ij} = -\sum_k \omega_{ik} \wedge \omega_{kj} + \tfrac{1}{2} \sum_{k,l} R_{ijkl} \omega_k \wedge \omega_l$$

where the functions R_{ijkl} are well defined on $\mathcal{F}$ and satisfy the symmetries

$$
R_{ijkl} = -R_{jikl} = -R_{ijlk}
$$
$$
\text{(85)} \qquad R_{ijkl} + R_{iklj} + R_{iljk} = 0.
$$

Just as in §3 of Chapter III, we shall regard the functions R_{ijkl} as the components of a function R which takes values in the vector space $\mathcal{K}_n \subset \Lambda^2(\mathbb{R}^n) \otimes \Lambda^2(\mathbb{R}^n)$ defined by the symmetries (85). Of course, the group $O(n)$ acts linearly on $\mathcal{K}_n$ in the obvious way and the map $R : \mathcal{F} \to \mathcal{K}_n$ is equivariant.

Let $\mathbb{E}^{n+r}$ be given its standard metric, and let $\mathcal{F}(\mathbb{E}^{n+r})$ denote the orthonormal frame bundle of $\mathbb{E}^{n+r}$. We shall denote elements of $\mathcal{F}(\mathbb{E}^{n+r})$ as $\mathbf{f} = (y; \mathbf{e}_1, \mathbf{e}_2, \ldots, \mathbf{e}_{n+r})$. Of course, the group $O(n+r)$ acts on the right on $\mathcal{F}(\mathbb{E}^{n+r})$ and makes it into a principal right fiber bundle over $\mathbb{E}^{n+r}$. We shall denote the tautological forms on $\mathcal{F}(\mathbb{E}^{n+r})$ by η_A and the associated Levi–Civita forms by $\eta_{AB} = -\eta_{BA}$. Since $\mathbb{E}^{n+r}$ is flat, these forms satisfy the structure equations

$$
d\eta_A = -\sum_B \eta_{AB} \wedge \eta_B
$$
$$
\text{(86)} \qquad d\eta_{AC} = -\sum_B \eta_{AB} \wedge \eta_{BC}.
$$

Let M be the manifold of 1-jets of isometries from N to $\mathbb{E}^{n+r}$. Thus, an element of M consists of a triple $m = (x, y, l)$ where $x \in N$, $y \in \mathbb{E}^{n+r}$, and $l : T_x N \to T_y \mathbb{E}^{n+r}$ is a linear map which is an isometry onto its image. We can embed M into $G_n(T(N \times \mathbb{E}^{n+r}))$ by letting $m = (x, y, l)$ correspond to the n-plane $\mathbb{E} = \{(v, l(v)) \mid v \in T_x N\}$ which lies in $T_{(x,y)}(N \times \mathbb{E}^{n+r})$. The canonical Pfaffian system on $G_n(T(N \times \mathbb{E}^{n+r}))$ then restricts to M to be the Pfaffian system I_0 of rank $n + r$ which we wish to study. We let J_0 be the canonical independence bundle. It contains I_0 and is of rank $2n + r$. In fact, J_0 is the bundle of semi-basic 1-forms for the projection $M \to N \times \mathbb{E}^{n+r}$.

The difficulty of working directly with the Pfaffian system I_0 on M is that there is no canonical set of generators for I_0 on M. To remedy this, consider the product manifold $\mathcal{F} \times \mathcal{F}(\mathbb{E}^{n+r})$. There is a canonical projection $q : \mathcal{F} \times \mathcal{F}(\mathbb{E}^{n+r}) \to M$ defined by letting $q(f, \mathbf{f})$ be the triple (x, y, l) where $x \in N$ is the base point of f, $y \in \mathbb{E}^{n+r}$ is the base point of $\mathbf{f}$, and $l : T_x N \to T_y \mathbb{E}^{n+r}$ is the linear map which satisfies $l(e_i) = \mathbf{e}_i$. Let $O(n) \times O(r)$ be the subgroup of $O(n+r)$ which preserves the n-plane spanned by the first n elements of a given orthonormal basis of $\mathbb{E}^{n+r}$. The group $O(n) \times O(r)$ acts on $\mathcal{F} \times \mathcal{F}(\mathbb{E}^{n+r})$ by the diagonal action in the $O(n)$-factor and in the standard way in the $O(r)$-factor. It is clear that the orbits of $O(n) \times O(r)$ are the fibers of q. Thus, $\mathcal{F} \times \mathcal{F}(\mathbb{E}^{n+r})$ has the structure of a principal right $O(n) \times O(r)$-bundle over M.

It is easy to see that the Pfaffian system I_0 on M pulls back up to $\mathcal{F} \times \mathcal{F}(\mathbb{E}^{n+r})$ to be spanned by the 1-forms $\{\eta_i - \omega_i \mid 1 \le i \le n\}$ and the 1-forms $\{\eta_a \mid n < a \le n + r\}$. Let $\mathcal{I}_0$ denote the differential system generated by I_0 on either M or $\mathcal{F} \times \mathcal{F}(\mathbb{E}^{n+r})$. The structure equations above give the formulas

$$(87) \qquad \left.\begin{aligned} d(\eta_i - \omega_i) &\equiv -\sum_j (\eta_{ij} - \omega_{ij}) \wedge \omega_j \\[2ex] d\eta_a &\equiv -\sum_j \eta_{aj} \wedge \omega_j \end{aligned}\right\} \quad \mathrm{mod}\ \mathcal{I}_0$$

which make clear the fact that, on $\mathcal{F} \times \mathcal{F}(\mathbb{E}^{n+r})$, the orbits of $O(n) \times O(r)$ are the Cauchy leaves of I_0. Moreover, the independence condition on M pulled up to $\mathcal{F} \times \mathcal{F}(\mathbb{E}^{n+r})$ can be represented by the n-form $\Omega = \omega_1 \wedge \omega_2 \wedge \cdots \wedge \omega_n$. To include the Cauchy characteristics, we shall define an augmented independence condition on $\mathcal{F} \times \mathcal{F}(\mathbb{E}^{n+r})$ by letting Ω_+ be the form of degree $n_+ = n + \frac{1}{2}n(n-1) + \frac{1}{2}r(r-1)$ obtained by wedging together the forms ω_i, $\{\omega_{ij} \mid i < j\}$, and the forms $\{\eta_{ab} \mid a < b\}$.

It is now easy to see that every n_+-dimensional integral element E_+ of $(\mathcal{I}_0, \Omega_+)$ on $\mathcal{F} \times \mathcal{F}(\mathbb{E}^{n+r})$ pushes down to M to be an n-dimensional integral element $E = q_*(E_+)$ of $(\mathcal{I}_0, \Omega)$. Conversely, for every point $(f, \mathbf{f}) \in \mathcal{F} \times \mathcal{F}(\mathbb{E}^{n+r})$ in the fiber over $m \in M$, the inverse image of an integral element $E \subset T_m M$ of $(\mathcal{I}_0, \Omega)$ is an integral element $E_+ = q_*^{-1}(E) \subset T_{(f,\mathbf{f})}\mathcal{F} \times \mathcal{F}(\mathbb{E}^{n+r})$ of $(\mathcal{I}_0, \Omega_+)$.

By the equations (87), the integral elements of $(\mathcal{I}_0, \Omega_+)$ based at a point $(f, \mathbf{f}) \in \mathcal{F} \times \mathcal{F}(\mathbb{E}^{n+r})$ are parametrized by a collection of $\frac{1}{2}(n^2 + n)r$ numbers $h = (h_{aij}) = (h_{aji})$ in the following way: For each such collection, the n_+-plane which is annihilated by the 1-forms of I_0 together with the 1-forms $\eta_{ij} - \omega_{ij}$ and the 1-forms $\eta_{ai} - h_{aij}\omega_j$ is an integral element of $(\mathcal{I}_0, \Omega_+)$ and, conversely, every integral element of $(\mathcal{I}_0, \Omega_+)$ is of this form. Thus, the space $V_n(\mathcal{I}_0, \Omega_+)$ of integral elements of $(\mathcal{I}_0, \Omega_+)$ is diffeomorphic to the product $\mathcal{F} \times \mathcal{F}(\mathbb{E}^{n+r}) \times (\mathbb{R}^r \otimes S^2(\mathbb{R}^n))$. Accordingly, we shall denote the elements of $V_n(\mathcal{I}_0, \Omega_+)$ by triples (f, f, h).

Now, the group $O(n) \times O(r)$ acts in the obvious way on the vector space $\mathbb{R}^r \otimes S^2(\mathbb{R}^n)$. If we let $O(n) \times O(r)$ act on $\mathcal{F} \times \mathcal{F}(\mathbb{E}^{n+r}) \times (\mathbb{R}^r \otimes S^2(\mathbb{R}^n))$ by the consequent natural diagonal action, then the quotient will clearly be the space of integral elements of $(\mathcal{I}_0, \Omega)$ on M. Thus,

$$M^{(1)} = (\mathcal{F} \times \mathcal{F}(\mathbb{E}^{n+r}) \times (\mathbb{R}^r \otimes S^2(\mathbb{R}^n)))/O(n) \times O(r).$$

Geometrically, we may represent the elements of $M^{(1)}$ as quadruples (x, y, l, H) where $(x, y, l) \in M$ is as before and $H : S^2(T_x N) \to (l(T_x N))^\perp$ is a linear map. Here, $(l(T_x N))^\perp$ is the r-dimensional vector space which is perpendicular to $l(T_x N)$ in $T_y \mathbb{E}^{n+r}$. As to be expected, the elements of $M^{(1)}$ are identifiable with the 2-jets of smooth maps which are isometries up to first order.

For simplicity, we shall denote by I (instead of $I_0^{(1)}$) the Pfaffian system which is generated on $\mathcal{F} \times \mathcal{F}(\mathbb{E}^{n+r}) \times (\mathbb{R}^r \otimes S^2(\mathbb{R}^n))$ by the 1-forms $\vartheta_i, \vartheta_a, \vartheta_{ij} = -\vartheta_{ji}$, and ϑ_{ai} where

$$\vartheta_i = \eta_i - \omega_i$$

$$\vartheta_a = \eta_a$$

(88)
$$\vartheta_{ij} = \eta_{ij} - \omega_{ij}$$

$$\vartheta_{ai} = \eta_{ai} - \sum_j h_{aij}\omega_j.$$

Of course, I is also well-defined as a Pfaffian system on $M^{(1)}$ and moreover, the Cauchy leaves of I on $\mathcal{F} \times \mathcal{F}(\mathbb{E}^{n+r}) \times (\mathbb{R}^r \otimes S^2(\mathbb{R}^n))$ are the orbits of the group $O(n) \times O(r)$. We let $\mathcal{I}$ denote the differential system generated by I on either $M^{(1)}$ or $\mathcal{F} \times \mathcal{F}(\mathbb{E}^{n+r}) \times (\mathbb{R}^r \otimes S^2(\mathbb{R}^n))$.

The structure equations of I are easily obtained in the form

(89)
$$\left. \begin{aligned} d\vartheta_i &\equiv d\vartheta_a \equiv 0 \\ d\vartheta_{ij} &\equiv \tfrac{1}{2} T_{ijkl}\omega_k \wedge \omega_l \\ d\vartheta_{ai} &\equiv -\sum_j \pi_{aij} \wedge \omega_j \end{aligned} \right\} \mod I$$

where the forms $\pi_{aij} = \pi_{aji}$ are given by the formula

(90)
$$\pi_{aij} = dh_{aij} - \sum_k [h_{akj}\omega_{ki} + h_{aik}\omega_{kj}] + \sum_b h_{bij}\eta_{ba}.$$

and the functions T_{ijkl} are given by the formulas

(91)
$$T_{ijkl} = \sum_a [h_{aik}h_{ajl} - h_{ail}h_{ajk}] - R_{ijkl}.$$

It is clear from (89) that there are no integral elements of $(\mathcal{I}, \Omega_+)$ at points of $\mathcal{F} \times \mathcal{F}(\mathbb{E}^{n+r}) \times (\mathbb{R}^r \otimes S^2(\mathbb{R}^n))$ where any of the functions T_{ijkl} are non-zero (the functions T_{ijkl} represent non-absorbable torsion).

The usual prescription for continuing the prolongation process at this point is to restrict to the subspace of $\mathcal{F} \times \mathcal{F}(\mathbb{E}^{n+r}) \times (\mathbb{R}^r \otimes S^2(\mathbb{R}^n))$ where all of the functions T_{ijkl} vanish. If we let $\mathcal{Z} \subset \mathcal{F} \times \mathcal{F}(\mathbb{E}^{n+r}) \times (\mathbb{R}^r \otimes S^2(\mathbb{R}^n))$ denote the locus of common zeros of the collection of functions

$$\mathcal{T} = \{T_{ijkl} \mid 1 \le i, j, k, l \le n\},$$

and let $q(\mathcal{Z}) \subset M^{(1)}$ denote its image in $M^{(1)}$, then $q(\mathcal{Z})$ represents the 2-jets of mappings of N to $\mathbb{E}^{n+r}$ which induce an isometry to first order and which satisfy the *Gauss equations* $T_{ijkl} = 0$. (Note that, from a given 2-jet of an immersion into Euclidean space, one can only compute the 1-jet of the induced metric. The Gauss equations demonstrate the remarkable fact that, even though one needs the the full 3-jet of the immersion to compute the full 2-jet of the metric, the Riemannian curvature tensor of the induced metric (which depends only on partial 2-jet information) can be computed using only the 2-jet of the immersion.)

Unfortunately, $\mathcal{Z}$ will not, in general, be a smooth manifold. To avoid this difficulty, we shall restrict our attention to a more manageable subspace. Let $Z \subset \mathcal{Z}$ denote the subspace consisting of the *ordinary* zeros of the collection $\mathcal{T}$. (Recall that if P is a smooth manifold and $\mathcal{P}$ is a collection of smooth functions on P, then a point $p \in P$ is called an ordinary zero of the collection $\mathcal{P}$ if there exists an integer k, an open neighborhood U of p, and a set of functions $f_1, f_2, \ldots, f_k \in \mathcal{P}$ whose differentials are linearly independent on U which have the property that an element $q \in U$ is a simultaneous zero of all of the functions in $\mathcal{P}$ if and only if it is is a zero of the functions $f_1, f_2, \ldots, f_k$. The integer k is, of course, the codimension of the zero locus of the functions in $\mathcal{P}$ at p.) Moreover, since we are only interested in integrals of $(\mathcal{I}, \Omega_+)$, we may as well restrict to $Z^* \subset X$, the open subset of Z on which the form Ω_+ is non-zero. Assuming that Z^* is non-empty (as it will be in many of the specific examples below), we can study the restriction of the differential system $(\mathcal{I}, \Omega_+)$ to Z^*. Note that since the collection $\mathcal{T}$ and the form Ω_+ are invariant (up to sign) under the action of the group $O(n) \times O(r)$, it follows that Z and Z^* are invariant under this action and thus are unions of its orbits. Let $Y \subset M^{(1)}$ be the image of Z under the quotient mapping. Then Y is a smooth submanifold of $M^{(1)}$. Moreover, if $Y^* \subset Y$ is the open subset where Ω does not vanish, then Y^* is clearly the image of Z^* under the quotient map. In fact, Z^* is an $O(n) \times O(r)$-bundle over Y^*.

In order to study the restriction of the Pfaffian system I to Z^* or Y^*, we shall need some information about the differentials of the functions in $\mathcal{T}$. Let us set

$$(92) \quad \tau_{ijkl} = dT_{ijkl} - \sum_m [T_{mjkl}\omega_{mi} + T_{imkl}\omega_{mj} + T_{ijml}\omega_{mk} + T_{ijkm}\omega_{ml}].$$

The structure equations (84) can be differentiated to yield the following formula for the derivative of R_{ijkl}

$$dR_{ijkl} = \sum_m [R_{mjkl}\omega_{mi} + R_{imkl}\omega_{mj} + R_{ijml}\omega_{mk} + R_{ijkm}\omega_{ml} + R_{ijklm}\omega_m].$$

where the functions R_{ijklm} are uniquely specified by this formula and represent the components of ∇R, the covariant derivative of the Riemann curvature tensor. If we regard $h_{ij} = (h_{aij})$ as an $\mathbb{R}^r$-valued function on $\mathcal{F} \times \mathcal{F}(\mathbb{E}^{n+r}) \times (\mathbb{R}^r \otimes S^2(\mathbb{R}^n))$ and use the standard inner product on $\mathbb{R}^r$, then the formula for T_{ijkl} can be simplified to the formula

$$T_{ijkl} = h_{ik} \cdot h_{jl} - h_{il} \cdot h_{jk} - R_{ijkl}.$$

It follows that if we let $\pi_{ij} = (\pi_{aij})$ denote the corresponding $\mathbb{R}^r$-valued 1-form, then the formula for the differentials of the functions T_{ijkl} can be written in the form

$$(93) \qquad \tau_{ijkl} = h_{ik} \cdot \pi_{jl} - h_{il} \cdot \pi_{jk} + \pi_{ik} \cdot h_{jl} - \pi_{il} \cdot h_{jk} - \sum_m R_{ijklm}\omega_m.$$

On Z, we have $dT_{ijkl} = \tau_{ijkl}$. If C is the codimension of Z in $\mathcal{F} \times \mathcal{F}(\mathbb{E}^{n+r}) \times (\mathbb{R}^r \otimes S^2(\mathbb{R}^n))$ at $(f, \mathbf{f}, h) \in Z$, then exactly C of the forms τ_{ijkl} are linearly independent at $(f, \mathbf{f}, h)$ and the vanishing of the C corresponding functions T_{ijkl} suffice to define Z in a neighborhood of $(f, \mathbf{f}, h)$. Moreover, since Z^* is an open subset of Z, it follows that, if $(f, \mathbf{f}, h) \in Z^*$, then these C 1-forms must be linearly independent from the 1-forms ω_i. In particular, at points of Z^*, the number of linearly independent 1-forms among the τ_{ijkl} is the same as the number of linearly independent 1-forms among the $\tau_{ijkl} \mod \{\omega_1, \ldots, \omega_n\}$. This latter rank clearly depends only on the "second fundamental form" h. This justifies the following terminology: We shall speak of Y^* as the space of 2-jets of isometric immersions with *ordinary second fundamental form*. The local isometric embeddings of N into $\mathbb{E}^{n+r}$ which correspond to integrals of I on Z^* will be referred to as ordinary isometric embeddings. We now proceed to investigate the ordinary isometric embeddings.

When we restrict to Z^*, the new relations on the forms in the coframing $\vartheta_i, \vartheta_a, \vartheta_{ij}, \vartheta_{ai}, \pi_{ij}, \omega_i, \omega_{ij}, \eta_{ab}$ will all be generated by setting the differentials of the T_{ijkl} equal to zero. Since, on Z, we have $dT_{ijkl} = \tau_{ijkl}$, it follows that the structure equations of I on Z^* are

$$(94) \qquad \left. \begin{aligned} d\vartheta_i &\equiv d\vartheta_a \equiv 0 \\[4pt] d\vartheta_{ij} &\equiv 0 \\[4pt] d\vartheta_{ai} &\equiv -\sum_j \pi_{aij} \wedge \omega_j \end{aligned} \right\} \mod I$$

where the relations on the forms π_{ij} are spanned by $\pi_{ij} = \pi_{ji}$ and

$$(95) \qquad h_{ik} \cdot \pi_{jl} - h_{il} \cdot \pi_{jk} + \pi_{ik} \cdot h_{jl} - \pi_{il} \cdot h_{jk} \equiv \sum_m R_{ijklm}\omega_m \quad \mod I.$$

Of course, one does not expect this system to be involutive in general. For one thing, we have not yet made any assumption about r, the embedding codimension. However, we can already gain some useful information about the isometric embedding problem by examining the characteristic variety of the symbol relations (95).

Assume that E is an integral element of $(\mathcal{I}, \Omega)$ at a point (x, y, l, H) of Y^*. We want to compute the condition that a covector $\xi \in E^*$ be characteristic. Let $(f, \mathbf{f}, h) \in Z^*$ be an element which lies over (x, y, l, H) and has the property that $\xi(q_*(v)) = \lambda \omega_n(v)$ for some non-zero real number λ and every $v \in E_+$ where $E_+ = q_*^{-1}(E)$ is the associated integral element of $(\mathcal{I}, \Omega_+)$ on Z^*. Since $O(n)$ acts transitively on the unit sphere in $\mathbb{R}^n$, such an element $(f, \mathbf{f}, h)$ clearly exists. The annihilator of E_+ is spanned by the 1-forms in I together with some $\mathbb{R}^r$-valued 1-forms

$$(96) \qquad \psi_{ij} = \psi_{ji} = \pi_{ij} - \sum_k h_{ijk}\omega_k,$$

where the vectors $h_{ijk} = h_{jik} = h_{ikj}$ in $\mathbb{R}^r$ satisfy the equations

$$(97) \qquad h_{ik} \cdot h_{jlm} - h_{il} \cdot h_{jkm} + h_{jl} \cdot h_{ikm} - h_{jk} \cdot h_{ilm} = R_{ijklm}(f).$$

The polar equations of the hyperplane $\omega_n = 0$ in E_+ are spanned by the 1-forms in I together with the 1-forms $\Psi = \{\psi_{ij} \mid i < n \text{ and } i \leq j\}$. It follows that this hyperplane is characteristic if and only if not all of the components of the 1-form ψ_{nn} can be obtained as linear combinations of the components of the forms in Ψ. Now, because Z^* is an open subset of the space of ordinary zeros of the functions T, aside from the symmetry $\psi_{ij} = \psi_{ji}$, the only relations among the forms ψ_{ij} are

$$(98) \qquad h_{ik} \cdot \psi_{jl} - h_{il} \cdot \psi_{jk} + h_{jl} \cdot \psi_{ik} - h_{jk} \cdot \psi_{il} = 0.$$

All of the relations in (98) which involve ψ_{nn} in a non-trivial way can be obtained by letting i and k be strictly less than n and setting j and l equal to n. This gives rise to the relations

$$(99) \qquad h_{ik} \cdot \psi_{nn} \equiv 0 \mod \Psi$$

for all pairs i and k which are strictly less than n. This set of relations implies $\psi_{nn} \equiv 0 \mod \Psi$ if and only if the set of vectors $\{h_{ij} \mid i, j < n\}$ spans the entire vector space $\mathbb{R}^r$. Thus, the hyperplane $\omega_n = 0$ in E^+ is characteristic if and only if, when we regard h as a quadratic form $h = \sum_{i,j} h_{ij}\omega_i \circ \omega_j$ with values in $\mathbb{R}^r$, there exists a non-zero vector $w \in \mathbb{R}^r$ so that $w \cdot h$ is a multiple of ω_n. Translating this into a statement about the original integral element E of $(\mathcal{I}, \Omega)$ based at $(x, y, l, H) \in Y^*$, we see that, after we make the natural identification $E \cong T_x N$, a covector $\xi \in E^*$ is characteristic if and only if there exists a non-zero vector $w \in (l(T_x N))^\perp$ so that $w \cdot H = \lambda \circ \xi$ for some $\lambda \in E^*$. We record this as the following fundamental proposition.

Proposition 4.1. *If E is an integral element of $(\mathcal{I}, \Omega)$ based at (x, y, l, H) in Y^*, then a covector $\xi \in E^*$ is characteristic if and only if there exists a non-zero vector $w \in (l(T_x N))^{\perp}$ so that $w \cdot H = \lambda \circ \xi$ for some $\lambda \in E^*$.*

Of course, Proposition 4.1 can be applied directly only to the cases where the system $(\mathcal{I}, \Omega)$ has been shown to be involutive. However, as discussed in Chapter V, the characteristic variety can only decrease in size when one prolongs. Thus, Proposition 4.1 serves as an important means of deriving an "upper bound" for the characteristic variety.

Proposition 4.1 also serves as motivation for the definition of a sort of characteristic variety associated to any "second fundamental form." Thus, let W be any r-dimensional Euclidean vector space and let V be any n-dimensional real vector space. Given an element $H \in W \otimes S^2(V^*)$, which we may regard as a quadratic form on V with values in W, we define $\Xi_H \subset \mathbb{P}V^*$ by the condition that $[\xi] \in \Xi_H$ if and only if there exists some non-zero $w \in W$ so that $w \cdot H = \lambda \circ \xi$ for some $\lambda \in V^*$ (note that we do not require λ to be non-zero). By tensoring with $\mathbb{C}$, we may define the associated complex variety $\Xi_H^{\mathbb{C}}$. It will also be useful to let $|H| \subset S^2(V^*)$ denote the linear subspace spanned by the quadratic forms $w \cdot H$ as w ranges over all of W.

As a first application of Proposition 4.1, we consider the "underdetermined" case when $r > \frac{1}{2}n(n-1)$.

Proposition 4.2. *If $r > \frac{1}{2}n(n-1)$, then $\Xi_H = \mathbb{P}V^*$ for all $H \in W \otimes S^2(V^*)$.*

Proof. Note first that if $\dim |H| < r$, then there exists a vector $w \in W$ so that $w \cdot H = 0$ and hence every covector is characteristic. On the other hand, if $\dim |H| = r$, then for every non-zero $\xi \in V^*$, the linear space $(\xi) = \{\lambda \circ \xi \mid \lambda \in V^*\}$ must have non-trivial intersection with $|H|$ for dimension reasons. Thus, again, every covector in V^* must be characteristic. $\square$

The "determined" case is also quite interesting:

Proposition 4.3. *If $r = \frac{1}{2}n(n-1)$, then for all $H \in W \otimes S^2(V^*)$, $\Xi_H^{\mathbb{C}}$ is either all of $\mathbb{P}(V^{\mathbb{C}})^*$ or else is an algebraic hypersurface in $\mathbb{P}(V^{\mathbb{C}})^*$ of degree n. Moreover, if $n > 2$, then the real characteristic variety Ξ_H is never empty.*

Proof. Again, if $\dim |H| < r$, then every covector is characteristic and there is nothing to prove. Thus, we may assume for the rest of the proof that $\dim |H| = r$. Let $h_1, h_2, \ldots, h_r$ be a basis of $|H| \subset S^2(V^*)$. Let $x^1, x^2, \ldots, x^n$ be a basis of V^* and let $\xi = \sum_i \xi_i x^i$ be a general element of V^* where we regard the ξ_i as variables. Since $S^2(V^*)$ is of dimension $r + n$, let Δ be a basis of $\Lambda^{r+n}(S^2(V^*))$. Then there exists a homogeneous polynomial $P(\xi_1, \xi_2, \ldots, \xi_n)$ of degree n (with real coefficients) so that

$$(100) \quad h_1 \wedge h_2 \wedge \cdots \wedge h_r \wedge (\xi \circ x^1) \wedge (\xi \circ x^2) \wedge \cdots \wedge (\xi \circ x^n) = P(\xi)\Delta.$$

Clearly, $[\xi] \in \mathbb{P}(V^{\mathbb{C}})^*$ is characteristic if and only if $P(\xi) = 0$. The stated properties of $\Xi_H^{\mathbb{C}}$ now follow immediately.

To show that Ξ_H is non-empty whenever $n > 2$, let us set $Q = S^2(V^*)/|H|$. Then Q a real vector space of dimension n. For any element b of $S^2(V^*)$, let us let $[\![b]\!] \in Q$ denote its reduction $\mod |H|$. Suppose now that Ξ_H were empty. Then for any two non-zero elements $\alpha, \beta \in V^*$, we must have $\alpha \circ \beta \notin |H|$ and thus $[\![\alpha \circ \beta]\!] \neq 0$. It follows that if we define $\mu : V^* \times V^* \to Q$ by $\mu(\alpha, \beta) = [\![\alpha \circ \beta]\!]$, then μ is a symmetric multiplication without zero divisors. Thus, if we choose any non-zero $e \in V^*$, then we may define a V^*-valued product $\alpha \circ \beta$ by letting $\alpha \circ \beta$ be the unique element of V^* which satisfies $\mu(e, \alpha \cdot \beta) = [\![\alpha \circ \beta]\!]$. This product defines the structure of a commutative (though not necessarily associative) algebra on V^*. By construction, this algebra has a unit e and has no zero divisors. By the commutativity of V^*, it follows that we may choose a basis of V^* in which all of the maps $m_\alpha : V^* \to V^*$ given by $m_\alpha(\beta) = \alpha \cdot \beta$ are simultaneously in (real) Jordan canonical form. From this, it immediately follows that these maps cannot all be invertible unless $n \leq 2$. $\qquad\square$

Note that in the case $n = 2$, we have $r = 1$. Thus $|H|$ consists of the multiples of a single quadratic form on V, say b. If b is positive (or negative) definite, then $b = \pm \xi \circ \bar{\xi}$ where ξ is a complex-valued 1-form on V. Thus, in this case, Ξ_H consists of the two (non-real) points $\{[\xi], [\bar{\xi}]\}$ in $\mathbb{P}(V^{\mathbb{C}})^*$. The case of isometric embedding when $n = 2$ and $r = 1$ has been discussed rather thoroughly in Chapters IV and V. We shall not discuss it further here.

It is interesting to remark that considerations from algebraic geometry allow us to compute the dimension and degree of $\Xi_H^{\mathbb{C}}$ for "generic" H. The dimension is easy: Since we have already treated the other cases, we may assume that $r < \frac{1}{2}n(n-1)$. Then for generic $H \in W \otimes S^2(V^*)$, we will have $\dim |H| = r$, and $|H|$ will be a generic r-plane in $S^2(V^*)$. The cone $\mathcal{C} \subset S^2(V^*)$ which consists of quadratic forms of rank 2 or less is a singular cone of dimension $2n - 1$. For dimension reasons, if $r + 2n - 1 \leq \frac{1}{2}n(n+1) = \dim S^2(V^*)$, then for generic H, $|H|^{\mathbb{C}} \cap \mathcal{C}^{\mathbb{C}} = \{0\}$. It then follows from Proposition 4.1 that if $r \leq \frac{1}{2}(n-1)(n-2)$ then, for generic H, $\Xi_H^{\mathbb{C}} = \emptyset$. If $r = k+1+\frac{1}{2}(n-1)(n-2)$ where $0 \leq k < n-1$, then the above general position argument shows that, again for generic H, $\dim \Xi_H^{\mathbb{C}} = k$.

The degree of $\Xi_H^{\mathbb{C}}$ is somewhat more difficult to compute. We shall not reproduce the argument here, instead we refer the reader to Bryant, Griffiths and Yang [1983], where it is shown that, for generic H,

$$\deg \Xi_H^{\mathbb{C}} = \binom{2n - 2 - k}{n - 1}.$$

Returning to the case $r = \frac{1}{2}n(n-1)$, let us say that an element $H \in W \otimes S^2(V^*)$ is *non-degenerate* if Ξ_H is not all of $\mathbb{P}V^*$. Note that the condition of non-degeneracy depends only on the subspace $|H| \subset S^2(V^*)$. In order to have non-degeneracy, $|H|$ must have dimension r. Moreover, as (100) shows, the coefficients of $P(\xi)$ are linear in the Plücker coordinates of the subspace $|H|$ in $G_r(S^2(V^*))$. It follows that the set of non-degenerate elements of $W \otimes S^2(V^*)$ is an open subset $\mathcal{U}$ in $W \otimes S^2(V^*)$. We can say more. Suppose that $H \in W \otimes S^2(V^*)$ is non-degenerate and that $\xi \in V^*$ is non-characteristic for H. Let $x^1, x^2, \ldots, x^n$ be a basis of V^* with the property that $\xi = x^n$. We may expand H in the form $H = \sum_{i,j} h_{ij} x^i \circ x^j$ where the $h_{ij} = h_{ji}$ are vectors in W. It follows immediately that the hypothesis that ξ not be characteristic is equivalent to the condition that the r vectors $\{h_{ij} \mid i \leq j < n\}$ be linearly independent. This leads to the following important observation.

Recall that $\mathcal{K}(V) \subset \Lambda^2(V^*) \otimes \Lambda^2(V^*)$, the space of Riemann curvature tensors on V is defined to be the kernel of the natural map from $\Lambda^2(V^*) \otimes \Lambda^2(V^*)$ to $V^* \otimes \Lambda^3(V^*)$ obtained by "skew symmetrizing on the last three indices." (When V is explicitly identified as $\mathbb{R}^n$, we use the already established notation $\mathcal{K}_n$ instead of $\mathcal{K}(\mathbb{R}^n)$.) It is well known that $\mathcal{K}(V)$ is actually a subspace of $S^2(\Lambda^2(V^*)) \subset \Lambda^2(V^*) \otimes \Lambda^2(V^*)$ and has dimension $n^2(n^2-1)/12$. There is a natural map $\gamma : W \otimes S^2(V^*) \to \mathcal{K}(V)$ which is defined in indices by setting

$$(101) \qquad \gamma(H)_{ijkl} = h_{ik} \cdot h_{jl} - h_{il} \cdot h_{jk}$$

where we have expanded H in indices as $H = \sum_{i,j} h_{ij} x^i \circ x^j$. It is important to note that, since no inner product on V is used in the definition of γ, this map is equivariant under the action of the group $O(W) \times GL(V)$. We shall return to this point later.

Referring to Lemma 3.10 of Chapter III, we have the proposition

Proposition 4.4. *If $r = \frac{1}{2}n(n-1)$ and $\mathcal{U} \subset W \otimes S^2(V^*)$ is the open set consisting of non-degenerate elements, then $\gamma : \mathcal{U} \to \mathcal{K}(V)$ is a surjective submersion.*

This leads to the following strengthening of Theorem 3.11 of Chapter III.

Proposition 4.5. *If $r = \frac{1}{2}n(n-1)$ and $\mathcal{Z} \subset \mathcal{F} \times \mathcal{F}(\mathbb{E}^{n+r}) \times (\mathbb{R}^r \otimes S^2(\mathbb{R}^n))$ is the set of zeros of the collection $\mathcal{T} = \{T_{ijkl}\}$, then $\mathcal{Z}_\mathcal{U} = \mathcal{Z} \cap (\mathcal{F} \times \mathcal{F}(\mathbb{C}^{n+r}) \times \mathcal{U})$ consists entirely of ordinary zeros of $\mathcal{T}$. Moreover, $\mathcal{Z}_\mathcal{U}$ is an open subset of Z^* on which the differential system $(\mathcal{I}, \Omega_+)$ is involutive with Cartan characters $s_p = \frac{1}{2}n(n-p)(n-p+1)$ for $p \leq n$ on the open subset $\mathcal{Z}_\mathcal{U}/(O(n) \times O(r)) \subset Y^*$.*

Proof. The function $T : \mathcal{F} \times \mathcal{F}(\mathbb{E}^{n+r}) \times (\mathbb{R}^r \otimes S^2(\mathbb{R}^n)) \to \mathcal{K}_n$ with components T_{ijkl} can be written in the form $T = \gamma - R$ where $\gamma : \mathbb{R}^r \otimes S^2(\mathbb{R}^n) \to \mathcal{K}_n$ is defined above and $R : \mathcal{F} \to \mathcal{K}_n$ is the Riemann curvature tensor. Since γ is a surjective submersion when restricted to $\mathcal{U} \subset \mathbb{R}^r \otimes S^2(\mathbb{R}^n)$, it follows that T is a surjective submersion when restricted to $\mathcal{F} \times \mathcal{F}(\mathbb{E}^{n+r}) \times \mathcal{U}$. Of course, this immediately implies that any zero of T which lies in $\mathcal{F} \times \mathcal{F}(\mathbb{E}^{n+r}) \times \mathcal{U}$ must be an ordinary zero of T. This establishes that $\mathcal{Z}_{\mathcal{U}} \subset Z$. Moreover, since T is a surjective submersion when restricted to any fiber of the form $\{f\} \times \{\mathbf{f}\} \times \mathcal{U}$, it follows that the projection $\mathcal{Z}_{\mathcal{U}} \to \mathcal{F} \times \mathcal{F}(\mathbb{E}^{n+r})$ is a surjective submersion. Thus, it follows that Ω_+ does not vanish on $\mathcal{Z}_{\mathcal{U}}$. In particular, $\mathcal{Z}_{\mathcal{U}} \subset Z^*$.

The proof of Theorem 3.11 can now easily be adapted to show that $\mathcal{Z}_{\mathcal{U}}$ is a open submanifold of the space of ordinary integral elements of the differential system with independence condition $(\mathcal{I}_{o,+}, \Omega_+)$ on $\mathcal{F} \times \mathcal{F}(\mathbb{E}^{n+r})$, where $\mathcal{I}_{o,+}$ is the differential ideal generated by the 1-forms in I_o together with the 1-forms $\vartheta_{ij} = \eta_{ij} - \omega_{ij}$. Moreover, that proof further shows that $(\mathcal{I}_{o,+}, \Omega_+)$ is involutive on $\mathcal{F} \times \mathcal{F}(\mathbb{E}^{n+r})$ with Cartan characters $\bar{s}_p = \frac{1}{2}n(n-p)$ for $p \leq n$ and $\bar{s}_p = 0$ for $n < p \leq n_+$.

Thus, the differential system $(\mathcal{I}, \Omega_+)$ on $\mathcal{Z}_{\mathcal{U}}$ is seen to be (an open subset of) the ordinary prolongation of the involutive system $(\mathcal{I}_{o,+}, \Omega_+)$. By Theorem 2.1 of Chapter VI, it follows that $(\mathcal{I}, \Omega_+)$ is involutive on $\mathcal{Z}_{\mathcal{U}}$ with Cartan characters $s_p = \frac{1}{2}n(n-p)(n-p+1)$ for $p \leq n$ and $s_p = 0$ for $n < p \leq n_+$. The remainder of the proposition now follows upon quotienting by the Cauchy leaves. $\square$

Combining Propositions 4.3–5, we arrive at the following result for isometric embedding in the determined case $r = \frac{1}{2}n(n-1)$.

Theorem 4.6. *The differential system for isometrically embedding a given metric on N^n into the Euclidean space of dimension $n + \frac{1}{2}n(n-1) = \frac{1}{2}n(n+1)$ in such a way that the second fundamental form is non-degenerate is an involutive differential system with independence condition $(\mathcal{I}, \Omega)$ on the manifold of 2-jets of immersions of N^n into $\mathbb{E}^{n(n+1)/2}$ which are infinitesimal isometries, satisfy the Gauss equations, and induce non-degenerate second fundamental forms. Moreover, the characters of this system are $s_p = \frac{1}{2}n(n-p)(n-p+1)$ for all $p \leq n$.*

Given such a non-degenerate isometric immersion, $u : N \to \mathbb{E}^{n(n+1)/2}$, the complex characteristic variety $\Xi_x^{\mathbb{C}} \subset \mathbb{P}(T_x^{\mathbb{C}}N)$ is an algebraic hypersurface of degree n. Moreover, if $n \geq 3$, then the real characteristic variety Ξ_x is non-empty for all $x \in N$.

Proof. Omitted. $\square$

Note that, as a consequence of the non-emptyness of the real characteristic variety when $n \geq 3$, it follows that the determined isometric embedding problem is never elliptic for $n \geq 3$, a result first noted by Tanaka [1973].

However, this does not mean that the isometric embedding problem in the smooth category is out of the reach of analysis when $n \geq 3$. In fact, in Bryant, Griffiths and Yang [1983], it is shown that a careful study of the characteristic variety when $n = 3$ can be coupled with the Nash–Moser Implicit Function Theorem to prove the existence of a smooth isometric embedding $N^3 \hookrightarrow \mathbb{E}^6$ in some neighborhood of any point $x \in N$ for which the Riemann curvature tensor is suitably non-degenerate. By applying more subtle results from the theory of P.D.E. of principal type, Yang and Goodman have been able to weaken this non-degeneracy assumption to the assumption that the Riemann curvature tensor be non-zero at x. At present, it is unknown whether there exists a local embedding on a neighborhood of a point where the Riemann curvature tensor vanishes. The analytical difficulties are similar to (but more complicated than) the difficulties one faces in trying to isometrically embed an abstract surface N^2 into $\mathbb{E}^3$ on a neighborhood of a point where the Gauss curvature vanishes. Also, Yang has some results on smooth solvability in the case $n = 4$. These results depend on the theory of P.D.E. of real principal type and show existence of an isometric embedding in a neighborhood of a point $x \in N^4$ where the metric satisfies some open condition on a finite jet of the metric at that point. For more details, see Goodman and Yang [1990]. In the cases $n \geq 5$, the analysis is complicated by the fact that the characteristic variety is generally no longer smooth (Bryant, Griffiths and Yang [1983]).

We now turn to the overdetermined cases $r < \frac{1}{2}n(n-1)$. In these cases, the Gauss equations represent non-trivial obstructions to isometric embedding. For example, when $n = 2$, isometric embeddings satisfying the condition $r = 0$ exist only when the Gauss curvature of the metric g vanishes identically. Since we have already dealt with this case in §1, we shall henceforth assume that $n > 2$.

A fundamental part of the problem is to study the mapping $\gamma : W \otimes S^2(V^*) \to \mathcal{K}(V)$. Since the dimensions of $S^2(V^*)$ and $\mathcal{K}(V)$ are $\frac{1}{2}n(n+1)$ and $n^2(n^2-1)/12$ respectively, it is clear that when n is large, the value of $r = \dim W$ must also be large if we are to have surjectivity of the map γ.

Let us make a detailed study of the case $n = 3$. Since, $\mathcal{K}(V)$ has dimension 6, it follows that $\mathcal{K}(V) = S^2(\Lambda^2(V^*))$. Thus, $\mathcal{K}(V)$ can be regarded as the space of quadratic forms on $\Lambda^2(V)$. The values of r which we will be interested in are $r = 1$ and $r = 2$.

In the case $r = 1$, the space $W \otimes S^2(V^*)$ is identified with $S^2(V^*)$, the space of quadratic forms on V. If $h \in S^2(V^*)$, we may diagonalize h in the form $h = \lambda_1(x^1)^2 + \lambda_2(x^2)^2 + \lambda_3(x^3)^2$, where x^1, x^2, x^3 is a basis of V^* and the λ_i are real numbers. It then follows that

$$\gamma(h) = \lambda_2\lambda_3(x^2 \wedge x^3)^2 + \lambda_3\lambda_1(x^3 \wedge x^1)^2 + \lambda_1\lambda_2(x^1 \wedge x^2)^2$$

as an element of $S^2(\Lambda^2(V^*)) = \mathcal{K}(V)$. Note that, as a quadratic form, $\gamma(h)$ has type $(3,0)$, $(1,2)$, $(1,0)$, $(0,1)$, or $(0,0)$. Conversely, any quadratic form on $\Lambda^2(V)$ whose type belongs to this list can be realized as $\gamma(h)$ for some $h \in S^2(V^*)$.

In fact, the map γ can be understood quite easily in terms of matrices. If we fix a basis x^1, x^2, x^3 of V^*, then we may identify elements h in $S^2(V^*)$ with 3×3 symmetric matrices in the usual way. Using the corresponding basis $x^2 \wedge x^3$, $x^3 \wedge x^1$, $x^1 \wedge x^2$ of $\Lambda^2(V^*)$, we may also identify elements K in $S^2(\Lambda^2(V^*)) = \mathcal{K}(V)$ with 3×3 symmetric matrices. Under these identifications, it is easy to see that the map γ becomes identified with the map Adj, which associates to each 3×3 symmetric matrix its adjoint matrix (i.e., the matrix of signed 2×2 minors). Since the identities

$$\det(Adj(h)) = (\det(h))^2$$

$$Adj(Adj(h)) = \det(h) \cdot h$$

hold for all 3×3 symmetric matrices, it follows easily that Adj is a 2-to-1 local diffeomorphism from the open set of invertible 3×3 symmetric matrices to the open set of 3×3 symmetric matrices with positive determinant. In particular, it follows that γ is a 2-to-1 local diffeomorphism from the open set of non-degenerate quadratic forms on V to the open set of quadratic forms on $\Lambda^2(V)$ with positive determinant. Thus, if h is a non-degenerate quadratic form on V, then the differential of γ at h is an isomorphism and hence is surjective.

In the case $r = 2$, we may let w_1, w_2 be an orthonormal basis of W. If we write $h \in W \otimes S^2(V^*)$ in the form $h = w_1 \otimes h_1 + w_2 \otimes h_2$ where h_1, h_2 belong to $S^2(V^*)$, then $\gamma(h) = \gamma(h_1) + \gamma(h_2)$. Since any non-zero quadratic form q on $\Lambda^2(V)$ can be written as a sum $q = q_1 + q_2$ where each q_i has positive determinant, it follows that when $r = 2$, the map $\gamma : W \otimes S^2(V^*) \to \mathcal{K}(V)$ is surjective and, moreover, if $\mathcal{O} \subset W \otimes S^2(V^*)$ is the open set on which the differential of γ is surjective, then $\gamma(\mathcal{O})$ contains all non-zero elements of $\mathcal{K}(V)$. In fact, $\mathcal{O}$ can be characterized as the set of $h \in W \otimes S^2(V^*)$ with the property that $|h|$ contains a non-degenerate quadratic form.

If $h \in W \otimes S^2(V^*)$ satisfies $\gamma(h) = 0$, then it is not difficult to see that h can be written in the form $h = w_1 \otimes (x^1)^2 + w_2 \otimes (x^2)^2$ where w_1, w_2 form an orthonormal basis of W and x^1, x^2 are elements of V^*. Since such an h does not belong to $\mathcal{O}$, it follows that $\gamma(\mathcal{O}) = \mathcal{K}(V) - \{0\}$.

Let us see how this analysis is reflected in the corresponding isometric embedding problems. First, the case $r = 1$ corresponds to the problem of isometrically embedding N^3 into $\mathbb{E}^4$. For simplicity, let us consider the case where the Riemann curvature tensor is non-degenerate at every $x \in N$ when regarded as a quadratic form on $\Lambda^2(T_x N)$. This corresponds to the assumption that $R : \mathcal{F} \to \mathcal{K}_3 \cong S^2(\Lambda^2(\mathbb{R}^3))$ satisfies $\det(R(f)) \neq 0$ for all $f \in \mathcal{F}$.

If $\det(R(f)) < 0$ for all $f \in \mathcal{F}$, then by our above discussion, the function $T(f, \mathbf{f}, h) = \gamma(h) - R(f)$ never vanishes on $\mathcal{F} \times \mathcal{F}(\mathbb{E}^4) \times (\mathbb{R}^1 \otimes S^2(\mathbb{R}^3))$. Thus $\mathcal{Z}$ is empty and hence there do not exist any local isometric embeddings of N into $\mathbb{E}^4$. This situation occurs, for example, when all of the sectional curvatures of N are negative since then, R is everywhere negative definite.

On the other hand, if $\det(R(f)) > 0$ for all $f \in \mathcal{F}$, then the above discussion shows that for each $f \in \mathcal{F}$, there exist exactly two solutions of the equation $\gamma(h) = R(f)$, each of which is the negative of the other. Let us write $h(f)$ for the unique element of $S^2(\mathbb{R}^3)$ which has positive determinant and which satisfies $\gamma(h(f)) = R(f)$. Then $\mathcal{Z}$ consists of the triples $(f, \mathbf{f}, \pm h(f))$. Since the 6 components of γ have linearly independent differentials along $\mathcal{Z}$, it follows that $Z^* = Z = \mathcal{Z}$. When we restrict to Z^*, the symbol relations (13) may be solved for the π_{ij} in the form

$$(102) \qquad \pi_{ij} \equiv (\det(R))^{-3/2} \sum_m R_{ijm}\omega_m \quad \mathrm{mod}\ I.$$

Here, the functions $R_{ijk} = R_{jik}$ are some universal polynomial expressions which are of degree 4 in the components R_{ijkl} and linear in the components R_{ijklm}. Referring to the structure equations (12), we see that the torsion of the system $(\mathcal{I}, \Omega_+)$ on Z^* vanishes if and only if $R_{ijk} = R_{ikj}$.

Let us define $T_{ijk} = R_{ijk} - R_{ikj}$. Due to the symmetry $R_{ijk} = R_{jik}$, there are only 8 independent functions among the T_{ijk}. It is easy to see that these functions are the components of a well-defined tensor T of rank 8 on N. On any neighborhood of a point where T is non-zero, there is no local isometric embedding of N into $\mathbb{E}^4$. On the other hand, if T vanishes identically (and $\det(R) > 0$) then the system I is a Frobenius system on Z^*. Since it is clear that the group of rigid motions of $\mathbb{E}^4$ acts transitively on the space of leaves of I, it follows that there exist local isometric embeddings of N into $\mathbb{E}^4$ and that they are unique up to rigid motions. Incidentally, even though T has 8 components, it can be shown that T satisfies a set of 3 linear conditions whose coefficients are linear in the components of R. Thus, the condition $T = 0$ is actually only 5 conditions on a metric g which satisfies $\det(R) > 0$. In fact, a little algebra shows that these 5 conditions can be expressed as the vanishing of a tensor of rank 5 whose components are cubic polynomials which are quadratic in the components of R and linear in the components of ∇R.

We now turn to the case $r = 2$. Let us begin by assuming that the metric g has the property that its Riemann curvature tensor R does not vanish identically at any point of N. As we have seen, when $r = 2$, the map $\gamma : W \otimes S^2(V^*) \to \mathcal{K}(V)$ restricts to become a surjective submersion $\gamma : \mathcal{O} \to \mathcal{K}(V) - \{0\}$. Let $\mathcal{O}^* \subset \mathcal{O}$ be the dense open subset consisting of those $h \in \mathcal{O}$ so that $|h|$ has dimension 2. It then follows that $\mathcal{Z}_{\mathcal{O}^*} = \mathcal{Z} \cap (\mathcal{F} \times \mathcal{F}(\mathbb{E}^5) \times \mathcal{O}^*)$ consists entirely of ordinary zeros of the collection $\mathcal{T}$ and that $\mathcal{Z}_{\mathcal{O}^*} \subset Z^*$. Since R is non-vanishing on $\mathcal{F}$, it is not difficult to see that $\mathcal{Z}_{\mathcal{O}^*}$ is a dense open subset of Z^* and that its projection onto $\mathcal{F} \times \mathcal{F}(\mathbb{E}^5)$ is a surjective submersion. If the sectional curvature of g is everywhere negative, then it can be shown that, in fact, $\mathcal{Z} = \mathcal{Z}_{\mathcal{O}^*} = Z^*$.

We now proceed to investigate the symbol and torsion of the system $(\mathcal{I}, \Omega_+)$ on $\mathcal{Z}_{\mathcal{O}^*}$. Of course, the symbol relations (95) play the crucial role.

Suppose that $(f, \mathbf{f}, h) \in \mathcal{Z}_{\mathcal{O}^*}$. Since $h \in \mathcal{O}^*$, it follows easily that there exists an orthonormal basis w_1, w_2 of $\mathbb{R}^2 = W$ so that, when we expand h in the form $h = w_1 \otimes h^1 + w_2 \otimes h^2$, the quadratic form h^2 is non-degenerate and not positive definite. Let $v \in \mathbb{R}^3$ be a null vector for h^2 which is not a null vector for h^1. Since the results of computing the ranks of polar equations, etc. for $\mathcal{I}$ will be the same at all points on a given $O(2) \times O(3)$-orbit in $\mathcal{Z}_{\mathcal{O}^*}$, we may assume that v is a multiple of the first element of the standard basis of $\mathbb{R}^3$. It follows that when we expand h in the form $h = \sum_{i,j} h_{ij} x^i \circ x^j$, the vector h_{11} is a non-zero multiple of w_1 and the quadratic form $h^2 = w_2 \cdot h$ is non-degenerate.

We now want to show that the symbol relations (95) and $\pi_{ij} = \pi_{ji}$ at $(f, \mathbf{f}, h) \in \mathcal{Z}_{\mathcal{O}^*}$ imply that $s_1' = 6$ and $s_p' = 0$ for all $p > 1$. To see this, it suffices to show that the reduced symbol relations allow us to express all of the components of π_{22}, π_{23}, and π_{33} as linear combinations of the forms in J ($=$ span of the forms in I and the forms ω^i) and the components of π_{11}, π_{12}, and π_{13}. If we let K denote the span of the forms in J and the components of π_{11}, π_{12} and π_{13}, then the symbol relations (13) imply

$$
(103) \qquad \left.
\begin{aligned}
h_{11} \cdot \pi_{22} &\equiv 0 \\
h_{11} \cdot \pi_{23} &\equiv 0 \\
h_{11} \cdot \pi_{33} &\equiv 0 \\
-h_{13} \cdot \pi_{22} + h_{12} \cdot \pi_{23} &\equiv 0 \\
-h_{13} \cdot \pi_{23} + h_{12} \cdot \pi_{33} &\equiv 0 \\
h_{33} \cdot \pi_{22} - 2h_{23} \cdot \pi_{23} + h_{22} \cdot \pi_{33} &\equiv 0
\end{aligned}
\right\} \quad \mathrm{mod}\ K
$$

(For example, the fourth relation is obtained by setting $(i, j, k, l) = (1, 2, 2, 3)$ in (95).) The first three relations in (103) show that the w_1-components of π_{22}, π_{23}, and π_{33} belong to K. Due to the non-degeneracy of h^2, the last three relations in (103) then show that the w_2-components of π_{22}, π_{23}, and π_{33} also belong to K. Since there are at least 6 linearly independent components among the π_{ij}, it follows that all of the components of π_{11}, π_{12}, and π_{13} are linearly independent. Thus, $s_1' = 6$ and $s_p' = 0$ for all $p > 1$, as claimed.

It follows by Cartan's test that the space of integral elements of $(\mathcal{I}, \Omega_+)$ based at any point of $\mathcal{Z}_{\mathcal{O}^*}$ is of dimension at most 6. We are now going to show that if the torsion of $(\mathcal{I}, \Omega_+)$ vanishes at $(f, \mathbf{f}, h) \in \mathcal{Z}_{\mathcal{O}^*}$, then there exists a 6-parameter family of integral elements of $(\mathcal{I}, \Omega_+)$ based at $(f, \mathbf{f}, h) \in \mathcal{Z}_{\mathcal{O}^*}$. To do this, we must show that the homogeneous system

$$
(104) \qquad h_{ik} \cdot h_{jlm} - h_{il} \cdot h_{jkm} + h_{jl} \cdot h_{ikm} - h_{jk} \cdot h_{ilm} = 0
$$

for the $\mathbb{R}^2$-valued unknowns $h_{ijk} = h_{jik} = h_{ikj}$ has a 6-parameter family of solutions. Note that by Cartan's test, this system of equations cannot have more than 6 linearly independent solutions. We shall show that, for an open subset of $h \in \mathcal{O}^*$, (104) has exactly 6 linearly independent solutions. By the principle of specialization and the upper bound on the space of solutions of (104) given by Cartan's test, this will imply that (104) has exactly 6 linearly independent solutions for all $h \in \mathcal{O}^*$.

Our argument will use the $GL(V)$-equivariance of the equations (104). For an open set of $h \in \mathbb{R}^2 \otimes S^3(\mathbb{R}^3)$, the corresponding two dimensional space of quadratic forms $|h|$ has the property that it contains no perfect squares (i.e., rank 1 quadratic forms) and the property that there exists a basis of $\mathbb{R}^3$ in which the elements of $|h|$ are diagonalized. Let x^1, x^2, x^3 be linear coordinates on $\mathbb{R}^3$ so that such an h can be written in the form

$$h = h_{11}(x^1)^2 + h_{22}(x^2)^2 + h_{33}(x^3)^2$$

where the vectors h_{ii} lie in $\mathbb{R}^2$. Note that the hypothesis that $|h|$ does not contain any rank 1 quadratic forms implies that the vectors h_{11}, h_{22}, h_{33} are pairwise linearly independent. The equations (104) may now be written in the form

$$
\begin{aligned}
h_{11} \cdot h_{23m} &= 0 \\
h_{22} \cdot h_{31m} &= 0 \\
h_{33} \cdot h_{12m} &= 0 \\
h_{22} \cdot h_{33m} - h_{33} \cdot h_{22m} &= 0 \\
h_{33} \cdot h_{11m} - h_{11} \cdot h_{33m} &= 0 \\
h_{11} \cdot h_{22m} - h_{22} \cdot h_{11m} &= 0
\end{aligned}
$$

(105)

for all m. A priori, this is 18 equations for the 20 unknown components of the 10 vectors h_{ijk}. However, at most 14 of these equations can be linearly independent. To see this, note that if we set $m = 1$ (resp. $2, 3$) in the first (resp. second, third) equations of (105) and use the symmetry of h_{ijk}, then we get the 3 equations $h_{kk} \cdot h_{123} = 0$. Due to the linear dependence of the three vectors h_{kk}, these 3 equations are linearly dependent. Also, if we set $m = 1$ in the fourth equation of (105), then we see that it is a linear combination of the second equation (with $m = 3$) and the third equation (with $m = 2$). Similarly, the fifth equation with $m = 2$ and the sixth equation with $m = 3$ are linear combinations of previous equations. Thus, there are at most 14 linearly independent equations in (105). It follows that the solution space of (105) must have dimension at least $20 - 14 = 6$. Since we have already seen that the solution space has dimension at most 6, we are done.

Our analysis so far of the case $r = 2$ has shown that the homogeneous symbol relations of the system $(\mathcal{I}, \Omega_+)$ on $\mathcal{Z}_{\mathcal{O}^*}$ are involutive and that the Cartan characters are $s_1' = 6$ and $s_p' = 0$ for all $p > 1$. Thus, if the torsion vanishes identically, then the system $(\mathcal{I}, \Omega_+)$ on $\mathcal{Z}_{\mathcal{O}^*}$ is involutive.

However, in general, the torsion does not vanish identically on $\mathcal{Z}_{\mathcal{O}^*}$. On heuristic grounds, this is obvious since the generic metric on N^3 cannot be isometrically embedded into $\mathbb{E}^5$. It is of some interest to see what the torsion of the system actually is. We claim that the unabsorbable torsion of the system $(\mathcal{I}, \Omega_+)$ on $\mathcal{Z}_{\mathcal{O}^*}$ takes values in a vector space of dimension 1. To see this, note that the components of ∇R satisfy the second Bianchi identity, namely

$$(106) \qquad\qquad R_{ijklm} + R_{ijmkl} + R_{ijlmk} = 0.$$

This implies that there are only 15 independent components of the covariant derivative of the Riemann curvature tensor. In other words, there is a subspace $\mathcal{K}^1(V) \subset \mathcal{K}(V) \otimes V^*$ of dimension 15 in which ∇R takes values (for general n, $\dim \mathcal{K}^1(V) = n^2(n^2 - 1)(n + 2)/24$, see Berger, Bryant and Griffiths [1983]). For $h \in \mathcal{O}^*$, define $\gamma_h : W \otimes S^3(V^*) \to \mathcal{K}^1(V)$ by the formula in indices

$$(107) \qquad (\gamma_h(p))_{ijklm} = h_{ik} \cdot p_{jlm} - h_{il} \cdot p_{jkm} + h_{jl} \cdot p_{ikm} - h_{jk} \cdot p_{ilm}$$

where $p = \sum_{ijk} p_{ijk} x^i \circ x^j \circ x^k$ belongs to $W \otimes S^3(V^*)$. We have already seen that γ_h has rank 14 for all $h \in \mathcal{O}^*$. It follows that there exists a linear functional $\lambda_h : \mathcal{K}^1(V) \to \mathbb{R}$ whose coefficients are polynomials in h of degree 14 so that $\ker \lambda_h = \operatorname{im} \gamma_h$. In particular, the torsion of $(\mathcal{I}, \Omega_+)$ is absorbable at $(f, \mathbf{f}, h)$ if and only if $\lambda_h(\nabla R(f)) = 0$.

It can happen that $\lambda_h(\nabla R(f))$ vanishes identically on $\mathcal{Z}_{\mathcal{O}^*}$. For example, if the Riemannian manifold (N^3, g) is locally symmetric, then $\nabla R \equiv 0$, so, a fortiori, $\lambda_h(\nabla R) \equiv 0$. A more general family of metrics for which this condition holds is the family of metrics induced on the non-degenerate quadratic hypersurfaces in $\mathbb{E}^4$. In fact, these examples, together with the corresponding metrics induced on the space-like portions of non-degenerate quadratic hypersurfaces in Minkowski space $\mathbb{M}^4$ exhaust the list of metrics with non-degenerate curvature on 3-manifolds on which this condition holds (see Berger, Bryant and Griffiths [1983]).

Thus, let us assume that g has the property $\det(R) \neq 0$ and that $\lambda_h(\nabla R) \equiv 0$. By our above description, such metrics are known to be real analytic in appropriate local coordinates. As we have shown, the system $(\mathcal{I}, \Omega)$ on $Y_{\mathcal{O}^*} = \mathcal{Z}_{\mathcal{O}^*}/O(2) \times O(3)$ is involutive. It follows that we may apply the Cartan–Kähler theorem to this system to conclude that such metrics can be locally isometrically embedded into $\mathbb{E}^5$ and that the general such embedding depends locally on 6 functions of 1 variable. Note that when $\det(R) > 0$, such metrics can be isometrically embedded into $\mathbb{E}^4$, but that they are rigid there.

More information about these embeddings can be obtained by studying the characteristic variety of the system $(\mathcal{I}, \Omega)$. For each integral element

E of $(\mathcal{I}, \Omega)$ the complex characteristic variety consists of 6 points. In fact, by our above computation of the complex characteristic variety $\Xi_H^{\mathbb{C}}$ for $H \in W \otimes S^2(V^*)$, these points are described as follows. If E is an integral element based at $(x, y, l, H) \in Y_{\mathcal{O}^*}$, then $|H| \subset S^2(E^*)$ has dimension 2 and contains at least one non-degenerate quadratic form. In the terminology of algebraic geometry, $|H|$ is called a *pencil* of quadrics. If $h_1 \in |H|$ is non-degenerate and $h_2 \in |H|$ is linearly independent from h_1, then the degenerate quadrics in $|H|$ are of the form $h_2 + th_1$ where t is a root of the cubic equation $\det(h_2 + th_1) = 0$. For each such root (counted with multiplicity) the quadric $h_2 + th_1$ factors in the form $\lambda \circ \mu$ where $\lambda, \mu \in (E^*)^{\mathbb{C}}$ are well defined up to scalar multiples. There are six such elements of $(E^*)^{\mathbb{C}}$ (counted with multiplicity). These give rise to the 6 points in $\mathbb{P}(E^*)^{\mathbb{C}}$ which constitute $\Xi_E^{\mathbb{C}}$.

For the generic pencil of quadrics $|H| \subset S^2(E^*)$, the base locus B consists of 4 points in $\mathbb{P}E^{\mathbb{C}}$ in general position except for the condition of being invariant under conjugation. Let us say that H is general if the base locus of $|H|$ consists of 4 points. Conversely, given a conjugation-invariant set B of 4 points in $\mathbb{P}E^{\mathbb{C}}$ with no 3 on a line, there is a unique real pencil of quadrics $|H_B|$ whose elements pass through all 4 points of B. The corresponding characteristic variety $\Xi_B \subset \mathbb{P}(\mathbb{E}^{\mathbb{C}})^*$ consists of the 6 lines which pass through 2 of the 4 points of B. Note that if all of the points of B are real, then all of the 6 points of Ξ_B are also real. If 2 of the points of B are real and the other 2 are non-real complex conjugates, then 2 of the 6 points of Ξ_B are real. If B consists of 2 pairs of non-real complex conjugate points, then 2 of the points of Ξ_B are real. In each case, note that Ξ_B contains at least 2 real points. Thus, it follows that the system $(\mathcal{I}, \Omega)$ is never elliptic on $Y_{\mathcal{O}^*} = \mathcal{Z}_{\mathcal{O}^*}/O(2) \times O(3)$.

Let us say that an immersion $u : N^3 \to \mathbb{E}^5$ is general if the induced second fundamental form H_x is general for all $x \in N$. We have seen that general isometric immersions exist locally and depend on 6 functions of 1 variable for the special case of $(N^3, g) = (S^3, \mathrm{can})$ since the canonical metric on S^3 satisfies $\nabla R \equiv 0$. We shall now show that the integrability of the characteristic variety obstructs the existence of a global general isometric immersion of (S^3, can) into $\mathbb{E}^5$.

Proposition 4.7. *There does not exist a general isometric immersion* $u : S^3 \to \mathbb{E}^5$ *when S^3 is given any one of the metrics induced by immersion as an ellipsoidal hyperquadric in $\mathbb{E}^4$.*

Proof. Suppose that such a u exists. Consider the associated integral $\hat{u} : S^3 \to Y_{\mathcal{O}^*}$ of $(\mathcal{I}, \Omega)$. Via $\hat{u}$, the characteristic variety of $(\mathcal{I}, \Omega)$ restricts to the projectivized complexified cotangent bundle of S^3 to become a submanifold $\Xi_{\hat{u}} \subset \mathbb{P}(T^*S^3)^{\mathbb{C}}$ whose base point projection $\Xi_{\hat{u}} \to S^3$ makes $\Xi_{\hat{u}}$ into a covering space of degree 6. By simple connectivity of S^3, it follows that $\Xi_{\hat{u}}$ is the disjoint union of 6 copies of S^3. Moreover, the base locus $B_{\hat{u}} \subset \mathbb{P}(TS^3)^{\mathbb{C}}$ dual to $\Xi_{\hat{u}}$ is a covering space $B_{\hat{u}} \to S^3$ of degree 4 and hence consists of the disjoint union of 4 copies of S^3. The fiber $(B_{\hat{u}})_x \subset \mathbb{P}(T_x S^3)^{\mathbb{C}}$ at each point $x \in S^3$ consists of 4 (distinct) points in general position subject only to the condition of being invariant under conjugation. It follows easily that the number of real points in $(B_{\hat{u}})_x$ is the same for all $x \in S^3$.

If all of the points in $(B_{\hat{u}})_x$ are real (for all x), then due to the fact that $GL(3, \mathbb{R})$ acts transitively on the set of quadruples of points in $\mathbb{RP}^2$ in general position, it follows easily that there exists a coframing η^1, η^2, η^3 on S^3 (not necessarily orthogonal) with the property that the 6 projectivized 1-forms

$$\{[\eta^1], [\eta^2], [\eta^3 + \eta^1 + \eta^2], [\eta^3 + \eta^1 - \eta^2], [\eta^3 - \eta^1 + \eta^2], [\eta^3 - \eta^1 - \eta^2]\}$$

are sections of $\Xi_{\hat{u}}$. By the involutivity of $(\mathcal{I}, \Omega)$ on $Y_{\mathcal{O}^*}$, any 1-form η for which $[\eta]$ is a section of $\Xi_{\hat{u}}$ is integrable, i.e., $\eta \wedge d\eta = 0$. Applying this integrability to each of the 6 1-forms above gives 6 equations which are equivalent to the 6 relations

$$(108) \qquad \eta^i \wedge d\eta^j + \eta^j \wedge d\eta^i = 0$$

for all i and j. It is easy to show that this is equivalent to the condition that there exist a 1-form λ so that $d\eta^i = \lambda \wedge \eta^i$ for all i. Differentiating this last relation gives the condition $d\lambda \wedge \eta^i = 0$ for all i. Since $d\lambda$ is a 2-form, it follows that $d\lambda = 0$. Due to the simple connectivity of S^3, it follows that there exists a function l on S^3 so that $\lambda = dl$. It then follows that the coframing $\tilde{\eta}^i = e^{-l}\eta^i$ satisfies $d\tilde{\eta}^i = 0$ for all i. Since S^3 clearly does not have any closed coframing, we have a contradiction.

In the case that 2 of the points in $(B_{\hat{u}})_x$ are real (for all x), then it is not difficult to see that there exists a coframing η^1, η^2, η^3 on S^3 (not necessarily orthogonal) with the property that the 6 projectivized 1-forms

$$\{[\eta^1], [\eta^2], [\eta^3 + i\eta^1 + \eta^2], [\eta^3 + i\eta^1 - \eta^2], [\eta^3 - i\eta^1 + \eta^2], [\eta^3 - i\eta^1 - \eta^2]\}$$

are sections of $\Xi_{\hat{u}}$. In the case that none of the points in $(B_{\hat{u}})_x$ are real (for all x), then it is not difficult to see that there exists a coframing η^1, η^2, η^3 on S^3 (not necessarily orthogonal) with the property that the 6 projectivized 1-forms

$$\{[\eta^1], [\eta^2], [\eta^3 + i\eta^1 + i\eta^2], [\eta^3 + i\eta^1 - i\eta^2], [\eta^3 - i\eta^1 + i\eta^2], [\eta^3 - i\eta^1 - i\eta^2]\}$$

are sections of $\Xi_{\hat{u}}$. In either case, applying the integrability of the characteristic variety shows that the equations (108) hold for this coframing. We have already seen that the equations (108) lead to a contradiction when the domain of the coframing is S^3 $\qquad\square$

Note that there do exist immersions of S^3 into $\mathbb{E}^5$ which are general. For example, if we regard $\mathbb{E}^5$ as the space of traceless symmetric 3×3 matrices and let $\Sigma \subset \mathbb{E}^5$ denote the set of such matrices whose eigenvalues are $\{1, -1, 0\}$, then the universal covering space of Σ is easily seen to be S^3 and it is not difficult to see that the second fundamental form of Σ is general at all points (use the fact that Σ is an orbit of $SO(3)$ under its natural irreducible action on $\mathbb{E}^5$). Of course, the induced metric on Σ, even though homogeneous, cannot be locally isometric to any of the "quadric" metrics for which the system $(\mathcal{I}, \Omega)$ is involutive on $Y_{\mathcal{O}^*}$.

As our final example in the case $r = 2$, let us consider the case where the metric g is flat. Then, the Riemann curvature satisfies $R \equiv 0$ on $\mathcal{F}$. Let us describe the set of ordinary zeros of the function T on $\mathcal{F} \times \mathcal{F}(\mathbb{E}^5) \times (\mathbb{R}^2 \otimes (S^2(\mathbb{R}^3)))$. Since $R \equiv 0$, it follows that $T(f, \mathbf{f}, h) = \gamma(h)$, and since the differential of γ is not surjective at any $h \in \mathbb{R}^2 \otimes S(\mathbb{R}^3)$ for which $\gamma(h) = 0$, we cannot directly apply the implicit function theorem to conclude that $T^{-1}(0) = \mathcal{F} \times \mathcal{F}(\mathbb{E}^5) \times \gamma^{-1}(0)$ has any smooth points. Nevertheless, by our previous discussion, we can parametrize $\gamma^{-1}(0)$ as follows. Let $O(2)$ denote the set of orthonormal bases (w_1, w_2) of $W = \mathbb{R}^2$. Then there exists a map $\mu : O(2) \times V^* \times V^* \to W \otimes S^2(V^*)$ defined by

$$\mu(w_1, w_2; x^1, x^2) = w_1 \otimes (x^1)^2 + w_2 \otimes (x^2)^2$$

whose image is precisely $\gamma^{-1}(0) \subset W \otimes S^2(V^*)$. If we let $\mathcal{D} \subset \mathbb{R}^5 \times \mathbb{R}^3 = V^* \times V^*$ denote the open set of pairs (x^1, x^2) of elements of V^* which satisfy $x^1 \wedge x^2 \neq 0$, then it is easy to see that the differential of μ has its maximum rank 7 precisely on the open subset $O(2) \times \mathcal{D}$. In fact, $\mu(O(2) \times \mathcal{D}) \subset \gamma^{-1}(0)$ consists of the set of $h \in \gamma^{-1}(0)$ for which $|h|$ is of dimension 2 and is a smooth submanifold of $W \otimes S^2(V^*)$ of dimension 7. Any point $h \in \gamma^{-1}(0)$ for which $|h|$ has dimension 1 or 0, is in the closure of $\mu(O(2) \times \mathcal{D})$ but is not a smooth point of $\gamma^{-1}(0)$. Thus, in order to see that $\mathcal{F} \times \mathcal{F}(\mathbb{E}^5) \times \mu(O(2) \times \mathcal{D})$ is the space of ordinary zeros of the collection $\mathcal{T}$, it suffices to show that the differentials of the functions in $\mathcal{T}$ contain at least 5 linearly independent 1-forms at every point of $\mathcal{F} \times \mathcal{F}(\mathbb{E}^5) \times \mu(O(2) \times \mathcal{D})$. To see this, note that we may take advantage of the $GL(V)$-equivariance of the map γ. Let $h \in \mu(O(2) \times \mathcal{D})$ and write h in the form $h = w_1 \otimes (x^1)^2 + w_2 \otimes (x^2)^2$. (The elements x^1, x^2 are not necessarily an orthonormal pair in V.) Making the appropriate $GL(V)$ change of basis for the 1-forms on $\mathcal{F} \times \mathcal{F}(\mathbb{E}^5) \times (W \otimes S^2(V^*))$, we can assume that h_{11} and h_{22} are an orthonormal basis of W and that all other h_{ij} are zero. Then a basis for the 1-forms τ_{ijkl} at $(f, \mathbf{f}, h)$ can be expressed in the form

$$
\begin{aligned}
&\tau_{2323} = h_{22} \cdot \pi_{33}, &\qquad &\tau_{3131} = h_{11} \cdot \pi_{33} \\
(109)\qquad &\tau_{1212} = h_{11} \cdot \pi_{22} + h_{22} \cdot \pi_{11}, &\qquad &\tau_{3112} = -h_{11} \cdot \pi_{23} \\
&\tau_{1223} = -h_{22} \cdot \pi_{13}, &\qquad &\tau_{2331} = 0.
\end{aligned}
$$

It follows that there exist 5 linearly independent 1-forms among the τ_{ijkl} at $(f, \mathbf{f}, h)$, as we wished to show. Thus, $Z = Z^* = \mathcal{F} \times \mathcal{F}(\mathbb{E}^5) \times \mu(O(2) \times \mathcal{D})$.

We are now going to show that the differential system $(\mathcal{I}, \Omega_+)$ is involutive on Z^* with Cartan characters $s_1 = 6$, $s_2 = 1$, and $s_p = 0$ for all $p > 2$. First, note that since $R \equiv 0$, we must have $\nabla R \equiv 0$ as well. Thus, the torsion of the differential system vanishes identically on Z^*. Examining the symbol relations, i.e., the right hand sides of (109), we see by inspection that there is a flag contained in each integral element which has the characters $s_1' = 6$, $s_2' = 1$, and $s_p' = 0$ for all $p > 2$. Direct calculation the the space of integral elements at each point of Z^* shows that there exists an 8-parameter family of integral elements at every point of Z^*. Thus, by Cartan's test, it follows that the system is in involution, as claimed. Note that applying Proposition 4.1 shows that the characteristic variety Ξ_H of an element $H = w_1 \otimes (x^1)^2 + w_2 \otimes (x^2)^2$ consists of the line $[\xi_1 x^1 + \xi_2 x^2]$ in $\mathbb{P}V^*$. Note that Ξ_H has degree and dimension 1 in accordance with the general theory. It follows that the submanifolds $N^3 \subset \mathbb{E}^5$ on which the induced metric is flat and whose second fundamental forms have rank 2 at each point depend on one function of two variables (in Cartan's terminology).

For further examples of isometric embedding for special metrics in codimensions below the natural embedding codimension, the reader may consult the aforementioned paper (Berger, Bryant and Griffiths [1983]) and its references to the work of Cartan. In particular, Cartan's study of $\gamma^{-1}(0) \subset \mathcal{K}_n$ constitutes his theory of "exteriorly orthogonal quadratic forms" which he used to study the isometric embeddings of $\mathbb{E}^n$ into $\mathbb{E}^{2n}$ and $\mathbb{H}^n$ into $\mathbb{E}^{2n-1}$ (here, $\mathbb{H}^n$ denotes hyperbolic n-space). In the latter problem, the analog of the equations (108), which are consequences of the integrability of the characteristic variety, can be used to prove the existence of "generalized Tschebysheff coordinates" which can be associated to any isometric embedding of $\mathbb{H}^n$ into $\mathbb{E}^{2n-1}$ (see Moore [1972]).

CHAPTER VIII

APPLICATIONS OF
COMMUTATIVE ALGEBRA AND
ALGEBRAIC GEOMETRY TO THE STUDY OF
EXTERIOR DIFFERENTIAL SYSTEMS

A linear Pfaffian differential system on a manifold M is given by sub-bundles

$$I \subset J \subset T^*(M)$$

such that

$$dI \subset \{J\}$$

where $\{J\} \subset \Omega^*(M)$ is the algebraic ideal generated by the sections of J. Setting

$$L = J/I$$

it follows that the exterior derivative induces a bundle mapping (cf. Section 5 of Chapter IV)

$$\overline{\delta} : I \to (T^*(M)/J) \otimes L.$$

Dualizing and using $(T^*(M)/J)^* \cong J^\perp$, this is equivalent to a bundle mapping

$$(1) \qquad\qquad \pi : J^\perp \to I^* \otimes L.$$

Locally, this mapping is given by the tableau matrix π as discussed in Chapter IV. Much of the discussion in the preceeding chapters has centered around fibrewise constructions, such as the symbol and characteristic variety, associated to the mapping (1). In this chapter we will isolate and considerably extend these discussions.

Given vector spaces W and V, a *tableau* has been defined to be a linear subspace

$$A \subset W \otimes V^*,$$

and the associated *symbol* has been defined to be

$$B = A^\perp \subset W^* \otimes V.$$

As explained in Chapter IV, the *first prolongation* $A^{(1)} \subset W \otimes S^2 V^*$ is defined by

$$A^{(1)} = \{P \in W \otimes S^2 V^* : v \lrcorner P \in A \text{ for all } v \in V\},$$

and A is said to be *involutive* in case equality holds in the inequality given by Cartan's test (cf. Proposition 3.6 in Chapter IV). This chapter will be an algebraic study of involutive tableau. Three of the basic results are stated in (2.4), (2.5) and (3.3) below.

One of the most useful exterior algebra facts is the Cartan lemma, and in Section 2 we begin with a generalization (Proposition 2.1) of this lemma that plays a critical role in the theory of exterior differential systems. This lemma leads naturally to the definition of the Spencer cohomology groups $H^{k,q}(A)$ of a tableau. The crucial role played by these groups in the development of the subject of overdetermined P.D.E. systems will be explained in Chapters IX and X. For us the main fact will be the characterization of involutive tableau as those for which all $H^{k,q}(A) = 0$, $k \geq 1$ (cf. the discussion in Section 1 of Chapter X). This result, which among other things implies that the prolongation of an involutive tableau is involutive, will be proved in Sections 2 and 3 below. Actually, we have chosen to also give a direct proof of the result that "A involutive $\Rightarrow A^{(1)}$ involutive" in Section 2, as among other things it shows how one is led naturally to the Spencer cohomology groups by purely differential system considerations. For another example of how cohomology naturally arises, it has been remarked in Chapter IV and will proved below that the torsion of a linear Pfaffian differential system lies in the family of vector spaces $H^{0,2}(A_x)$, where A_x is the tableau lying over $x \in M$.

In Section 3 we dualize Cartan's test for involution. This is done by introducing a graded module M_A naturally associated to a tableau A, and the condition that A be involutive is seen to be that M_A admit a quasi-regular sequence. The latter condition is, by more or less standard commutative algebra, equivalent to the vanishing result

$$H_{k,q}(M_A) = 0, \quad k \geq 1$$

for the Koszul homology groups of the module M_A. These Koszul homology groups then turn out to be dual to the Spencer cohomology groups, thus establishing the equivalence of involutivity and the vanishing of Spencer cohomology. At the end of Section 3, this characterization of involutivity is used to prove the fact that, given a tableau A, there is a q_0 such that the prolongations $A^{(q)}$ of A are involutive for $q \geq q_0$, a result used in the proof of the Cartan–Kuranishi theorem. In fact, we prove the stronger result that q_0 depends only on the sequence of numbers $\dim A^{(q)}$; this requires a localization argument and is related to the construction of Hilbert schemes in algebraic geometry.

In Section 4, we further pursue the use of Koszul homology by showing that the above vanishing result leads to a natural definition of what is meant by an involutive module, and then it is shown that involutive modules have canonical free resolutions where the maps are homogeneous of degree one.

This translates into statements such as: the compatibility equations for an involutive, overdetermined linear P.D.E. system are of first order (see Theorem 1.8, Chapter X), and so forth. The results in this section will be used in Section 6 when Guillemin's normal form is discussed.

In Section 5 we introduce what amounts to the micro-localization of a linear Pfaffian differential system. The concepts an involutive sheaf and of the characteristic sheaf of a tableau are introduced, and more or less standard algebro-geometric results are used to prove a number of results about these, culminating in the proof of Theorem 3.15 in Chapter V and the proof of Proposition 3.10 below, which is the essential algebraic step in the proof of Theorem 3.1 in Chapter VI.

The formal introduction of homological methods is due to Spencer [1961]. Most all of the results in this chapter were found in the early and middle 1960's and are due to Guillemin, Quillen, Spencer and Sternberg with crucial input coming from Mumford and Serre. In addition to Spencer's paper cited above, we would like to call attention to Singer and Sternberg [1965], Guillemin and Sternberg [1964], Quillen [1964] and Guillemin [1968] where much of the theory first appeared.

§1. Involutive Tableaux.

We begin by introducing some notations. Let W be a vector space with basis $\{w_a\}$, V a vector space with basis $\{v_i\}$ and dual basis $\{x^i\}$, $S^q V^*$ the qth symmetric product of V^*, and $W \otimes S^q V^*$ the W-valued polynomials of the form

$$P = P_I^a w_a \otimes x^I$$

where $I = (i_1, \ldots, i_q)$ runs over multi-indices of length q and $x^I = x^{i_1} \ldots x^{i_q}$. By

$$\frac{\partial P}{\partial x^i} = v_i \lrcorner P$$

we mean the formal derivative treating W as constants. We repeat and generalize some definitions from Chapter IV.

Definition 1.1. i) A linear subspace $A \subset W \otimes V^*$ will be called a *tableau*. More generally, a subspace

$$A \subset W \otimes S^{p+1} V^*$$

will be called a *tableau of order p*. ii) Given a tableau of order p, we inductively define the qth *prolongation*

$$A^{(q)} \subset W \otimes S^{p+q+1} V^*$$

by $A^{(0)} = A$ and

$$A^{(q)} = \{P \in W \otimes S^{p+q+1}V^* : \frac{\partial P}{\partial x^i} \in A^{(q-1)} \text{ for all } i\}.$$

To motivate this definition in the case $p = 0$ of an ordinary tableau, we suppose that the linear Pfaffian system $(\mathcal{I}, \Omega)$ has no integrability conditions, and we denote its first prolongation by $(\mathcal{I}^{(1)}, \Omega)$ on the manifold $M^{(1)}$. If A_x denotes the tableau of $(\mathcal{I}, \Omega)$ at a typical point $x \in M$, then the fibre of

$$\pi : M^{(1)} \to M$$

over x is an affine linear space whose associated vector space is $A_x^{(1)}$ (cf. (125) in Chapter IV).

Now $(\mathcal{I}^{(1)}, \Omega)$ is again a linear Pfaffian differential system whose tableau at a point y with $\pi(y) = x$ is $A_x^{(1)}$. Assuming again that there are no integrability conditions, the second prolongation is a linear Pfaffian differential system $(\mathcal{I}^{(2)}, \Omega)$ over the manifold $M^{(2)}$ where

$$\pi^{(1)} : M^{(2)} \to M^{(1)}$$

is a family of affine linear spaces whose associated linear space $z \in M^{(2)}$ is $A_x^{(2)}$ where $(\pi^{(1)} \circ \pi)(z) = x$.

In summary, *in the absence of integrability conditions the fibres of the prolongations*

$$\pi^{(q-1)} : M^{(q)} \to M^{(q-1)}$$

are affine linear spaces whose associated vector spaces are the prolongations $A^{(q)}$ *of a tableau* A. More precisely, if $x_0, x_1, x_2, \ldots$ is a sequence of points $x_q \in M^{(q)}$ with $\pi^{(q-1)}(x_q) = x_{q-1}$, then $A_{x_q} = A_{x_0}^{(q)}$.

Returning to the general discussion, if we denote by $S^+V^* = \bigoplus_{q \geq 1} S^q V^*$ the maximal ideal in the polynomial algebra $SV^* = \bigoplus_{q \geq 1} S^q V^*$ and define the *total prolongation* to be

$$\mathbf{A} = \bigoplus_{q \geq 0} A^{(q)} \subset W \otimes S^+V^*,$$

then it is clear that

(2)   **A** *is the largest graded subspace of* $W \otimes S^+V^*$

        that is closed under differentiation and satisfies

        $\mathbf{A} \cap (W \otimes S^{p+1}V^*) = A$, $\mathbf{A} \cap (W \otimes S^q V^*) = 0$

        for $q \leq p$.

We note that the grading on A is shifted by $p+1$ from that on S^+V^*; that is $A^{(q)} \subset W \otimes S^{p+q+1}V^*$.

Another useful characterization of prolongations is based on the following observation: If we consider each of

$$S^{p+1}V^* \otimes S^qV^*, \ S^pV^* \otimes S^{q+1}V^*, \ \text{and} \ S^{p+q+1}V^*$$

as subspaces of $\otimes^{p+q+1}V^*$, then

$$(3) \qquad (S^{p+1}V^* \otimes S^qV^*) \cap (S^pV^* \otimes S^{q+1}V^*) = S^{p+q+1}V^*,$$

and consequently

$$(4) \qquad A^{(q)} = (A \otimes S^qV^*) \cap (W \otimes S^{p+q+1}V^*).$$

We shall now recall and extend the concept of involution for a tableau. Referring to a definition from Chapter IV we have the subspaces

$$A_i^{(q)} = \{P \in A^{(q)} : \frac{\partial P}{\partial x^1} = \cdots = \frac{\partial P}{\partial x^i} = 0\},$$

where $x^1, \ldots, x^n$ is assumed to be a general basis of V^*. Remark that

$$(5) \qquad A_i^{(q)} = (A_i)^{(q)}$$

where $A_i = A_i^{(0)}$, and that

$$A_n^{(q)} = (0), \ A_0^{(q)} = A^{(q)}.$$

The proof of Proposition 3.6 in Chapter IV works equally for a general tableau and gives:

Proposition 1.2. *We have*

$$\dim A^{(1)} \leq \dim A + \dim A_1 + \cdots + \dim A_{n-1}$$

with equality holding if, and only if, the maps

$$(6) \qquad \frac{\partial}{\partial x^i} : A_{i-1}^{(1)} \to A_{i-1}$$

are surjective for $i = 1, \ldots, n$.

If we define the characters $s_1, \ldots, s_n$ of a tableau A by (cf. Definition 3.5 in Chapter IV)

$$s_1 + \cdots + s_k = \dim A - \dim A_k,$$

then the inequality in Proposition 1.2 is

$$\dim A^{(1)} \leq s_1 + 2s_2 + \cdots + ns_n.$$

Here and throughout this chapter we have dropped the primes since these are the only characters which we shall consider, and we shall also set $s = \dim W$ instead of using s_0 as was done in Chapter IV.

Definition 1.3. A tableau of order p is *involutive* if

$$\dim A^{(1)} = \dim A + \dim A_1 + \cdots + \dim A_{n-1}.$$

Being involutive is equivalent to the maps (6) being surjective. This is equivalent to the equality

$$\dim A^{(1)} = s_1 + 2s_2 + \cdots + ns_n$$

in Cartan's test, which agrees with Definition 3.7 in Chapter IV.

The following are the basic properties of involutive tableaux:

(7)   *Every prolongation of an involutive tableau is involutive.*

(8)   *If A is any tableau, then there is a q_0 such that the*

           prolongations $A^{(q)}$ are involutive for $q \geq q_0$.

These will be proven later in this chapter. For the moment we shall use (7) to prove the relations (which we have already encountered in Chapter III)

$$(9) \qquad \begin{aligned} s_n^{(1)} &= s_n \\ s_{n-1}^{(1)} &= s_n + s_{n-1} \\ &\;\;\vdots \qquad\quad \vdots \\ s_1^{(1)} &= s_n + \cdots + s_1 \end{aligned}$$

where the $s_k^{(1)}$ are the characters of $A^{(1)}$. As a corollary we have that

(10)   *The **character** l, i.e., the largest l such that $s_l \neq 0$, and*

           *the **Cartan integer** s_l are invariant under prolongation*

           of an involutive tableau.

Proof of (9). By definition

$$s_1^{(1)} + \cdots + s_k^{(1)} = \dim A^{(1)} - \dim A_k^{(1)}.$$

On the other hand, by the surjectivity of the maps (6) all the tableaux A_k are involutive. Hence

$$\dim A_k^{(1)} = \dim A_k + \cdots + \dim A_{n-1}$$
$$\Rightarrow \dim A^{(1)} - \dim A_k^{(1)} = \dim A + \cdots + \dim A_{k-1}$$
$$\Rightarrow s_1^{(1)} + \cdots + s_k^{(1)} = \dim A + \cdots + \dim A_{k-1}$$
$$\Rightarrow s_k^{(1)} = \dim A_{k-1}$$
$$= s_k + \cdots + s_n.$$

Example 1.4. Let $A \subset W \otimes V^*$ be a tableau, $B = A^\perp \subset W^* \otimes V$ be the symbol relations, and

$$B^\lambda = B_a^{\lambda i} w_a^* \otimes v_i$$

a basis for B. In the jet space $J^1(V, W)$ with coordinates (x^i, z^a, p_i^a) let M be defined by the equations

$$B_a^{\lambda i} p_i^a = 0.$$

The restriction to M of the contact system

$$\begin{cases} dz^1 - p_i^a dx^i = 0 \\ dx^1 \wedge \cdots \wedge dx^n \neq 0 \end{cases}$$

corresponds to the linear, homogeneous constant coefficient P.D.E. system

$$(11) \qquad B_a^{\lambda i} \frac{\partial z^a(x)}{\partial x^i} = 0.$$

We observe that

> *The total prolongation* **A** *is the space of formal power series*
> *solutions, with zero constant term, to* (11).

This is immediate from

$$\frac{\partial^q}{\partial x^I}\left(B_a^{\lambda i} \frac{\partial z^a(x)}{\partial x^i}\right) = B_a^{\lambda i} \frac{\partial^{q+1} z^a(x)}{\partial x^I \partial x^i}.$$

The involutivity of the tableau A is equivalent to the involutivity of the Pfaffian differential system associated to (11)—cf. Proposition 3.8 in Chapter IV.

More generally, let $A \subset W \otimes S^{p+1}V^*$ be a tableau of order p. We may identify $W^* \otimes SV$ with the constant coefficient differential operators on $W \otimes SV$; thus

$$(w_a^* \otimes v_I)(P_j^b w_b \otimes x^J) = \frac{\partial}{\partial x^I}(P_j^a x^J).$$

Giving A is equivalent to giving its annihilator $B = A^\perp \subset W^* \otimes S^{p+1}V$, so that a tableau of order p gives a linear, homogeneous constant coefficient P.D.E. system of order $p+1$ and vice versa.

Roughly speaking, a general linear Pfaffian differential system has the two aspects consisting of its tableau and torsion. To be involutive means that

i) for each x, the tableau is involutive (i.e., the constant coefficient system corresponding to x—like freezing the leading coefficients in a P.D.E.—is involutive); and

ii) the torsion vanishes (i.e., the integralibility conditions are satisfied— cf. Theorem 5.16 in Chapter IV). As we shall explain below, the tableau of $(\mathcal{I}, \Omega)$ influences both the tableau and torsion of $(\mathcal{I}^{(1)}, \Omega)$.

§2. The Cartan–Poincaré Lemma, Spencer Cohomology.

One of the most useful facts in exterior algebra is the familiar

Cartan Lemma: *Let V be a vector space and suppose there is a quadratic relation*

$$\sum_i w_i \wedge v_i = 0, \quad w_i, v_i \in V,$$

where the v_i are linearly independent. Then

$$\begin{cases} w_i = \sum_j a_{ij} v_j & where \\ a_{ij} = a_{ji} \end{cases}.$$

We shall give a generalization of the result that plays a crucial role in our theory of exterior differential systems.

Let U and V be vector spaces and suppose given a linear map

$$\Omega : U \to V$$

with adjoint

$$\Omega^* : V^* \to U^*.$$

We set

$$C^{p,q} = S^p V^* \otimes \Lambda^q U^*$$

and define a boundary operator

$$\delta_\Omega : C^{p,q} \to C^{p-1,q+1}$$

by the rule

$$(12) \quad \delta_\Omega(v_1^* \otimes \cdots \otimes v_p^* \otimes \psi) = \sum_\alpha v_1^* \otimes \cdots \otimes \hat{v}_\alpha^* \otimes \cdots \otimes v_p^* \otimes \Omega^*(v_\alpha^*) \wedge \psi$$

where $v_1^*, \ldots, v_p^* \in V^*$ and $\psi \in \Lambda^q U^*$. It is clear that $\delta_\Omega^2 = 0$, and we denote the resulting cohomology by

$$H^{p,q}(\Omega) = \ker\{C^{p,q} \xrightarrow{\delta_\Omega} C^{p-1,q+1}\}/\delta_\Omega C^{p+1,q-1}.$$

Proposition 2.1 (The Cartan–Poincaré lemma). *We have the isomorphism*

$$H^{p,q}(\Omega) \cong S^p(\ker \Omega^*) \otimes \Lambda^q(\operatorname{coker} \Omega^*).$$

We will give the proof in two steps.

Step One. Suppose that Ω is an isomorphism and use it to identify U with V. If we choose linear coordinates $x^1, \ldots, x^n$ on U, the elements in $C^{p,q}$ are

$$\varphi = \sum_{i_1 < \cdots < i_q} \varphi_{i_1 \ldots i_q}(x) dx^{i_1} \wedge \cdots \wedge dx^{i_q}$$

where $\varphi_{i_1 \ldots i_q}(x) \in S^p V^*$ is a polynomial of degree p. Briefly, $C^{p,q}$ consists of polynomial differential forms having polynomial degree p and exterior degree q. Thinking of Ω as the identity map, we have by (12)

$$\delta_\Omega(x^{j_i} \cdots x^{j_p} dx^{i_1} \wedge \cdots \wedge dx^{i_q}) = \sum_\alpha x^{j_1} \cdots \hat{x}^{j_\alpha} \cdots x^{j_p} dx^{j_\alpha} \wedge dx^{i_1} \wedge \cdots \wedge dx^{i_q}.$$

This implies that

$$\delta_\Omega(\varphi) = d\varphi$$

is the usual exterior derivative. We must show that

$$(13) \qquad H^{p,q}(\Omega) = \begin{cases} 0 & p + q > 0 \\ \mathbb{R} & p = q = 0. \end{cases}$$

Let

$$e = \sum_i x^i \partial/\partial x^i$$

be the Euler vector field. For $\varphi \in C^{p,q}$, Euler's theorem on homogeneous forms implies that

$$(14) \qquad \mathcal{L}_e(\varphi) = (p + q)\varphi,$$

where $\mathcal{L}_e$ denotes the Lie derivative along e. Combining (14) with the Cartan family formula gives the homotopy relation

$$(p+q)\varphi = i(e)d\varphi + di(e)\varphi,$$

and this implies (13).

Step Two. In the general case, we may choose bases for U, V so that Ω has the matrix

$$\begin{pmatrix} \begin{array}{ccc|c} 1 & & 0 & 0 \\ & \ddots & & \\ 0 & & 1 & \\ \hline 0 & & & 0 \end{array} \end{pmatrix}$$

Thus, in terms of suitable linear coordinates

$$\begin{cases} x^i, y^\alpha \text{ on } V \\ u^\alpha, w^\lambda \text{ on } U \end{cases}$$

we will have

$$\begin{cases} \Omega^*(dx^i) = 0 \\ \Omega^*(dy^\alpha) = du^\alpha. \end{cases}$$

Using multi-index notations, such as $I = \{i_1, \ldots, i_q\}$ where $i_1 < \cdots < i_q$, we may write a typical element in $C^{p,q}$ as

$$(15) \qquad \tilde{\varphi} = \sum_{I,A} \varphi_{IA}(x,y)du^I \wedge dw^A$$

where $\varphi_{IA}(x,y)$ is a polynomial in x, y and

$$\begin{cases} \deg \varphi_{IA}(x,y) = p \\ |I| + |A| = q. \end{cases}$$

We shall identify $\tilde{\varphi}$ in (15) with the expression

$$(16) \qquad \varphi = \sum_{I,A} \varphi_{IA}(x,y)dy^I \wedge dw^A.$$

When this is done,

$$\delta_\Omega(\varphi) = \sum_{I,A} \frac{\partial \varphi_{IA}(x,y)}{\partial y^\alpha} dy^\alpha \wedge dy^I \wedge dw^A.$$

In other words, δ_Ω is the exterior derivative with respect to the y variables, treating the x and w variables as parameters. This suggest that we set

$$C^{r,s,\rho,\sigma} = \begin{cases} \text{set of } \varphi \text{ given by (16) where} \\ \varphi_{IA}(x,y) \text{ has degree } r \text{ in } x \text{ and degree } s \text{ in } y, \\ \text{and where } |I| = \rho,\ |A| = \sigma. \end{cases}$$

Then

$$\delta_\Omega : C^{r,s,\rho,\sigma} \to C^{r,s-1,\rho+1,\sigma}$$

and with the obvious notation

$$(17) \qquad H^{p,q}(\Omega) \cong \bigoplus_{\substack{r+s=p \\ \rho+\sigma=q}} H^{r,s,\rho,\sigma}.$$

On the other hand, the proof of Step 1 gives

$$(18) \qquad H^{r,s,\rho,\sigma} = \begin{cases} 0 & \text{unless } s = \rho = 0 \\ C^{r,s,\rho,\sigma} & \text{when } s = \rho = 0. \end{cases}$$

Combining (17) and (18) gives the result.

We will use the Cartan–Poincaré lemma in the following form:

Corollary 2.2. *Suppose that* $\omega^1, \ldots, \omega^n \in U^*$ *are linearly independent 1-forms on a vector space* U *and that* $\varphi_{i_1 \ldots i_q} \in \Lambda^r U^*$ *are* r-*forms* $(r > 0)$ *that satisfy the conditions*

$$(19) \qquad \begin{cases} \varphi_{i_1 \ldots i_q} \text{ is symmetric in } i_1, \ldots, i_q \\ \sum_i \varphi_{i_1 \ldots i_{q-1} i} \wedge \omega^i = 0. \end{cases}$$

Then there exist $\psi_{j_1 \ldots j_{q+1}} \in \Lambda^{r-1} U^*$ *that satisfy*

$$(20) \qquad \begin{cases} \psi_{j_1 \ldots j_{q+1}} \text{ is symmetric in } j_1, \ldots, j_{q+1} \\ \sum_i \psi_{j_1 \ldots j_q j} \wedge \omega^j = \varphi_{j_1 \ldots j_q}. \end{cases}$$

Proof. In this case $V = \mathbb{R}^n$ and

$$\Omega = U \to V$$

given by

$$\Omega(u) = (\omega^1(u), \ldots, \omega^n(u)), \quad u \in U,$$

is surjective. In particular

$$(21) \qquad H^{q,r}(\Omega) = 0 \text{ when } r > 0.$$

The conditions (19) are

$$\begin{cases} \varphi \in S^q V^* \otimes \Lambda^r U^* \\ \delta_\Omega(\varphi) = 0, \end{cases}$$

and the conditions (20) are

$$\begin{cases} \psi \in S^{q+1} V^* \otimes \Lambda^{r-1} U^* \\ \delta_\Omega(\psi) = \varphi. \end{cases}$$

Thus the corollary is equivalent to (21). $\qquad\qquad\square$

When $r = q = 1$, this corollary is the usual *Cartan lemma*.

When Ω is an isomorphism, the Cartan–Poincaré lemma is the Poincaré lemma for polynomial differential forms.

We shall give a variant of this discussion of the Cartan–Poincaré lemma. Let V, W be vector spaces and set

$$C^{k,q} = W \otimes S^k V^* \otimes \Lambda^q V^*.$$

Choosing bases $\{w_a\}$ for W and $\{x^i\}$ for V^* we may think of $\varphi \in C^{k,q}$ as a W-valued polynomial differential form

$$\varphi = \sum_{|I|=k,\, |J|=q} w_a \otimes \varphi^a_{IJ} x^I dx^J.$$

We define

$$\delta : C^{k,q} \to C^{k-1,q+1}$$

to be the usual exterior differentiation treating the w_a as constants. Denoting the resulting cohomology by $H^{k,q}$ we have from (13)

$$(22) \qquad\qquad H^{k,q} = \begin{cases} W & k = q = 0 \\ 0 & \text{otherwise.} \end{cases}$$

Now let $A \subset W \otimes V^*$ be a tableau with prolongations $A^{(p)} \subset W \otimes S^{p+1} V^*$, and define $C^{k,q}(A) \subset C^{k,q}$ by

$$(23) \qquad\qquad C^{k,q}(A) = \begin{cases} A^{(k-1)} \otimes \Lambda^q V^* & k \geq 1 \\ W \otimes \Lambda^q V^* & k = 0. \end{cases}$$

The defining property of prolongations given by equation (2) in §1 above implies that

$$\delta : C^{k,q}(A) \to C^{k-1,q+1}(A),$$

and we denote by $H^{k,q}(A)$ the resulting cohomology:

$$(24) \qquad H^{k,q}(A) = \ker\{\delta : C^{k,q}(A) \to C^{k-1,q+1}(A)\}/\delta C^{k+1,q-1}(A).$$

Definition 2.3. The $H^{k,q}(A)$ are the *Spencer cohomology groups* associated to the tableau A.

With the correspondence in notation $A^{(p)} \leftrightarrow g_{k+p}$, this definition coincides with that in Chapter IX.

For us their importance resides in the following result, the first half of which will be proved in a moment and the remainder in §3.

Theorem 2.4. *If A is involutive, then*

$$H^{k,q}(A) = 0 \quad k \geq 1, \; q \geq 0.$$

The converse is also true.

In general, if

$$A \subset W \otimes S^{p+1}V^*$$

is a tableau of order p we define $C^{k,q}(A) \subset C^{k+p,q}$ by

$$(25) \qquad C^{k,q}(A) = \begin{cases} A^{(k-1)} \otimes \Lambda^q V^* & k \geq 1 \\ W \otimes S^p V^* \otimes \Lambda^q V^* & k = 0. \end{cases}$$

This agrees with (23) when $p = 0$, which is the case of an ordinary tableau. Again by the defining property of prolongations we may define the Spencer cohomology groups by the same formula (24).

We will now prove the following proposition, the first statement of which gives one half of Theorem 2.4, and the second statement of which gives a result pertaining to Chapter VI. The complete proof of Theorem 2.4 will be given later.

Proposition 2.5. *Let A be a tableau of order p.*
(i) *If A is involutive, then*

$$H^{k,q}(A) = 0 \text{ for } k \geq 1, \; q \geq 0.$$

(ii) *If A is involutive, then the prolongations $A^{(q)}$ are involutive for $q \geq 1$.*

We note that (ii) follows from (26) below and Theorem 2.4; however, the proof of Theorem 2.4 is somewhat lengthy and so we will first give a direct proof of (ii).

As an application of Proposition 2.5, we let $A \subset W \otimes V^*$ be an ordinary tableau and picture its cohomology as coming from the diagram

$$
\begin{array}{ccccc}
A^{(2)} & A^{(2)} \otimes V^* & A^{(2)} \otimes \Lambda^2 V^* \\
& \searrow & \searrow \\
A^{(1)} & A^{(1)} \otimes V^* & A^{(1)} \otimes \Lambda^2 V^* \\
& \searrow & \searrow \\
A & A \otimes V^* & A \otimes \Lambda^2 V^* \\
& \searrow & \searrow \\
W & W \otimes V^* & W \otimes \Lambda^2 V^*.
\end{array}
$$

For the 1st prolongation its Spencer cohomology comes from the diagram

$$
\begin{array}{ccccc}
A^{(3)} & A^{(3)} \otimes V^* & A^{(3)} \otimes \Lambda^2 V^* \\
& \searrow & \searrow \\
A^{(2)} & A^{(2)} \otimes V^* & A^{(2)} \otimes \Lambda^2 V^* \\
& \searrow & \searrow \\
A^{(1)} & A^{(1)} \otimes V^* & A^{(1)} \otimes \Lambda^2 V^* \\
& \searrow & \searrow \\
W \otimes V^* & W \otimes V^* \otimes V^* & W \otimes V^* \otimes \Lambda^2 V^*.
\end{array}
$$

Comparing these two it is clear that

$$(26) \qquad H^{k,q}(A^{(1)}) \cong H^{k+1,q}(A), \quad k \geq 1.$$

In particular, we have from Proposition 2.5 above:

(27)   *If A is involutive, then*

$$H^{k,q}(A^{(1)}) = (0), \quad k \geq 1.$$

Returning to our general discussion, the proof of Proposition 2.5 will follow from the two assertions

(28)   *A involutive $\Rightarrow H^{p,q}(A) = 0$ for $p \geq 1$, $q \geq 0$.*

(29)   *$H^{p,q}(A) = 0$ for $p \geq 1$ and statement* (ii) *in*

                 Proposition 2.5 when $\dim V \leq n - 1$ *together imply*

                 (ii) *in Proposition 2.5 when* $\dim V = n$.

Our proof of (28) is due to Sternberg, and the idea is this: *Using the surjectivity of the maps* (6) *in* §1 *above, a standard proof of the Poincaré lemma carries over verbatim to give $H^{p,q}(A) = 0$ for $p \geq 1$, $q \geq 1$ (the case $p \geq 1$, $q = 0$ must be treated separately and will be left to the reader).*

In fact, the exposition will be clearer if we just give this standard inductive proof of the usual Poincaré lemma and leave it for the reader to simply observe that the argument also establishes (29).

Let U be the closed cube $\{x \in \mathbb{R}^n : |x^i| \leq 1\}$ and let $\varphi \in \Omega^q(U)$ be a closed C^∞ q-form (the coefficients of φ are assumed smooth in a neighborhood of U). If $q \geq 1$ we want to find $\eta \in \Omega^{q-1}(U)$ satisfying

$$d\eta = \varphi.$$

Suppose that φ involves only the differentials $dx^1, \ldots, dx^k$. The construction of η will be by descending induction on k. Namely, we will inductively find η_k such that

$$\varphi - d\eta_k \text{ involves only } dx^1, \ldots, dx^{k-1}.$$

Then $\eta = \eta_1$ will be our required form. To find η_k we write

$$\varphi = \varphi' + \varphi'' \wedge dx^k$$

where φ', φ'' involve only $dx^1, \ldots, dx^{k-1}$. From

$$0 = d\varphi = d\varphi' + d\varphi'' \wedge dx^k,$$

by looking at the coefficients of $dx^k \wedge dx^l$ where $l > k$ we see that

$$\frac{\partial \varphi''}{\partial x^l} = 0, \qquad l > k.$$

Here, the derivatives of a form mean the derivatives of its coefficients. By elementary calculus, we may find a $(q-1)$-form η_k involving only $dx^1, \ldots, dx^{k-1}$ and satisfying

$$(30) \qquad \begin{cases} \frac{\partial \eta_k}{\partial x^k} = \varphi'' \\ \frac{\partial \eta_k}{\partial x^l} = 0 \quad \text{for } l > k. \end{cases}$$

In other words, if $C_k^\infty(U)$ are the C^∞ functions in U that only depend on $x^1, \ldots, x^k$, then the mappings

$$(31) \qquad \partial/\partial x^k : C_k^\infty(U) \to C_k^\infty(U)$$

are surjective. In fact, given $f \in C_k^\infty(U)$ the function

$$g(x^1, \ldots, x^k) = \int_{-1}^{x^k} f(x^1, \ldots, x^{k-1}, t)dt$$

satisfies $g \in C_k^\infty(U)$ and $\partial g/\partial x^k = f$. We note the similarity between (31) and (6).

Now consider

$$\psi = \varphi - d\eta_k$$
$$= \varphi - \varphi'' \wedge dx^k + \text{ terms involving only } dx^1, \ldots, dx^{k-1}.$$

By (30) the form ψ involves only $dx^1, \ldots, dx^{k-1}$, and finding it completes the induction step in our proof of the usual Poincaré lemma. $\square$

As suggested above, the proof of (28) is the same with (6) in §1 replacing (31).

We now turn to the proof of (ii) in Proposition 2.5. The result is due to Cartan; a direct proof is given in Singer and Sternberg [1965] and, as we have noted, the result follows in a general way from (26) and Theorem 2.4. We shall give the argument for an ordinary tableau $A \subset W \otimes V^*$, the proof for a general tableau being essentially the same. Thus we must show that:

$$(32) \qquad A \text{ involutive } \Rightarrow A^{(1)} \text{ involutive}.$$

By our inductive strategy (29) we may assume
   (i) that (32) is true when $\dim V \le n - 1$;

(ii) that (i) in Proposition 2.5 is true when $\dim V \leq n$. We must then prove (32) when $\dim V = n$.

By definition and Proposition 1.2 above we must show that there exists a basis $x^1, \ldots, x^n$ for V^* such that the maps

$$(33_i) \qquad\qquad \partial/\partial x^i : A^{(2)}_{i-1} \to A^{(1)}_{i-1}$$

are surjective for $i = 1, \ldots, n$. By the involutivity of A there exists a basis such that the maps

$$(34) \qquad\qquad \partial/\partial x^i : A^{(1)}_{i-1} \to A_{i-1}, \quad i = 1, \ldots, n,$$

are all surjective, and this is the basis for V^* we shall use.

Next we note that A_1 is itself an involutive tableau in $n - 1$ variables. By our induction assumption we may then assume that the maps (33_i) are surjective for $i = 2, \ldots, n$. It remains to prove that

$$\partial/\partial x^1 : A^{(2)} \to A^{(1)}$$

is surjective.

Let $Q \in A^{(1)} \subset W \otimes S^2 V^*$. We want to find $P \in A^{(2)} \subset W \otimes S^3 V^*$ satisfying

$$(35) \qquad\qquad \frac{\partial P}{\partial x^1} = Q.$$

The idea is to use the surjectivity of (34) to solve for the derivatives of this equation. Then the vanishing of cohomology will allow us to "integrate" this solution.

Thus consider

$$\frac{\partial Q}{\partial x^j} \in A \subset W \otimes V^*.$$

By (34) when $i = 1$ we may find $T_j \in A^{(1)}$ with

$$(36) \qquad\qquad \frac{\partial T_j}{\partial x^1} = \frac{\partial Q}{\partial x^j}.$$

Consider the 1-form

$$T = T_j dx^j \in A^{(1)} \otimes \Lambda^1 V^*.$$

Its exterior derivative satisfies

$$dT \in A \otimes \Lambda^2 V^*$$

$$\frac{\partial}{\partial x^1}(dT) = d\left(\frac{\partial T}{\partial x^1}\right) = d^2 Q = 0,$$

where the derivatives of a form mean the derivatives of its coefficients. Denote by $\overline{\varphi}$ the restriction of a form whose coefficients are functions of $x^2, \ldots, x^n$ to the subspace $x^1 = 0$ (thus $\varphi \to \overline{\varphi}$ means to set $dx^1 = 0$). Then, since $d\overline{\varphi} = \overline{d\varphi}$,

$$d(\overline{dT}) = 0.$$

We may thus consider $\overline{dT}$ as a class in

$$H^{1,2}(A_1) = 0$$

by our induction assumption (ii). Consequently, we may find

$$S = \sum_{\alpha=2}^{n} S_\alpha(x^2, \ldots, x^n)dx^\alpha = \overline{S}$$

with $S_\alpha \in A_1^{(1)}$ and

$$\overline{dT} = dS.$$

It follows that

(37) $$\qquad\qquad d(T - S) = dx^1 \wedge U$$

where

$$U = \sum_{\alpha=2}^{n} U_\alpha(x^2, \ldots, x^n)dx^\alpha = \overline{U}, \quad U_\alpha \in A_1.$$

Taking exterior derivatives of both sides of (37) gives

$$0 = dx^1 \wedge dU.$$

It follows that

$$U \in H^{1,1}(A_1) = 0,$$

again by our induction assumption. Thus

$$U = -dW, \quad \text{where } W \in A_1^{(1)}$$

$$\Rightarrow d(T - S) = d(W\,dx^1).$$

Setting

$$R = S + W\,dx^1 = \sum_{i=1}^{n} R_j(x^2, \ldots, x^n)dx^j,$$

where $R_j \in A_1^{(1)}$, we see that

$$d(T - R) = 0.$$

Then, by the usual Poincaré lemma for homogeneous polynomials differential forms there exists P such that

$$\begin{cases} \text{(i)} & T - R = dP \\ \text{(ii)} & P \in W \otimes S^3 V^*. \end{cases}$$

The first equation implies that

$$P \in A^{(2)},$$

and we claim that (35) is satisfied. In fact

$$\frac{\partial}{\partial x^j}\left(\frac{\partial P}{\partial x^1} - Q\right) = \frac{\partial}{\partial x^1}(T_j - R_j) - \frac{\partial Q}{\partial x^j}$$

$$= \frac{\partial T_j}{\partial x^i} - \frac{\partial Q}{\partial x^j}$$

$$= 0$$

by (36). $\square$

We have now proved that the prolongations of an involutive tableau are involutive. One point of putting the argument here is that, whether one uses the language or not, the proof unavoidedly uses Spencer cohomology.

Discussion. In Chapter VI, Theorem 2.1 we have proved that the prolongation of an involutive differential system $(\mathcal{I}, \Omega)$ is again involutive. In case $(\mathcal{I}, \Omega)$ is a linear Pfaffian system we may give an alternate proof based on Proposition 2.5 above as follows:

To show that $(\mathcal{I}^{(1)}, \Omega)$ is involutive on the manifold $M^{(1)}$ of integral elements of $(\mathcal{I}, \Omega)$, we must show that

i) for each point $y \in M_1^{(1)}$, the tableau $A_y^{(1)}$ is involutive.

ii) the integrability conditions for $(\mathcal{I}^{(1)}, \Omega)$ are satisfied.

Now (i) follows from (ii) in Proposition 2.5, and so we must establish (ii). We have seen in Chapter IV and will recall below that the integrability conditions of a linear Pfaffian differential system live in a family of quotient vector spaces. These equivalence classes were called the *torsion* of the linear Pfaffian system, and what we shall prove is that

(38)   *The torsion of $(\mathcal{I}, \Omega)$ lives naturally in the family*

              of vector spaces $H^{0,2}(A_x)$, and

(39)   *The torsion of $(\mathcal{I}^{(1)}, \Omega)$ lives naturally in the family*

              of vector spaces $H^{0,2}(A_x^{(1)}) \cong H^{1,2}(A_x)$ (cf. (26)).

Since $(\mathcal{I}, \Omega)$ is assumed to be involutive, its torsion vanishes. Then the torsion of $(\mathcal{I}^{(1)}, \Omega)$ vanishes by the vanishing of cohomology assertion (i) in Proposition 2.5. This, at least in outline form, is the proof for linear Pfaffian systems of the fact that $(\mathcal{I}^{(1)}, \Omega)$ is involutive if $(\mathcal{I}, \Omega)$ is.

We will now discuss how (38) and (39) are established, and for this we begin with a standard homological construction. Let $A \subset W \otimes V^*$ be a tableau and set

$$B = A^{\perp} \subset W^* \otimes V$$

so that we have an exact sequence of vector spaces

$$0 \to A \to W \otimes V^* \to B^* \to 0.$$

We define vector spaces $B^{*(k)}$ by

$$0 \to A^{(k)} \to W \otimes S^{k+1}V^* \to B^{*(k)} \to 0$$

and set

$$C^{k,q}(B) = \begin{cases} B^{*(k-1)} \otimes \Lambda^q V^* & k \geq 1 \\ 0 & k = 0. \end{cases}$$

Then we have an exact sequence of complexes

$$0 \to C^{\cdot,\cdot}(A) \to C^{\cdot,\cdot} \to C^{\cdot,\cdot}(B) \to 0$$

where $C^{\cdot,\cdot}(A) = \bigoplus_{k,q} C^{k,q}(A)$, $C^{\cdot,\cdot} = \bigoplus_{k \geq 0}(W \otimes S^{k+1}V^*)$, and $C^{\cdot,\cdot}(B) = \bigoplus_{k,q} C^{k,q}(B)$. Associated to this is a long exact cohomology sequence, and using (13) above this gives

$$H^{p,q}(B) \xrightarrow{\sim} H^{p-1,q+1}(A), \quad p \geq 1 \text{ and } q \geq 1.$$

In particular, we have

$$(40) \qquad \begin{cases} \text{(i)} \;\; H^{0,2}(A) \cong H^{1,1}(B) \\ \text{(ii)} \;\; H^{0,2}(A^{(1)}) \cong H^{1,2}(A) \cong H^{2,1}(B). \end{cases}$$

To establish (38) we use equation (51) of Chapter IV to write the structure equations of $(\mathcal{I}, \Omega)$ as

$$d\theta^a \equiv A^a_{\varepsilon i}\pi^{\varepsilon} \wedge \omega^i + \frac{1}{2}c^a_{ij}\omega^i \wedge \omega^j \quad \mathrm{mod}\ \{I\}$$

where $\{I\}$ is the *algebraic* ideal in $\Omega^* M$ generated by the sections of $I \subset T^* M$. Under a change

$$\pi^{\varepsilon} \to \pi^{\varepsilon} + p^{\varepsilon}_j \omega^j$$

we have

$$(41) \qquad c_{ij}^a \to c_{ij}^a + \frac{1}{2}\left(A_{\varepsilon i}^a p_j^\varepsilon - A_{\varepsilon j}^a p_i^\varepsilon\right).$$

Recalling that

$$C^{0,2}(A) = W \otimes \Lambda^2 V^*$$

$$C^{1,1}(A) = A \otimes V^*$$

$$H^{0,2}(A) = W \otimes \Lambda^2 V^* / \delta(A \otimes V^*)$$

and that A_x is spanned by the

$$A_{x,\varepsilon} = A_{\varepsilon i}^a(x) w_a \otimes v_i^*,$$

may easily compute from (41) that the cocycle

$$\frac{1}{2}\, c_{ij}^a(x) w_a \otimes v_i^* \wedge v_j^* \in C^{0,2}(A_x)$$

gives a class $[c](x)$ in $H^{0,2}(A_x)$ whose vanishing is necessary and sufficient
for the existence of integral elements lying over $x \in M$ (cf. Proposition 5.14
in Chapter IV for the complete argument here).

If we write the structure equations in dual form as (cf. equation (83) of
Chapter IV)

$$\begin{cases} d\theta^a \equiv \pi_i^a \wedge \omega^i \mod \{I\} \\ B_a^{\lambda i}\pi_i^a \equiv C_j^\lambda \omega^j \mod \{I\} \end{cases}$$

where the

$$B^\lambda = B_a^{\lambda i} w_a^* \otimes v_i$$

give a basis for the symbol relations in the annihilator

$$B_x = A_x^\perp,$$

then the π_i^a are determined modulo I up to a substitution

$$\pi_i^a \to \pi_i^a + p_{ij}^a \omega^j, \quad p_{ij}^a = p_{ji}^a.$$

Under such a substitution

$$C_j^\lambda \to C_j^\lambda + B_a^{\lambda i} p_{ij}^a,$$

so that the equivalence class

$$[C_j^\lambda] \in B_x^* \otimes V^* / \delta(W \otimes S^2 V^*) = H^{1,1}(B_x)$$

is well-defined. If we denote this equivalence class by $[C](x)$, then it is easy to verify that:

$$[c](x) = [C](x) \ \textit{under the isomorphism} \ \text{(i)} \ \textit{in} \ (40).$$

We will now recall from equation (124) of Chapter IV the structure equations of the first prolongation $(\mathcal{I}^{(1)}, \Omega)$

$$(42) \qquad \left\{ \begin{array}{ll} \text{(i)} & d\theta^a \equiv 0 \mod \{I^{(1)}\} \\[4pt] \text{(ii)} & d\theta^a_i \equiv \pi^a_{ij} \wedge \omega^j \mod \{I^{(1)}\} \ \text{where} \\[4pt] \text{(iii)} & \pi^a_{ij} = \pi^a_{ji} \\[4pt] \text{(iv)} & B^{\lambda i}_a \pi^a_{ij} \equiv C^\lambda_{jk} \omega^k \mod \{I^{(1)}\}. \end{array} \right.$$

We will see that the equivalence class

$$[C^\lambda_{jk}] \in B^{*(1)}_x \otimes V^* / \delta B^{*(2)}_x$$

is well-defined, that the coboundary $\delta(C^\lambda_{jk}) = 0$, and that the vanishing of the resulting cohomology class $[C^{(1)}](x) \in H^{2,1}(B_x)$ is the necessary and sufficient condition for the existence of integral elements for $(\mathcal{I}^{(1)}, \Omega)$ lying over x. Referring to (28) and (ii) in (39), we will then see that, in the involutive case, this vanishing is automatically satisfied.

In summary, *the involutivity of the tableau of* $(\mathcal{I}, \Omega)$ *gives both the involutivity of the tableau and vanishing of the torsion of the prolonged system* $(\mathcal{I}^{(1)}, \Omega)$.

From equation (126) of Chapter IV we have

$$C^\lambda_{jk} \omega^j \wedge \omega^k \equiv 0 \mod \{I^{(1)}\},$$

which is just $\delta(C^\lambda_{jk}) = 0$. Using (iv) in (31) it is now straightforward to show that writing $C^{(1)}$ as a coboundary is equivalent to being able to absorb the integrability conditions into the π^a_{ij}, so that (iv) becomes

$$B^{\lambda i}_a \pi^a_{ij} \equiv 0 \mod \{I^{(1)}\}.$$

In this way we have now established that $(\mathcal{I}^{(1)}, \Omega)$ is involutive.

§3. The Graded Module Associated to a Tableau; Koszul Homology.

We want to finish laying the basis for the proof of Theorem 2.4 above and for the Cartan–Kuranishi prolongation theorem in Chapter VI. The algebraic basis for both of these comes by studying a certain graded SV-module

M_A associated to a tableau A. A very interesting confluence occurs in that Cartan's test for involution dualizes into the condition that M_A admit a quasi-regular sequence, and such modules are standard fodder for the cannons of homological algebra. Before embarking on the formal discussion, we remark on why this should be so: The dual of

$$\partial/\partial x^i : S^q V^* \to S^{q-1} V^*$$

is the multiplication

$$v_i : S^{q-1} V \to S^q V.$$

Thus, Cartan's test in form of the <u>surjectivity</u> of the differentiation maps given by (6) above dualizes to the <u>injectivity</u> of suitable multiplication maps.

Let $A \subset W \otimes V^*$ be a tableau and

$$\mathbf{A} = \bigoplus_{q \geq 0} A^{(q)} \subset W \otimes S^+ V^*$$

the total prolongation of A. Dually, we set

$$\begin{cases} B = A^\perp \subset W^* \otimes V \\ B_q = A^{(q)\perp} \subset W^* \otimes S^{q+1} V \\ \mathbf{B} = \oplus_{q \geq 0} B_q \subset W^* \otimes S^+ V. \end{cases}$$

For $P \in W \otimes S^{p+1} V^*$ and $m \in W^* \oplus S^p V \cong (W \otimes S^p V^*)^*$ we have

$$(43) \qquad \left\langle m, \frac{\partial P}{\partial x^i} \right\rangle = \langle v_i m, P \rangle$$

where $v_i \in V$ is a basis with dual basis $x^i \in V^*$ and $\langle \, , \, \rangle$ is the pairing between dual vector spaces. It then follows from (2) above that:

$(44) \qquad \mathbf{B} \subset W^* \otimes S^+ V$ *is the graded SV-submodule of*

$W^* \otimes S^+ V$ *generated by* $B \subset W^* \otimes V.$

Here, $W^* \otimes SV$ is the obvious free SV-module and $W \otimes S^+ V$ the submodule corresponding to the maximal ideal in SV. We remark that since $B = B_0$, the grading on $\mathbf{B}$ is shifted by one from that induced by the natural grading on $W^* \otimes SV$. To rectify this we shall introduce the standard *shift notation*: if $M = \bigoplus_q M_q$ is a graded SV-module, then we define the new graded SV-module $M^{[p]}$ by

$$(M^{[p]})_l = M_{p+l}.$$

With this notation the inclusion

$$\mathbf{B}^{[-1]} \to W^* \otimes SV$$

is a homogeneous SV-module mapping of degree zero. We define the graded SV-module M_A to be the quotient, so that we have

$$(45) \qquad 0 \to \mathbf{B}^{[-1]} \to W^* \otimes SV \to M_A \to 0.$$

We note that

$$(46) \qquad M_{A,q} = \begin{cases} W^* & q = 0 \\ A^{(q-1)*} & q \geq 1. \end{cases}$$

Definition 3.1. i) $\mathbf{B}$ is the *symbol module* associated to the tableau A, and ii) M_A is the *graded SV-module* associated to the tableau A.

To explain how Cartan's test dualizes, we need to recall some essentially standard commutative algebra definitions. Let $M = \bigoplus_{q \geq 0} M_q$ be a non-negatively graded SV-module.

Definition 3.2. i) The element $v \in V$ is *quasi-regular* if the kernel K in

$$0 \to K \to M \xrightarrow{v} M \to 0$$

of multiplication by v has no elements in positive degree; i.e., $K^+ = (0)$.

ii) The *sequence* $v_1, \ldots, v_k$ of elements of V is *quasi-regular* for M if v_j is quasi-regular for $M/(v_1, \ldots, v_{j-1})M$ for $j = 1, \ldots, k$.

iii) The module M is *quasi-regular* if there is a basis $v_1, \ldots, v_n$ for V that is quasi-regular for M.

The usual definitions of regular element, regular sequence, and regular module are the same as above but where multiplication has no kernel.

The grading on the quotient module is the obvious one:

$$(M/(v_1, \ldots, v_{j-1})M)_q = M_q/(v_1 M_{q-1} + \cdots + v_{j-1} M_{q-1}).$$

The point of all this is the following:

Proposition 3.3. *The tableau $A \subset W \otimes V^*$ is involutive if, and only if, the associated graded module M_A is quasi-regular.*

Proof. By Proposition 1.2 and Definition 1.3 above, the involutivity of A is equivalent to the existence of a basis $x^1, \ldots, x^n \in V^*$ such that each mapping

$$(47) \qquad \partial/\partial x^i : A_{i-1}^{(q+1)} \to A_{i-1}^{(q)}$$

is surjective for all $q \geq 0$. Recalling that by definition

$$\mathbf{A}_{i-1} = \{P \in \mathbf{A} : \frac{\partial P}{\partial x^1} = \cdots = \frac{\partial P}{\partial x^{i-1}} = 0\}$$

it follows from (43) and (44) that

$$A_{i-1}^{(q)\perp} = (v_1, \ldots, v_{i-1}) \cdot M_{A,q}$$
$$\cap \qquad\qquad \cap$$
$$A^{(q)*} \cong \qquad M_{A,q+1},$$

and therefore

$$(A_{i-1}^{(q)})^* \cong A^{(q)*}/A_{i-1}^{(q)\perp}$$
$$\cong M_{A,q+1}/(v_1, \ldots, v_{i-1}) \cdot M_{A,q}.$$

The surjectivity of (47) for $q \geq 0$ is equivalent to the injectivity of the dual mapping

$$(47^*) \quad v_i : M_{A,q+1}/(v_1, \ldots, v_{i-1}) \cdot M_{A,q} \to M_{A,q+2}/(v_1, \ldots, v_{i-1}) \cdot M_{A,q+1}$$

for $q \geq 0$. This implies the proposition. $\qquad\square$

It is an remarkable coincidence of previously unrelated historical terminology that *the condition that the constant coefficient Pfaffian differential system associated to the tableau A have a regular integral flag is equivalent to M_A being quasi-regular in the sense of commutative algebra.*

We shall now quickly extend this discussion to a higher order tableau. Let

$$A \subset W \otimes S^{p+1}V^*$$

be a tableau of order p with prolongations

$$A^{(q)} \subset W \otimes S^{p+q+1}V^*.$$

We set

$$B = A^\perp \subset W^* \otimes S^{p+1}V$$

$$B_q = A^{(q)\perp} \subset W^* \otimes S^{p+q+1}V$$

$$\mathbf{B} = \bigoplus_{q \geq 0} B_q, \quad B_0 = B.$$

Using the shift notation we define the graded SV-module M_A associated to A by the exact SV-module sequence

$$(48) \qquad 0 \to \mathbf{B}^{[-1]} \to (W^* \otimes SV)^{[p]} \to M_A \to 0.$$

Then

$$M_{A,q} = \begin{cases} W^* \otimes S^p V & q = 0 \\ A^{(q)*} & q \geq 1. \end{cases}$$

The reason for the choice of $M_{A,0}$ will be discussed below. Remark on the obvious but interesting point that

$$(W^* \otimes SV)^{[p]} = (W^* \otimes S^p V) \oplus (W^* \otimes S^{p+1}V) \oplus \ldots$$

is not a free SV-module when $p \geq 1$. This will be further discussed below.

For now the important observation is that *the statement and proof of Proposition 3.3 carry over verbatim to a higher order tableau.*

It is a well-known result in commutative algebra that the condition for a graded SV-module to admit a regular sequence is expressed by the vanishing of suitable Koszul homology groups. Moreover, the same proof works for quasi-regular sequences. We shall now explain this.

Let $M = \bigoplus_q M_q$ be a graded SV-module with module mappings

$$S^p V \otimes M_q \to M_{p+q}.$$

Set

$$C_{p,q} = M_p \otimes_{\mathbb{C}} \Lambda^q V$$

and define a boundary operator

$$(49) \qquad\qquad \partial : C_{p,q} \to C_{p+1,q-1}$$

by the formula

$$(50) \quad \partial(m \otimes v_{i_1} \wedge \cdots \wedge v_{i_q}) = \sum_\alpha (-1)^{\alpha+1} v_{i_\alpha} \cdot m \otimes v_{i_1} \wedge \cdots \wedge \hat{v}_{i_\alpha} \wedge \cdots \wedge v_{i_q}.$$

It is clear that $\partial^2 = 0$, and we set

$$(51) \qquad H_{p,q}(M) = \ker\{\partial : C_{p,q} \to C_{p+1,q-1}\}/\partial C_{p-1,q+1}.$$

Definition 3.4. $H_{p,q}(M)$ are the *Koszul homology groups* of the graded SV-module M.

The Koszul homology groups will be used in this and some of the following sections. For the moment the following is the relevant property:

Proposition 3.5. *The following are equivalent:*
   i) $H_{+,n}(M) = \cdots = H_{+,n-q}(M) = 0$;
   ii) *there is a quasi-regular sequence of length q, say $v_1, \ldots, v_q$, for M;*
   iii) *every generic sequence of length q is quasi-regular for M.*

We denote by S^+V the maximal ideal in SV and shall begin by establishing the

Lemma 3.6. *The following are equivalent:*

i) $H_{+,n}(M) = 0$

ii) *for* $m \in M$, $S^+ V \cdot m = 0 \Rightarrow m \in M_0$

iii) *there exists a* $v \in V$ *such that* $v \cdot m = 0 \Rightarrow m \in M_0$

iv) *for generic* $v \in V$, $v \cdot m = 0 \Rightarrow m \in M_0$.

Proof. We begin by showing the equivalence of i) and ii). The top end of the complex that computes Koszul homology is

$$0 \to H_{\cdot,n}(M) \to M \otimes \Lambda^n V \xrightarrow{\partial} M \otimes \Lambda^{n-1} V,$$

where by the boundary formula (50)

$$\partial(m \otimes v_1 \wedge \cdots \wedge v_n) = \sum_{\alpha=1}^{n} (-1)^{\alpha+1} v_\alpha \cdot m \otimes v_1 \wedge \cdots \wedge \hat{v}_\alpha \wedge \cdots \wedge v_n.$$

From this the equivalence of i) and ii) is clear.

The equivalence of ii)–iv) is also pretty clear. What is obvious is that iv) $\Rightarrow$ iii) $\Rightarrow$ ii), and so we must prove that ii) $\Rightarrow$ iv). Let $J_0 \subset M_0$ be a vector space complement to $\{M \in M_0 : v \cdot m = 0 \text{ for all } v \in V\}$ and set $M' = J_0 \oplus M^+$. Then M' is a graded SV-module with the property:

$$(52) \qquad\qquad v \cdot m = 0 \text{ for all } v \in V \Leftrightarrow m = 0.$$

Referring to §1, n^0 1 of Bourbaki [1961], with the terminology employed there we have that

$$SV \notin \text{Ass}(M').$$

Moreover, by Corollary 2 in §1, n^0 1 of Bourbaki [1961], the condition that the multiplication

$$P : M' \to M'$$

by $P \in SV$ be injective is that $P \notin I$ for any prime ideal $I \in \text{Ass}(M')$. By the corollary to Theorem 2 in §1, n^0 4 of Bourbaki [1961], the set $\text{Ass}(M')$ is a finite set $I_1, \ldots, I_k$ of proper prime ideals. Then $SV \backslash (I_1 \cup \cdots \cup I_k)$ consists of the elements P such that multiplication by P is injective. Each $I_j \cap V$ is a proper linear subspace, and $V \backslash ((I_1 \cap V) \cup \cdots \cup (I_k \cap V))$ is the open dense set of elements that are generic in the sense of (iv). $\qquad \square$

We next need the following trivial but basic

Lemma 3.7. *Multiplication by* $v \in V$ *induces the zero map*

$$H_{p,q}(M) \xrightarrow{v} H_{p+1,q}(M).$$

Proof. If $\varphi \in M_p \otimes \Lambda^q V$ is a cycle, then $\varphi \wedge v \in M_p \otimes \Lambda^{q+1} V$ and by (50)

$$(53) \qquad \partial(\varphi \wedge v) = (\partial\varphi) \wedge v + (-1)^q v \cdot \varphi.$$

This implies the lemma. $\qquad\square$

Proof of Proposition 3.5. If v is quasi-regular then we have exact sequences of SV-modules

$$\begin{cases} 0 \to J_0 \to M \xrightarrow{v} vM \to 0, & J_0 \subset M_0, \\ 0 \to vM \to M \to M/vM \to 0, \end{cases}$$

where in the first sequence v is a module map homogeneous of degree one, while all other maps are homogeneous of degree zero. By standard reasoning these give long exact homology sequences

$$\begin{cases} \to H_{p,q}(J_0) \to H_{p,q}(M) \xrightarrow{v} H_{p+1,q}(vM) \xrightarrow{\partial} H_{p+1,q-1}(J_0) \to \\ \to H_{p+1,q}(vM) \to H_{p+1,q}(M) \to H_{p+1,q}(M/vM) \to H_{p+2,q-1}(vM) \to . \end{cases}$$

Using Lemma 3.7 and the fact that $H_{p+1,q-1}(J_0) = 0$ for $p \geq 0$, this pair of sequences combines to give

$$0 \to H_{p+1,q}(M) \to H_{p+1,q}(M/vM) \to H_{p+1,q-1}(M) \to 0$$

for $p \geq 0$. Now take $q = n$ and use Lemma 3.6 to obtain

$$(54) \qquad H_{p+1,n}(M/vM) \cong H_{p+1,n-1}(M), \quad p \geq 0.$$

If there is a $\tilde{v} \in V$ that is quasi-regular for M/vM, then we conclude from Lemma 3.6 that

$$(55) \qquad H_{p+1,n-1}(M) = 0, \quad p \geq 0.$$

Conversely, if (55) holds then by (54) and the lemma we may find $\tilde{v} \in V$ that is quasi-regular for M/vM. Continuing in this way with an obvious descending induction gives the proposition. $\qquad\square$

Corollary 3.8. *The tableau A (of any order) p is involutive if, and only if,*

$$H_{p,q}(M_A) = 0, \quad \text{for } p \geq 1,\ q \geq 0.$$

Completion of the Proof of Theorem 2.4. We must show that the vanishing of suitable cohomology implies that the tableau A is involutive. By (43) and the definition of M_A, the complexes of vector spaces

$$\cdots \to A^{(p)} \otimes \Lambda^q V^* \xrightarrow{\delta} A^{(p-1)} \otimes \Lambda^{q+1} V^* \to \cdots$$
$$\cdots \leftarrow M_{A,p} \otimes \Lambda^q V \leftarrow M_{A,p-1} \otimes \Lambda^{q+1} V \leftarrow \cdots$$

and mutually dual. It follows that *Spencer cohomology is dual to Koszul homology*, i.e.,

$$(56) \qquad\qquad H^{p,q}(A) \cong H_{p,q}(M_A)^*.$$

Under this duality the above corollary translates into the implication

$$H^{p,q}(A) = 0 \text{ for } p \geq 1, \ q \geq 0 \Rightarrow A \text{ involutive.}$$

$\square$

Remark. Let $\mathbf{D} = \bigoplus_{q \geq 0} D_q$ denote the algebra of constant coefficient differential operators on $W \otimes SV^*$. Then $\mathbf{D} \cong \operatorname{Hom}(W, W) \otimes SV^*$ acts on the graded vector space $\mathbf{A} = \bigoplus_{q \geq 0} A^{(q)}$. The Cartan family formula gives

$$\frac{\partial \varphi}{\partial x^i} = \mathcal{L}_{\partial/\partial x^i}(\varphi) = \frac{\partial}{\partial x^i} \lrcorner \, \delta\varphi + \delta\left(\frac{\partial}{\partial x^i} \lrcorner \, \varphi\right)$$

where $\varphi \in \mathbf{A} \otimes \Lambda V^* \subset W \otimes SV^* \otimes \Lambda V^*$ and where $\partial\varphi/\partial x^i$ means the derivative of the polynomial coefficients of the exterior monomials $dx^{i_1} \wedge \cdots \wedge dx^{i_q}$ treating the vector space W as constants. It follows that

$$\delta\varphi = 0 \Rightarrow \frac{\partial \varphi}{\partial x^i} = \delta(\partial/\partial x^i \lrcorner \, \varphi)$$

$$\Rightarrow \mathbf{D}^+ \text{ acts trivially on } \bigoplus_q H^q(\mathbf{A})$$

where $\mathbf{D}^+$ is the ideal of positive degree operators and $H^q(\mathbf{A}) = \bigoplus_p H^{p,q}(A)$. By (56) this is the dual statement to Lemma 3.7 above. Of course, we could dualize completely our use of Koszul homology and work entirely with Spencer cohomology, but this would fail to make clear the connection between involutivity and more standard notions from commutative algebra.

In order to establish the prolongation theorem in Chapter VI we need to prove the

Proposition 3.9. *Let A be a tableau. Then there is a k_0 such that the prolongations $A^{(k)}$ are involutive for $k \geq k_0$.*

We will actually prove a stronger result. To state it, we recall that there is a function, the *Hilbert function* $P_A(q)$ of the graded module M_A, such that

$$\dim A^{(q)} = P_A(q)$$

for all q. The result we shall actually need is given by the

Proposition 3.10. *Let $A \subset W \otimes V^*$ be a tableau and $P_A(q)$ the Hilbert function of the graded module M_A. Then there is a k_0 depending on $\dim W$ and $P_A(q)$ such that $A^{(k)}$ is involutive for $k \geq k_0$.*

We will now prove Proposition 3.9. Then we shall prove Proposition 3.10 at the end of §5 after we have discussed localization.

By Theorem 2.4, it will suffice to show that there is a k_0 such that all

$$(57) \qquad H^{p,q}(A^{(k)}) = 0, \quad p \geq 1,\ q \geq 0 \text{ and } k \geq k_0.$$

By the definition of Spencer cohomology we have (cf. (26) above)

$$(58) \qquad H^{p,q}(A^{(k)}) \cong H^{p+k,q}(A), \quad p \geq 1 \text{ and } q \geq 0.$$

Combining (56)–(58) we see that Proposition 3.9 follows from (and in fact is equivalent to) the

Proposition 3.9′. *Let M_A be the graded module associated to a tableau. Then there exists a p_0 such that*

$$H_{p,q}(M_A) = 0 \text{ for } p \geq p_0,\ q \geq 0.$$

In fact, this proposition is valid for any finitely generated graded SV-module M, and it is this more general result that we shall prove. For this we set

$$C_{p,q} = M_p \otimes \Lambda^q V$$

$$C_q = \bigoplus_p C_{p,q},$$

and define a graded SV-module structure on C_q by

$$v \cdot (m \otimes v_{i_1} \wedge \cdots \wedge v_{i_q}) = (v \cdot m) \otimes v_{i_1} \wedge \cdots \wedge v_{i_q}$$

where $v, v_i \in V$ and $m \in M$. The boundary mappings (49) induce

$$(59) \qquad \partial : C_q \to C_{q-1},$$

and it is immediate from (50) that (59) *is a mapping of graded SV-modules, homogeneous of degree one.* As a consequence, assuming that $M.$ is finitely generated the following are all finitely generated SV-modules

$$\ker \partial,\ \text{image}\,\partial,\ H_q(M) = \bigoplus_p H_{p,q}(M).$$

Moreover, and this is the crucial point, by Lemma 3.7 above *the maximal ideal S^+V acts trivially on the SV-module $H_q(M)$.* From this we infer that $H_q(M)$ is a finitely generated $SV/S^+V = \mathbb{C}$ module, i.e., it is a finite dimensional vector space. $\qquad \square$

In the next section we shall discuss the interpretation of Koszul homology $H_{p,q}(M)$ as it pertains to resolutions of M by free modules. It will turn out that Proposition 3.9 is really just the statement that a sub-module of a finitely generated SV^* module is itself finitely generated as this latter statement pertains to the relations among the generators of M, and then the relations among the relations, and so forth, i.e., it is equivalent to the statement that M has a finite resolution by finitely generated free modules.

§4. The Canonical Resolution of an Involutive Module.

In the preceeding section we defined the graded module M_A associated to a tableau, and then we used more or less standard commutative algebra to relate the existence of quasi-regular sequences for M_A to the vanishing of Koszul homology. There the motivation was to complete the proof of Theorem 2.4, among other things providing the conceptual basis for the result that the prolongations of an involutive tableau are involutive.

Another use of Koszul homology is in the construction of resolutions of graded modules, and we would now like to pursue the implications of this for the theory of differential systems. Before doing this we will try to give some motivation by the following

Example 4.1. We identify $W^* \otimes SV$ with constant coefficient linear differential operators on $W^* \otimes SV^*$, so that by definition

$$(w_a^* \otimes v_I)(P_J^b w_b \otimes x^J) = \frac{\partial}{\partial x^I}(P_J^a x^J).$$

As discussed in Chapter IV, giving a tableau $A \subset W \otimes V^*$ is equivalent to giving the linear homogeneous constant coefficient P.D.E. system

$$(60) \qquad\qquad D^\lambda u(x) = 0$$

where

$$D^\lambda = B_a^{\lambda i} w_a^* \partial/\partial x^i$$

$$u = w_a u^a(x)$$

$$D^\lambda u = B_a^{\lambda i} \frac{\partial u^a(x)}{\partial x^i},$$

and where the

$$B^\lambda = B_a^{\lambda i} w_a^* \otimes v_i$$

give a basis for $B = A^\perp \subset W^* \otimes V$. The symbol module $\mathbf{B}$ corresponds to the algebra of constant coefficient differential operators on $W \otimes SV^*$ generated by the operators D^λ. In particular, the solutions to (60) satisfy

$$Du = 0$$

for all $D \in \mathbf{B}$.

More generally we consider the inhomogeneous linear P.D.E. system

$$(61) \qquad\qquad D^\lambda u(x) = f^\lambda(x).$$

Any relation

$$e_\lambda B^\lambda = 0, \quad e_\lambda \in SV$$

on the generators of the symbol module gives an *integrability condition*

$$E_\lambda f^\lambda(x) = 0$$

on the f^λ, where $E_\lambda \in \mathbb{C}[\partial/\partial x^1, \ldots, \partial/\partial x^n]$ corresponds to $e_\lambda \in \mathbb{C}[v_1, \ldots, v_n]$ when we set $v_i = \partial/\partial x^i$. In this way, not only the generators but also the relations of the symbol module enter into the theory. Once we agree to study the relations of $\mathbf{B}$, we may as well go ahead and study entire resolutions.

In fact, returning to the general discussion, it is well-known that in some ways the most important properties of a finitely generated SV-module M are given by its generators and relations, and then the generators and relations of its relations, etc. In brief, one wants to find *resolutions* of M by free SV-modules $\mathbf{E} = E \otimes SV$, where E is a finite-dimensional vector space. It is also well-known that any M has an essentially canonical minimal such resolution where the vector spaces E are appropriate Koszul homology groups of M. It has been proved above that the involutivity of A is equivalent to the vanishing of certain of the Koszul homology groups of M_A, which then turns out to be equivalent to the property that the canonical resolution be especially simple (this was conjectured by Guillemin and Sternberg and proved by Serre; see Guillemin and Sternberg [1964]). Here we shall recall some of the definitions and elementary facts, together with a derivation of the canonical resolution of an involutive module.

We set $S = SV$ and consider an arbitrary finitely generated, graded S-module $M = \bigoplus_{k \in \mathbb{Z}} M_k$. For simplicity of exposition we shall assume that M is *non-negatively graded* in the sense that $M_k = 0$ for $k < 0$. The action of S on M is given by vector space mappings $S_j \otimes M_k \to M_{j+k}$ satisfying the customary conditions. We let $S^+ = \bigoplus_{k \geq 1} S^k V$ be the maximal ideal of S; then $M/S^+ M$ is a finite dimensional vector space whose dimension equals the minimal number of generators of M as an S-module.

Associated to M are its Koszul homology groups $H_{k,q}(M)$, whose definition was given in the preceeding section. Here we shall give a number of remarks concerning these groups (cf. Green [1989a] for a general discussion of Koszul homology and its relationship to algebraic geometry):

$$(62) \qquad H_{k,0}(M) \cong M_k/V \cdot M_{k-1}$$

$$\cong \{\text{new generators of } M \text{ in degree } k\}.$$

In particular

$$(63) \qquad \left\{ \begin{array}{l} M \text{ is generated} \\ \text{in degree zero} \end{array} \right\} \Leftrightarrow H_{k,0}(M) = 0 \text{ for all } k > 0.$$

$$(64) \qquad H_{k,n}(M) \cong \mathrm{Ann}_V(M_k) = \{v \in V \mid v \cdot M_k = 0\}.$$

Of importance is the *shift mapping*, which we recall associates to a graded module M a new graded module $M^{[p]}$ defined by

$$(65) \qquad (M^{[p]})_q = M_{p+q}.$$

It is then clear from the definitions that

$$(66) \qquad H_{k,q}(M^{[p]}) \cong H_{k+p,q}(M) \text{ for } q \geq 0.$$

An exact sequence of graded S-modules

$$(67) \qquad 0 \to M' \to M \to M'' \to 0$$

where all maps are homogeneous of degree zero, gives a long exact homology sequence

$$(68) \qquad \to H_{k-1,q+1}(M'') \xrightarrow{\partial} H_{k,q}(M') \to H_{k,q}(M) \to H_{k,q}(M'')$$

$$\xrightarrow{\partial} H_{k+1,q-1}(M') \to \ldots \ .$$

In case the module maps in (67) are homogeneous of degrees other than zero, then we still have an exact sequence (68) with a shift in indices given by using the shift mapping to make the maps homogeneous of degree zero.

　　An example when this shift occurs is given by the exact S-module sequence

$$0 \to \mathbf{B} \to W^* \otimes SV \to M_A \to 0$$

where the first map is homogeneous of degree $+1$. Here, A is an ordinary tableau, $\mathbf{B}$ is its symbol module, and M_A is the associated graded SV-module. The sequence (45) above

$$0 \to \mathbf{B}^{[-1]} \to W^* \otimes S \to M_A \to 0$$

has all maps homogeneous of degree zero, and (68) together with (66) gives
(70)

$$\cdots \to H_{k,q}(W^* \otimes SV) \to H_{k,q}(M_A) \xrightarrow{\partial} H_{k,q-1}(\mathbf{B}) \to H_{k+1,q-1}(W^* \otimes SV) \to \cdots$$

(71) $\qquad$ M is a *free module* generated in degree zero (i.e.,

$\qquad\qquad$ $M \cong M_0 \otimes SV$) if, and only if,

$$(72) \qquad H_{k,q}(M) = 0 \text{ for } (k,q) \neq (0,0).$$

This property will be proved when we establish Proposition 4.3 below.

We recall our assumption that all modules are non-negatively graded; thus $M = \bigoplus_{q \geq 0} M_q$.

Definition 4.2. i) A graded S-module M is *involutive* if

$$(73) \qquad H_{k,q}(M) = 0 \text{ for } k \geq 1 \text{ and all } q \geq 0$$

ii) A graded S-module $M = \bigoplus_{k \geq k_0} M_k$ is k_0-*involutive* if

$$H_{k,q}(M) = 0 \text{ for } k \neq k_0 \text{ and all } q \geq 1.$$

In the latter case it follows that

$$\tilde{M} = M^{[k_0]}$$

is involutive in the usual sense.

We may reformulate the discussion in the preceeding section by the statement:

$$(74) \qquad \textit{The tableau } A \textit{ is involutive if, and only if, the associated}$$
$$\textit{graded module } M_A \textit{ is involutive.}$$

From (70) and (72) we have the following:

$$(75) \qquad \textit{The tableau } A \textit{ is involutive if, and only if, the symbol}$$
$$\textit{module } \mathbf{B}, \textit{ is involutive.}$$

By applying the standard construction of a minimal free resolution of a graded S-module in terms of its Koszul homology groups, we shall derive the following result due to Serre (see Guillemin and Sternberg [1964]):

Proposition 4.3. *Let M be an involutive SV-module. Then there is a free resolution*

$$(76) \qquad 0 \to E_n \xrightarrow{\varphi_n} E_{n-1} \to \cdots \to E_1 \xrightarrow{\varphi_1} E_0 \xrightarrow{\varphi_0} M \to 0$$

where $\deg \varphi_0 = 0$, $\deg \varphi_i = 1$, for $i \geq 1$ and $E_i = H_{0,i}(M) \otimes SV$. The converse is also true.

We shall call (76) the *canonical resolution of an involutive module* (it is canonical in the sense that it is minimal in the sense explained below). Before deriving it we shall make a few remarks.

$$(77) \qquad \textit{An involutive module is generated in degree zero, i.e.,}$$
$$M = S \cdot M_0.$$

This follows from the definition together with (63).

Next we consider an exact sequence

$$0 \to M' \to M \to M'' \to 0$$

of graded SV-modules and degree zero maps.

(78)   *If M, M'' are involutive and $M_0 \to M_0''$ is an isomorphism, then M' is 1-involutive. Moreover, there are canonical short exact sequences*

$$(79) \qquad 0 \to H_{0,q}(M) \to H_{0,q}(M'') \xrightarrow{\partial} H_{1,q-1}(M') \to 0.$$

This is clear from our definitions and (68).

For 1-involutive modules M' we shall use the notation

$$\tilde{M}' = M'^{[1]}.$$

Then under the hypotheses of (19), $\tilde{M}'$ is involutive and (79) is

$$(80) \qquad 0 \to H_{0,q}(M) \to H_{0,q}(M'') \to H_{0,q-1}(\tilde{M}') \to 0.$$

We may now easily give the construction of the canonical resolution of an involutive module. By (62) and (63)

$$\begin{cases} H_{0,0}(M) \cong M_0, \text{ and} \\ M \text{ is generated in degree zero.} \end{cases}$$

Setting $E_0 = H_{0,0}(M) \otimes SV$, this gives a short exact sequence

$$0 \to N \to E_0 \to M \to 0$$

to which (78) and (79) apply. Thus $\tilde{N}$ is involutive, and by (71) and (80)

$$H_{0,q}(M) \cong H_{0,q-1}(\tilde{N}), \quad q \geq 1.$$

We now repeat the construction replacing M by $\tilde{N}$, and continue.

Note that if M is free and $H_{0,1}(M'') = 0$ in (77), then $M' = 0$ and so $M \cong M'$ is free. In other words, if N is involutive and $H_{0,1}(N) = 0$, then N is free. This implies that the resolution process terminates after at most n steps.

We have now established the constructive half of Proposition 4.3; since we will not use the converse the proof of this will not be given (in any case it is standard). $\qquad\square$

Discussion. We shall, without proofs, put Proposition 4.3 in a general context. For this we let

$$(i) \qquad 0 \to E_m \xrightarrow{\varphi_m} E_{n-1} \to \cdots \to E_1 \xrightarrow{\varphi_1} E_0 \xrightarrow{\varphi_0} M, \quad \deg\varphi_i = 0,$$

be a resolution of a finitely generated graded S-module M by free modules

$$E_p = \bigoplus_q (B_{p,q} \otimes S^{[-q]}),$$

where the $B_{p,q}$ are finite dimensional vector spaces whose dimensions $b_{p,q}$ give the number of $S^{[-q]}$'s occurring in E_p. The maps $\varphi_i : E_i \to E_{i-1}$ are normalized to be of degree zero and are given by a matrix whose entries break into blocks corresponding to maps

$$(ii) \qquad \varphi_{i,k,l} : B_{i,k} \otimes S^{[-k]} \to B_{i-1,l} \otimes S^{[-l]}.$$

Each such block is clearly given by a $b_{i,k} \times b_{i-1,l}$ matrix whose individual entries are homogeneous polynomials of degree $k - l$. We shall say that the resolution (i) is *minimal* in case the entries in each $\varphi_{i,k,l}$ have strictly positive degree.

To explain this, it is clear that a non-zero block (ii) where $k = l$ gives a redundancy in (i) in the sense that in terms of suitable bases for $B_{i,k}$ and $B_{i-1,k}$ the matrix of $\varphi_{i,k,l}$ will have the form

$$\begin{pmatrix} I_\alpha & 0 \\ 0 & 0 \end{pmatrix},$$

where I_α is the $\alpha \times \alpha$ identity matrix and where the I_α piece induces isomorphism between free sub-modules of E_i and E_{i-1} which may then be deleted from (i). It follows that every M has a minimal free resolution (i), and this resolution may be seen to be essentially unique. In fact, a general result given for example in Green [1989a] is that

$$(iii) \qquad H_{p,q}(M) \cong B_{q,q+p}.$$

We will briefly discuss this result.

We recall that $M = \bigoplus_{q \geq 0} M_q$ is assumed to be non-negatively graded. Then (iii) implies that

$$(iv) \qquad B_{p,q} = 0 \text{ for } q < p.$$

In other words, the generators of M (which correspond to the $B_{0,q}$) are in non-negative degrees, the relations among the generators (which correspond to the $B_{1,q}$) are themselves generated in degrees at least one, and so forth. Using (iv) we may write

$$\text{(v)} \qquad E_q = \bigoplus_{p \geq 0}(C_{p,q} \otimes S^{[-q-p]})$$

where by (iii)

$$\text{(vi)} \qquad C_{p,q} \cong H_{p,q}(M).$$

Now the simplest modules, the free modules, are those for which

$$C_{p,q} = 0, \quad q \neq 0.$$

By (vi) these are characterized by

$$H_{p,q}(M) = 0, \quad q \neq 0.$$

The next simplest modules, at least from our point of view, are those for which

$$\text{(vii)} \qquad C_{p,q} = 0, \quad p \neq 0.$$

By (vi) and the Definition 4.2 these are just the involutive modules, and clearly (v) above is equivalent to Proposition 4.3.

Generalizing slightly the above, we may see from (vi) and the definition that a finitely generated module M is k_0-involutive if, and only if, its minimal resolution (i) has

$$\text{(viii)} \qquad \begin{cases} E_p = B_p \otimes S^{[-k_0-p]} \text{ where} \\ B_p = H_{k_0,p}(M) \end{cases}$$

Example 4.4. The *truncation* S_{k_0} of S is defined by

$$(S_{k_0})_q = \begin{cases} S^q V & q \geq k_0 \\ 0 & q < k_0. \end{cases}$$

It is easy to verify that

$$H_{p,q}(S_{k_0}) = 0 \text{ for } p \neq k_0.$$

The minimal resolution of M may be constructed using Young symmetrizers (cf. Green [1989a]). When $k_0 = 1$ the truncation is the maximal ideal S^+V.

Resolving S^+V is equivalent to resolving $SV/S^+V = \mathbb{C}$, and this is provided by the standard Koszul resolution

$$\ldots \xrightarrow{\partial} S \otimes \Lambda^2 V \xrightarrow{\partial} S \otimes V \xrightarrow{\partial} S \to \mathbb{C} \to 0, \quad \deg \partial = 1,$$

obtained by dualizing the polynomial de Rham complex

$$0 \to \mathbb{C} \to SV^* \xrightarrow{d} SV^* \otimes V^* \xrightarrow{d} SV^* \otimes \Lambda^2 V^* \xrightarrow{d} \ldots$$

and applying Proposition 2.1 above.

The inhomogeneous P.D.E. system whose symbol module is S_{k_0} is

$$\frac{\partial^{k_0} u(x)}{\partial x^I} = f_I(x), \quad |I| = k_0.$$

The involutivity of this system implies in particular that the compatibility conditions for this system are all of 1st order (see the discussion below). When $k_0 = 1$ the system is

$$\frac{\partial u(x)}{\partial x^i} = f_i(x),$$

and the compatibility conditions are obviously

$$\frac{\partial f_i(x)}{\partial x^j} = \frac{\partial f_j(x)}{\partial x^i}.$$

The remainder of this section will be a series of remarks, some of which are of interest in themselves and some of which are for later use.

(81)   A piece of the canonical resolution is (where we set
$S^m V = 0$ for $m < 0$)

$$\tag{82}
\begin{aligned}
0 \to H_{0,n}(M) \otimes S^m V &\to H_{0,n-1}(M) \otimes S^{m+1} V \to \ldots \\
\to H_{0,1}(M) \otimes S^{m+n-1} V &\to H_{0,0}(M) \otimes S^{m+n} V \to M_{m+n} \to 0.
\end{aligned}$$

Of importance below will be the explicit description of the maps

$$\varphi_i : H_{0,i}(M) \to H_{0,i-1}(M) \otimes V$$

occurring in the resolution (76). For this we now denote the Koszul boundary by

$$\partial_V : \Lambda^q V \to \Lambda^{q-1} V \otimes V$$

where $\partial_V(v_{i_1} \wedge \cdots \wedge v_{i_q}) = \sum_\alpha (-1)^{\alpha+1} v_{i_1} \wedge \cdots \wedge \hat{v}_{i_\alpha} \wedge \cdots \wedge v_{i_q} \otimes v_{i_\alpha}$. Recalling that

$$H_{0,i}(M) = \ker\{\partial : M_0 \otimes \Lambda^i V \to M_1 \otimes \Lambda^{i-1}V\}^1$$
$$\cap$$
$$M_0 \otimes \Lambda^i V$$

we infer the commutative diagram

$$\begin{array}{ccc} H_{0,i}(M) & \overset{\varphi_i}{\longrightarrow} & H_{0,i-1}(M) \otimes V \\ \cap & & \cap \\ M_0 \otimes \Lambda^i V & \overset{1 \otimes \partial_V}{\longrightarrow} & M_0 \otimes \Lambda^{i-1}V \otimes V. \end{array}$$

For $\xi \in M_0 \otimes \Lambda^i V$, the cycle condition $\partial \xi = 0$ is that the composite map

$$M_0 \otimes \Lambda^i V \overset{1 \otimes \partial_V}{\longrightarrow} M_0 \otimes \Lambda^{i-1}V \otimes V \to M_1 \otimes \Lambda^{i-1}V$$

applied to ξ be zero, i.e., identifying $\Lambda^{i-1}V \otimes V$ with $V \otimes \Lambda^{i-1}V$,

$$(1 \otimes \partial_V)(\xi) \in (\text{degree one relations for } M) \otimes \Lambda^{i-1}V$$
(83)
$$\cap$$
$$(M_0 \otimes V) \otimes \Lambda^{i-1}V.$$

This implies that $(1 \otimes \partial_V)(\xi)$ lies in the subspace $H_{0,i-1}(M) \otimes V$ of $M_0 \otimes \Lambda^{i-1}V \otimes V$ and that φ_i is induced by $1 \otimes \partial_V$.

(84)   In case $M = M_A$ for involutive tableau $A \subset W \otimes V^*$ with symbol relations $B \subset W^* \otimes V$, the first few pieces of the canonical resolution (82) are (using (83))

$$\begin{array}{ccc} 0 \to H_{0,0}(M_A) & \cong (M_A)_0 \to 0; \\ \| & \| \\ W^* & W^* \end{array}$$

$$\begin{array}{ccccccc} 0 \to H_{0,1}(M_A) & \to & H_{0,0}(M_A) \otimes V & \to & (M_A)_1 & \to 0 \\ \| & & \| & & \| \\ 0 \to \quad B & \to & W^* \otimes V & \to & A^* & \to 0, \end{array}$$

and (more interestingly)

(85)
$$\begin{array}{ccccccccc} 0 \to H_{0,2}(M_A) \to & H_{0,1}(M_A) \otimes V & \to & H_{0,0}(M_A) \otimes S^2 V & \to & (M_A)_2 \to 0 \\ \cap & \| & & \| & & \| \\ W^* \otimes \Lambda^2 V \to & B \otimes V & \to & W^* \otimes S^2 V & \to & A^{(1)*} \to 0. \end{array}$$

---

[1] For $i = 1$ this just gives that

$$H_{0,1}(M) \cong \ker(M_0 \otimes V \to M_1)$$
$$\cong \text{degree one relations for } M.$$

Referring to Example 4.1 above, we let

$$D^\lambda = B_a^{\lambda i} w_a^* \otimes v_i \in W^* \otimes V$$

be a basis for B. Then by (85), elements of $H_{0,2}(M_A)$ are of the form

$$q = \frac{1}{2} q_a^{ij} w_a^* \otimes \partial/\partial x^i \wedge \partial/\partial x^j \in W^* \otimes \Lambda^2 V$$

where

$$q_a^{ij} + q_a^{ji} = 0$$

and

$$q_a^{ij} = m_\lambda^j B_a^{\lambda i}.$$

Since

$$H_{0,2}(M_A) \cong H_{0,1}(\mathbf{B}) \cong \left\{ \begin{array}{l} \text{relations among the} \\ \text{generators } D^\lambda \text{ of } B \end{array} \right\},$$

it follows that:

(86)   *All relations among the $D^\lambda = B_a^{\lambda i} w_a^* \otimes \partial/\partial x^i$*

        are generated by linear relations of the form

$$m_\lambda^j \partial/\partial x^j \, D^\lambda = 0.$$

(Remarkably, this fact is found in Cartan [1953].)[2]

Referring now to the inhomogeneous, constant coefficient linear P.D.E. system (cf. (61) above)

$$D^\lambda u(x) = f^\lambda(x),$$

we see that the *compatibility conditions on the $f^\lambda(x)$* are generated by the 1st order equations

$$m_\lambda^j \frac{\partial f^\lambda(x)}{\partial x^i} = 0.$$

(87)   *The dual of the graded piece (82) of the canonical*

        resolution of the involutive module M_A associated

        to an involutive tableau A is (setting

        $q = m + n$)

---

[2] Referring to (85), the kernel $H_{0,1}(B)$ of $B \otimes V \to W^* \otimes \Lambda^2 V$ is identified with all $m_\lambda^j D^\lambda \otimes \partial/\partial x^j \in B \otimes V \subset W^* \otimes V \otimes V$ that lie in $W^* \otimes \Lambda^2 V$; this is just the condition $m_\lambda^j B_a^{\lambda i} + m_\lambda^i B_a^{\lambda i} = 0$, which is clearly equivalent to $m_\lambda^j \partial/\partial x^j \, D^\lambda = 0$. The reference to Cartan is page 1045 in the 1984 edition of his collected works.

$$(88) \qquad \begin{aligned} 0 \to A^{(q)} \to H^{0,0}(A) \otimes S^{q+1}V^* \to \\ H^{0,1}(A) \otimes S^q V^* \to H^{0,2}(A) \otimes S^{q-1}V^* \to \dots \end{aligned}$$

Although we do not need it for our work here we remark that (88) is essentially the symbol sequence of the Spencer complex associated to an involutive linear P.D.E. system (cf. Chapter X).

For later purposes it is useful to make explicit the maps in (88). First note that, by the definitions,

$$(89) \qquad H^{0,i}(A) = W \otimes \Lambda^i V^* / \delta(A \otimes \Lambda^{i-1}V^*).$$

We claim that δ induces

$$(90) \qquad \left(\frac{W \otimes \Lambda^i V^*}{\delta(A \otimes \Lambda^{i-1}V^*)} \right) \otimes S^k V^* \xrightarrow{\ \overline{\delta}\ } \left(\frac{W \otimes \Lambda^{i+1} V^*}{\delta(A \otimes \Lambda^i V^*)} \right) \otimes S^{k-1}V^*,$$

which is the map induced by exterior differentiation $\delta : W \otimes S^k V^* \otimes \Lambda^i V^* \to W \otimes S^{k-1}V^* \otimes \Lambda^{i+1}V^*$ (thinking of the $S^k V^*$ as polynomial functions).

Proof. Let $\psi \in A \otimes \Lambda^{i-1}V^*$ and $P \in S^k V^*$. Then, from the definition of $\overline{\delta}$, since $\delta\psi \in W \otimes \Lambda^i V^*$ are constant coefficient differential forms

$$\overline{\delta}(\delta\psi \otimes P) = \sum_i (-1)^i \delta\psi \wedge dx^j \otimes \frac{\partial P}{\partial x^j}$$

$$= \sum_j (-1)^i \delta(\psi \wedge dx^j) \otimes \frac{\partial P}{\partial x^j} \in (\delta(A \otimes \Lambda^i V^*)) \otimes S^{k-1}V^*.$$

$\qquad\qquad\qquad\qquad\qquad\qquad\qquad\qquad\qquad\qquad\qquad\qquad\qquad \square$

From this and (83) it follows that:

(91) *The maps in the long exact sequence (88) are induced by $\overline{\delta}$ in (90).*

(92) Finally, we want to discuss the "1st derived part" of an involutive module M. Recall from (64) that

$$H_{k,n}(M) = 0 \quad k \geq 1$$

$$H_{0,n}(M) = \{m \in M_0 : v \cdot m = 0 \text{ for all } v \in V\}$$

$$= W_0^*$$

where the last equation is a definition of W_0^*. Setting $N = M/W_0^*$ we have an exact sequence of graded SV-modules and degree zero maps

$$(93) \qquad\qquad 0 \to W_0^* \to M \to N \to 0$$

where N has the following properties (both of which come from the exact Koszul homology sequence of (93)):

$$\begin{cases} \text{(i)} & N \text{ is involutive} \\ \text{(ii)} & H_{0,n}(N) = 0. \end{cases}$$

Moreover, the sequence (93) gives

$$(94) \qquad \begin{cases} \text{(i)} & 0 \to W_0^* \to W^* \to N_0 \to 0 \quad \text{(degree zero)} \\ \text{(ii)} & M_k \xrightarrow{\sim} N_k \quad (k \geq 1). \end{cases}$$

From (i) we have an intrinsic subspace

$$W_1 = W_0^{*\perp} \subset W.$$

Now suppose that $M = M_A$ for an involutive tableau A. From (ii) in (94) in the case $k = 1$ we have

$$(95) \qquad\qquad A \subset W_1 \otimes V^* \subset W \otimes V^*.$$

Definition 4.5. i) We shall say that two tableau $A_i \subset W_i \otimes V^*$, $i = 1, 2$ are *equivalent* if there is a vector space W and inclusions $W_i \subset W$ such that $A_1 = A_2$ as subspaces of $W \otimes V^*$.

ii) We shall say that two *symbol mappings* $\sigma_i : W_i \otimes V^* \to U_i$ are *equivalent* in case the tableau $A_i = \ker \sigma_i$ are equivalent.

To see what this means, suppose that $W_2 = W$ and we have the situation (95). Choose a direct sum decomposition

$$W = W_0 \oplus W_1$$

and basis for W compatible with this composition. Then the tableau matrix $\|\pi_i^a\|$ for A has the form

$$\left\| \begin{array}{c} 0 \\ \pi_i^\rho \end{array} \right\|$$

where $\|\pi_i^\rho\|$ is the tableau matrix for $A \subset W_1 \otimes V^*$. In the language of linear Pfaffian differential systems, assuming involutivity we easily see that *the 0-block* (corresponding to $W_0^* \subset W^*$) *gives the 1st derived subsystem of our Pfaffian differential system.* Thus:

For an involutive, linear Pfaffian differential system the 1st derived system corresponds to the subspaces

$$H_{0,n}(M_A) \subseteq W.$$

Below we shall introduce a refinement of the characteristic variety called the characteristic sheaf, and in the next section shall prove that

(96)   *the characteristic sheaf of an involutive linear Pfaffian*

differential system uniquely determines the symbol

mapping up to equivalence.

There are examples to show that the corresponding statement for the characteristic variety is false.

§5. Localization; the Proofs of Theorem 3.2 and Proposition 3.8.

We continue with our purely algebraic discussion centered around the algebraic properties of an involutive tableau $A \subset W \otimes V^*$.

It is well-known that a major advance in commutative algebra occurred by localizing, or by what is essentially equivalent, by the use of sheaf theory. It is also well-known that a major advance in linear P.D.E. theory occurred by microlocalizing in the cotangent bundle.

It is therefore reasonable to adapt these two techniques to the theory of linear Pfaffian differential systems in the expectation that they may prove useful in the study of geometric problems. That is what we shall do in this section, which will be broken into a number of discussions.

In order to simplify notation, we make the convention for this section: *Unless mentioned otherwise, all vector spaces will be assumed to be complex.*

Preliminaries. For a complex vector space V we set $P = PV^*$ with homogeneous coordinate ring $S = SV = \bigoplus_{k \geq 0} S^k V$ and with $\mathcal{O}$ denoting the structure sheaf on P. We will use the well known F.A.C. "dictionary", cf. Serre [1955] and Hartshorne [1977].

$$(97) \qquad \left\{ \begin{array}{c} \text{coherent sheaves of} \\ \mathcal{O}\text{-modules and sheaf} \\ \text{maps over } P \end{array} \right\} \leftrightarrow \left\{ \begin{array}{c} \text{graded } S\text{-modules} \\ \text{and } S\text{-module} \\ \text{maps} \end{array} \right\}.$$

We remark that all modules are assumed to be of finite type. The italics around the word dictionary signify that the correspondence (97) is not a bijection. In the study of involutive, linear Pfaffian differential systems what will be lost are the graded modules associated to the 1st derived system part of the tableau; this will exactly correspond to the equivalence relation introduced in Definition 4.5 above.

We will denote by $\mathcal{O}(q)$ the usual sheaf of "locally holomorphic homogeneous functions of degree q" on P. It is standard that

$$H^0(P, \mathcal{O}(q)) = S^q V$$

(this holds for all q if we agree to set $S^q V = 0$ for $q < 0$). Given a coherent sheaf $\mathcal{F}$ on P the corresponding graded S-module is

$$\begin{cases} F = \bigoplus_q F_q \\ F_q = H^0(P, \mathcal{F}(q)) \end{cases}$$

where $\mathcal{F}(q) = \mathcal{F} \otimes_{\mathcal{O}} \mathcal{O}(q)$. The module structure

$$S^p V \otimes F_q \rightarrow F_{p+q}$$

is obtained from the sheaf pairing

$$\mathcal{O}(p) \otimes_{\mathcal{O}} \mathcal{F}(q) \rightarrow \mathcal{F}(p+q)$$

by passing to global sections.

The inverse map

$$\text{graded } S\text{-module } F \rightarrow \text{coherent sheaf } \mathcal{F}$$

is obtained by *localization*. The following are some of its basic properties:

i) If $\mathbf{E} = E \otimes_{\mathbb{C}} S$ is a free S-module, then the localization $\mathcal{E}$ is the trivial vector bundle with fibre E (in general we shall identify holomorphic vector bundles and locally free sheaves);

ii) If the graded module F localizes to $\mathcal{F}$, then the shift $F^{[q]}$ localizes to $\mathcal{F}(q)$;

iii) exact module sequences go to exact sheaf sequences (but not quite conversely); and

iv) $F_q = H^0(P, \mathcal{F}(q))$ for $q \geq q_0(F)$.

To construct the localization, we may use the fact that F has finite free resolution and use i). Remark that a module map

$$\varphi : \mathbf{E} \rightarrow \mathbf{F}$$

between free modules $\mathbf{E} = E \otimes S$ and $\mathbf{F} = F \otimes S$ will by definition be homogeneous of some degree $q \geq 0$; it is given by an element $\varphi \in \text{Hom}(E, F) \otimes S^q V$, which we may think of as a global section of $\text{Hom}(\mathcal{E}(-q), \mathcal{F})$. Thus φ localizes to

$$\varphi : \mathcal{E}(-q) \rightarrow \mathcal{F},$$

and in this way we know what is meant by the maps appearing in a free resolution.

A basic general fact is that *the dictionary* (97) *is exact modulo finite dimensional vector spaces.* This is illustrated by the following

Example 5.1. Let $P_1, \ldots, P_m \in S^q V$ be homogeneous forms of degree q on P. There is then an exact sequence of graded S-modules

$$(98) \qquad \bigoplus^m S \xrightarrow{\varphi} S^{[q]} \to Q \to 0$$

where the maps have degree zero and

$$\begin{cases} \varphi(Q_1, \ldots, Q_m) = \sum_i Q_i P_i, \\ Q = \operatorname{coker} \varphi \end{cases} .$$

The image $I = \varphi(\bigoplus^m S)$ is (a shift of) the homogeneous ideal $\{P_1, \ldots, P_m\}$ generated by the P_r's. The sheaf sequence corresponding to (98) is

$$(99) \qquad \bigoplus^m \mathcal{O} \xrightarrow{\varphi} \mathcal{O}(q) \to \mathcal{Q} \to 0$$

where $\mathcal{Q}$ is a coherent sheaf supported on the algebraic variety $Z \subset P$ defined by the ideal I.

In particular, if $Z = \emptyset$ then $\mathcal{Q} = (0)$ and (99) is

$$\bigoplus^m \mathcal{O} \xrightarrow{\varphi} \mathcal{O}(q) \to 0.$$

By the general theory (Theorem A in Serre [1955] to be specific), the induced mapping

$$\bigoplus^m H^0(P, \mathcal{O}(p)) \to H^0(P, \mathcal{O}(p+q))$$

is surjective for $p \geq p_0$; this is a special case of *Hilbert's nullstellensatz*. In this case $Q = S/I$ is a finite dimensional vector space that is "lost" in the dictionary (97). The simplest special case is when

$$P_i = \xi_i$$

so that

$$Q = \mathbb{C}[\xi_1, \ldots, \xi_n] / \{\xi_1, \ldots, \xi_n\} \cong \mathbb{C}.$$

This example also explains why the finiteness Theorem 3.12 in Chapter V should be true. Namely, consider the linear homogeneous P.D.E. system for one unknown function

$$(100) \qquad P_r(D)u(x) = 0, \quad r = 1, \ldots, m$$

where $P_r(D) \in \mathbb{C}[\partial/\partial x^1, \ldots, \partial/\partial x^n]$ is the constant coefficient operator corresponding to $P_r \in S^q V$. If the complex characteristic variety $\{[\xi] \in P :$

all $P_r(\xi) = 0\}$ of (100) is empty, then by the nullstellensatz we have that $\xi^\alpha \in \{P_1(\xi), \ldots, P_m(\xi)\}$ for $|\alpha| \geq p_0 + q$; this gives

$$D^\alpha u(x) = 0 \text{ for } |\alpha| \geq p_0 + q.$$

Thus $u(x)$ is a polynomial of degree at most $p_0 + q$, and therefore the solution space to (100) is finite dimensional. In fact, the solution space is naturally isomorphic to $Q^*(= $ dual vector space to Q). In the special case when $P_1 = \xi_1, \ldots, P_n = \xi_n$ we obtain only the constant functions. It will come out of our discussion that this is the only *involutive* system (100) with empty characteristic variety. This is a special case of Corollary 3.11 to Theorem 3.6 in Chapter V.

The Characteristic Sheaf. We consider a graded S-module M that has a presentation

$$(101) \qquad\qquad U^* \otimes S \xrightarrow{\varphi_1} W^* \otimes S \xrightarrow{\varphi_0} M \to 0$$

where U, W are finite dimensional vector spaces and φ_i has degree i for $i = 0, 1$ (we may think of φ_1 as a matrix whose entries are linear functions). The localization of (101) will be denoted by

$$(102) \qquad\qquad \mathcal{U}^*(-1) \xrightarrow{\varphi_1} \mathcal{W}^* \xrightarrow{\varphi_0} \mathcal{M} \to 0$$

where $\mathcal{U}^*, \mathcal{W}^*$ are the trivial bundles with respective fibres U^*, W^*.

Definition 5.2. We call $\mathcal{M}$ the *characteristic sheaf* of M and support$(\mathcal{M}) = \Xi$ the *characteristic variety* of M.

It may be proved that both these definitions are independent of the particular presentation (101).

To explain our motivation for this terminology, we let $A \subset W \otimes V^*$ be a tableau and set

$$U = W \otimes V^*/A \cong B^*$$

so that we have the symbol mapping

$$\sigma : W \otimes V^* \to U.$$

Let M_A be the graded S-module associated to A as given by Definition 3.1 above.

Definition 5.3. We call $\mathcal{M}_A$ the *characteristic sheaf of the tableau* and $\Xi_A = $ support$(\mathcal{M}_A)$ the *characteristic variety of the tableau.*

In this special case (101) is

$$B \otimes S \xrightarrow{\sigma^*} W^* \otimes S \to M_A \to 0,$$

which is just the definition of M_A, and the localization (102) is

$$\mathcal{B}^*(-1) \xrightarrow{\sigma^*} \mathcal{W}^* \to \mathcal{M}_A \to 0.$$

For each $[\xi] \in P$ we define

$$\sigma_\xi : W \to U$$

by

$$\sigma_\xi(w) = \sigma(w \otimes \xi).$$

Clearly, σ_ξ is defined only up to non-zero multiples. In intrinsic terms, if $L_\xi \subset V^*$ is the line corresponding to $[\xi] \in P$, then

$$\sigma_\xi = \sigma \mid W \otimes L_\xi.$$

This induces the mapping

$$W \to U \otimes L_\xi^*$$

which dualizes to

$$B \otimes L_\xi \to W^*.$$

Now L_ξ is the fibre of $\mathcal{O}(-1)$ at $[\xi]$ so that the last mapping is

$$B(-1)_{[\xi]} \xrightarrow{\sigma_\xi^*} W_{[\xi]}^*,$$

where the subscript denotes the fibres of the various bundles at $[\xi] \in P$. Our conclusion is that

$$\sigma_\xi^* \text{ is the mapping } \sigma^* \text{ localized at } [\xi].$$

Consequently, if we denote by $\mathcal{F}_{[\xi]} = \mathcal{F}/m_{[\xi]} \cdot \mathcal{F}$ the fibre of a coherent sheaf $\mathcal{F}$ at $[\xi]$ where $m_{[\xi]} \subset \mathcal{O}_{[\xi]}$ is the maximal ideal, then

$$(\mathcal{M}_A)_{[\xi]} = (\ker \sigma_\xi)^*.$$

Moreover,

$$\text{support } \mathcal{M}_A = \Xi_A$$

is the characteristic variety of the tableau, defined set-theoretically as

$$\Xi_A = \{[\xi] \in P : \ker \sigma_\xi \neq 0\}.$$

We may summarize by saying that

the characteristic sheaf of a tableau contains not only the information of where the symbol fails to be injective, but by how much it fails to be injective.

Involutive Sheaves. Returning to the general discussion we give the following

Definition 5.4. An *involutive sheaf* is a coherent sheaf $\mathcal{F}$ that has a presentation

$$(103) \qquad 0 \to \mathcal{F}_k(-k) \xrightarrow{\psi_k} \mathcal{F}_k(-(k-1)) \xrightarrow{\psi_{k-1}} \cdots \to$$
$$\mathcal{F}_1(-1) \xrightarrow{\psi_1} \mathcal{F}_0 \xrightarrow{\psi_0} \mathcal{F} \to 0$$

for some $k \leq n$ and where $\mathcal{F}_i = F_i \otimes_{\mathbb{C}} \mathcal{O}$ is a trivial vector bundle.

We shall give two remarks concerning this definition.

The first is that an involutive module has been defined (cf. Definition 4.2 above) to be one whose Koszul homology groups have a certain vanishing property. Now associated to a coherent sheaf are both its cohomology groups $H^i(\mathcal{F}(q)) = H^i(P, \mathcal{F}(q))$ and Koszul groups $K^i_{p,q}(\mathcal{F})$, the latter defined to be the middle homology of the 3-term complex

$$H^i(\mathcal{F}(p-1)) \otimes \Lambda^{q+1}V \to H^i(\mathcal{F}(p)) \otimes \Lambda^q V \to H^i(\mathcal{F}(p+1)) \otimes \Lambda^{q-1}V$$

(these are the usual Koszul groups for the graded S-module $\bigoplus_q H^i(\mathcal{F}(q))$). We may then equivalently define an involutive sheaf by the vanishing conditions

$$\begin{cases} H^i(\mathcal{F}(q)) = 0 & \text{for } i \neq 0,\, q \geq 0 \\ K^0_{p,q}(\mathcal{F}) = 0 & \text{for } p \geq 1,\, q \geq 0. \end{cases}$$

For our purposes, however, it is more convenient to take the existence of a presentation (103) as our definition.

Our second remark is that the presentation (103) is *not* unique. For example we consider the localized Koszul complex

$$(104) \quad 0 \to \Lambda^n V \otimes \mathcal{O}(-n) \xrightarrow{\partial} \ldots \xrightarrow{\partial} \Lambda^2 V \otimes \mathcal{O}(-2) \xrightarrow{\partial} V \otimes \mathcal{O}(-1) \xrightarrow{\partial} \mathcal{O} \to 0$$

obtained by choosing a basis $v_1, \ldots, v_n$ for $V = H^0(P, \mathcal{O}(1))$ and defining, as always,

$$\partial(v_{i_1} \wedge \cdots \wedge v_{i_q} \otimes f) = \sum_{\alpha=1}^q (-1)^{\alpha+1} v_{i_1} \wedge \cdots \wedge \hat{v}_{i_\alpha} \wedge \cdots \wedge v_{i_q} \otimes v_{i_\alpha} \cdot f,$$

for $1 \leq i_1, \ldots, i_q \leq n$ and $f \in \mathcal{O}(-q)$. Although not strictly necessary for our purposes, it may be shown that:

(105)   *For $\mathcal{F}$ involutive, any two resolutions (103)*

            differ by a direct sum of resolutions (104).

(106)   *If we define (103) to be minimal in case*

$$H^0(P, \mathcal{F}_0) \xrightarrow{\sim} H^0(P, \mathcal{F})$$

            is an isomorphism, then any involutive $\mathcal{F}$ has a

            unique minimal resolution (103) with $k \leq n - 1$.

We next observe that

$$(107) \qquad \textit{The characteristic sheaf of an involutive}$$
$$\textit{module } M \textit{ is an involutive sheaf.}$$

In fact, the localization of the canonical resolution (cf. Proposition 4.3 above)

$$0 \to E_n \xrightarrow{\varphi_n} E_{n-1} \xrightarrow{\varphi_{n-1}} \cdots \to E_1 \xrightarrow{\varphi_1} E_0 \xrightarrow{\varphi_0} M \to 0$$

of M gives a presentation

$$0 \to \mathcal{E}_n(-n) \xrightarrow{\varphi_n} \mathcal{E}_{n-1}(-(n-1)) \xrightarrow{\varphi_{n-1}} \cdots \to \mathcal{E}_1(-1) \xrightarrow{\varphi_1} \mathcal{E}_0 \xrightarrow{\varphi_0} \mathcal{M} \to 0$$

of the type (103).

We now shall prove the converse:

Proposition 5.5. *An involutive sheaf is the characteristic sheaf of an involutive module.*

Proof. What we want to do is take global sections of (103) tensored with $\mathcal{O}(r)$ for each $r \geq 0$. In order for this to work, we need (as always) a vanishing theorem.

Lemma 5.6. *Let $\mathcal{F}$ be an involutive sheaf. Then*
    i) $H^i(P, \mathcal{F}(q)) = 0$ *for* $i \geq 1$, $q \geq 0$;
    ii) $\mathcal{F}$ *is generated by its global sections; and*
    iii) $F_n^* \cong \ker\{H^0(P, \mathcal{F}_0) \to H^0(P, \mathcal{F})\}$ *(this is understood to be zero if* $k < n$ *in (103)).*

Proof. Although one may prove this result directly by induction on k, the argument is clearer if we use spectral sequences. For each r we obtain from (103) the long exact sheaf sequence

$$(108) \qquad \begin{aligned} 0 \to \mathcal{F}_k(-k+r) \to \mathcal{F}_{k-1}(-(k-1)+r) \to \cdots \to \\ \mathcal{F}_1(-1+r) \to \mathcal{F}_0(r) \to \mathcal{F}(r) \to 0 \end{aligned}$$

We may view (108) as a *complex of sheaves* whose *cohomology sheaves* are zero. Associated to any complex of sheaves are two spectal sequences both having the same abutment. Since the cohomology sheaves of (108) are trivial, one spectral sequence has its E_1 terms equal to zero and therefore the abutment is trivial. If we label the terms in the complex (108) with $\mathcal{F}_k(-k+r)$ corresponding to the index 0 and $\mathcal{F}(r)$ to the index $k+1$, then the other spectral sequence has

$$\begin{cases} E_1^{p,q} = H^q(P, \mathcal{F}_{k-p}(-(k-p)+r)) & 0 \leq p \leq k \\ E_1^{k+1,q} = H^q(P, \mathcal{F}(r)) \end{cases}.$$

We shall use the well-known fact that

$$(109) \qquad H^q(P, \mathcal{O}(s)) = \begin{cases} 0 & 1 \leq s \leq n-2 \text{ and all } q \\ 0 & q = n-1 \text{ and } s \geq -(n-1) \\ \mathbb{C} & q = n-1 \text{ and } s = -n \end{cases}.$$

For $r \geq 0$ the only possible non-zero terms in the spectral sequence are therefore

$$E_1^{p,0} = H^0(P, \mathcal{F}_{k-p}(-(k-p)+r)) \cong F_{k-p} \otimes S^{-k+p+r}V$$

$$E_1^{k+1,0} = H^0(P, \mathcal{F}(r))$$

$$E_1^{0,n-1} = H^{n-1}(P, \mathcal{F}_n(-n)) \cong F_n^* \text{ in case } k = n, \ r = 0$$

(here we recall our convention that $S^t V = (0)$ for $t < 0$). It follows that

i) $H^q(P, \mathcal{F}(r)) = 0$ for $q \geq 1$, $r \geq 0$;

ii) all differentials $d_t = 0$ for $t \geq 2$, except that when $k = n$, $r = 0$ we have a short exact sequence

$$0 \to F_n^* \xrightarrow{d_n} F_0 \xrightarrow{d_1} H^0(P, \mathcal{F}) \to 0;$$

iii) we have an exact sequence induced by the maps d_1

$$0 \to F_k \otimes S^{-k+r}V \to F_{k-1} \otimes S^{-k+1+r}V \to \cdots$$
$$\to F_1 \otimes S^{-1+r}V \to F_0^r \otimes S^r V \to H^0(P, \mathcal{F}(r)) \to 0$$

where $F_0^r = F_0$ for $r \geq 1$ and $F_0^0 = F_0/d_n F_n^*$ in case $k = n$, $r = 0$. The lemma is now clear.

To prove the proposition we define the graded S-module

$$F = \bigoplus_q F_q \text{ where}$$

$$F_q = H^0(P, \mathcal{F}(q)) \quad q \neq 0 \text{ and}$$

$$F_0 = H^0(P, \mathcal{F}) \oplus F_n.$$

Then from the long exact sequence in iii) just above we infer that F has a resolution

$$0 \to F_n \xrightarrow{\psi_n} F_{n-1} \xrightarrow{\psi_{n-1}} \cdots F_1 \xrightarrow{\psi_1} F_0 \xrightarrow{\psi_0} F \to 0$$

where $\deg \psi_0 = 0$ and $\deg \psi_i = 1$ for $i \geq 1$. In this way every involutive sheaf is the characteristic sheaf of an involutive module. $\qquad \square$

As discussed at the end of §4 above, any involutive module M has a unique decomposition

$$M = N \oplus H_{0,n}(M) \text{ where}$$

$$M_q = N_q \quad q \geq 1 \text{ and}$$

$$H_{0,n}(N) = 0.$$

The composite map

$$\begin{pmatrix} \text{involutive} \\ \text{modules} \end{pmatrix} \to \begin{pmatrix} \text{involutive} \\ \text{sheaves} \end{pmatrix} \to \begin{pmatrix} \text{involutive} \\ \text{modules} \end{pmatrix}$$

is given by

$$M \to N.$$

This is a consequence of statement iii) in the lemma above.

The above proof and discussion have the following corollary:

(110)　　　　i) *If $\mathcal{F}$ is an involutive sheaf, then the sheaf Euler characteristic satisfies*

(111)　　　　$$\chi(P, \mathcal{F}(q)) = H^0(P, \mathcal{F}(q)), \quad q \geq 0.$$

　　　　ii) *If M is an involutive module with characteristic sheaf $\mathcal{M}$, then*

(112)　　　　$$M = \bigoplus_{q \geq 0} H^0(P, \mathcal{M}(q)) \oplus H_{0,n}(M)$$

　　　　iii) *If $\mathcal{M}$ is an involutive sheaf, then the maps*

$$H^0(P, \mathcal{M}) \otimes S^q V \to H^0(P, \mathcal{M}(q))$$

are surjective for $q \geq 0$.

We are now ready to prove a number of results that have been previously stated above. We begin with

Proof of (96) *above.* Let $\mathcal{M}$ be an involutive sheaf and set

$$W^* = H^0(P, \mathcal{M})$$

$$V = H^0(P, \mathcal{O}(1))$$

$$A^* = H^0(P, \mathcal{M}(1)).$$

Define

$$\mu : W^* \otimes V \to A^*$$

to be the mapping on global sections induced by the sheaf mapping $\mathcal{M} \otimes_{\mathcal{O}} \mathcal{O}(1) \to \mathcal{M}(1)$. By iii) above μ is surjective, and setting $B = \ker \mu$ we have an exact vector space sequence

$$0 \to B \to W^* \otimes V \xrightarrow{\mu} A^* \to 0.$$

The dual of this is

$$0 \to A \to W \otimes V^* \xrightarrow{\sigma} B^* \to 0,$$

where A is our desired tableau and σ is a symbol mapping. By our discussion, if we start with the symbol σ_1 of an involutive tableau, construct the associated characteristic sheaf, and then construct from this characteristic sheaf the symbol σ as above, σ_1 differs from σ by a trivial symbol; i.e., σ_1 is equivalent to σ in the sense of Definition 4.5 above. $\qquad\square$

Proof of Theorem 3.6 in Chapter V. Let $A \subset W \otimes V^*$ be an involutive tableau with characters $s_1, \ldots, s_n$ and 1st prolongation $A^{(1)} \subset W \otimes S^2 V^*$. Then, respectively by definition and by Cartan's test,

$$(113_0) \qquad \dim A = s_1 + s_2 + \cdots + s_n$$

$$(113_1) \qquad \dim A^{(1)} = s_1 + 2s_2 + \cdots + ns_n.$$

The characters $s_1^{(1)}, \ldots, s_n^{(1)}$ of $A^{(1)}$ are uniquely determined by the relations

$$(114_1) \qquad s_1^{(1)} + \cdots + s_k^{(1)} = \dim A^{(1)} - \dim A_k^{(1)}, \quad k = 0, \ldots, n.$$

It follows that

$$(115_1) \qquad \dim A^{(1)} = s_1^{(1)} + \cdots + s_n^{(1)},$$

and also, as noted in equation (9) above,

$$(116_1) \qquad
\begin{aligned}
s_n^{(1)} &= s_n \\
s_{n-1}^{(1)} &= s_n + s_{n-1} \\
&\;\;\vdots \\
s_1^{(1)} &= s_n + s_{n-1} + \cdots + s_1.
\end{aligned}$$

Note that if A has character l and Cartan integer $\sigma = s_l$, then $A^{(1)}$ has the same character and Cartan integer. This will remain true for all the prolongations.

Now, by Proposition 2.5 above, $A^{(1)} \subset W \otimes S^2 V^*$ is again an involutive tableau (of order one) with 1st prolongation $A^{(2)} \subset W \otimes S^3 V^*$. By Cartan's test

$$\dim A^{(2)} = s_1^{(1)} + 2s_2^{(1)} + \cdots + ns_n^{(1)},$$

which by (116_1) gives

$$(113_2) \quad \dim A^{(2)} = s_1 + 3s_2 + \cdots + (n(n-1/2)s_{n-1} + (n(n+1)/2)s_n.$$

The characters $s_1^{(2)}, \ldots, s_n^{(2)}$ are given by formula (114_2), and then the relations (115_2) and (116_2) are valid. In general we will use (113_q)–(116_q) to denote the formulas (113)–(116) corresponding to $A^{(q)}$. In general the characters $s_i^{(q)}$ of $A^{(q)} \subset W \otimes S^{q+1} V^*$ are given by (114_q) and then (115_q) and (116_q) are satisfied. Recursively we therefore obtain our main formula

$$(113_q) \qquad\qquad \dim A^{(q)} = \sum_{k=1}^n \binom{k+q-1}{q} s_k.$$

Now let M_A be the involutive module associated to A with characteristic sheaf $\mathcal{M}_A$. Then for $q \geq 1$, by definition and by (112)

$$A^{(q)*} \cong (M_A)_{q+1} = H^0(P, \mathcal{M}(q+1)).$$

Combining this with (111) and (113_q) gives the following expression for the sheaf Euler characteristic

$$\chi(P, \mathcal{M}(q+1)) = \sum_{k=1}^n \binom{k+q-1}{q} s_k.$$

To complete the proof we shall compare the two sides of this formula for large q.

If the tableau A has character l and Cartan integer σ, thus $s_l = \sigma$ and $s_{l+1} = \cdots = s_n = 0$, then

$$(117) \qquad \sum_{k=1}^n \binom{k+q-1}{q} s_k = \frac{\sigma q^{l-1}}{(l-1)!} + \text{(lower order terms in } q\text{)}.$$

On the other hand we may evaluate the sheaf Euler characteristic $\chi(P, \mathcal{M}(q+1))$ by the Riemann–Roch formula (Fulton and Lang [1985]). (Of course, there are more elementary methods but this is perhaps the clearest conceptually.) If $\text{supp}\, \mathcal{M} = \Xi$ then it follows from that formula that

$$\chi(P, \mathcal{M}(q+1)) = \int_\Xi P(\mathcal{M}, q)$$

where $P(\mathcal{M}, q)$ involves the Chern classes of $\mathcal{M}$ and $\omega = c_1(\mathcal{O}(1))$ as they appear in the Chern character of $\mathcal{M}(q)$ twisted by the Todd class of Ξ. Explicitly

$$(118) \qquad P(\mathcal{M}, q) = \frac{\kappa q^{m-1} \omega^{m-1}}{(m-1)!} + (\text{lower order terms in } q)$$

where $m - 1 = \dim \Xi$ and κ is the fibre dimension of $\mathcal{M}$ over a general point of Ξ. From the preceeding four equations we infer first that $m = l$ and secondly that

$$\sigma = \kappa \int_\Xi \omega^{m-1} = \kappa \delta$$

where $\delta = \deg \Xi$. $\qquad\qquad\qquad\qquad\qquad\qquad\qquad\qquad\qquad\qquad\square$

We now turn to the proof of Proposition 3.10. From Corollary 3.8 what must be proved is this:[3]

$$(119) \qquad\qquad \textit{Let M be a quotient of a free module}$$

$$W^* \otimes SV. \textit{ Then there exists a}$$

$$p_0, \textit{ depending on } \dim W^* \textit{ and the Hilbert}$$

$$\textit{function } P_M(q) \textit{ of } M, \textit{ such that the Koszul}$$

$$\textit{homology groups}$$

$$H_{p,q}(M) = 0 \quad \text{for} \quad p \geq p_0,\ 0 \leq q \leq n = \dim V.$$

Referring to the discussion following the proof of Proposition 4.3 (cf. (i) and (iii) in that discussion), the statement (119) is an assertion about the minimal free resolution of M. More precisely, it is equivalent to

$$(120) \qquad\qquad\qquad B_{p,q} = 0 \quad \text{if} \quad q - p \geq p_0.$$

If R is the module of relations defined by

$$0 \to R \to W^* \otimes SV \to M \to 0,$$
$$\|$$
$$E_0$$

---

[3]We would like to thank Mark Green for help with this argument. His paper Green [1989a] serves as a general reference on Koszul groups, and Green [1989b] has a discussion related to finding an effectively computable bound for p_0. A weaker version of this appears in Goldschmidt [1968a] and in Goldschmidt [1974].

then clearly (120) for M implies the same statement for R, so that (119) follows from (and is in fact equivalent to) the assertion:

(121)   *Let $R \subset W^* \otimes SV$ be a sub-module*

          of a free module. Then there is a p_0,

          depending only on $\dim W^*$ *and the Hilbert*

          function $P_R(q) = \dim W^*.\ P_{SV}(q) - P_M(q)$,

          such that

$$H_{p,q}(R) = 0 \quad \text{for} \quad p \geq p_0,\ 0 \leq q \leq n.$$

We will prove (121) by a localization argument. Given a graded module we denote by $\mathcal{R}$ the corresponding sheaf and by $\bar{R} = \oplus \bar{R}_q$, $\bar{R}_q = H^0(\mathcal{R}(q))$ the module associated to the sheaf $\mathcal{R}$. $\bar{R}$ is usually called the *saturation* of R. There is always a module mapping $R \to \bar{R}$, and when R is a submodule of a free module this mapping is injective.

Proof. Suppose that $R \subset W^* \otimes SV$. Then $\mathcal{R} \subset \mathcal{W}^*$ where $\mathcal{W}^*$ is the trivial bundle with fibre W^*. It follows on the one hand that $\bar{R} \subset \overline{W^* \otimes SV}$, while on the other hand we infer from (109) that $\overline{W^* \otimes SV} = W^* \otimes SV$. From the commutative diagram

$$
\begin{array}{ccc}
 & 0 & \\
 & \downarrow & \\
0 \to R \to & W^* \otimes SV & \\
\downarrow & \quad \downarrow\approx & \\
0 \to \bar{R} \to & \overline{W^* \otimes SV} &
\end{array}
$$

we infer that $R \to \bar{R}$ is injective. $\qquad\qquad \square$

The main step in our proof of (121) is to show that a weaker statement is true when $R = \bar{R}$ is saturated. More precisely, recall that by definition the *Hilbert polynomial* $\chi_R(q)$ of a graded module R is the unique polynomial such that

$$\chi_R(q) = P_R(q)$$

for large q. Then it is clear that

$$\chi_R(q) = \sum_i (-1)^i \dim H^i(\mathcal{R}(q))$$

is the Euler characteristic of the twists of the localization of R. The weaker version of (121) is this:

(122)   *Let $R \subset W^* \otimes SV$ be a saturated*

          sub-module of a free module. Then there exists

          a p_0, depending only on $\dim W^*$ *and on*

          the coefficients of the Hilbert polynomial

          $\chi_R(q)$, such that

$$H_{p,q}(R) = 0 \quad \text{for} \quad p \geq p_0, \ 0 \leq q \leq n.$$

Assuming this result we will complete the proof of (121), and then we shall prove (122).

We consider the exact sequence

$$(123) \qquad\qquad 0 \to R \to \bar{R} \to \bar{R}/R \to 0.$$

Since $\bar{R}/R$ is a finite dimensional vector space over $\mathbb{C}$, it is clear that

$$\chi_R(q) = \chi_{\bar{R}}(q).$$

Thus we may apply (122) to the saturated module $\bar{R}$ to conclude that

$$(124) \qquad H_{p,q}(\bar{R}) = 0 \quad \text{for} \quad p \geq p_1 = p_1(\dim W^*, P_R(q)), \ 0 \leq q \leq n$$

where p_1 depends on $\dim W^*$ and $\chi_{\bar{R}}(q) = \chi_R(q)$. On the other hand, let p_2 be an integer such that

$$P_R(q) = P_{\bar{R}}(q) \quad \text{for} \quad q \geq p_2.$$

We will see below that there is a q_0 depending only on $\chi_{\bar{R}}(q) = \chi_R(q)$, and therefore only on $P_R(q)$, such that $\chi_{\bar{R}}(q) = P_{\bar{R}}(q)$ for $q \geq q_0$. Thus p_2 may be chosen to satisfy $p_2 \geq q_0$ and $P_R(q) = \chi_R(q)$ for $q \geq p_2$, and therefore $p_2 = p_2(P_R(q))$ may be assumed to also depend only on the Hilbert function of R. Then

$$(\bar{R}/R)_p = 0 \quad \text{for} \quad p \geq p_2(P_R(q)).$$

But then clearly

$$(125) \qquad\qquad H_{p,q}(\bar{R}/R) = 0 \quad \text{for} \quad p \geq p_2(P_R(q)).$$

Combining (124) and (125) and using the long exact homology sequence of (123) gives the desired result (121). It remains to give the

Proof of (122). We shall use the following.

Definition 5.7. A coherent sheaf $\mathcal{R}$ on $P = \mathbb{P}V^*$ is said to be m-regular in case

$$H^i(\mathcal{R}(m - i)) = 0 \quad \text{for} \quad i > 0.$$

The smallest m such that $\mathcal{R}$ is m-regular is called the regularity $m(\mathcal{R})$ of $\mathcal{R}$.

Lemma 5.8. *Let*

$$\begin{cases} 0 \to E_n \to E_{n-1} \to \cdots \to E_0 \to R \to 0 \\ E_p = \bigoplus_q B_{p,q} \otimes S^{[-q]} \end{cases}$$

be the minimal resolution of a saturated module R. Then for the localization $\mathcal{R}$ the regularity

$$m(\mathcal{R}) = \max\{q - p : B_{p,q} \neq 0\}.$$

Proof. We will prove the result in the first non-trivial case when the minimal resolution has two terms; the general case is proved by an analogous argument with spectral sequences replacing the long exact cohomology sequence—cf. Theorem 2.3 in Green [1989a].

We first note that the sheaf $\mathcal{O}$ on $\mathbb{P}^{n-1}$ is 0-regular but is not (-1)-regular; this follows from (109) above. In general, by using shifts we may reduce to considering the case of 0-regularity.

Suppose that

$$(125) \qquad 0 \to \bigoplus_p B_{1,p} \otimes S[-p] \xrightarrow{\varphi} \bigoplus_q B_{0,q} \otimes S[-q] \to R \to 0$$

is the minimal resolution of R. Writing $\varphi = \bigoplus_{p,q} \varphi_{p,q}$ where $\varphi_{p,q} : B_{1,p} \otimes S[-p] \to B_{0,q} \otimes S[-q]$ is given by a matrix of homogeneous polynomials of degree $p - q$, it follows trivially that $\varphi_{p,q} = 0$ for $p < q$ and by minimality that $\varphi_{p,p} = 0$.

Suppose first that $\max\{q - p : B_{p,q} \neq 0\} = 0$. Then by localizing (125) we have

$$(126) \quad 0 \to \bigoplus_p B_{1,p} \otimes \mathcal{O}(-p - i) \to \bigoplus_q B_{0,q} \otimes \mathcal{O}(-q - i) \to \mathcal{R}(-i) \to 0$$

where $B_{0,q} = 0$ for $q > 0$, $B_{1,p} = 0$ for $p > 1$. Then only non-trivial piece of the exact cohomology sequence is

$$(127) \quad \begin{aligned} 0 \to H^{n-2}(\mathcal{R}(-i)) \to \bigoplus_p H^{n-1}(B_{1,p} \otimes \mathcal{O}(-p - i)) \to \\ \bigoplus_q H^{n-1}(B_{0,q} \otimes \mathcal{O}(-q - i)) \to H^{n-1}(\mathcal{R}(i)) \to 0 \end{aligned}$$

Taking $i = n - 2$ we have $H^{n-1}(B_{1,p} \otimes \mathcal{O}(-p - n - 2)) = 0$ since $B_{1,p} = 0$ for $p > 1$, and taking $i = n - 1$ we have $H^{n-1}(B_{0,q} \otimes \mathcal{O}(-q - (n - 1))) = 0$ since $B_{0,q} = 0$ for $q > 0$. Thus $\mathcal{R}$ is 0-regular.

Conversely, suppose that $\mathcal{R}$ is 0-regular. From the lemma below we see that $\mathcal{R}$ is m-regular for $m \geq 0$. Let p_0 be the largest integer such that $B_{1,p_0} \neq 0$. Taking $i = n - p_0$ in (127) and using that $\varphi_{p_0,q} = 0$ for $q \geq p_0$ we obtain

$$0 \to H^{n-2}(\mathcal{R}(-n + p_0)) \xrightarrow{\sim} B_{1,p_0} \otimes H^{n-1}(\mathcal{O}(-n)) \to 0.$$

For $p_0 \geq 2$ we have $-n + p_0 = -(n-2) + m$ where $m \geq 0$ and so $H^{n-2}(\mathcal{R}(-n + p_0)) = 0$. Thus $B_{1,p} = 0$ for $p \geq 2$. Similarly, let q_0 be the largest integer such that $B_{0,q} \neq 0$. If $q_0 \geq 1$ then take $i = n - q_0$ in (127) and use that $\varphi_{p,q_0} = 0$ for $p \geq 1$ to obtain

$$0 \to B_{0,q_0} \otimes H^{n-1}(\mathcal{O}(-n)) \xrightarrow{\sim} H^{n-1}(\mathcal{R}(-n + q_0)) \to 0.$$

Writing $-n + q_0 = -(n-1) + m$ where $m \geq 0$, this last term is zero by the m-regularity of $\mathcal{R}$. Thus $m(\mathcal{R}) = 0$ as desired. $\square$

Referring to (iii) in the discussion following the proof of Proposition 4.2, we have

$$(128) \qquad m(\mathcal{R}) = \max\{p : H_{p,q}(R) \neq 0 \text{ for some } q\}.$$

Thus our desired result (122) follows from the following.

Proposition 5.9. *Let $\mathcal{R} \subset \mathcal{W}^*$ be a coherent subsheaf of the trivial sheaf $\mathcal{W}^*$ on P. Then there exists a polynomial $f(s, a_i)$ that depends on $s = \dim W^*$ and the coefficients a_i of the Hilbert polynomial $\chi(R(q)) = \sum a_i \binom{q}{i}$ such that the regularity*

$$m(\mathcal{R}) \leq f(s, a_i).$$

Proof (cf. Mumford [1966], Chapter XIV). We shall give the argument when $\mathcal{R} \subset \mathcal{O}$; the general case is the same. Then $\mathcal{R}$ is a sheaf of ideals on $P \cong \mathbb{P}^{r-1}$. The proof will be by induction, and we choose a general hyperplane $H \subset P$ defined by $h \in H^0(\mathcal{O}_P(1))$. Then we have the sequence

$$0 \to \mathcal{R}(-1) \xrightarrow{h} \mathcal{R} \to \mathcal{R} \otimes_{\mathcal{O}_P} \mathcal{O}_H \to 0$$
$$\|$$
$$\mathcal{R}_H$$

which is injective on the left. A local argument shows that the sequence is in fact exact and that $\mathcal{R}_H \subset \mathcal{O}_H$ is a subsheaf. Tensoring with $\mathcal{O}_P(m+1)$ we obtain

$$(129) \qquad 0 \to \mathcal{R}(m) \to \mathcal{R}(m+1) \to \mathcal{R}_H(m+1) \to 0,$$

and therefore

$$\chi(\mathcal{R}_H(m+1)) = \chi(\mathcal{R}(m+1)) - \chi(\mathcal{R}(m))$$
$$= \sum_{i=0}^{n-1} a_i \left[\binom{m+1}{i} - \binom{m}{i} \right]$$
$$= \sum_{i=0}^{n-2} a_{i+1} \binom{m}{i}.$$

The induction assumption applies to $\mathcal{R}_H$, so we may assume that it is $g(a_i)$-regular for a suitable polynomial g depending only on n; we set $m_1 = g(a_i)$. Then the exact cohomology sequences of (129) give

$$(130') \qquad 0 \to H^0(\mathcal{R}(m)) \to H^0(\mathcal{R}(m+1)) \xrightarrow{\rho_{m+1}} H^0(\mathcal{R}_H(m+1)) \to$$
$$H^1(\mathcal{R}(m)) \to H^1(\mathcal{R}(m+1)) \to 0$$

for $m \geq m_1 - 2$, and

$$(130'') \qquad 0 \to H^i(\mathcal{R}(m)) \to H^i(\mathcal{R}(m+1)) \to 0$$

for $i \geq 2$, $m \geq m_1 - 2$. Since $H^i(\mathcal{R}(m+1)) = 0$ for m sufficiently large and $i \geq 1$ this last sequence gives

$$(131) \qquad H^i(\mathcal{R}(m)) = 0 \qquad i \geq 2, \ m \geq m_1 - 2.$$

Turning to H^1, from $(130')$ we see that if $m \geq m_1 - 2$ then either ρ_{m+1} is surjective or else $h^1(\mathcal{R}(m)) > h^1(\mathcal{R}(m+1))$. Since $h^1(\mathcal{R}(m)) < \infty$ there is an m_2 such that $m_2 \geq m_1$ and ρ_{m_2} is surjective. In a moment we will show that ρ_{m_2+1} is also surjective, from which it follows that, for $m \geq m_1$, $h^1(\mathcal{R}(m))$ is strictly decreasing as a function of m until it reaches zero. Then clearly

$$\mathcal{R} \text{ is } [m_1 + h^1(\mathcal{R}(m_1))]\text{-regular.}$$

But

$$h^1(\mathcal{R}(m_1)) = h^0(\mathcal{R}(m_1)) - \chi(\mathcal{R}(m_1))$$
$$\leq h^0(\mathcal{O}_P(m_1)) - \chi(\mathcal{R}(m_1))$$
$$= h(a_i, m_1)$$
$$= f(a_i)$$

where h and f are suitable polynomials.

It remains to show that ρ_{m_2+1} is surjective. By the lemma below,

$$H^0(\mathcal{O}_P(1)) \otimes H^0(\mathcal{R}_H(m_2)) \to H^0(\mathcal{R}_H(m_2+1))$$

is surjective. Thus the horizontal composite map in the commutative diagram

$$H^0(\mathcal{O}_P(1)) \otimes H^0(\mathcal{R}(m_2)) \xrightarrow{1 \otimes \rho_{m_2}} H^0(\mathcal{O}_P(1)) \otimes H^0(\mathcal{R}_H(m_2)) \to H^0(\mathcal{R}_H(m_2+1))$$
$$\downarrow$$
$$H^0(\mathcal{R}(m_2+1)) \xrightarrow{\quad \rho_{m_2+1} \quad}$$

is surjective, and it then follows that ρ_{m_2+1} is surjective. $\qquad \square$

We will be done once we prove the

Lemma 5.10. *If $\mathcal{R}$ is m-regular, then for $k > m$*

(i) $H^0(\mathcal{O}_P(1)) \otimes H^0(\mathcal{R}(k)) \to H^0(\mathcal{R}(k+1))$ *is onto; and*

(ii) $\mathcal{R}$ *is k-regular.*

Proof. From the exact sequence (129) for $m - i - 1$ we obtain

$$H^i(\mathcal{R}(m - i)) \to H^i(\mathcal{R}_H(m - i)) \to H^{i+1}(\mathcal{R}(m - i - 1)),$$

and so $\mathcal{R}_H$ is m-regular. We may apply an induction hypothesis to conclude (i) and (ii) for $\mathcal{R}_H$. From the exact sequence (129) for $m - i$

$$H^{i+1}(\mathcal{R}(m - i - 1)) \to H^{i+1}(\mathcal{R}(m - i)) \to H^{i+1}(\mathcal{R}_H(m - i)), \qquad i \geq 0,$$

and from (ii) for $\mathcal{R}_H$ we see that $\mathcal{R}$ is $(m + 1)$-regular. From the diagram

$$
\begin{array}{ccccc}
H^0(\mathcal{R}(k{-}1))\otimes H^0(\mathcal{O}_P(1)) & \xrightarrow{\alpha} & H^0(\mathcal{R}_H(k{-}1))\otimes H^0(\varphi_P(1)) \to H^1(\mathcal{R}(k{-}2))\otimes H^0(\mathcal{O}_P(1)) \\
\big\downarrow{\scriptstyle\mu} & & \big\downarrow{\scriptstyle\beta} \\
H^0(\mathcal{R}(k{-}1)) \xrightarrow{h} H^0(\mathcal{R}(k)) \to & & H^0(\mathcal{R}_H(k))
\end{array}
$$

we see that α is surjective for $k > m$, and by (i) for $\mathcal{R}_H$, β is also surjective for $k > m$. But then $H^0(\mathcal{R}(k))$ is spanned by image μ + image h, while clearly image $h \subseteq$ image μ. $\qquad\square$

§6. Proof of Theorem 3.15 in Chapter V; Guillemin's Normal Form.

It is clear that Theorem 3.15 in Chapter V is a purely algebraic result dealing with a tableau $A \subset W \otimes V^*$. Accordingly, we shall reformulate the result in a purely algebraic manner and then prove the reformulated version.

Setting

$$U = W \otimes V^*/A$$

we consider the symbol mapping

$$\sigma : W \otimes V^* \to U$$

and define

$$
(132) \qquad
\begin{aligned}
\Xi_p &= \{\Omega \in G_{n-p}(V^*) : \sigma : W \otimes \Omega \to U \text{ fails to be injective}\} \\
&= \{\Omega \in G_{n-p}(V^*) : A \cap W \otimes \Omega \neq 0\}.
\end{aligned}
$$

It is clear that

$$\Xi_{n-1} = \Xi_A \subset \mathbb{P}V^*$$

is the characteristic variety of the tableau. Using the projective duality isomorphism

$$G_{n-p}(V^*) \xrightarrow{\approx} G_p(V)$$
$$\downarrow \qquad\qquad \downarrow$$
$$\Omega \qquad \rightarrow \quad \Omega^{\perp}$$

we define in the obvious way

$$\Xi_p^{\perp} \subset G_p(V).$$

On the other hand, in Definition 3.14 in Chapter V we have defined

$$\Lambda_p \subset G_p(V),$$

and we shall show that:

(133)   *If the tableau A has character l, then*

$$\Xi_p^{\perp} = \Lambda_p \ \text{for } l \leq p \leq n-1.$$

Proof. We let $\{w_a\}$, $\{v_i\}$ and

$$z_\varepsilon = A_{\varepsilon i}^a w_a \otimes v_i^*$$

be bases for W, V, and A respectively. Setting

$$\pi_i^a = A_{\varepsilon i}^a z_\varepsilon^* \in A^*$$

we consider the tableau matrix

$$\pi = \left\| \begin{matrix} \pi_1^1 & \cdots & \pi_n^1 \\ \vdots & & \\ \pi_1^s & \cdots & \pi_n^s \end{matrix} \right\|.$$

Using the additional index ranges

$$\begin{cases} 1 \leq \lambda,\ \mu \leq l \\ l+1 \leq \rho,\ \sigma \leq n, \end{cases}$$

the character l of the tableau A is the smallest integer with the following property: If the basis $\{v_i\}$ is chosen generically, then

(134) $$\pi_\rho^a \equiv 0 \operatorname{modulo}\{\pi_\lambda^b\}.$$

That is, *the $\pi_\rho^a \in A^*$ in the last $n - l$ columns of π should be linear combinations of the π_λ^b in the first l columns.* Clearly (134) is equivalent to

$$(135) \qquad \text{all } \pi_\lambda^b(\psi) = 0 \Rightarrow \text{all } \pi_\rho^a(\psi) = 0, \text{ for any } \psi \in A.$$

Now let $\Omega \in G_{n-l}(V^*)$ and choose a basis $\{v_i\}$ for V so that

$$\Omega = \text{span}\{v_\rho^*\},$$

or equivalently

$$\Omega^\perp = \text{span}\{v_\lambda\}.$$

Recalling that if $s_1, \ldots, s_n$ denote the characters of A and $s(\Omega)$ denotes the dimension of $\text{span}\{\pi_\lambda^b\} \subset A^*$,

$$\Omega \in \Lambda_l \Leftrightarrow s(\Omega) < s_1 + \cdots + s_l$$

$$\Leftrightarrow (134) \text{ fails to be true}$$

$$\Leftrightarrow \left\{ \begin{array}{c} \text{there exists } 0 \neq \psi \in A \text{ with} \\ \text{all } \pi_\lambda^b(\psi) = 0 \text{ but some } \pi_\rho^a(\psi) \neq 0 \end{array} \right\}.$$

But then

$$(136) \qquad 0 \neq \pi_i^a(\psi)w_a \otimes v_i^* \in A \cap W \otimes \Omega.$$

Conversely, any element of A is of this form for a suitable ψ and the condition that (136) lie in $W \otimes \Omega$ is that all $\pi_\lambda^b(\psi) = 0$. This proves (133) when $p = l$, and it is clear that the same argument works for $l \le p \le n-1$. $\square$

We note that (133) does not depend on the involutivity of A.

Using (133), it is a small and straightforward exercise to reformulate Theorem 3.15 in Chapter V as follows:

Theorem 6.1. *For an involutive tableau $A \subset W \otimes V^*$ of character l, the following are equivalent conditions on $\Omega \in G_{n-l}(V^*)$:*
  i) $A \cap W \otimes \Omega \neq 0$
  ii) *for some line $L_\xi \subset \Omega$*

$$A \cap W \otimes L_\xi \neq 0.$$

Here, L_ξ is the line in V^* corresponding to $[\xi] \in \mathbb{P}V^*$, and $A \cap W \otimes L_\xi \neq 0$ is just our condition (132) that $[\xi] \in \Xi_A$. On the other hand, $A \cap W \otimes \Omega \neq 0$ is the condition (132) that $\Omega \in \Xi_l$. It is clear that even without assuming that A is involutive, ii) $\Rightarrow$ i), and what we have to do is to show that i) $\Rightarrow$ ii). This requires involutivity and amounts to showing that

If $A \cap W \otimes \Omega \neq 0$, then $A \cap W \otimes \Omega$ contains a non-zero decomposable vector $w \otimes \xi$. Equivalently, we must show that:

$$(137) \qquad \mathbb{P}\Omega \cap \Xi_A = \emptyset \Rightarrow \sigma : W \otimes \Omega \to U \text{ is injective.}$$

We shall formulate a stronger result than (137), and shall then prove this stronger result by a localization argument.

Let M be an involutive graded S-module with canonical resolution

$$0 \to \mathbf{E}_n \to \mathbf{E}_{n-1} \to \cdots \to \mathbf{E}_1 \to \mathbf{E}_0 \to M \to 0$$

having localization

$$(138) \qquad \begin{aligned} 0 \to \mathcal{E}_n(-n) &\xrightarrow{\psi_n} \mathcal{E}_{n-1}(-(n-1)) \xrightarrow{\psi_{n-1}} \cdots \to \\ \mathcal{E}_1(-1) &\xrightarrow{\psi_1} \mathcal{E}_0 \xrightarrow{\psi_0} \mathcal{M} \to 0. \end{aligned}$$

Here $\mathcal{M}$ is the characteristic sheaf of M and $\mathcal{E}_i$ is the trivial vector bundle with fibre $E_i = H_{0,i}(M)$ where $\mathbf{E}_i = E_i \otimes SV$. For any k-plane $\Omega \in G_k(V^*)$ we may restrict all sheaves to $\mathbb{P}\Omega \cong \mathbb{P}^{k-1}$, and then

$$H^0(\mathbb{P}\Omega, \mathcal{E}_i(q)) \cong E_i \oplus S^q\Omega^*.$$

Hence there are induced maps

$$E_i \otimes S^q\Omega^* \xrightarrow{\psi_i} E_{i-1} \otimes S^{q+1}\Omega^*, \quad q \geq 0.$$

Proposition 6.2. *If $\mathbb{P}\Omega \cap \operatorname{supp}\mathcal{M} = \emptyset$, then for each $l \geq 1$ the sequence*

$$(139) \qquad E_l \to E_{l-1} \otimes \Omega^* \to \cdots \to E_1 \otimes S^{l-1}\Omega^* \to E_0 \otimes S^l\Omega^* \to 0$$

is exact.[4]

The case $l = 1$ is the exact sequence

$$E_1 \to E_0 \otimes \Omega^* \to 0.$$

In case $M = M_A$ is the graded module associated to be involutive tableau A, then $\operatorname{supp}\mathcal{M} = \Xi_A$ is the characteristic variety of the tableau and the above sequence is

$$B \xrightarrow{\sigma^*} W^* \otimes \Omega^* \to 0,$$

---

[4]The l used in this proof has nothing to do with the character of the tableau A.

which dualizes to

$$0 \to W \otimes \Omega \xrightarrow{\sigma} U.$$

Thus, Proposition 6.2 in the case $l = 1$ implies (137).

Proof of Proposition 6.2. We set

$$\Xi = \operatorname{supp} \mathcal{M}$$

and will prove the exactness of (139) first where $\mathbb{P}\Omega$ is a point in V^* and then in general.

$\dim \Omega = 1$. In this case $\mathbb{P}\Omega$ is a point $[\xi] \in \mathbb{P}V^* \backslash \Xi$. Denoting by $\mathcal{O}_\xi$ the local ring of $\mathcal{O}_{\mathbb{P}V^*}$ at $[\xi]$, the exact stalk sequence of (138) at $[\xi]$ is the following exact sequence of $\mathcal{O}_\xi$-modules

$$(140) \quad 0 \to \mathcal{E}_n(-n)_\xi \to \mathcal{E}_{n-1}(-(n-1))_\xi \to \cdots \to \mathcal{E}_1(-1)_\xi \to \mathcal{E}_{0\xi} \to 0.$$

Here we are using that

$$\mathcal{M}_\xi = 0 \text{ if } [\xi] \notin \operatorname{supp} \mathcal{M}.$$

Now $\mathcal{E}_i(-q)$ is a vector bundle whose fibre over $[\xi] \in \mathbb{P}V^*$ is $E_i \otimes L_\xi^{-q}$. Thus

$$\mathcal{E}_i(q)_\xi = E_i \otimes_{\mathbb{C}} L_\xi^{-q} \otimes_{\mathbb{C}} \mathcal{O}_\xi,$$

and from Nakayama's lemma it follows that (140) gives the exact sequence of vector spaces

$$0 \to E_n \otimes L_\xi^{-n} \to E_{n-1} \otimes L_\xi^{-(n-1)} \to \cdots \to E_1 \otimes L_\xi^{-1} \to E_0 \to 0.$$

Tensoring this sequence with L_ξ^l gives the exactness of (139) for all l when $\dim \Omega = 1$.

This case is due to Quillen [1964]. According to the discussion centered around (87) and (88) above it may be rephrased as follows: *for non-characteristic covectors, the symbol sequence of the Spencer sequence associated to an involutive linear* P.D.E. *system is exact.*

Remark. If we set

$$h_i = \dim E_i = \dim H_{0,i}(M),$$

then from

$$E_2 \otimes L_\xi^{-2} \to E_1 \otimes L_\xi^{-1} \to E_0 \to 0$$

we infer that

$$(141) \qquad\qquad h_2 \geq h_1 - h_0.$$

This inequality, a special case of which is due to Cartan (cf. the reference in footnote 2), has the following interpretation for an involutive, constant coefficient linear P.D.E. system

$$(142) \qquad B_a^{\lambda i} \frac{\partial u^a(x)}{\partial x^i} = f^\lambda(x).$$

Here $1 \le a \le s =$ number of unknown functions and $1 \le \lambda \le t =$ number of equations. We have

$$h_0 = s, \quad h_1 = t,$$

and so if the system (142) is *determined* or *underdetermined* then (141) doesn't say anything. However, suppose we are in the *overdetermined* case

$$t > s.$$

Setting

$$D^\lambda = B_a^{\lambda i} w_a^* \otimes \partial/\partial x^i,$$

h_2 is the number of linearly independent, 1st order constant coefficient relations (cf. (86) above)

$$(143) \qquad m_\lambda^j \partial/\partial x^j D^\lambda = 0.$$

From (141) we have that:

The number of relations (143) is $\ge t - s$.

As discussed in §4 above, the relations (143) give the 1st order compatibility conditions

$$m_\lambda^j \frac{\partial f^\lambda(x)}{\partial x^j} = 0$$

for the formal solvability of (142). By involutiveness, all the compatibility conditions are 1st order.

Returning to the proof of Proposition 6.2 in general, we suppose that $\dim \Omega = k$ so that $\mathbb{P}\Omega \cong \mathbb{P}^{k-1}$. Over $\mathbb{P}\Omega$ we have the exact complex

$$(144) \quad 0 \to \mathcal{E}_n(-n+l) \to \mathcal{E}_{n-1}(-n+l+1) \to \cdots \to \mathcal{E}_1(l-1) \to \mathcal{E}_0(l) \to 0$$

where $\mathcal{E}_k(m)$ is $E_k \otimes \mathcal{O}_{\mathbb{P}\Omega}(m)$. For $-n \le j \le 0$ we set $\mathcal{F}_j = \mathcal{E}_{-j}(j+l)$ so that (144) becomes the exact complex[5]

$$(145) \qquad 0 \to \mathcal{F}_{-n} \to \mathcal{F}_{-n+1} \to \cdots \to \mathcal{F}_{-1} \to \mathcal{F}_0 \to 0.$$

---

[5]This argument is similar to the proof of Proposition 5.5 above except that we have chosen to use a different indexing convention.

We note that by (109) above

$$(146) \qquad H^q(\mathbb{P}\Omega, \mathcal{F}_j) = \begin{cases} 0 & q \neq 0, \ k-1 \\ 0 & q = k-1, \ j \geq -k-l-1 \\ 0 & q = 0, \ j \leq -l-1 \end{cases}.$$

Associated to the complex of sheaves (145) are two spectral sequences both abutting to the hypercohomology of the complex of sheaves. Since (145) is exact one of these has E_1 term equal to zero. Thus the other spectral sequence abuts to zero.

Now by our indexing convention both spectral sequences are in the second quadrant. The one with non-zero E_1 term has

$$E_1^{p,q} = H^q(\mathbb{P}\Omega, \mathcal{F}_p)$$

where $-n \leq p \leq 0$ and $0 \leq q \leq k-1$. By (146) the only non-zero terms are

$$\begin{cases} E_1^{p,k-1}, & p \leq -k-l \\ E_1^{p,0}, & p \geq -l. \end{cases}$$

From this we see that the only possible non-zero differentials are

$$\begin{aligned} d_1 &: E_1^{p,0} \to E_1^{p+1,0} & p \geq -l \\ d_1 &: E_1^{p,k-1} \to E_1^{p+1,k-1} & p \leq -k-l-1 \\ d_k &: E_k^{-k-l,k-1} \to E_k^{-l,0}. \end{aligned}$$

It follows that

$$E_2^{p,q} = 0 \text{ unless } \begin{cases} p = -k-l, & q = k-1 \\ p = -l, & q = 0 \end{cases}$$

$$E_2 = \cdots = E_k$$

$$E_{k+1} = 0.$$

In particular, the complex

$$H^0(\mathbb{P}\Omega, \mathcal{F}_{-l}) \xrightarrow{d_1} H^0(\mathbb{P}\Omega, \mathcal{F}_{-l+1}) \to \cdots \to H^0(\mathbb{P}\Omega, \mathcal{F}_{-1}) \to H^0(\mathbb{P}\Omega, \mathcal{F}_0) \to 0$$

is exact. This implies Proposition 6.2. $\qquad\square$

Guillemin's Normal Form. Victor Guillemin established a "normal form" for the Spencer complex of an involutive linear P.D.E. system (cf. Guillemin [1968]). This result is closely related to the special case $l = 2$ of Proposition 6.2 and so we should like to discuss it here.

Let $A \subset W \otimes V^*$ be an involutive tableau, considered as the image of an injective linear mapping

$$\pi : A \to W \otimes V^*.$$

In terms of bases $\{w_a\}$ for W and $\{x^i\}$ for V^* we write

$$\pi = \pi_i^a w_a \otimes x^i$$

where $\pi_i^a \in A^*$. We will establish a certain normal form for the symbol relations on the π_i^a's. This normal form will in fact correspond to the symbol relations when the tableau is put in the normal form given by equation (90) in Chapter IV. Referring to that discussion, the forms π_i^a for $a < s_i$ are given by linear equations

$$(147) \qquad\qquad \pi_i^a = \sum_{\substack{b \leq s_j \\ j \leq i}} B_{ib}^{aj} \pi_j^b.$$

Involutivity of the tableau will have strong commutation properties on matrices derived from the $\|B_{ib}^{aj}\|$ above. For example, when $s_1 = \cdots = s_l = s$ and $s_{l+1} = \cdots = s_n = 0$, the above equations reduce to

$$(147') \qquad\qquad \pi_\rho^a = B_{\rho b}^{a\lambda} \pi_\lambda^b$$

where $1 \leq \lambda \leq l$ and $l + 1 \leq \rho, \sigma \leq n$. We have seen in section 5 of Chapter IV that involutivity is equivalent to the commutation relations

$$[B_\rho(\xi), B_\sigma(\xi)] = 0 \quad \text{for all } \xi$$

where

$$B_\rho(\xi) = \|B_{\rho b}^{a\lambda} \xi_\lambda\|.$$

There will be an analogous statement in the general case, one that will be given following a general discussion.

The geometric picture is this: If A has character l then the complex characteristic variety Ξ_A has dimension $l - 1$, and therefore by a generic linear projection may be realized as a finite branched covering over a $\mathbb{P}^{l-1}$. We shall then explore how this representation may be used to at least partially normalize the relations (147).

To carry this out we let $\Omega \subset V^*$ be a maximal, non-characteristic subspace. Then $\dim V^* = n$, $\dim \Omega = n - l$ and

$$\mathbb{P}\Omega \cap \Xi_A = \emptyset.$$

Set $E = \Omega^\perp \subset V$ so that $E^* \cong V^*/\Omega$. Then $\mathbb{P}E^* \cong \mathbb{P}^{l-1}$ and by linear projection there is a diagram

$$
\begin{array}{ccc}
\mathbb{P}V^* \backslash \mathbb{P}\Omega & \supset & \Xi_A \\
\downarrow \tilde{\omega}' & & \downarrow \tilde{\omega} \\
\mathbb{P}E^* & = & \mathbb{P}E^*.
\end{array}
$$

where the map $\Xi_A \xrightarrow{\tilde{\omega}} \mathbb{P}E^*$ is a finite branched covering map. If $\xi \in \mathbb{P}E^*$ and we set (the following has intrinsic meaning)

$$\{\Omega, \xi\} = \operatorname{span} \Omega, \xi,$$

then $\{\Omega, \xi\} \cong \mathbb{P}^{n-l}$ and we have

$$
\begin{cases}
\overline{\tilde{\omega}'^{-1}(\xi)} = \{\Omega, \xi\} \\
\tilde{\omega}^{-1}(\xi) = \{\Omega, \xi\} \cdot \Xi_A.
\end{cases}
$$

Choose our basis $\{x^i\}$ for V^* so that $\Omega = \operatorname{span}\{x^{l+1}, \ldots, x^n\}$. We keep the additional index ranges

$$
\begin{cases}
1 \leq \lambda, \ \mu \leq l \\
l+1 \leq \rho, \ \sigma \leq n
\end{cases}
$$

and write points in V^* as

$$\zeta = (\xi, \eta)$$

$$= \xi_\lambda x^\lambda + \eta_\rho x^\rho;$$

i.e., in homogeneous coordinates

$$\zeta = [\xi_1, \ldots, \xi_l; \eta_{l+1}, \ldots, \eta_n]$$

(from now on we drop the bracket around points ξ in a projective space). Then with this notation

$$\tilde{\omega}(\zeta) = \xi.$$

Since Ω is non-characteristic, we have

$$(148) \qquad\qquad A \cap W \otimes \Omega = 0.$$

Elements in $W \otimes \Omega$ are of the form $\psi = \psi^a_\rho w_a \otimes x^\rho$. Thus (148) is equivalent to (cf. (135) above)

$$\pi^a_\lambda(\psi) = 0 \Rightarrow \pi^a_\rho(\psi) = 0, \quad \psi \in A.$$

In other words, the basis $\{x^i\}$ for V^* has the property that $\Omega = \text{span}\{x^{l+1}, \ldots, x^n\}$ is non-characteristic if, and only if, the forms π^a_i are all linear combinations of the π^a_λ for $1 \leq \lambda \leq l$, in which case we have the relations (147) among the symbol relations.

We now recall from §5 of Chapter IV that, in the involutive case, we may assume that

$$\pi^a_i = 0 \quad \text{for} \quad a > s_1.$$

Thus we may as well assume that $s_1 = s$, and then the equations (147) give in particular relations (147'). We shall derive a general commutation property of the matrices $B_\rho(\xi)$. For this we let $\xi \in E^*$ and consider expressions

$$(149) \qquad w \otimes \xi + w_\rho \otimes x^\rho \in A \cap (W \otimes \{\Omega, \xi\}).$$

We note that, by (148), the tensor $w_\rho \otimes x^\rho \in W \otimes \Omega$ is uniquely determined by the property that $w \otimes \xi + w_\rho \otimes x^\rho \in A$, and we define

$$W_\xi = \{w \in W : \text{ there exists an expression (149)}\}.$$

Clearly $W_\xi \subset W$ is a linear subspace, and we shall show that for any tensor (149)

$$(150) \qquad\qquad B_\rho(\xi)w = w_\rho$$

where $B_\rho(\xi)$ is the linear transformation associated to the matrix (147').

Proof. If $w = \mu^a w_a \in W_\xi$, then for some $\psi \in A$ we have

$$\pi(\psi) = w \otimes \xi + w_\rho \otimes x^\rho$$

$$= \mu^a \xi_\lambda w_a \otimes x^\lambda + \mu^a_\rho w_a \otimes x^\rho$$

where we have set $w_\rho = \mu^a_\rho w_a$ and the ξ_λ are the components of ξ. This gives

$$\pi^a_\lambda(\psi) = \mu^a \xi_\lambda$$

$$\pi^a_\rho(\psi) = \mu^a_\rho.$$

But then (147') gives

$$\mu^a_\rho = B^{a\lambda}_{\rho b} \mu^b \xi_\lambda,$$

i.e.,

$$B_\rho(\xi)w = w_\rho.$$

We shall now show that

Proposition 6.3. *With the above notations*

$$(151) \qquad\qquad B_\rho(\xi)W_\xi \subseteq W_\xi$$

$$(152) \qquad\qquad [B_\rho(\xi), B_\sigma(\xi)]|_{W_\xi} = 0.$$

Proof. We shall use the exact sequence (139), or rather its dual. We recall the dual sequence (88) in §3 above of the canonical resolution of an involutive module, which using the definition of the Spencer cohomology groups may be written as

$$0 \to A^{(q)} \to W \otimes S^{q+1}V^* \xrightarrow{\bar\delta} \left(\frac{W \otimes V^*}{A}\right) \otimes S^q V^* \xrightarrow{\bar\delta} \left(\frac{W \otimes \Lambda^2 V^*}{\delta(A \otimes V^*)}\right) \otimes S^{q-1}V^*.$$

Here, for $\psi \in W \otimes \Lambda^i V^*$ and $P \in S^{q+1-i}V^*$ the mapping $\bar\delta$ is given by

$$(153) \qquad \bar\delta(\psi \otimes P) = \sum_i \psi \wedge dx^i \otimes \frac{\partial P}{\partial x^i} \quad \mathrm{mod}\, \delta(A \otimes \Lambda^i V^*)$$

(cf. the proof of (90) in §3 above). For $\Omega \subset V^*$ a non-characteristic subspace, the dual of (139) above when $l = 2$ is

$$(154) \qquad 0 \to W \otimes S^2\Omega \xrightarrow{\bar\delta} \left(\frac{W \otimes V^*}{A}\right) \otimes \Omega \xrightarrow{\bar\delta} \left[\frac{W \otimes \Lambda^2 V^*}{\delta(A \otimes V^*)}\right].$$

By (153) we have for $\psi \in W \otimes \Lambda^i V^*$ and $P(x)$ depending only on the last $(n - l)$-variables, i.e., $P \in S\Omega$,

$$\bar\delta(\psi \otimes P(x^{l+1}, \ldots, x^n)) = \sum_\rho \psi \wedge dx^\rho \otimes \frac{\partial P}{\partial x^\rho}.$$

Suppose now that

$$w \otimes \xi + w_\rho \otimes x^\rho \in A \subset W \otimes V^*.$$

Taking the exterior product with ξ gives

$$w_\rho \otimes x^\rho \wedge \xi \equiv 0 \quad \mathrm{mod}\, \delta(A \otimes V^*)$$

(here we are using (153) above when $P = \xi$). But

$$w_\rho \otimes x^\rho \wedge \xi = -\bar\delta((w_\rho \otimes \xi) \otimes x^\rho)$$

where $(w_\rho \otimes \xi) \otimes x^\rho \in W \otimes V^* \otimes \Omega$. By the exactness of (154) we have

$$(w_\rho \otimes \xi) \otimes x^\rho = \bar{\delta}(\tfrac{1}{2} w_{\rho\sigma} x^\rho x^\sigma) \in \left[\frac{W \otimes V^*}{A}\right] \otimes \Omega$$

where

$$w_{\rho\sigma} = w_{\sigma\rho} \in W.$$

It follows that

$$\begin{cases} w_\rho \otimes \xi \equiv w_{\rho\sigma} x^\sigma \text{ modulo } A \\ w_{\rho\sigma} = w_{\sigma\rho} \end{cases}.$$

The first of these equations gives (151) and the second gives (152). $\square$

We may now complete our description of the branched covering $\tilde{\omega} : \Xi_A \to \mathbb{P}E^*$. Given $\xi \in PE^*$ there are points

$$[\xi, \eta(\xi)] \in \Xi_A \cap \tilde{\omega}^{-1}(\xi).$$

Here, we are writing $\eta(\xi)$ to express the fact that via the finite branched covering mapping $\Xi_A \to \mathbb{P}E^*$, the inverse image of a point $\xi \in \mathbb{P}E^*$ is a finite number of points $[\xi, \eta(\xi)]$ whose η-coordinates are algebraic functions of ξ. Thus, setting $\zeta = (\xi, \eta(\xi))$ there is a non-zero vector $w \in W$ with

$$\sigma_\zeta(w) = 0.$$

This gives that

$$w \otimes \xi + w \otimes \eta(\xi) \in A \cap \{\Omega, \xi\},$$

and so $w \in W_\xi$. Referring to (150), we have

$$B_\rho(w) = \eta_\rho(\xi) w$$

where the $\eta_\rho(\xi)$ are the components of $\eta(\xi)$. Thus, the commuting linear transformations $B_\rho(\xi)|_{W_\xi}$ may be put in Jordan normal form with the last $n - l$ coordinates on each sheet giving the eigenvalues for the common eigenvector of these transformations.

At this stage, one needs to be careful in treating the non-semisimple parts of the $B_\rho(\xi)$'s. Moreover, one must worry about the characteristic ideal and not just the characteristic variety.[6] We refer to Gabber [1981] for further references and discussion.

---

[6] Again it is interesting to note in Cartan [1953] attention is drawn to the technical difficulties encountered when there is an essential non-diagonal piece to the $B_\rho(\xi)$'s (cf. footnote 9 in Chapter V).

§7. The Graded Module Associated to a Higher Order Tableau.

It is well known that a system of higher order P.D.E. may be rewritten as a (much larger) 1st order P.D.E. system. In general, however, it is preferable to treat the higher order system directly. Similarly, in many geometric examples the tableau of a linear Pfaffian differential system looks like the tableau of a higher order system—we have called these a tableau of order p. It is desirable to adapt the formalism—characteristic variety, graded module associated to a tableau, etc.—to a tableau of order p, and this is what we shall do in this section. For reasons to appear below, we shall make the notation shift $p \to q - 1$.

We consider a *tableau of order $q - 1$*

$$A \subset W \otimes S^q V^*$$

with prolongations

$$A^{(k)} \subset W \otimes S^{k+q} V^*$$

where (cf. (2)–(4) above)

$$(155) \qquad A^{(k)} = W \otimes S^{k+q} V^* \cap A \otimes S^k V^*.$$

We set

$$\mathbf{A} = \bigoplus_{k \geq 0} A^{(k)} \subset W \otimes SV^*.$$

By definition the *symbol* of A is

$$B = A^\perp \subset W^* \otimes S^q V,$$

and the *symbol module* is, again by definition,

$$\mathbf{B} = \bigoplus_{k \geq 0} B_k$$

where

$$B_k = A^{(k)\perp} \subset W^* \otimes S^{k+q} V.$$

As before (the case $q = 1$), $\mathbf{B}$ is the SV-submodule of $W^* \otimes SV$ generated by $B = B_0$ with an appropriate shift in grading.

Example 7.1. We make the identification

$$SV \cong \mathbb{C}[\partial/\partial x^1, \ldots, \partial/\partial x^n]$$

by the mapping $v_i \to \partial/\partial x^i$. Given a tableau $A \subset W \otimes S^q V^*$ of order $q - 1$ with symbol $B \subset W^* \otimes S^q V$, we choose a basis

$$D^\lambda = B_a^{\lambda I} w_a^* \otimes \partial^I / \partial x^I, \quad |I| = q,$$

for B, and thereby establish a 1–1 correspondence between tableaux of order $q - 1$ and q^{th} order, constant coefficient homogeneous linear P.D.E. systems

$$D^\lambda u(x) = 0.$$

The symbol module is just the sub-module of $W^* \otimes \mathbb{C}[\partial/\partial x^1, \ldots, \partial/\partial x^n]$ generated by the D^λ's.

Returning to the general discussion we recall our "shift" notations

$$(W^* \otimes SV)_k^{[q-1]} = W^* \otimes S^{k+q-1}$$

$$(W^* \otimes SV)^{[q-1]} = \bigoplus_{k \geq 0} (W^* \otimes SV)_k^{[q-1]}$$

and give the following:

Definition 7.2. Given a tableau $A \subset W \otimes S^q V^*$, we define the *associated graded module M_A* by the exact sequence

$$(156) \qquad\qquad 0 \to \mathbf{B} \to (W^* \otimes SV)^{[q-1]} \to M_A \to 0,$$

where the 1st map is homogeneous of degree one and the 2nd is homogeneous of degree zero.

It follows that

$$M_A = \bigoplus_{k \geq 0} (M_A)_k$$

where

$$(M_A)_k = \begin{cases} A^{(k-1)*} & k \geq 1 \\ W^* \otimes S^{q-1}V & k = 0. \end{cases}$$

When $q = 1$ this coincides with the graded module associated to an ordinary tableau introduced in §3 above. Generalizing (57) above we have

(157)

> *The tableau $A \subset W \otimes S^q V^*$ is involutive if, and only if,*
>
> *the associated graded module M_A is involutive in the*
>
> *sense of Definition 4.1 above.*

The proof of (57) is based on (and, in fact, is equivalent to) the surjectivity of the maps $\partial/\partial x^i : A_{i-1}^{(q+1)} \to A_{i-1}^{(q)}$, $q \geq 0$. According to Cartan's test, this surjectivity is by definition the same as involutiveness *for a tableau of any order*. Consequently, the proof of (57) given above carries over pretty much verbatim to give a proof of (157). The "pretty much" refers to the fact that some care must be taken in the definition of the degree zero piece of M_A. This will be discussed now.

The simplest involutive module is $(W^* \otimes SV)^{[q-1]}$. However, this is not a free module if $q \geq 2$. To explain this we remark that $(W^* \otimes SV)^{[q-1]}$ corresponds to the empty P.D.E. system. However, when $q \geq 2$ there are compatibility conditions that functions u_I^α, $|I| = q-1$, be $(q-1)$-jets; these conditions correspond to the non-freeness of $(W^* \otimes SV)^{[q-1]}$ as expressed by[7]

$$H_{0,1}((W^* \otimes SV)^{[q-1]}) \neq 0 \text{ if } q \geq 2.$$

The meaning of "correspond to" will be elaborated on below.

Example 7.3. We consider a constant coefficient linear P.D.E. system for one unknown function

$$(158) \qquad B^{\lambda I} \frac{\partial^q u(x)}{\partial x^I} = f^r(x), \quad \lambda = 1, \ldots, t.$$

The symbol module $\mathbf{B}$ of (158) is generated by the homogeneous polynomials

$$P^\lambda(\xi) = B^{\lambda I} \xi_I \in S^q V.$$

If (158) is involutive, then it follows from the exact homology sequence of (156) that $\mathbf{B}$ is an involutive S-module. Let

$$(159) \qquad \cdots \to \mathbf{E}_1 \xrightarrow{\varphi_1} \mathbf{E}_0 \xrightarrow{\varphi_0} \mathbf{B} \to 0$$

be its canonical resolution where $\mathbf{E}_i = E_i \otimes_{\mathbb{C}} S$ and φ_0 has degree zero while φ_i has degree one for $i \geq 1$. The localization of (159) is

$$(160) \qquad \cdots \to \mathcal{E}_1(-q-1) \xrightarrow{\varphi_1} \mathcal{E}_0(-q) \xrightarrow{\varphi_0} \mathcal{T} \to 0$$

where $\mathcal{T} \subset \mathcal{O}$ is the sheaf of ideals of the characteristic variety $\Xi = \{\xi : P^\lambda(\xi) = 0\}$. Over a point $\xi \notin \Xi$, the fibre sequence of (163) is

$$\to E_1 \otimes L_\xi^{-(q+1)} \to E_0 \otimes L_\xi^{-q} \to \mathbb{C} \to 0.$$

Setting $h_i = \dim E_i$ $(= \dim H_{0,i}(\mathbf{B}))$ it follows that

$$(161) \qquad h_1 \geq h_0 - 1.$$

Now

$$h_0 = \text{ number of equations (158)}$$

---

while h_1 is the number of independent linear compatibility conditions

$$(162) \qquad m_\lambda^j \frac{\partial}{\partial x^j}\left(B^{\lambda I}\frac{\partial^q}{\partial x^I}\right) = 0.$$

Setting

$$Q_\lambda(\xi) = m_\lambda^j \xi_j$$

(162) is equivalent to

$$(163) \qquad Q_\lambda(\xi)P^\lambda(\xi) = 0, \quad \deg Q_\lambda(\xi) = 1.$$

We therefore conclude from (164) that:

(164)   *If* (158) *is involutive, then there are at least* $t-1$
          independent linear relations (163) *among the*
          polynomials $P^\lambda(\xi)$.

This result is due to Cartan in the case of three independent variables (his proof applies to the general case).

We remark that (164) remains true for any involutive tableau whose associated constant coefficient linear P.D.E. system is (158). A special case of this result was given in Example 7.2 in Chapter IV.

Returning to the general discussion, the involutivity of M_A is expressed by

$$(165) \qquad H_{k,q}(M_A) = 0 \text{ for } k \geq 1.$$

Using this we may repeat verbatim the construction of the canonical free resolution of M_A. However, in contrast to the case $q = 1$ the resolution does not begin with $(W^* \otimes SV)^{[q-1]}$ (which in any case is not free if $q \geq 2$), but with the free module $\tilde{W}^* \otimes SV$ where $\tilde{W}^* = W^* \otimes S^{q-1}V$ $(= H_{0,0}(M_A))$. What is happening is that the homological formalism is leading us to the symbol algebra underlying the treatment of a higher order P.D.E. system as a large 1st order system.

Definition 7.4. Given a tableau $A \subset W \otimes S^q V^*$ and $q \geq 2$, we define

$$\begin{cases} \tilde{W} = W \otimes S^{q-1}V^*, \text{ and} \\ \tilde{A} \subset \tilde{W} \otimes V^* \end{cases}$$

to be the image of A under the natural inclusion $W \otimes S^q V^* \subset W \otimes S^{q-1}V^* \otimes V^*$.

Example 7.5. In the case $q = 2$ we let

$$B_a^{\lambda ij} \frac{\partial^2 u^a(x)}{\partial x^i \partial x^j} = 0$$

be the 2nd order linear homogeneous constant coefficient system corresponding to $A \subset W \otimes S^2 V^*$. Then the 1st order system

$$\frac{\partial u_i^a(x)}{\partial x^j} - \frac{\partial u_j^a(x)}{\partial x^i} = 0$$

$$B_a^{\lambda ij} \frac{\partial u_i^a(x)}{\partial x^j} = 0$$

corresponds to $\tilde{A} \subset (W \otimes V^*) \otimes V^*$.

Returning to the general discussion, we have the following result (which would be interesting only if it were false, since it would then say that we have the wrong formalism):

Proposition 7.6. *There is a natural isomorphism*

$$M_{\tilde{A}} \xrightarrow{\sim} M_A$$

of graded SV-modules.

This gives us another proof of the previously noted

Corollary 7.7. *A is an involutive tableau of order $q - 1$ if, and only if, $\tilde{A}$ is an involutive tableau in the usual sense.*

Proof of Proposition 7.6. Using (3) and (4) above we have as subspaces of $W \otimes \underbrace{V^* \otimes \cdots \otimes V^*}_{k+q}$,

$$\tilde{A}^{(k)} = \tilde{W} \otimes S^{k+1}V^* \cap \tilde{A} \otimes S^k V^*$$

$$= W \otimes S^{q-1}V^* \otimes S^{k+1}V^* \cap A \otimes S^k V^*$$

where $A \subset W \otimes S^q V^*$. It follows from this and (155) that

$$\tilde{A}^{(k)} = A^{(k)}.$$

More precisely, $\tilde{A}^{(k)}$ is the image of $A^{(k)}$ under the natural inclusion

$$(166) \qquad \begin{array}{c} j_k : W \otimes S^{q+k}V^* \subset W \otimes S^{q-1}V^* \otimes S^{k+1}V^* \\ \| \\ \tilde{W} \otimes S^{k+1}V^* \end{array} \quad .$$

We now define the obvious degree zero graded vector space mapping μ by the diagram

$$\tilde{W}^* \otimes SV \xrightarrow{\mu} (W^* \otimes SV)^{[q-1]}$$
$$\| j^*$$
$$W^* \otimes S^{q-1}V \otimes SV$$

where $j^* = \bigoplus_{k \geq 0} j_k^*$. A basic observation, whose straightforward verification we omit, is that j^* is induced by the multiplication

$$S^{q-1}V \otimes SV \to (SV)^{[q-1]}.$$

This induces a commutative diagram

$$
\begin{array}{ccc}
\tilde{W}^* \otimes SV & \to M_{\tilde{A}} \to 0 \\
\downarrow \mu & \downarrow j^* \\
(W^* \otimes SV)^{[q-1]} & \to M_A \to 0
\end{array}
$$

where μ is now a graded module mapping and where $j^* = \bigoplus_{k \geq 0} j_k^*$ with

$$j_k^* : \tilde{A}^{(k)*} \xrightarrow{\sim} A^{(k)*}$$

being induced by the dual of (166). The module isomorphism $j^* : M_{\tilde{A}} \xrightarrow{\sim} M_A$ is the one promised in the proposition. $\qquad\square$

Finally, we shall relate the graded module $M_{A^{(q)}}$ associated to the qth prolongation $A^{(q)} \subset W \otimes S^{q+1}V^*$ of a tableau $A \subset W \otimes V^*$ to M_A. Denoting by $M^+ = \bigoplus_{q \geq 1} M_q$ the positively graded sub-module of a graded module M, the simple answer is given by the

Proposition 7.8. *There is a natural isomorphism*

$$M_{A^{(q)}}^+ \cong (M_A)^{[q]+}.$$

Corollary 7.9. *If A is involutive, then so are its prolongations $A^{(q)}$.*

Corollary 7.10. *The characteristic varieties Ξ of A and $\Xi^{(q)}$ of $A^{(q)}$ coincide.*

Both of these results have been proved above. Since as noted in §3 above, Corollary 7.9 is essentially a homological result, the present proof is a natural way of establishing it.

We shall give a sketch of the proof in the case $q = 1$. The point is that we have

$$A^{(1)} \subset A \otimes V^* \subset W \otimes V^* \otimes V^*$$

$$A^{(1)(k)} \subset A \otimes S^{k+1}V^* \subset W \otimes \underbrace{V^* \otimes \cdots \otimes V^*}_{k+2}$$

$$A^{(k+1)} \subset W \otimes S^{k+2}V^* \subset W \otimes \underbrace{V^* \otimes \cdots \otimes V^*}_{k+2}$$

and, as subspaces of $W \otimes \underbrace{V^* \otimes \cdots \otimes V^*}_{k+2}$,

$$A^{(1)(k)} = A^{(k+1)}.$$

$\square$

Finally, we remark on the characteristic variety of a higher order tableau. Let

$$A \subset W \otimes V^{*(p+1)}$$

be a tableau of order p given as the kernel of a symbol mapping

$$\sigma : W \otimes V^{*(p+1)} \to U.$$

For $0 \neq \xi \in V^*$ we define

$$\sigma_\xi : W \to U$$

by

$$\sigma_\xi(w) = \sigma(w \otimes \underbrace{\xi \otimes \cdots \otimes \xi}_{p+1}),$$

and then we define the characteristic variety of A by

$$(167) \qquad \Xi_A = \{\xi \in \mathbb{P}V^* : \ker \sigma_\xi \neq (0)\}.$$

To justify this we consider A as an ordinary tableau A_1 by the inclusion

$$W \otimes V^{*(p+1)} \subset (W \otimes \underbrace{V^* \otimes \cdots \otimes V^*}_{p}) \otimes V^*.$$

Setting $W_1 = W \otimes V^* \otimes \cdots \otimes V^*$ we may give A_1 as the kernel of a suitable mapping

$$\sigma_1 : W_1 \otimes V^* \to U_1.$$

As in the proof of (12) in §3 of Chapter V we may show that:

$$(168) \qquad \textit{There is a natural isomorphism}$$

$$\ker \sigma_\xi \cong \ker \sigma_{1,\xi}.$$

In particular, the characteristic varieties of A and A_1 coincide and are given by

$$\Xi_A = \{\xi \in V^* : W \otimes \xi^{p+1} \cap A \neq 0\}$$

where ξ^{p+1} is the $(p+1)^{st}$ symmetric product of ξ.

CHAPTER IX

PARTIAL DIFFERENTIAL EQUATIONS

In this chapter and the next, we present an introduction to the theory of overdetermined systems of partial differential equations, both linear and non-linear, as it has been developed over the last twenty five years. Rather than giving complete proofs, we have preferred in general to present many examples illustrating the various methods used in the theory.

The modern theory of these systems was initially undertaken by Matsushima [1953, 1954-55] and Kuranishi [1957, 1961, 1962] within the framework of exterior differential systems. The first major result was the Cartan–Kuranishi prolongation theorem (Kuranishi [1957]). Using Ehresmann's theory of jets, Spencer [1962] introduced fundamental new tools for the theory of overdetermined systems in order to study deformations of pseudogroup structures. In particular, to linear equations, he associated certain complexes of differential operators, namely the so-called naive and sophisticated Spencer sequences (see Example 1.13, Chapter X). Intrinsic constructions of these Spencer complexes were given by Bott [1963] and investigated by Quillen [1964]. The formal theory of overdetermined systems was then systematically studied by Goldschmidt [1967a, 1967b, 1968a, 1968b, 1970b, 1972a, 1974]; for linear equations, introductory accounts are contained in Malgrange [1966-67], Spencer [1969] and Goldschmidt [1970a].

This chapter is devoted to the basic existence theorem of Goldschmidt [1967b] for systems of non-linear partial differential equations. This result consists of two parts. First, it provides conditions which guarantee the existence of sufficiently many formal solutions for an arbitrary system. Then for an analytic system satisfying these conditions, it gives us the convergence of formal solutions and thus the existence of local solutions. We show how this theorem can be used to prove the existence of solutions for two systems involving the Ricci curvature.

§1. An Integrability Criterion.

Let V and V' be finite-dimensional vector spaces and let U be an open subset of $\mathbb{R}^n$. Consider the system of non-linear partial differential equations of order k

$$(1) \qquad \Phi(x, D^\alpha u) = 0$$

for the unknown V-valued function u on U, where $x \in U$ and Φ is a V'-valued function, and $\alpha = (\alpha_1, \ldots, \alpha_n)$ ranges over all multi-indices of norm $\leq k$. If $l \geq 0$, we say that a V-valued function u_0 on a neighborhood of $x_0 \in U$ is an infinitesimal solution of (1) of order $k + l$ at x_0 if

$$D^\beta \Phi(x, D^\alpha u_0)|_{x=x_0} = 0,$$

for all multi-indices β of norm $\leq l$. Clearly, the conditions for u_0 to be an infinitesimal solution of order $k + l$ at x_0 are in fact imposed only on its Taylor series

$$p = \sum_{0 \leq |\alpha| \leq k+l} a_\alpha \frac{(x - x_0)^\alpha}{\alpha!}$$

at x_0 of order $k + l$, with $a_\alpha = (D^\alpha u_0)(x_0)$, which we call a formal solution of (1) of order $k + l$ at x_0. If $m \geq l$, we say that the polynomial

$$q = \sum_{0 \leq |\beta| \leq k+m} b_\beta \frac{(x - x_0)^\beta}{\beta!}$$

of degree $k + m$ extends the polynomial p if its coefficients of order $\leq k + l$ agree with those of p. Let R_k denote the set of all formal solutions of (1) of order k.

If Φ is analytic, we are interested in finding a convergent power series solution of (1) on a neighborhood of x_0. Thus we first seek formal power series solutions of (1) at x_0. In particular, given a formal solution of (1) of order $k + l$, we wish to extend it to a formal solution of higher order. The aim of the existence theory of Goldschmidt [1967b] is to provide sufficient conditions under which a formal solution of order k can be extended to a formal power series solution, and in the analytic case to an analytic solution. One such condition is *formal integrability*: it requires that, for all $l \geq 0$, every formal solution of (1) of order $k + l$ can be extended to a formal solution of order $k + l + 1$. However, verifying directly the formal integrability of an equation is extremely tedious and difficult. We now present sufficient conditions for formal integrability, which in general can be effectively verified.

Let u_0 be an infinitesimal solution of (1) of order k at x_0 and let p be the corresponding formal solution of order k at x_0. If β is a multi-index of norm k, we denote by

$$\sigma_\beta(\Phi)_p : V \to V'$$

the derivative of Φ, considered as a function of the independent variables $(x, D^\alpha u)$, in the direction $D^\beta u$ at $(x_0, (D^\alpha u_0)(x_0))$. We denote by $T^*_{x_0}$ the cotangent space of $\mathbb{R}^n$ at x_0 and by $S^m W$ the m-th symmetric power of a subspace W of $T^*_{x_0}$; we write ξ^m for the m-th symmetric power of an element ξ of $T_{x_0}^*$. The symbol $\sigma(\Phi)_p$ and its first prolongation $\sigma_1(\Phi)$ of Φ at p are the unique linear mappings

$$\sigma(\Phi)_p : S^k T_{x_0}^* \otimes V \to V',$$

$$\sigma_1(\Phi)_p : S^{k+1} T_{x_0}^* \otimes V \to T_{x_0}^* \otimes V'$$

determined by

$$\sigma(\Phi)_p(\xi^k \otimes v) = \sum_{|\alpha|=k} (\sigma_\alpha(\Phi)_p v) \cdot \xi^\alpha,$$

$$\sigma_1(\Phi)_p(\xi^{k+1} \otimes v) = (k+1) \sum_{|\alpha|=k} \xi \otimes (\sigma_\alpha(\Phi)_p v) \cdot \xi^\alpha,$$

for $\xi = \sum_{j=1}^{n} \xi_j dx^j \in T_{x_0}^*$ and $v \in V$, where $\alpha = (\alpha_1, \ldots, \alpha_n)$ and

$$\xi^\alpha = \xi_1^{\alpha_1} \cdot \ldots \cdot \xi_n^{\alpha_n}.$$

We also call the kernel $g_{k,p}$ of the mapping $\sigma(\Phi)_p$ the symbol of Φ at p and we set $g_{k+1,p} = \text{Ker } \sigma_1(\Phi)_p$. Associated to the subspace $g_{k,p}$ of $S^k T_{x_0}^* \otimes V$ are the Spencer cohomology groups $H^{k+l,j}(g_{k,p})$, with $l, j \geq 0$, which will be defined in §2. Let $\{\eta_1, \ldots, \eta_n\}$ be a basis of $T_{x_0}^*$; we say that $\{\eta_1, \ldots, \eta_n\}$ is a quasi-regular basis for g_k at p if

$$\dim g_{k+1,p} = \dim g_{k,p} + \sum_{j=1}^{n-1} \dim(g_{k,p} \cap (S^k W_j \otimes V)),$$

where W_j is the subspace of $T_{x_0}^*$ generated by $\eta_1, \ldots, \eta_j$. The existence of a quasi-regular basis for g_k at p is equivalent to the vanishing of all these Spencer cohomology groups (see Theorem 2.14).

The criterion of Goldschmidt [1967b] for the existence of formal solutions may be stated as follows. The formal solution p of order k can be extended to a formal solution of infinite order if:

(i) the mapping Φ in the variables $(x, D^\alpha u)$, with $0 \leq \alpha \leq k$, is of constant rank in a neighborhood of $(x_0, (D^\alpha u_0)(x_0))$;

(ii) for all x in a neighborhood of x_0, there exists a formal solution of (1) of order k at x;

(iii) in a neighborhood of x_0, every formal solution of (1) of order k can be extended to a formal solution of order $k+1$;

(iv) for all $q \in R_k$ in a neighborhood of p, the rank of the linear mapping $\sigma_1(\Phi)_q$ is independent of q;

(v) for all $q \in R_k$ in a neighborhood of p, the symbol $g_{k,q}$ at q is 2-acyclic, i.e. the cohomology groups $H^{k+l,2}(g_{k,q})$ vanish for $l \geq 0$.

The condition (v) can be replaced by the stronger condition of involutivity which need only be verified at p:

(vi) there exists a quasi-regular basis of $T^*_{x_0}$ for g_k at p.

In fact, conditions (i)–(v) imply the existence of sufficiently many formal solutions extending p, and, whenever Φ is a real-analytic function, the existence of an analytic solution u of (1) in a neighborhood of x_0 satisfying $(D^\beta u)(x_0) = (D^\beta u_0)(x_0)$, for all $0 \leq |\beta| \leq k$.

§2. Quasi-Linear Equations.

In this section, we give an intrinsic version of the basic existence theorem of Goldschmidt [1967b]; for simplicity, here we mainly restrict our attention to quasi-linear equations.

We assume that all objects and mappings are differentiable of class C^∞. In general, we do not require that the dimensions of the different components of a differentiable manifold be the same, and we allow the rank of a vector bundle over a manifold Y to vary over the different components of Y. Let X be a differentiable manifold of dimension n, whose tangent and cotangent bundles we denote by T and T^* respectively. If k is a non-negative integer, we let $\bigotimes^k T^*$, $S^k T^*$ and $\bigwedge^k T^*$ be the k-th tensor, symmetric and exterior powers of T^*, respectively. We shall identify $S^k T^*$ and $\bigwedge^k T^*$ with sub-bundles of $\bigotimes^k T^*$ by means of the injective mappings

$$S^k T^* \to \bigotimes^k T^*, \qquad \bigwedge^k T^* \to \bigotimes^k T^*$$

sending the symmetric product $\beta_1 \cdot \ldots \cdot \beta_k$, with $\beta_1, \ldots, \beta_k \in T^*$, into

$$\sum_{\sigma \in \mathfrak{S}_k} \beta_{\sigma(1)} \otimes \cdots \otimes \beta_{\sigma(k)},$$

and the exterior product $\beta_1 \wedge \cdots \wedge \beta_k$ into

$$\sum_{\sigma \in \mathfrak{S}_k} \operatorname{sgn} \sigma \cdot \beta_{\sigma(1)} \otimes \cdots \otimes \beta_{\sigma(k)},$$

where $\mathfrak{S}_k$ is the group of permutations of $\{1, \ldots, k\}$ and $\operatorname{sgn} \sigma$ is the signature of the element σ of $\mathfrak{S}_k$.

If Y, Z are differentiable manifolds and $\rho_1 : Z \to X$, $\rho_2 : Z \to Y$ are mappings, and if F_1 is a vector bundle over X and F_2 is a vector bundle over Y, we denote by $F_1 \otimes_Z F_2$ the vector bundle $\rho_1^{-1} F_1 \otimes \rho_2^{-1} F_2$ over Z;

if E is a vector bundle over X, we shall sometimes also denote by E the vector bundle $\rho_1^{-1}E$ over Z induced by ρ_1.

A fibered manifold E over X is a manifold together with a surjective submersion $\pi : E \to X$. A submanifold F of E is said to be a fibered submanifold if $\pi|_F : F \to X$ is a fibered manifold. We denote by $\mathcal{E}$ the sheaf of sections of E over X. Recall that two sections s and s' of E over a neighborhood V of $x_0 \in X$ have the same k-jet at x_0 if $s(x_0) = s'(x_0)$ and if in some, and hence in all, local coordinate systems the Taylor series of s and s' agree up through order k. The class determined by s at x_0 will be denoted by $j_k(s)(x_0)$, and we write $\pi(j_k(s)(x_0)) = x_0$. The set $J_k(E)$ of all such k-jets together with the projection π is a fibered manifold over X, and $x \mapsto j_k(s)(x)$ is a section of $J_k(E)$ over V, which we call the k-jet $j_k(s)$ of the section s; often $J_k(E)$ is called the bundle of k-jets of sections of E. If $m \geq k$ and $p \in J_m(E)$, we let $\pi_k(p)$ be the element of $J_k(E)$ that it determines; thus $\pi_k j_m(s)(x_0) = j_k(s)(x_0)$. We also know that $\pi_k : J_m(E) \to J_k(E)$ is a fibered manifold. We shall identify $J_0(E)$ with E.

Let $e \in E$ with $\pi(e) = x_0$; then there is an open neighborhood U of e and diffeomorphisms

$$\varphi : U \to \mathbb{R}^n \times \mathbb{R}^m, \qquad \psi : \pi U \to \mathbb{R}^n$$

such that the diagram

$$
\begin{array}{ccc}
U & \xrightarrow{\ \varphi\ } & \mathbb{R}^n \times \mathbb{R}^m \\
\downarrow{\scriptstyle \pi} & & \downarrow{\scriptstyle \mathrm{pr}_1} \\
\pi U & \xrightarrow{\ \psi\ } & \mathbb{R}^n
\end{array}
$$

commutes, where pr_1 is the projection onto the first factor. We obtain corresponding coordinate systems $(x^1, \ldots, x^n, y^1, \ldots, y^m)$ for E on U and $(x^1, \ldots, x^n)$ for X on πU. A standard local coordinate system for $J_k(E)$ on $\pi_0^{-1}(U)$ is

$$(x^i, y^j, y^j_\alpha),$$

where $1 \leq i \leq n$, $1 \leq j \leq m$ and $\alpha = (\alpha_1, \ldots, \alpha_n)$ ranges over all multi-indices satisfying $1 \leq |\alpha| \leq k$. If s is a section of E over a neighborhood of x with $s(x) \in U$, then

$$y^j_\alpha(j_k(s)(x)) = D^\alpha y^j(s(x)),$$

$$y^j(j_k(s)(x)) = y^j(s(x)).$$

If E is a vector bundle, recall that $J_k(E)$ is also a vector bundle; we have

$$j_k(s)(x_0) + j_k(s')(x_0) = j_k(s + s')(x_0),$$

$$a j_k(s)(x_0) = j_k(as)(x_0),$$

for $a \in \mathbb{R}$. The morphism of vector bundles

$$\epsilon : S^k T^* \otimes E \to J_k(E)$$

determined by

$$\epsilon(((df_1 \cdot \ldots \cdot df_k) \otimes s)(x)) = j_k\left(\left(\textstyle\prod_{i=1}^{k} f_i\right) \cdot s\right)(x),$$

where $f_1, \ldots, f_k$ are real-valued functions on X vanishing at $x \in X$ and s is a section of E over X, is well-defined since $\prod_{i=1}^{k} f_i$ vanishes to order $k-1$ at x. For $k < 0$, we set $J_k(E) = 0$. One easily verifies that the sequence

$$0 \to S^k T^* \otimes E \xrightarrow{\epsilon} J_k(E) \xrightarrow{\pi_{k-1}} J_{k-1}(E) \to 0$$

is exact, for $k \geq 0$.

Let $\pi : E \to X$ and $\pi' : E' \to X$ be fibered manifolds over X; a mapping $\varphi : E \to E'$ is a morphism of fibered manifolds over X if $\pi' \circ \varphi = \pi$.

We return to the study of equation (1). Let $\tilde{V}, \tilde{V}'$ be the trivial vector bundles $U \times V$, $U \times V'$ respectively. The function Φ determines a morphism $\varphi : J_k(\tilde{V}) \to \tilde{V}'$ of fibered manifolds over U; in fact

$$\Phi(x, D^\alpha u) = \varphi(j_k(\tilde{u})(x)),$$

where $x \in U$ and $\tilde{u}$ is the graph of the V-valued function u on U. We identify the set R_k of formal solutions of (1) of order k with

$$\{p \in J_k(\tilde{V}) \mid \varphi(p) = 0\}.$$

The solutions of (1) depend only on R_k. We reserve the terminology of differential equations for such R_k satisfying the additional regularity condition that it be a fibered submanifold of the jet bundle. More precisely, we have:

Definition. A (non-linear) partial differential equation R_k of order k on E is a fibered submanifold of $\pi : J_k(E) \to X$. A solution s of R_k is a section of E such that $j_k(s)$ is a section of R_k.

Let F be an open fibered submanifold of $J_k(E)$ and $\varphi : F \to E'$ be a morphism of fibered manifolds over X; let s' be a section of E' over X. We set

$$(2) \qquad R_k = \mathrm{Ker}_{s'} \, \varphi = \{p \in F \mid \varphi(p) = s'(\pi(p))\}.$$

If

$$(3) \qquad s'(X) \subset \varphi(F),$$

the fibers of R_k are non-empty. If (3) holds and

$$(4) \qquad \varphi \text{ has locally constant rank,}$$

then, according to Proposition 2.1 of Goldschmidt [1967b], R_k is a fibered submanifold of $J_k(E)$ and a partial differential equation. The l-th prolongation of φ is the morphism

$$p_l(\varphi) : \pi_k^{-1}F \to J_l(E')$$

defined on the open subset $\pi_k^{-1}F$ of $J_{k+l}(E)$ by

$$p_l(\varphi)(j_{k+l}(s)(x)) = j_l(\varphi \circ j_k(s))(x),$$

for all $x \in X$ and $s \in \mathcal{E}_x$, with $j_k(s)(x) \in F$. We set $p_0(\varphi) = \varphi$. We consider the subset

$$R_{k+l} = \operatorname{Ker}_{j_l(s')} p_l(\varphi)$$

of $\pi_k^{-1}F$; the natural projection $\pi_{k+l} : J_{k+l+1}(E) \to J_{k+l}(E)$ sends R_{k+l+1} into R_{k+l}, for all $l \geq 0$. If conditions (3) and (4) hold, then R_{k+l} depends only on R_k and is called the l-th prolongation of R_k (see §3). We remark that any partial differential equation of order k on E can be written locally (in $J_k(E)$) in the form (2) with φ of constant rank (see Goldschmidt [1967b]).

Let $(x^1, \ldots, x^n, y^1, \ldots, y^m)$ be the coordinate system on the open subset U of E considered above, where $(x^1, \ldots, x^n)$ is a coordinate system for X, and let $(x^1, \ldots, x^n, z^1, \ldots, z^p)$ be a similar coordinate system for E' on an open subset U' of E', with $\pi'U' = \pi U$. We consider the standard coordinate systems on the jet bundles. If $\varphi(F \cap \pi_0^{-1}U) \subset U'$, the morphism φ is determined by the p functions

$$\varphi^r(x^i, y^j, y^j_\alpha), \qquad 1 \leq |\alpha| \leq k,$$

on $F \cap \pi_0^{-1}U$ equal to $z^r \circ \varphi$. The first prolongation $p_1(\varphi)$ of φ is then determined by

$$\varphi^r(x^i, y^j, y^j_\alpha)$$

and

$$\frac{\partial \varphi^r}{\partial x^k}(x^i, y^j, y^j_\alpha) + \sum_{l=1}^m \frac{\partial \varphi^r}{\partial y^l}(x^i, y^j, y^j_\alpha)y^l_{\epsilon_k} + \sum_{\substack{1 \leq |\beta| \leq k \\ 1 \leq l \leq m}} \frac{\partial \varphi^r}{\partial y^l_\beta}(x^i, y^j, y^j_\alpha)y^l_{\beta+\epsilon_k},$$

with $1 \leq r \leq p$, $1 \leq k \leq n$, where ϵ_k is the multi-index whose k-th entry is equal to one and whose other entries are equal to zero.

Example 2.1. Let $F = E' = J_k(E)$ and φ be the identity mapping of $J_k(E)$. Then $p_l(\varphi)$ is the canonical imbedding

$$\lambda_l : J_{k+l}(E) \to J_l(J_k(E))$$

sending $j_{k+l}(s)(x)$ into $j_l(j_k(s))(x)$, for $x \in X$ and $s \in \mathcal{E}_x$.

We again consider equation (1); if s' is the zero-section of $\tilde{V}'$, then $R_k = \text{Ker}_{s'}\, \varphi$, and, for $l \geq 0$, we may identify R_{k+l} with the set of formal solutions of (1) of order $k + l$.

We now return to the situation considered above. If R_k is a differential equation, we therefore call an element of R_{k+l} a formal solution of R_k of order $k + l$, and an element of

$$R_\infty = \text{pr}\,\lim R_{k+l}$$

a formal solution of R_k (of infinite order). Given a formal solution of R_k of order $m \geq k$, we seek conditions which will insure that it can be extended to a formal solution. One such condition is:

(5)   the mappings $\pi_{k+l} : R_{k+l+1} \to R_{k+l}$ are surjective, for all $l \geq 0$.

Spencer [1962] formulated the so-called δ-Poincaré estimate, which was proved by Ehrenpreis, Guillemin and Sternberg [1965], and later by Sweeney [1967], and which gives the convergence of power series solutions for analytic partial differential equations satisfying condition (5). Malgrange [1972] (Appendix) realized that this estimate is essentially equivalent to the "privileged neighborhood theorem" of Grauert [1960] and used it together with the method of majorants to prove directly the following existence theorem for analytic differential equations. An adaptation of the proof of a result of Douady [1966] yields the required theorem of Grauert.

Theorem 2.2. *Suppose that X is a real-analytic manifold, that E, E' are real-analytic fibered manifolds and that $\varphi : F \to E'$ is a real-analytic morphism and s' is an analytic section of E'. Let $x_0 \in X$ and $l \geq 0$. If $\pi_{k+m} : R_{k+m+1,x_0} \to R_{k+m,x_0}$ is surjective for all $m \geq l$, then given $p \in R_{k+l,x_0}$ there exists an analytic section s of E over a neighborhood U of x_0 such that $j_{k+l}(s)(x_0) = p$ and $j_k(s)(x) \in R_k$ for all $x \in U$.*

If conditions (3) and (4) hold, the section s given by the theorem is a solution of the differential equation R_k.

We present below sufficient conditions for (5) to hold which involve only a finite number of prolongations of R_k.

We now assume that E and E' are vector bundles over X. We say that the morphism of fibered manifolds $\varphi : F \to E'$ is *quasi-linear* if there exists a morphism of vector bundles

$$\sigma(\varphi) : S^k T^* \otimes E \to E'$$

over $\pi_{k-1}F$ such that

$$\varphi(p + \epsilon u) = \varphi(p) + \sigma(\varphi)_{\pi_{k-1}p}(u),$$

for all $p \in F$, $u \in S^k T^* \otimes E$, with $p + \epsilon u$ belonging to F. Here ϵ is the monomorphism $S^k T^* \otimes E \to J_k(E)$ and the vector bundles $S^k T^* \otimes E$ and E' are considered as induced vector bundles over $\pi_{k-1}F$, via the mapping π. If φ is quasi-linear, the mapping $\sigma(\varphi)$ is uniquely determined by φ and is called the *symbol* of φ.

If φ is quasi-linear and $\sigma(\varphi)$ is an epimorphism, and if

$$F + \epsilon(S^k T^* \otimes E) \subset F,$$

then it is easily seen that φ is a surjective submersion; thus under these hypotheses, conditions (3) and (4) hold and so (2) is a differential equation.

Example 2.3. Let E be the vector bundle $S^2 T^*$ and consider the fibered submanifold $S^2_+ T^*$ of E whose sections are the positive-definite symmetric 2-forms on X. Let F be the open fibered submanifold $J_2(S^2_+ T^*)$ of $J_2(E)$. A section g of $S^2_+ T^*$ is a Riemannian metric on X and we consider the Levi–Civita connection ∇^g of g and the Riemann curvature tensor $\mathcal{R}(g)$ of g, which is the section of $\bigwedge^2 T^* \otimes \bigwedge^2 T^*$ determined by

$$\mathcal{R}(g)(\xi_1, \xi_2, \xi_3, \xi_4) = g(\xi_4, (\nabla^g_{\xi_1} \nabla^g_{\xi_2} - \nabla^g_{\xi_2} \nabla^g_{\xi_1} - \nabla^g_{[\xi_1, \xi_2]})\xi_3),$$

for $\xi_1, \xi_2, \xi_3, \xi_4 \in \mathcal{T}$. In fact, according to the first Bianchi identity, $\mathcal{R}(g)$ is a section of the sub-bundle G of $\bigwedge^2 T^* \otimes \bigwedge^2 T^*$ consisting of those elements θ of $\bigwedge^2 T^* \otimes \bigwedge^2 T^*$ which satisfy the relation

$$\theta(\xi_1, \xi_2, \xi_3, \xi_4) + \theta(\xi_2, \xi_3, \xi_1, \xi_4) + \theta(\xi_3, \xi_1, \xi_2, \xi_4) = 0,$$

for all $\xi_1, \xi_2, \xi_3, \xi_4 \in T$; according to Lemma 3.1 of Gasqui and Goldschmidt [1983], G is equal to the image of the morphism of vector bundles

$$\tau : S^2 T^* \otimes S^2 T^* \to \bigwedge^2 T^* \otimes \bigwedge^2 T^*$$

defined by

$$\tau(u)(\xi_1, \xi_2, \xi_3, \xi_4) = \frac{1}{2}\{u(\xi_1, \xi_3, \xi_2, \xi_4) + u(\xi_2, \xi_4, \xi_1, \xi_3)$$
$$- u(\xi_1, \xi_4, \xi_2, \xi_3) - u(\xi_2, \xi_3, \xi_1, \xi_4)\},$$

for all $u \in S^2 T^* \otimes S^2 T^*$ and $\xi_1, \xi_2, \xi_3, \xi_4 \in T$. Let $E' = G$ and let

$$\Phi : J_2(S^2_+ T^*) \to G$$

be the morphism of fibered manifolds over X sending $j_2(g)(x)$ into $\mathcal{R}(g)(x)$, for $x \in X$. Then Φ is quasi-linear and its symbol

$$\sigma(\Phi) : S^2 T^* \otimes S^2 T^* \to G$$

over $J_1(S^2_+ T^*)$ is determined by τ; in fact, in terms of the local coordinate expression for the curvature of a metric, it is easily seen that

$$\Phi(j_2(g)(x) + \epsilon u) = \mathcal{R}(g)(x) + \tau u,$$

for $g \in S^2_+ T^*_x$ and $u \in (S^2 T^* \otimes S^2 T^*)_x$, with $x \in X$. Since τ is an epimorphism onto G, we see that Φ is a surjective submersion; hence if R is a section of G over X, then

$$N_2 = \operatorname{Ker}_R \Phi$$

is a differential equation, whose solutions are the Riemannian metrics g satisfying the equation

$$\mathcal{R}(g) = R.$$

Example 2.4. Suppose that $n \geq 3$. If g is a Riemannian metric on X, we denote by

$$\mathrm{Tr}_g : \bigwedge^2 T^* \otimes \bigwedge^2 T^* \to T^* \otimes T^*$$

the morphism defined by

$$(\mathrm{Tr}_g u)(\xi_1, \xi_2) = \sum_{i=1}^{n} u(t_i, \xi_1, t_i, \xi_2),$$

for $x \in X$, $u \in (\bigwedge^2 T^* \otimes \bigwedge^2 T^*)_x$ and $\xi_1, \xi_2 \in T_x$, where $\{t_1, \ldots, t_n\}$ is an orthonormal basis of T_x. It is well-known that

$$(6) \qquad\qquad \mathrm{Tr}_g(G) = S^2 T^*$$

(see for example Gasqui [1982]). The Ricci curvature $\mathrm{Ric}(g)$ of g is the section of $S^2 T^*$ equal to $-\mathrm{Tr}_g \mathcal{R}(g)$. Now as in Example 2.3, let $E = S^2 T^*$ and $F = J_2(S^2_+ T^*)$. We set $E' = S^2 T^*$ and let

$$\varphi : J_2(S^2_+ T^*) \to S^2 T^*$$

be the morphism of fibered manifolds over X sending $j_2(g)(x)$ into $\mathrm{Ric}(g)(x)$, for $x \in X$. Since the morphism Φ of Example 2.3 is quasi-linear, we see that this morphism φ is also quasi-linear and that its symbol

$$\sigma(\varphi) : S^2 T^* \otimes S^2 T^* \to S^2 T^*$$

over $J_1(S_+^2 T^*)$ sends $(j_1(g)(x), u)$ into $-\mathrm{Tr}_g \tau(u)$, for $u \in (S^2 T^* \otimes S^2 T^*)_x$; in fact,

$$\varphi(j_2(g)(x) + \epsilon u) = \mathrm{Ric}(g)(x) - \mathrm{Tr}_g \tau(u),$$

for $u \in (S^2 T^* \otimes S^2 T^*)_x$, with $x \in X$. According to (6), $\sigma(\varphi)$ is an epimorphism, and so we see that φ is a surjective submersion. Hence if R is a section of $S^2 T^*$ over X, then

$$N_2 = \mathrm{Ker}_R \, \varphi$$

is a differential equation, whose solutions are the Riemannian metrics g satisfying the equation

$$(7) \qquad\qquad\qquad \mathrm{Ric}(g) = R.$$

Example 2.5. Suppose that $n \geq 3$ and let $E = S^2 T^*$, $F = J_2(S_+^2 T^*)$ and $E' = S^2 T^*$ as in Example 2.4. Let $\lambda \in \mathbb{R}$ and

$$\psi_\lambda : J_2(S_+^2 T^*) \to S^2 T^*$$

be the morphism of fibered manifolds over X sending $j_2(g)(x)$ into $\mathrm{Ric}(g)(x) - \lambda g(x)$, for $g \in S^2 T_x^*$, with $x \in X$. Since the morphism φ of Example 2.4 is quasi-linear, we see that ψ_λ is also quasi-linear and that its symbol

$$\sigma(\psi_\lambda) : S^2 T^* \otimes S^2 T^* \to S^2 T^*$$

over $J_1(S_+^2 T^*)$ is equal to $\sigma(\varphi)$. Therefore ψ_λ is a surjective submersion; if 0 is the zero-section of $S^2 T^*$, then

$$N_2^\lambda = \mathrm{Ker}_0 \, \psi_\lambda$$

is a differential equation whose solutions are the Riemannian metrics g satisfying the identity

$$\mathrm{Ric}(g) = \lambda g,$$

and are Einstein metrics.

We now return to the situation we were considering before the above examples.

Let

$$\Delta_{l,k} : S^{k+l} T^* \to S^l T^* \otimes S^k T^*$$

be the natural inclusion. Let $\rho : Y \to X$ be a fibered manifold; if

$$\psi : S^k T^* \otimes E \to E'$$

is a morphism of vector bundles over Y, where $S^k T^* \otimes E$ and E' are considered as induced vector bundles over Y via the mapping ρ, the l-th prolongation

$$(\psi)_{+l} : S^{k+l} T^* \otimes E \to S^l T^* \otimes E'$$

of ψ is the morphism of vector bundles over Y equal to the composition

$$S^{k+l} T^* \otimes E \xrightarrow{\Delta_{l,k} \otimes \mathrm{id}} S^l T^* \otimes S^k T^* \otimes E \xrightarrow{\mathrm{id} \otimes \psi} S^l T^* \otimes E'.$$

If φ is quasi-linear, the l-th prolongation of $\sigma(\varphi)$ (over $\pi_{k-1} F$) is denoted by $\sigma_l(\varphi)$. The following result is given by Goldschmidt [1967b], §5, and is easily verified using the standard local coordinates on the jet bundles.

Proposition 2.6. *If the morphism* $\varphi : F \to E'$ *is quasi-linear, then, for* $l \geq 1$, *so is the morphism*

$$p_l(\varphi) : \pi_k^{-1} F \to J_l(E')$$

and its symbol is determined by $\sigma_l(\varphi)$; *we have*

$$(8) \qquad p_l(\varphi)(p + \epsilon u) = p_l(\varphi)(p) + \epsilon \sigma_l(\varphi)_{\pi_{k-1}p}(u),$$

for all $p \in J_{k+l}(E)$, $u \in S^{k+l} T^* \otimes E$, *with* $\pi_k p \in F$.

Example 2.7. Let E, E' be arbitrary vector bundles over X; assume that $F = J_k(E)$ and that $\varphi : J_k(E) \to E'$ is a morphism of vector bundles. Then

$$D = \varphi \circ j_k : \mathcal{E} \to \mathcal{E}'$$

is a linear differential operator of order k. In fact, any linear differential operator $\mathcal{E} \to \mathcal{E}'$ of order k is obtained in this way. The l-th prolongation

$$p_l(\varphi) = p_l(D) : J_{k+l}(E) \to J_l(E')$$

of φ is a morphism of vector bundles over X. The symbol

$$\sigma(\varphi) : S^k T^* \otimes E \to E'$$

of φ is the morphism $\varphi \circ \epsilon$ of vector bundles over X. The l-th prolongation of the symbol of φ is the morphism of vector bundles $\sigma_l(\varphi) = (\sigma(\varphi))_{+l}$ over X. Then the diagram

$$
\begin{array}{ccc}
S^{k+l} T^* \otimes E & \xrightarrow{\ \sigma_l(\varphi)\ } & S^l T^* \otimes E' \\
\downarrow{\scriptstyle \epsilon} & & \downarrow{\scriptstyle \epsilon} \\
J_{k+l}(E) & \xrightarrow{\ p_l(\varphi)\ } & J_l(E')
\end{array}
$$

commutes. We set

$$p(D) = p_0(D) = \varphi, \qquad \sigma(D) = \sigma_0(D) = \sigma(\varphi)$$

and

$$\sigma_l(D) = \sigma_l(\varphi),$$

for $l \geq 0$. A linear differential equation of order k on E is a sub-bundle R_k of $J_k(E)$. If φ has locally constant rank, then $R_k = \operatorname{Ker} \varphi$ is a linear differential equation, and its l-th prolongation $R_{k+l} = \operatorname{Ker} p_l(\varphi)$ is a vector bundle with variable fiber which depends only on R_k.

We now describe the first obstruction to the integrability of the non-linear equations determined by the morphism of fibered manifolds φ, which is assumed to be quasi-linear. Let W be the cokernel of the morphism $\sigma_1(\varphi)$, which is a vector bundle over $\pi_{k-1}F$ with variable fiber; if $\nu : T^* \otimes E' \to W$ is the natural projection, the sequence

$$(9) \qquad S^{k+1}T^* \otimes E \xrightarrow{\ \sigma_1(\varphi)\ } T^* \otimes E' \xrightarrow{\ \nu\ } W \to 0$$

is exact. We define a mapping

$$\Omega : R_k \to W$$

as follows. If $p \in R_{k,x}$, with $x \in X$, let $q \in J_{k+1}(E)$ with $\pi_k q = p$; then by (8) and the exactness of (9), the element

$$(10) \qquad \Omega(p) = \nu \epsilon^{-1}(p_1(\varphi)q - j_1(s')(x))$$

of $W_{\pi_{k-1}p}$ is well-defined, since $\varphi(p) = s'(x)$. We denote by 0 the zero-section of W; the following result is given by Proposition 2.1 of Goldschmidt [1972a].

Proposition 2.8. *The sequence*

$$R_{k+1} \xrightarrow{\ \pi_k\ } R_k \mathrel{\mathop{\rightrightarrows}^{\Omega}_{0 \circ \pi_{k-1}}} W$$

is exact, i.e.

$$\pi_k R_{k+1} = \{p \in R_k \mid \Omega(p) = 0\}.$$

Proof. Let $p \in R_{k,x}$, with $x \in X$, and let q be an element of $J_{k+1}(E)$ satisfying $\pi_k q = p$. If $q \in R_{k+1}$, according to (10) we see that $\Omega(p) = 0$. Conversely, if $\Omega(p) = 0$, then by the exactness of (9) there exists $u \in (S^{k+1}T^* \otimes E)_x$ such that

$$\sigma_1(\varphi)_{\pi_{k-1}p} u = -\epsilon^{-1}(p_1(\varphi)q - j_1(s')(x)).$$

Then, by (8), we have

$$p_1(\varphi)(p + \epsilon u) = j_1(s')(x)$$

and the element $q' = p + \epsilon u$ of $J_{k+1}(E)$ belongs to R_{k+1} and satisfies $\pi_k q' = p$.

Thus Ω represents the obstruction to the surjectivity of $\pi_k : R_{k+1} \to R_k$. The techniques for computing this first obstruction were first applied to equations arising in the theory of Lie pseudogroups by Goldschmidt [1972a, 1972b]. Subsequently, Gasqui [1975, 1979a, 1979b, 1982] studied the first obstruction and formal integrability questions for several other equations.

Examples 2.4 and *2.5* (continued). Let g be a Riemannian metric on X. If $\mathbf{1}$ denotes the trivial real line bundle over X, we consider the trace mappings

$$\mathrm{Tr}_g^0 : S^2 T^* \to \mathbf{1},$$

$$\mathrm{Tr}_g^1 : T^* \otimes S^2 T^* \to T^*$$

defined by

$$\mathrm{Tr}_g^0 u = \sum_{i=1}^n u(t_i, t_i),$$

$$(\mathrm{Tr}_g^1 v)(\xi) = \sum_{i=1}^n v(t_i, t_i, \xi),$$

for $x \in X$, $u \in S^2 T_x^*$, $v \in (T^* \otimes S^2 T^*)_x$, $\xi \in T_x$, where $\{t_1, \ldots, t_n\}$ is an orthonormal basis of T_x. The Bianchi operator

$$B_g : S^2 T^* \to T^*$$

of g is the first-order linear differential operator defined by

$$B_g u = \ \mathrm{Tr}_g^1 \nabla^g u - \tfrac{1}{2} d \, \mathrm{Tr}_g^0 u,$$

for $u \in S^2 T^*$. Clearly, the symbol

$$\sigma(B_g) : T^* \otimes S^2 T^* \to T^*$$

of B_g is equal to $\mathrm{Tr}_g^1 - \tfrac{1}{2} \mathrm{id} \otimes \mathrm{Tr}_g^0$. Since $\nabla^g g = 0$, we see that $B_g g = 0$. We recall that the Ricci curvature $\mathrm{Ric}(g)$ of g satisfies the Bianchi identity

$$(11) \qquad\qquad\qquad B_g \mathrm{Ric}(g) = 0.$$

The following algebraic result is proved by Gasqui [1982] using decompositions of $O(n)$-modules into irreducible submodules.

Lemma 2.9. *The sequence of vector bundles over X*

$$S^3T^* \otimes S^2T^* \xrightarrow{(\mathrm{Tr}_g \circ \tau)+1} T^* \otimes S^2T^* \xrightarrow{\sigma(B_g)} T^* \to 0$$

is exact.

Therefore the sequence of vector bundles

$$(12) \qquad S^3T^* \otimes S^2T^* \xrightarrow{\sigma_1(\varphi)} T^* \otimes S^2T^* \xrightarrow{\nu} T^* \to 0$$

over $J_1(S_+^2T^*)$ is exact, where ν sends $(j_1(g)(x), u)$ into $\sigma(B_g)u$, for $g \in S_+^2T_x^*$, $u \in (T^* \otimes S^2T^*)_x$, with $x \in X$. We now compute the first obstruction to the integrability of the equation N_2^λ. Let $p = j_2(g)(x) \in N_2^\lambda$, with $x \in X$ and $g \in S_+^2T_x^*$; then $(\mathrm{Ric}(g) - \lambda g)(x) = 0$. By (10), (11) and the exactness of (12), we have

$$\Omega(p) = \nu \epsilon^{-1} j_1(\mathrm{Ric}(g) - \lambda g)(x)$$

$$= \sigma(B_g)\epsilon^{-1} j_1(\mathrm{Ric}(g) - \lambda g)(x)$$

$$= B_g(\mathrm{Ric}(g) - \lambda g)(x) = 0,$$

since $B_g g = 0$. Therefore if N_3^λ denotes the first prolongation of N_2^λ, the mapping $\pi_2 : N_3^\lambda \to N_2^\lambda$ is surjective, by Proposition 2.8.

The computation of the first obstruction Ω for the equation N_2 of Example 2.4 is quite similar. Namely, if $p = j_2(g)(x) \in N_2$, with $x \in X$ and $g \in S_+^2T_x^*$, then $(\mathrm{Ric}(g) - R)(x) = 0$ and

$$\Omega(p) = \nu \epsilon^{-1} j_1(\mathrm{Ric}(g) - R)(x)$$

$$= \sigma(B_g)\epsilon^{-1} j_1(\mathrm{Ric}(g) - R)(x)$$

$$= B_g(\mathrm{Ric}(g) - R)(x) = -(B_g R)(x).$$

Thus by Proposition 2.8, we see that $p \in \pi_2 N_3$ if and only if $(B_g R)(x) = 0$. For a general section R, the mapping $\pi_2 : N_3 \to N_2$ is not surjective (see DeTurck [1981, 1982]). In fact, a solution g of (7) is also a solution of the equation $B_g R = 0$, and this will be taken into account in the next example.

Example 2.10. If $h \in S^2T^*$, let

$$h^\flat : T \to T^*$$

be the mapping determined by

$$h(\xi, \eta) = \langle \eta, h^\flat(\xi) \rangle,$$

for $\xi, \eta \in T$. If h is non-degenerate, that is, if $h^\flat$ is an isomorphism, we denote by

$$h^\sharp : T^* \to T$$

the inverse of $h^\flat$. We suppose that $n \geq 3$ and consider the objects of Example 2.4. The proofs of the following algebraic lemmas due to DeTurck [1981] can be proved using the methods of Gasqui [1982] involving decompositions into irreducible $O(n)$-modules; here g denotes a Riemannian metric on X.

Lemma 2.11. *The mapping*

$$S^2 T^* \otimes S^2 T^* \xrightarrow{(\mathrm{Tr}_g \circ \tau) \oplus \sigma_1(B_g)} S^2 T^* \oplus (T^* \otimes T^*)$$

is an epimorphism of vector bundles.

Lemma 2.12. *The sequence of vector bundles*

$$S^3 T^* \otimes S^2 T^* \xrightarrow{(\mathrm{Tr}_g \circ \tau)+_1 \oplus \sigma_2(B_g)} (T^* \otimes S^2 T^*) \oplus (S^2 T^* \otimes T^*) \xrightarrow{\mu_g} T^* \to 0$$

is exact, where μ_g sends $u \oplus v$, with $u \in T^* \otimes S^2 T^*$ and $v \in S^2 T^* \otimes T^*$, into $\sigma(B_g)u$.

Let h be a section of $S^2 T^*$. We define a section $L^g(h)$ of $S^2 T^* \otimes T$ by

$$(13) \qquad g(L^g(h)(\xi,\eta),\zeta) = \frac{1}{2}\{(\nabla^g_\xi h)(\eta,\zeta) + (\nabla^g_\eta h)(\zeta,\xi) - (\nabla^g_\zeta h)(\xi,\eta)\},$$

for $\xi, \eta, \zeta \in T$. If $\omega \in \bigotimes^k T^*$, let $L^g(h)\omega$ be the element of $\bigotimes^{k+1} T^*$ given by
(14)

$$(L^g(h)\omega)(\xi,\xi_1,\ldots,\xi_k) = -\sum_{j=1}^{k} \omega(\xi_1,\ldots,\xi_{j-1},L^g(h)(\xi,\xi_j),\xi_{j+1},\ldots,\xi_k),$$

for $\xi, \xi_1, \ldots, \xi_k \in T$. By means of (14), it is easily seen that $L^g(h)R$ is a section of $T^* \otimes S^2 T^*$ satisfying the relation

$$(15) \qquad \langle \xi, \sigma(B_g)(L^g(h)R) \rangle = -R((\mathrm{Tr}^0_g \otimes \mathrm{id})L^g(h),\xi),$$

for $\xi \in T$. From (13), it follows directly that

$$(16) \qquad B_g(h) = g^\flat(\mathrm{Tr}^0_g \otimes \mathrm{id})L^g(h).$$

If R is a non-degenerate section of $S^2 T^*$, from (15) and (16) we deduce that

$$(17) \qquad B_g(h) = - g^\flat \cdot R^\sharp(\sigma(B_g)(L^g(h)R)).$$

Let $x \in X$ and assume that $h(x) = 0$; then $g + h$ is a Riemannian metric on a neighborhood U of x and

$$\nabla^{g+h} - \nabla^g : T \to T^* \otimes T$$

is a differential operator of order zero on U which arises from a section of $S^2 T^* \otimes T$. In fact, we have

$$\nabla^{g+h} - \nabla^g = L^g(h)$$

at x; the verification of this relation is essentially the same as that of identity (4.8) of Gasqui and Goldschmidt [1983]. Thus

$$(\nabla^{g+h} R)(x) = (\nabla^g R)(x) + (L^g(h)R)(x),$$

and so

$$(18) \qquad\qquad B_{g+h}(R) = B_g(R) + \sigma(B_g)(L^g(h)R)$$

at x.

We now assume that R is a non-degenerate section of $S^2 T^*$. We consider the morphism of fibered manifolds

$$\psi_R : J_1(S^2_+ T^*) \to T^*$$

defined by

$$\psi_R(j_1(g)(x)) = (g^\flat \cdot R^\sharp)(B_g(R))(x).$$

According to (18) and (17), we have

$$\psi_R(j_1(g + h)(x)) = \psi_R(j_1(g)(x)) - B_g(h)(x)$$
$$= \psi_R(j_1(g)(x)) - \sigma(B_g)\epsilon^{-1} j_1(h)(x).$$

Thus ψ_R is quasi-linear and its symbol

$$\sigma(\psi_R) : T^* \otimes S^2 T^* \to T^*$$

over $S^2_+ T^*$ sends $(g(x), u)$ into $-\sigma(B_g)u$, for $u \in (T^* \otimes S^2 T^*)_x$, and is surjective by Lemma 2.9. Therefore, according to Proposition 2.6, the morphism of fibered manifolds

$$\Psi = \varphi \oplus p_1(\psi_R) : J_2(S^2_+ T^*) \to S^2 T^* \oplus J_1(T^*)$$

is quasi-linear, and its symbol

$$\sigma(\Psi) : S^2 T^* \otimes S^2 T^* \to S^2 T^* \oplus J_1(T^*)$$

over $J_1(S_+^2 T^*)$ at $p = j_1(g)(x)$ is determined by

$$-\{(\mathrm{Tr}_g \circ \tau) \oplus \sigma_1(B_g)\} : (S^2 T^* \otimes S^2 T^*)_x \to (S^2 T^* \oplus (T^* \otimes T^*))_x;$$

in fact,

$$\sigma(\Psi)(j_1(g)(x), u) = -(\mathrm{Tr}_g \tau(u) \oplus \epsilon \sigma_1(B_g) u),$$

for $u \in (S^2 T^* \otimes S^2 T^*)_x$. Let $x \in X$ and $g_0 \in S_+^2 T_x^*$. Since the symbol of ψ_R is surjective, there exists $p \in J_1(S_+^2 T^*)_x$ with $\psi_R(p) = 0$ and $\pi_0(p) = g_0$. By Lemma 2.11, the image of $\sigma(\Psi)$ is equal to the vector bundle over $J_1(S_+^2 T^*)$ induced from $S^2 T^* \oplus \epsilon(T^* \otimes T^*)$; hence it is easily seen that there exists $q \in J_2(S_+^2 T^*)$ satisfying $\pi_1(q) = p$ and $\Psi(q) = R(x) \oplus 0$ and that Ψ is a submersion. Therefore

$$N_2' = \mathrm{Ker}_{R \oplus 0} \Psi$$

is a differential equation of order 2 satisfying

$$(19) \qquad\qquad \pi_0 N_2' = S_+^2 T^*.$$

Its solutions are the same as of those of equation (7). The diagram

$$
\begin{array}{ccc}
J_3(S_+^2 T^*) & \xrightarrow{\;p_1(\Psi)\;} & J_1(S^2 T^*) \oplus J_1(J_1(T^*)) \\[2pt]
\Big\uparrow{\scriptstyle\,\mathrm{id}} & & \Big\uparrow{\scriptstyle\,\mathrm{id} \oplus \lambda_1} \\[2pt]
J_3(S_+^2 T^*) & \xrightarrow{\;p_1(\varphi) \oplus p_2(\psi_R)\;} & J_1(S^2 T^*) \oplus J_2(T^*)
\end{array}
$$

is commutative, where λ_1 is injective; hence the first prolongation N_3' of N_2' is equal to

$$\mathrm{Ker}_{j_1(R) \oplus 0}\, p_1(\varphi) \oplus p_2(\psi_R).$$

By Proposition 2.6, we have

$$(p_1(\varphi) \oplus p_2(\psi_R))(p + \epsilon u)$$

$$= (p_1(\varphi) \oplus p_2(\psi_R))(p) - ((\epsilon(\mathrm{Tr}_g \circ \tau)_{+1} u) \oplus (\epsilon \sigma_2(B_g) u)),$$

for $p = j_3(g)(x) \in J_3(S_+^2 T^*)$, $u \in (S^3 T^* \otimes S^2 T^*)_x$. Therefore by Lemma 2.12, the first obstruction

$$\Omega' : N_2' \to T^*$$

for the equation N_2' is well-defined by

$$(20) \qquad \Omega'(p) = \mu_g(\epsilon^{-1} j_1(\mathrm{Ric}(g) - R)(x) \oplus \epsilon^{-1} j_2(g^\flat \cdot R^\sharp(B_g(R)))(x)),$$

for $p = j_2(g)(x) \in N_2'$, with $x \in X$ and $g \in S_+^2 T_x^*$ satisfying

$$(\mathrm{Ric}(g) - R)(x) = 0$$

and

$$(21) \qquad j_1(g^\flat \cdot R^\sharp(B_g(R)))(x) = 0.$$

It is easily verified that the sequence

$$N_3' \xrightarrow{\;\pi_2\;} N_2' \underset{0\circ\pi}{\overset{\Omega'}{\rightrightarrows}} T^*$$

is exact. Moreover, by (20), (11) and (21), we have

$$\Omega'(p) = \sigma(B_g)\epsilon^{-1}j_1(\mathrm{Ric}(g) - R)(x)$$
$$= (B_g(\mathrm{Ric}(g) - R))(x) = -(B_g R)(x) = 0.$$

Thus, the mapping $\pi_2 : N_3' \to N_2'$ is surjective.

For $k \geq 0$, we denote by

$$\delta = \Delta_{1,k} : S^{k+1}T^* \to T^* \otimes S^k T^*$$

the natural inclusion; we have

$$\delta(\beta_1 \cdot \ldots \cdot \beta_{k+1}) = \sum_{i=1}^{k+1} \beta_i \otimes \beta_1 \cdot \ldots \cdot \hat{\beta}_i \cdot \ldots \cdot \beta_{k+1},$$

for all $\beta_1, \ldots, \beta_{k+1} \in T^*$, where the symbol $\hat{}$ above a letter means that it is omitted. We extend δ to a morphism of vector bundles

$$\delta : \textstyle\bigwedge^j T^* \otimes S^{k+1}T^* \to \bigwedge^{j+1}T^* \otimes S^k T^*$$

sending $\omega \otimes u$ into $(-1)^j \omega \wedge \delta u$, for all $\omega \in \bigwedge^j T^*$ and $u \in S^{k+1}T^*$. If we set $S^l T^* = 0$, for $l < 0$, the Poincaré lemma for forms with polynomial coefficients implies that the sequence

$$(22) \quad 0 \to S^k T^* \xrightarrow{\;\delta\;} T^* \otimes S^{k-1}T^* \xrightarrow{\;\delta\;} \textstyle\bigwedge^2 T^* \otimes S^{k-2}T^* \xrightarrow{\;\delta\;} \cdots$$
$$\to \textstyle\bigwedge^n T^* \otimes S^{k-n}T^* \to 0$$

is exact, for $k \geq 1$.

Let $\rho : Y \to X$ be a fibered manifold and let

$$\psi : S^k T^* \otimes E \to E'$$

be a morphism of vector bundles over Y, where $S^k T^* \otimes E$ and E' are considered as induced vector bundles over Y via the mapping ρ. Let g_k be the kernel of ψ, which is a vector bundle over Y with variable fiber. For $l \geq 0$, the kernel g_{k+l} of the l-th prolongation $(\psi)_{+l}$ of ψ is equal to

$$(S^{k+l} T^* \otimes E) \cap (S^l T^* \otimes g_k)$$

and is called the l-th prolongation of g_k. We set $g_{k+l} = S^{k+l} T^* \otimes E$, considered as a vector bundle over Y, for $l < 0$. It is easily seen that the diagram

$$(23) \quad
\begin{array}{ccc}
\bigwedge^j T^* \otimes S^{k+l+1} T^* \otimes E & \xrightarrow{\ \mathrm{id} \otimes (\psi)_{+(l+1)}\ } & \bigwedge^j T^* \otimes S^{l+1} T^* \otimes E' \\[4pt]
\Big\downarrow{\scriptstyle \delta} & & \Big\downarrow{\scriptstyle \delta} \\[4pt]
\bigwedge^{j+1} T^* \otimes S^{k+l} T^* \otimes E & \xrightarrow{\ \mathrm{id} \otimes (\psi)_{+l}\ } & \bigwedge^{j+1} T^* \otimes S^l T^* \otimes E'
\end{array}$$

commutes, and so the morphism δ induces by restriction mappings

$$\delta : \bigwedge^j T^* \otimes g_{k+l+1} \to \bigwedge^{j+1} T^* \otimes g_{k+l};$$

the cohomology of the complexes

$$(24) \quad 0 \to g_m \xrightarrow{\delta} T^* \otimes g_{m-1} \xrightarrow{\delta} \bigwedge^2 T^* \otimes g_{m-2} \xrightarrow{\delta} \cdots \to \bigwedge^n T^* \otimes g_{m-n} \to 0$$

is the *Spencer cohomology* of g_k. We denote by $H^{m-j,j}(g_k)$ the cohomology of (24) at $\bigwedge^j T^* \otimes g_{m-j}$. We say that g_k is r-acyclic if $H^{m,j}(g_k) = 0$, for all $m \geq k$ and $0 \leq j \leq r$, and that g_k is involutive if it is n-acyclic. It is easily seen that g_k is always 1-acyclic.

The following theorem asserts that all but a finite number of these cohomology groups vanish, whenever there is an integer d such that $\dim E_x \leq d$ for all $x \in X$ (see Quillen [1964], Sweeney [1968]).

Theorem 2.13. *If there is an integer d for which $\dim E_x \leq d$, for all $x \in X$, then there exists an integer k_0 depending only on n, k and d such that*

$$H^{m,j}(g_k) = 0,$$

for all $m \geq k_0$, $j \geq 0$.

Let $x \in X$ and $\{t_1, \ldots, t_n\}$ be a basis of T_x. If $\{\alpha_1, \ldots, \alpha_n\}$ is the basis of T_x^* dual to $\{t_1, \ldots, t_n\}$, then we denote by $S^k T^*_{x,\{t_1,\ldots,t_j\}}$ the subspace of $S^k T_x^*$ generated by the symmetric products $\alpha_{i_1} \cdot \ldots \cdot \alpha_{i_k}$, with $j+1 \leq i_1 \leq \cdots \leq i_k \leq n$. If $y \in Y$, with $\rho(y) = x$, we set

$$g_{k,y,\{t_1,\ldots,t_j\}} = g_{k,y} \cap (S^k T^*_{x,\{t_1,\ldots,t_j\}} \otimes E_x).$$

We say that $\{t_1, \ldots, t_n\}$ is a quasi-regular basis for g_k at y if

$$\dim g_{k+1,y} = \dim g_{k,y} + \sum_{j=1}^{n-1} \dim g_{k,y,\{t_1,\ldots,t_j\}}.$$

The following criterion for the involutivity of g_k is due to Serre (see Guillemin and Sternberg [1964], Appendix; see also §§2,3, Chapter VIII).

Theorem 2.14. *The following conditions are equivalent:*
  (i) *there exists a quasi-regular basis of T_x for g_k at y;*
  (ii) $H^{m,j}(g_k)_y = 0$, *for all* $m \geq k$, $j \geq 0$.

An elementary argument, due to Sternberg and based on É. Cartan's proof of the Poincaré lemma, shows that condition (i) of the above theorem implies (ii). Any basis of T_x is quasi-regular for the sub-bundle T^* of $S^1 T^*$; since $S^{k+1} T^*$ is its k-th prolongation, this argument proves that the sequence (22) is exact (see §2, Chapter VIII).

Using the preceding theorem, it is easily seen that:

Lemma 2.15. *If g_{k+1} is a vector bundle over Y and g_k is involutive at $y_0 \in Y$, then g_k is involutive for all y in a neighborhood of y_0.*

We again consider the morphism $\varphi : F \to E'$, where F is an open fibered submanifold of $J_k(E)$. Assume that φ is quasi-linear and suppose that conditions (3) and (4) hold; then R_k is a differential equation. The symbol g_k of R_k and the l-th prolongation g_{k+l} of the symbol of R_k are the vector bundles with variable fiber over R_k whose fibers at $p \in R_k$ are

$$\tag{25} g_{k,p} = \text{Ker } \sigma(\varphi)_{\pi_{k-1}p},$$

$$\tag{26} g_{k+l,p} = \text{Ker } \sigma_l(\varphi)_{\pi_{k-1}p} = (g_k)_{+l,p};$$

they depend only on R_k and not on φ (see §3).

We say that the differential equation R_k is *formally integrable* if:
  (i) g_{k+l+1} is a vector bundle over R_k, for all $l \geq 0$;
  (ii) $\pi_{k+l} : R_{k+l+1} \to R_{k+l}$ is surjective, for all $l \geq 0$.

The following result (Goldschmidt [1967b], Theorem 8.1) is the formal part of the basic existence theorem for quasi-linear morphisms; it gives us a criterion for formal integrability. Together with Theorem 2.2, it provides us with the existence of analytic solutions for analytic quasi-linear partial differential equations.

Theorem 2.16. *Assume that (3) and (4) hold. If*
  (A) g_{k+1} *is a vector bundle over R_k,*
  (B) $\pi_k : R_{k+1} \to R_k$ *is surjective,*
  (C) g_k *is 2-acyclic,*

then R_k is formally integrable.

In fact, if (3), (4), (A) and (B) hold, Goldschmidt [1967b] constructs the second obstruction to the integrability of R_k, which is a mapping

$$\kappa : R_{k+1} \to H^{k,2}(g_k)$$

over R_k, and then proves that the sequence

$$R_{k+2} \xrightarrow{\pi_{k+1}} R_{k+1} \underset{0 \circ \pi_k}{\xrightarrow{\kappa}} H^{k,2}(g_k)$$

is exact. Thus the higher obstructions to integrability lie in the cohomology groups $H^{k+l,2}(g_k)$ and condition (C) then implies that the mappings $\pi_{k+l} : R_{k+l+1} \to R_{k+l}$ are surjective for $l > 0$.

A complete proof of Theorem 2.16 for linear equations will be given in Chapter X (Theorem 1.6); in particular, in the course of this proof, the mapping κ will be constructed.

According to Theorem 2.14, we may replace condition (C) in the above theorem by:

(C′) *for all $p \in R_k$, there exists a quasi-regular basis of $T_{\pi(p)}$ for g_k at* p.

We remark that conditions (A), (B) and (C′) are of "finite type", in the sense that they involve only φ and $\sigma(\varphi)$ and their first prolongations.

From Theorems 2.13 and 2.16, Goldschmidt [1967b] deduces the following version of the Cartan–Kuranishi prolongation theorem (Kuranishi [1957]), which asserts that the condition of formal integrability is of "finite type": to determine whether R_k is formally integrable, we need examine only a finite number of prolongations of φ.

Theorem 2.17. *Assume that X is connected and that (3) and (4) hold. There exists an integer $k_0 \geq k$ depending only on n, k and the rank of E such that, if*
(i) *g_{k+l+1} is a vector bundle over R_k, for all $0 \leq l \leq k_0 - k$,*
(ii) *$\pi_{k+l} : R_{k+l+1} \to R_{k+l}$ is surjective, for all $0 \leq l \leq k_0 - k$,*
then R_k is formally integrable.

We now show how Theorems 2.16 and 2.2 give us the existence of solutions for the equations of Examples 2.5 and 2.10 in the analytic case.

Examples 2.5 and 2.10 (continued). For the differential equations N_2^λ and N_2' we have verified condition (B). The involutivity of the symbols of N_2^λ and N_2' will be a consequence of the following lemma of Gasqui [1982] and DeTurck [1981].

Lemma 2.18. *Let g be a Riemannian metric on X and $x \in X$. An orthonormal basis of T_x is quasi-regular for the kernels of the morphisms*

$$\text{(27)} \qquad\qquad \text{Tr}_g \circ \tau : S^2 T^* \otimes S^2 T^* \to S^2 T^*,$$

$$\text{(28)} \qquad (\text{Tr}_g \circ \tau) \oplus \sigma_1(B_g) : S^2 T^* \otimes S^2 T^* \to S^2 T^* \oplus (T^* \otimes T^*)$$

at x.

We remark that the first prolongations of the kernels of the morphisms (27) and (28) are equal to the kernels of $(\text{Tr}_g \circ \tau)_{+1}$ and $(\text{Tr}_g \circ \tau)_{+1} \oplus \sigma_2(B_g)$ respectively.

Let g be a Riemannian metric on X and $x \in X$; set $p = j_2(g)(x)$. If $p \in N_2^\lambda$ (resp. N_2'), then the fiber of the symbol of N_2^λ (resp. N_2') is the kernel of (27) (resp. (28)) at x. Thus by Lemma 2.18 condition (C$'$) holds for N_2^λ and N_2', while condition (A) for these equations is a consequence of Lemmas 2.9 and 2.12. Therefore, by Theorem 2.16, N_2^λ and N_2' are formally integrable.

Assume that X is a real-analytic manifold. Then ψ_λ and N_2^λ are analytic. If R is an analytic non-degenerate section of $S^2 T^*$, then Ψ and N_2' are analytic. The following result of DeTurck [1981] is now a direct consequence of Theorem 2.2 and (19); its proof outlined here is a variant of the one given by DeTurck.

Theorem 2.19. *Let X be a real-analytic manifold of dimension $n \geq 3$ and let R be an analytic non-degenerate section of $S^2 T^*$. If $x \in X$ and $g_0 \in S_+^2 T_x^*$, there exists an analytic Riemannian metric g on a neighborhood of x such that*

$$g(x) = g_0, \quad \text{Ric}(g) = R.$$

The following theorem is due to Gasqui [1982].

Theorem 2.20. *Let X be a real-analytic manifold of dimension $n \geq 3$ and $x \in X$. Let g_0 be a Riemannian metric on X and $R_0 \in G_x$ such that*

$$-\text{Tr}_{g_0} R_0 = \lambda g_0(x),$$

with $\lambda \in \mathbb{R}$. Then there exists an analytic Riemannian metric g on a neighborhood of x such that

$$g(x) = g_0, \quad \mathcal{R}(g)(x) = R_0, \quad \text{Ric}(g) = \lambda g.$$

Proof. Since the morphism Φ of Example 2.3 is quasi-linear and its symbol is surjective, we see that there is an element p of $J_2(S_+^2 T^*)_x$ satisfying $\pi_0(p) = g_0(x)$ and $\Phi(p) = R_0$. From our hypothesis on R_0, we see that $p \in N_2^\lambda$. Because N_2^λ is formally integrable, Theorem 2.2 gives us an analytic solution g of N_2^λ over a neighborhood of x such that $j_2(g)(x) = p$.

§3. Existence Theorems.

We now briefly show how the results of §2 can be generalized to arbitrary systems of partial differential equations. As the equations are in general no longer quasi-linear, we must consider the structure of affine bundle which the jet bundles possess.

We no longer assume that E and E' are vector bundles, but continue to suppose that they are fibered manifolds over X. We denote by $V(E)$ the bundle of vectors tangent to the fibers of $\pi : E \to X$.

According to Proposition 5.1 of Goldschmidt [1967b], for $k \geq 1$ the jet bundle $J_k(E)$ is an affine bundle over $J_{k-1}(E)$ modeled on the vector bundle $S^k T^* \otimes_{J_{k-1}(E)} V(E)$ over $J_{k-1}(E)$. In fact, if $p \in J_{k-1}(E)$ with $\pi(p) = x$, the vector space $S^k T_x^* \otimes V_{\pi_0 p}(E)$ considered as an additive group acts freely and transitively on the fiber of $J_k(E)$ over p; for $u \in S^k T_x^* \otimes V_{\pi_0 p}(E)$, we denote by $u + q$ the image of the element q of $J_k(E)_p$ under the action of u. In terms of the standard local coordinate system for $J_k(E)$ considered in §2, this action can be described as follows. If $\alpha = (\alpha_1, \ldots, \alpha_n)$ is a multi-index of norm k, we consider the section

$$dx^\alpha = (dx^1)^{\alpha_1} \cdot \ldots \cdot (dx^n)^{\alpha_n}$$

of $S^k T^*$. If an element q of $J_k(E)$ satisfies $\pi_0 q \in U$ and $\pi(q) = x$, and if (x^i, y^j, y_α^j) are the coordinates of q and

$$u = \sum_{|\alpha|=k} a_\alpha^j (dx^\alpha)(x) \otimes \frac{\partial}{\partial y^j}(\pi_0 q),$$

the coordinates of $u + q$ are (x^i, y^j, z_α^j), where

$$z_\alpha^j = \begin{cases} y_\alpha^j + a_\alpha^j, & \text{if } |\alpha| = k, \\ y_\alpha^j, & \text{if } 1 \leq |\alpha| < k. \end{cases}$$

An intrinsic definition of this action is given by Goldschmidt [1967b], §5. If E is a vector bundle, then $V_{\pi_0 p}(E)$ is canonically isomorphic to E_x and so

$$S^k T_x^* \otimes V_{\pi_0 p}(E) \simeq S^k T^* \otimes E_x;$$

in this case, the action of $S^k T_x^* \otimes V_{\pi_0 p}(E)$ on $J_k(E)$ can be described in terms of the vector bundle structure of $J_k(E)$ and the morphism ϵ:

$$u + q = \epsilon(u) + q,$$

where on the left-hand side u is considered as an element of $S^k T_x^* \otimes V_{\pi_0 p}(E)$, while on the right-hand side u is viewed as an element of $(S^k T^* \otimes E)_x$.

From the affine bundle structure of $J_k(E)$, we obtain a morphism of vector bundles

$$\mu : S^k T^* \otimes_{J_k(E)} V(E) \to V(J_k(E)),$$

sending (p, u), with $p \in J_k(E)$ and $u \in S^k T^*_{\pi(p)} \otimes V_{\pi_0 p}(E)$, into the tangent vector

$$\frac{d}{dt}(p + tu)|_{t=0},$$

where $t \in \mathbb{R}$. It is easily seen that the sequence

$$0 \to S^k T^* \otimes_{J_k(E)} V(E) \xrightarrow{\mu} V(J_k(E)) \xrightarrow{\pi_{k-1*}} \pi_{k-1}^{-1} V(J_{k-1}(E)) \to 0$$

of vector bundles over $J_k(E)$ is exact (see Goldschmidt [1967b], Proposition 5.2). We shall identify $S^k T^* \otimes_{J_k(E)} V(E)$ with its image in $V(J_k(E))$ under the mapping μ.

Let F be an open fibered submanifold of $J_k(E)$ and let $\varphi : F \to E'$ be a morphism of fibered manifolds over X. The mapping $\varphi_* : V(F) \to V(E')$ induces a morphism

$$\varphi_* = \varphi_* \circ \mu : S^k T^* \otimes_F V(E) \to V(E'),$$

which we also denote by $\sigma(\varphi)$; let $\sigma_l(\varphi)$ be the composition

$$S^{k+l} T^* \otimes_F V(E) \xrightarrow{\Delta_{l,k} \otimes \, \mathrm{id}} S^l T^* \otimes S^k T^* \otimes_F V(E) \xrightarrow{\mathrm{id} \, \otimes \varphi_*} S^l T^* \otimes_{E'} V(E').$$

According to Proposition 5.6 of Goldschmidt [1967b], for $l \geq 1$ the mapping

$$p_l(\varphi) : \pi_k^{-1} F \to J_l(E')$$

is a morphism of affine bundles over $p_{l-1}(\varphi)$ whose associated morphism of vector bundles is induced by $\sigma_l(\varphi)$; in other words,

$$(29) \qquad p_l(\varphi)(u + p) = \sigma_l(\varphi)_{\pi_k p}(u) + p_l(\varphi)p,$$

for all $p \in J_{k+l}(E)$, with $\pi_k p \in F$, and $u \in S^{k+l} T^*_{\pi(p)} \otimes V_{\pi_0 p}(E)$. This formula is easily verified using the standard local coordinates on jet bundles. If E, E' are vector bundles and φ is quasi-linear, then $\sigma_l(\varphi)$ can be identified with the mapping of §2 denoted there by $\sigma_l(\varphi)$; moreover, Proposition 2.6 can be deduced from formula (29).

Let R_k be a differential equation of order k on E. The l-th prolongation of R_k is the subset R_{k+l} of $J_{k+l}(E)$ determined by the equality

$$\lambda_l R_{k+l} = J_l(R_k) \cap \lambda_l J_{k+l}(E),$$

where $J_l(R_k)$ is considered as a subset of $J_l(J_k(E))$. The projection $\pi_{k+l}:$ $J_{k+l+1}(E) \to J_{k+l}(E)$ sends R_{k+l+1} into R_{k+l}. The symbol of R_k is the sub-bundle with varying fiber

$$g_k = V(R_k) \cap (S^k T^* \otimes_{R_k} V(E))$$

of $(S^k T^* \otimes_{J_k(E)} V(E))|_{R_k}$, which is equal to the kernel of the morphism of vector bundles

$$(S^k T^* \otimes_{J_k(E)} V(E))|_{R_k} \to (V(J_k(E))|_{R_k})/V(R_k)$$

over R_k. Let g_{k+l} be the l-th prolongation of g_k; it is a sub-bundle with varying fiber of $S^{k+l} T^* \otimes_{R_k} V(E)$.

If s' is a section of E' over X and R_k is given by (2), and if conditions (3) and (4) are satisfied, then the l-th prolongation R_{k+l} of R_k is equal to $\mathrm{Ker}_{j_l(s')} p_l(\varphi)$ and

$$g_k = (\mathrm{Ker}\ \sigma(\varphi))|_{R_k}, \quad g_{k+l} = (\mathrm{Ker}\ \sigma_l(\varphi))|_{R_k};$$

thus in this case, our definition of R_{k+l} coincides with the one given in §2. Using the identity (29), the first obstruction $\Omega : R_k \to W$ to the integrability of R_k, where W is a vector bundle with variable fiber over R_k, can be constructed in a way similar to that of §2; then we still have

$$\pi_k R_{k+1} = \{p \in R_k \mid \Omega(p) = 0\}$$

(see Goldschmidt [1972a], Proposition 2.1). If moreover E, E' are vector bundles and φ is quasi-linear, the morphisms $\sigma(\varphi)$ and $\sigma_l(\varphi)$ can be identified with the mappings of §2, while g_k and g_{k+l} are given by (25) and (26).

We say that a differential equation R_k of order k on E is formally integrable if:

   (i) g_{k+l+1} is a vector bundle over R_k, for all $l \geq 0$;

   (ii) $\pi_{k+l} : R_{k+l+1} \to R_{k+l}$ is surjective, for all $l \geq 0$.

If R_k is formally integrable, then for $l \geq 0$, according to Proposition 7.2 of Goldschmidt [1967b], R_{k+l} is a fibered submanifold of $J_{k+l}(E)$ and $\pi_{k+l} :$ $R_{k+l+1} \to R_{k+l}$ is an affine sub-bundle of $\pi_{k+l} : J_{k+l+1}(E)|_{R_{k+l}} \to R_{k+l}$, whose associated vector bundle is the vector bundle $\pi_k^{-1} g_{k+l+1}$ over R_{k+l} induced from g_{k+l+1} by $\pi_k : R_{k+l} \to R_k$; this last statement implies that for each $p \in R_{k+l}$ the fiber of R_{k+l+1} over p is an affine subspace of the fiber of $J_{k+l+1}(E)$ over p whose associated vector space is $g_{k+l+1,\pi_k p}$.

Theorem 2.16 can be viewed as a special case of Theorem 8.1 of Goldschmidt [1967b], which we now state as

Theorem 3.1. *Let R_k be a differential equation of order k on E. If*

   (A) *g_{k+1} is a vector bundle over R_k,*

   (B) *$\pi_k : R_{k+1} \to R_k$ is surjective,*

   (C) *g_k is 2-acyclic,*

then R_k is formally integrable.

Again as for Theorem 2.16, according to Theorem 2.14, we may replace condition (C) in the above theorem by:

   (C′) *for all $p \in R_k$, there exists a quasi-regular basis of $T_{\pi(p)}$ for g_k at p.*

Theorem 2.17 also holds for a differential equation of order k on E, with $k_0 \geq k$ depending only on n, k and the dimension of E.

Theorem 2.2 now provides us with the existence of analytic solutions:

Theorem 3.2. *Suppose that X is a real-analytic manifold and that E is a real-analytic fibered manifold. If R_k is an analytic formally integrable differential equation of order k on E, then for all $p \in R_{k+l}$ there exists an analytic solution s of R_k on a neighborhood of $x = \pi(p)$ such that $j_{k+l}(s)(x) = p$.*

We now present the intrinsic formulation of the criterion of §1 for the existence of analytic solutions of analytic equations; it is a direct consequence of Theorems 3.1 and 2.2.

Theorem 3.3. *Assume that X is a real-analytic manifold and that E, E' are real-analytic fibered manifolds. Let F be an open fibered submanifold of $J_k(E)$ and let $\varphi : F \to E'$ be an analytic morphism of fibered manifolds over X and s' be an analytic section of E' over X. Let*

$$R_k = \mathrm{Ker}_{s'}\, \varphi, \qquad R_{k+1} = \mathrm{Ker}_{j_1(s')}\, p_1(\varphi),$$

$$g_k = (\mathrm{Ker}\, \sigma(\varphi))|_{R_k}, \quad g_{k+1} = (\mathrm{Ker}\, \sigma_1(\varphi))|_{R_k},$$

and let $p \in R_k$. If there exists a neighborhood U of p in F such that:

   (i) *φ has constant rank on U,*

   (ii) *$s'(\pi U) \subset \varphi(F)$,*

   (iii) *$\pi_k : R_{k+1} \cap \pi_k^{-1}(U) \to R_k \cap U$ is surjective,*

   (iv) *$g_{k+1}\,|_{(U \cap R_k)}$ is a vector bundle,*

   (v) *$H^{k+l,2}(g_k)|_{(U \cap R_k)} = 0$, for all $l \geq 0$,*

then there exists an analytic section s of E over a neighborhood V of $x_0 = \pi(p)$ such that $j_k(s)(x_0) = p$ and $j_k(s)(x) \in R_k$ for all $x \in V$.

According to Lemma 2.15, the condition (v) can be replaced by the stronger condition:

   (vi) *there exists a quasi-regular basis of T_{x_0} for g_k at p.*

Kuranishi [1967] proves the existence of analytic solutions of analytic partial differential equations by the method of Cartan–Kähler, using the Cauchy–Kowalewski theorem, under these assumptions (i)–(iv) and (vi).

CHAPTER X

LINEAR DIFFERENTIAL OPERATORS

In this chapter, we consider only linear systems of partial differential equations, and use the notation and terminology introduced in Chapter IX. In general, if $D : \mathcal{E} \to \mathcal{F}$ is a linear differential operator, where E, F are vector bundles over the manifold X, and if f is a section of F, the inhomogeneous equation

$$Du = f$$

is not solvable for a section u of E unless f satisfies a requisite compatibility condition. Indeed, certain conditions must be imposed on the formal power series expansion $j_\infty(f)(x)$ of f at $x \in X$ in order that it may be written as $j_\infty(Du)(x)$, for some section u of E. Under certain regularity assumptions on D, they can be expressed in terms of a differential operator $P : \mathcal{F} \to \mathcal{B}$ of finite order, where B is a vector bundle over X. This operator is called the compatibility condition for D and is obtained by repeatedly differentiating the equation. We then obtain a complex of differential operators

$$\mathcal{E} \xrightarrow{\ D\ } \mathcal{F} \xrightarrow{\ P\ } \mathcal{B}$$

which is exact at the formal power series level: the formal power series expansion $j_\infty(f)(x)$ of f at x can be written in the form $j_\infty(Du)(x)$, for some section u of E, if and only if Pf vanishes to infinite order at x. For example, the inhomogeneous equation $du = f$, where u is a real-valued function and f is a 1-form on X, is not solvable for u unless $df = 0$.

In Section 1, we present a complete proof of the formal existence theorem of Goldschmidt [1967a] for homogeneous linear systems (Theorem 1.6), and existence results for the compatibility condition of a linear differential operator, as well as existence theorems for analytic differential operators. We also construct formally exact complexes of differential operators, whose vector bundles can be explicitly described under specific hypotheses; these include the sophisticated Spencer sequence (see Theorems 1.9 and 1.11, Examples 1.10, 1.12 and 1.13).

Section 2 is devoted to various examples of these complexes, while Section 3 is concerned with exactness results for our complexes under the additional assumption of ellipticity.

§1. Formal Theory and Complexes.

Let E, F be vector bundles over the differentiable manifold X of dimension n. We denote by $C^\infty(U, E)$ the space of sections of the vector bundle E over an open subset U of X. Let $\varphi : J_k(E) \to F$ be a morphism of vector bundles and let $D = \varphi \circ j_k : \mathcal{E} \to \mathcal{F}$ be the corresponding differential operator of order k. We consider the morphisms associated to φ and D in Example 2.7 of Chapter IX. Let

$$J_\infty(E) = \mathrm{pr}\ \lim J_m(E)$$

be the bundle of jets of infinite order of sections of E, and set $p_\infty(D) = \mathrm{pr}\ \lim p_m(D)$. If $k = 1$, then

$$(1) \qquad\qquad D(fs) = \sigma(D)(df \otimes s) + fDs,$$

where f is a real-valued function on X and s is a section of E over X. Indeed, this formula holds if f is a constant function; on the other hand, if $x \in X$ and $f(x) = 0$, then it is valid at x, according to the definitions of ϵ and $\sigma(D)$.

Let R_k be the kernel of φ and R_{k+l} the kernel of $p_l(\varphi)$. The symbol g_k of (R_k, φ) is the kernel of $\sigma(\varphi) : S^k T^* \otimes E \to F$ and its l-th prolongation g_{k+l} is the kernel of the morphism of vector bundles

$$\sigma_l(\varphi) = (\sigma(\varphi))_{+l} : S^{k+l} T^* \otimes E \to S^l T^* \otimes F$$

over X. For $l < 0$, we write $R_{k+l} = J_{k+l}(E)$ and $g_{k+l} = S^{k+l} T^* \otimes E$; then the sequence

$$0 \to g_{k+l} \xrightarrow{\epsilon} R_{k+l} \xrightarrow{\pi_{k+l-1}} R_{k+l-1}$$

is exact. We recall that, if φ has locally constant rank, R_k is a linear differential equation and that the l-th prolongation R_{k+l} of R_k is determined by the equality

$$\lambda_l R_{k+l} = J_l(R_k) \cap \lambda_l J_{k+l}(E).$$

If R_k is a vector bundle, we call g_k the symbol of the equation R_k. If R_{k+l} is a vector bundle, it is easily verified that the diagram

$$
\begin{array}{ccccccc}
0 & \longrightarrow & R_{k+l+m} & \longrightarrow & J_{k+l+m}(E) & \xrightarrow{\ p_{l+m}(\varphi)\ } & J_{l+m}(F) \\
& & & & \downarrow{\scriptstyle \mathrm{id}} & & \downarrow{\scriptstyle \lambda_m} \\
0 & \longrightarrow & (R_{k+l})_{+m} & \longrightarrow & J_{k+l+m}(E) & \xrightarrow{\ p_m(p_l(\varphi))\ } & J_m(J_l(F))
\end{array}
$$

is commutative and exact, where λ_m is injective; therefore the m-th prolongation $(R_{k+l})_{+m}$ of the equation R_{k+l} is equal to R_{k+l+m}. It follows that the m-th prolongation $(g_{k+l})_{+m}$ of g_{k+l} is equal to g_{k+l+m}. We say that R_k is formally integrable if:

(i) R_{k+l} is a vector bundle for all $l \geq 0$;

(ii) $\pi_{k+l} : R_{k+l+1} \to R_{k+l}$ is surjective for all $l \geq 0$.

We recall the following lemma of Goldschmidt [1967a] which we will require later.

Lemma 1.1. *If*

$$E' \xrightarrow{\psi'} E \xrightarrow{\psi''} E''$$

is an exact sequence of vector bundles over X, then the kernel of ψ' and the cokernel of ψ'' are both vector bundles.

Let B be a vector bundle over X. If $D' : \mathcal{F} \to \mathcal{B}$ is a differential operator of order l, we say that the sequence

$$(2) \qquad \mathcal{E} \xrightarrow{D} \mathcal{F} \xrightarrow{D'} \mathcal{B}$$

is formally exact if the sequence

$$J_\infty(E) \xrightarrow{p_\infty(D)} J_\infty(F) \xrightarrow{p_\infty(D')} J_\infty(B)$$

is exact.

Example 1.2. Suppose that $X = \mathbb{R}^n$ and that (2) is a complex of constant coefficient differential operators. If U is an open convex subset of $\mathbb{R}^n$, according to the Ehrenpreis–Malgrange theorem (see Ehrenpreis [1970], Malgrange [1963] and Hörmander [1973], §7.6), the sequence

$$C^\infty(U, E) \xrightarrow{D} C^\infty(U, F) \xrightarrow{D'} C^\infty(U, B)$$

is exact.

Unfortunately, formal exactness is not a good concept for operators with variable coefficients, even for analytic operators, as we shall see below with Example 1.5. It shall be replaced by stronger conditions (see Theorem 1.4).

Lemma 1.3. *Let $\varphi : J_k(E) \to F$, $\psi : J_l(F) \to B$ be morphisms of vector bundles over X, and set $D = \varphi \circ j_k$, $D' = \psi \circ j_l$. Let $m_0 \geq 0$ and assume that the sequences of vector bundles*

$$(3) \qquad J_{k+l+m}(E) \xrightarrow{p_{l+m}(\varphi)} J_{l+m}(F) \xrightarrow{p_m(\psi)} J_m(B)$$

are exact, for all $m \geq m_0$. Then the sequence (2) is formally exact, R_{k+l+m} is a vector bundle for all $m \geq m_0$ and $N_{l+m_0} = \text{Ker } p_{m_0}(\psi)$ is a formally integrable differential equation of order $l + m_0$ on F. Moreover, if $R_{k+l+m_0} = \text{Ker } p_{l+m_0}(\varphi)$ is formally integrable, the sequences

$$(4) \quad S^{k+l+m+1}T^* \otimes E \xrightarrow{\sigma_{l+m+1}(\varphi)} S^{l+m+1}T^* \otimes F \xrightarrow{\sigma_{m+1}(\psi)} S^{m+1}T^* \otimes B$$

are exact for all $m \geq m_0$.

Proof. The first assertion of the lemma is a direct consequence of Corollary 2, §3, n^0 5 of Bourbaki [1965], since finite-dimensional vector spaces are artinian. From Lemma 1.1, it follows that R_{k+l+m} and $N_{l+m} = \operatorname{Ker} p_m(\psi)$ are vector bundles, for $m \geq m_0$; the exactness of the sequences (3) gives us also the surjectivity of $\pi_{l+m} : N_{l+m+1} \to N_{l+m}$, for $m \geq m_0$. For $m \geq m_0$, we consider the commutative diagram (5). If R_{k+l+m_0} is formally integrable, its columns are exact; by means of this diagram, the exactness of the sequences (3) then implies the exactness of (4), for $m \geq m_0$.

If X is a real-analytic manifold and E is a real-analytic vector bundle over X, we denote by $\mathcal{E}_\omega$ the sheaf of analytic sections of E.

Theorem 1.4. *Suppose that X is a real-analytic manifold, and that E, F, B are real-analytic vector bundles over X. Let $\varphi : J_k(E) \to F$, $\psi : J_l(F) \to B$ be real-analytic morphisms of vector bundles over X, and set $D = \varphi \circ j_k$, $D' = \psi \circ j_l$. Let $m_0 \geq 0$; if the sequences (3) are exact for all $m \geq m_0$ and $R_{k+l+m_0} = \operatorname{Ker} p_{l+m_0}(\varphi)$ is formally integrable, then the sequence*

$$\mathcal{E}_\omega \xrightarrow{D} \mathcal{F}_\omega \xrightarrow{D'} \mathcal{B}_\omega$$

is exact.

Proof. Let f be an analytic section of F over a neighborhood U of $x \in X$ satisfying $D'f = 0$. From the exactness of the sequences (3), we deduce that, for $m \geq l + m_0$,

$$N_{k+m} = \operatorname{Ker}_{j_m(f)} p_m(\varphi)|_U$$

is an affine sub-bundle of the vector bundle $J_{k+m}(E)|_U$ whose associated vector bundle is $R_{k+m}|_U$. By means of the commutativity of diagram (5) and the exactness of the sequence (4), with $m \geq m_0$, given by Lemma 1.3, we easily see that $\pi_{k+l+m} : N_{k+l+m+1} \to N_{k+l+m}$ is surjective for all $m \geq m_0$. According to Theorem 2.2, Chapter IX, there exists an analytic section s of E over a neighborhood of x such that $j_{k+l+m_0}(s)$ is a section of N_{k+l+m_0}; then $Ds = f$ on this neighborhood.

Goldschmidt [1968a] shows that one can replace the hypothesis "R_{k+l+m_0} is formally integrable" in the above theorem by the weaker condition:

$\pi_m : R_{m+r} \to R_m$ has locally constant rank for all $m \geq k+l+m_0$, $r \geq 0$.

The following example shows that these conditions can not be weakened in an essential way.

$$
\begin{array}{ccccccccc}
 & & 0 & & 0 & & 0 & & 0 \\
 & & \downarrow & & \downarrow & & \downarrow & & \downarrow \\
0 \to & g_{k+l+m+1} & \to & S^{k+l+m+1}T^* \otimes E & \xrightarrow{\sigma_{l+m+1}(\varphi)} & S^{l+m+1}T^* \otimes F & \xrightarrow{\sigma_{m+1}(\psi)} & S^{m+1}T^* \otimes B \\
 & \downarrow{\scriptstyle \epsilon} & & \downarrow{\scriptstyle \epsilon} & & \downarrow{\scriptstyle \epsilon} & & \downarrow{\scriptstyle \epsilon} \\
0 \to & R_{k+l+m+1} & \to & J_{k+l+m+1}(E) & \xrightarrow{p_{l+m+1}(\varphi)} & J_{l+m+1}(F) & \xrightarrow{p_{m+1}(\psi)} & J_{m+1}(B) \\
 & \downarrow{\scriptstyle \pi_{k+l+m}} & & \downarrow{\scriptstyle \pi_{k+l+m}} & & \downarrow{\scriptstyle \pi_{l+m}} & & \downarrow{\scriptstyle \pi_m} \\
0 \to & R_{k+l+m} & \to & J_{k+l+m}(E) & \xrightarrow{p_{l+m}(\varphi)} & J_{l+m}(F) & \xrightarrow{p_m(\psi)} & J_m(B) \\
 & \downarrow & & \downarrow & & \downarrow & & \downarrow \\
 & & 0 & & 0 & & 0 & & 0
\end{array}
$$

$$(5)$$

Example 1.5. Suppose that $X = \mathbb{R}$ with its standard coordinate x, and that $E = F$ is the trivial real line bundle. We identify a section of E with a real-valued function on X. Consider the analytic first-order differential operator $D : \mathcal{E} \to \mathcal{E}$ given by

$$Df = x^2 \frac{df}{dx} - f,$$

where $f \in \mathcal{E}$. Let $\varphi : J_1(E) \to E$ be the morphism of vector bundles satisfying $D = \varphi \circ j_1$, and $R_{l+1} = \operatorname{Ker} p_l(\varphi)$. For $l \geq 0$, it is easily verified that R_{l+1} is a vector bundle of rank 1 on X and that $\pi_0 : R_{l+1} \to E$ is an isomorphism on $X - \{0\}$; therefore R_1 is formally integrable on $X - \{0\}$. However $\dim g_{l+1,x} = 1$ for $x = 0$, and so $\pi_l : R_{l+1} \to R_l$ is not surjective at $x = 0$ and does not have constant rank on X. On the other hand, the sequence

$$0 \to R_{l+1} \to J_{l+1}(E) \xrightarrow{\; p_l(\varphi) \;} J_l(E) \to 0$$

is exact, for all $l \geq 0$; thus the sequences (3) corresponding to the complex

$$(6) \qquad\qquad\qquad \mathcal{E} \xrightarrow{\; D \;} \mathcal{E} \to 0$$

are exact for all $m \geq 0$, and this complex is formally exact by Lemma 1.3. However the sub-complex

$$(7) \qquad\qquad\qquad \mathcal{E}_\omega \xrightarrow{\; D \;} \mathcal{E}_\omega \to 0$$

of (6) is not exact. Indeed, we see that $R_{\infty,x} = 0$ for $x = 0$. Hence there exists a unique formal solution of infinite order at 0 of the equation

$$(8) \qquad\qquad\qquad Df = -x;$$

in fact, if f is a solution of the equation (8) on a neighborhood of 0, then the Taylor series of f at 0 is

$$\sum_{n=0}^{\infty} n!\, x^{n+1}.$$

Therefore there does not exist a real-analytic function f on a neighborhood of 0 satisfying (8) near 0, and (7) is not exact. One can also see that the complex (6) itself is not exact.

     We now give the version of Theorem 2.16, Chapter IX for linear equations (Goldschmidt [1967a], Theorem 4.1).

$$
\begin{array}{ccccccccc}
& & 0 & & 0 & & 0 & & 0 \\
& & \downarrow & & \downarrow & & \downarrow & & \downarrow \\
0 \longrightarrow & g_{k+l+1} \longrightarrow & S^{k+l+1}T^* \otimes E & \xrightarrow{\ \sigma_{l+1}(\varphi)\ } & S^{l+1}T^* \otimes F & \xrightarrow{\ \sigma_l(\psi)\ } & S^l T^* \otimes B \\
& \downarrow{\scriptstyle \epsilon} & \downarrow{\scriptstyle \epsilon} & & \downarrow{\scriptstyle \epsilon} & & \downarrow{\scriptstyle \epsilon} \\
0 \longrightarrow & R_{k+l+1} \longrightarrow & J_{k+l+1}(E) & \xrightarrow{\ p_{l+1}(\varphi)\ } & J_{l+1}(F) & \xrightarrow{\ p_l(\psi)\ } & J_l(B) \\
& \downarrow{\scriptstyle \pi_{k+l}} & \downarrow{\scriptstyle \pi_{k+l}} & & \downarrow{\scriptstyle \pi_l} & & \downarrow{\scriptstyle \pi_{l-1}} \\
0 \longrightarrow & R_{k+l} \longrightarrow & J_{k+l}(E) & \xrightarrow{\ p_l(\varphi)\ } & J_l(F) & \xrightarrow{\ p_{l-1}(\psi)\ } & J_{l-1}(B) \\
& & \downarrow & & \downarrow & & \downarrow \\
& & 0 & & 0 & & 0
\end{array}
$$

$$(9)$$

Theorem 1.6. *Let $\varphi : J_k(E) \to F$ be a morphism of vector bundles. If*
   (A) R_{k+1} *is a vector bundle,*
   (B) $\pi_k : R_{k+1} \to R_k$ *is surjective,*
   (C) g_k *is 2-acyclic,*
then R_k is a formally integrable linear differential equation.

Proof. According to (A), the morphism $p_1(\varphi)$ is of locally constant rank and so its cokernel B is a vector bundle. We denote by $\psi : J_1(F) \to B$ the natural projection; we set $p_{-1}(\psi) = 0$. For $l \geq 0$, we consider the commutative diagram (9). Its columns are exact, and its rows are complexes and are exact at $S^{k+l+1}T^* \otimes E$, $J_{k+l+1}(E)$ and $J_{k+l}(E)$. We denote by h_l the cohomology of the top row at $S^{l+1}T^* \otimes F$. If $l = 0$, it follows from (B) and the definition of ψ that the top row is exact, i.e. $h_0 = 0$.

Lemma 1.7. *Let $l \geq 0$. If $h_l = 0$, then we have an isomorphism*

$$(10) \qquad\qquad h_{l+1} \to H^{k+l,2}(g_k).$$

Proof. According to the commutativity of the diagram (23) of Chapter IX, the diagram (11) commutes and is exact, except perhaps for its first column at $\bigwedge^2 T^* \otimes g_{k+l}$ and its first row at $S^{l+2}T^* \otimes F$; hence it gives us a natural isomorphism (10).

We now return to the proof of Theorem 1.6. From diagram (9) with $l = 1$, we obtain an exact sequence

$$R_{k+2} \xrightarrow{\pi_{k+1}} R_{k+1} \xrightarrow{\Omega} h_1;$$

if $p \in R_{k+1}$, then $\Omega(p)$ is the cohomology class of $\epsilon^{-1}p_{l+1}(\varphi)q$ in h_1, where $q \in J_{k+2}(E)$ satisfies $\pi_{k+1}q = p$; by means of the isomorphism (10) with $l = 0$, we therefore have an exact sequence

$$R_{k+2} \xrightarrow{\pi_{k+1}} R_{k+1} \xrightarrow{\kappa} H^{k,2}(g_k).$$

Since $H^{k,2}(g_k) = 0$, we see that $\pi_{k+1} : R_{k+2} \to R_{k+1}$ is surjective. More generally, by induction on $l \geq 1$, from diagram (9), Lemma 1.7 and (C), we simultaneously obtain the surjectivity of $\pi_{k+l} : R_{k+l+1} \to R_{k+l}$ and the exactness of the second row of (9). Moreover, by Lemma 1.1 and the exactness of the top row of (9), g_{k+l} is a vector bundle for $l \geq 1$; the exactness of the sequences

$$0 \to g_{k+l+1} \xrightarrow{\epsilon} R_{k+l+1} \xrightarrow{\pi_{k+l}} R_{k+l} \to 0$$

and (A) now imply that R_{k+l} is a vector bundle for all $l \geq 0$.

$$
(11)\quad
\begin{array}{ccccccc}
& & 0 & & 0 & & 0 & & 0 \\
& & \downarrow & & \downarrow & & \downarrow & & \downarrow \\
0 \to & g_{k+l+2} & \to & S^{k+l+2}T^* \otimes E & \xrightarrow{\sigma_{l+2}(\varphi)} & S^{l+2}T^* \otimes F & \xrightarrow{\sigma_{l+1}(\psi)} & S^{l+1}T^* \otimes B \\
& \downarrow{\scriptstyle\delta} & & \downarrow{\scriptstyle\delta} & & \downarrow{\scriptstyle\delta} & & \downarrow{\scriptstyle\delta} \\
0 \to & T^* \otimes g_{k+l+1} & \to & T^* \otimes S^{k+l+1}T^* \otimes E & \xrightarrow{\mathrm{id}\otimes\sigma_{l+1}(\varphi)} & T^* \otimes S^{l+1}T^* \otimes F & \xrightarrow{\mathrm{id}\otimes\sigma_l(\psi)} & T^* \otimes S^l T^* \otimes B \\
& \downarrow{\scriptstyle\delta} & & \downarrow{\scriptstyle\delta} & & \downarrow{\scriptstyle\delta} \\
0 \to & \bigwedge^2 T^* \otimes g_{k+l} & \to & \bigwedge^2 T^* \otimes S^{k+l}T^* \otimes E & \xrightarrow{\mathrm{id}\otimes\sigma_l(\varphi)} & \bigwedge^2 T^* \otimes S^l T^* \otimes F \\
& \downarrow{\scriptstyle\delta} & & \downarrow{\scriptstyle\delta} \\
0 \to & \bigwedge^3 T^* \otimes g_{k+l-1} & \to & \bigwedge^3 T^* \otimes S^{k+l-1}T^* \otimes E
\end{array}
$$

We thus have completed the proof of Theorem 1.6 and, if we denote by $D' : \mathcal{F} \to \mathcal{B}$ the first-order differential operator $\psi \circ j_1$, in the process, by Lemma 1.3, we have also proved the following

Theorem 1.8. *Let $\varphi : J_k(E) \to F$ be a morphism of vector bundles and let D be the differential operator $\varphi \circ j_k$ of order k. Assume that conditions (A), (B) and (C) of Theorem 1.6 hold. Then there exist a vector bundle B and a first-order linear differential operator $D' : \mathcal{F} \to \mathcal{B}$ such that the sequence*

$$\mathcal{E} \xrightarrow{D} \mathcal{F} \xrightarrow{D'} \mathcal{B}$$

is formally exact; moreover the sequences

$$J_{k+l+1}(E) \xrightarrow{p_{l+1}(D)} J_{l+1}(F) \xrightarrow{p_l(D')} J_l(B)$$

are exact, for all $l \geq 0$.

Again, as in Theorems 2.16 and 3.1 of Chapter IX, according to Theorem 2.14, Chapter IX, we may replace condition (C) by:

(C′) *for all $x \in X$, there exists a quasi-regular basis of T_x for g_k at x.*

Similarly, using Theorem 2.13, Chapter IX, we have the corresponding version of Theorem 2.17, Chapter IX, which is given by Goldschmidt [1968a], Theorem 4.2.

The following generalization of Theorem 1.8 is given by Goldschmidt [1968a], Theorem 3 and [1967a], Theorem 4.4.

Theorem 1.9. *Let $\varphi : J_k(E) \to F$ be a morphism of vector bundles and let $D = \varphi \circ j_k$.*

(i) *Assume that X is connected and that there is an integer $l_0 \geq 0$ such that R_{k+l} is a vector bundle for all $l \geq l_0$. Then there exists a complex*

$$(12) \qquad \mathcal{E} \xrightarrow{D} \mathcal{B}_0 \xrightarrow{P_1} \mathcal{B}_1 \xrightarrow{P_2} \mathcal{B}_2 \to \cdots \to \mathcal{B}_{j-1} \xrightarrow{P_j} \mathcal{B}_j \xrightarrow{P_{j+1}} \cdots ,$$

where B_j is a vector bundle and $B_0 = F$, and where $P_j : \mathcal{B}_{j-1} \to \mathcal{B}_j$ is a linear differential operator of order l_j, which is formally exact; moreover, if $r_0 = 0$ and $r_j = l_1 + l_2 + \cdots + l_j$, for $j \geq 1$, the sequences

$$(13)$$
$$J_{k+m}(E) \xrightarrow{p_m(D)} J_m(B_0) \xrightarrow{p_{m-r_1}(P_1)} J_{m-r_1}(B_1) \xrightarrow{p_{m-r_2}(P_2)} J_{m-r_2}(B_2) \to \cdots$$
$$\to J_{m-r_{j-1}}(B_{j-1}) \xrightarrow{p_{m-r_j}(P_j)} J_{m-r_j}(B_j) \to \cdots$$

are exact at $J_{m-r_j}(B_j)$ for $m \geq r_{j+1}$ and $j \geq 0$.

(ii) *Let*

$$0 = r_0 < r_1 < \cdots < r_j < r_{j+1} < \cdots$$

be integers such that R_{k+r_1-1} is formally integrable and

$$H^{k+r_j+m-j,j+1}(g_k) = 0$$

for all $j \geq 1$ and $m \geq 0$. Then there exists a complex (12), where B_j is a vector bundle and $B_0 = F$, and where $P_j : \mathcal{B}_{j-1} \to \mathcal{B}_j$ is a linear differential operator of order $l_j = r_j - r_{j-1}$, for $j \geq 1$, which is formally exact; moreover the sequences (13) are exact at $J_{m-r_j}(B_j)$ for $m \geq r_{j+1}$ and $j \geq 0$.

If the hypotheses of Theorem 1.9,(i) are satisfied and if the mappings $\pi_{k+l} : R_{k+l+1} \to R_{k+l}$ are of constant rank for all $l \geq l_0$, the cohomologies of two different sequences (12) are isomorphic and depend, up to isomorphisms, only on R_{k+l_0} (see Goldschmidt [1968a]). The sequences of type (12) were first introduced by Kuranishi [1964].

We now give a brief outline of the proof of Theorem 1.9,(ii). Since $p_{r_1}(\varphi)$ has locally constant rank, its cokernel B_1 is a vector bundle. We denote by $\psi_1 : J_{r_1}(F) \to B_1$ the natural projection and set $P_1 = \psi_1 \circ j_{r_1}$. As R_{k+r_1-1} is formally integrable and $H^{k+r_1-1+m,2}(g_k) = 0$, for all $m \geq 0$, an argument similar to the one given above to prove the exactness of the sequences of vector bundles of Theorem 1.8 tells us that the sequences

$$J_{k+r_1+m}(E) \xrightarrow{p_{r_1+m}(\varphi)} J_{r_1+m}(F) \xrightarrow{p_m(\psi_1)} J_m(B_1)$$

are exact for all $m \geq 0$. If $j \geq 2$ and P_i, B_i are defined for $1 \leq i \leq j-1$ and if the sequences (13) are exact at $J_{m-r_i}(B_i)$ for $m \geq r_{i+1}$ and $0 \leq i \leq j-1$, then, according to Lemma 1.1, $p_{l_j}(P_{j-1})$ has locally constant rank and so its cokernel B_j is a vector bundle; we denote by $\psi_j : J_{l_j}(B_{j-1}) \to B_j$ the natural projection and set $P_j = \psi_j \circ j_{l_j}$. If $p_m = l_j + m + 1$, $q_m = p_m + l_{j-1}$, one shows that the sequences

$$S^{q_m}T^* \otimes B_{j-2} \xrightarrow{\sigma_{p_m}(P_{j-1})} S^{p_m}T^* \otimes B_{j-1} \xrightarrow{\sigma_{m+1}(P_j)} S^{m+1}T^* \otimes B_j$$

are exact for $m \geq 0$. Then one deduces that the sequences

$$J_{l_{j-1}+l_j+m}(B_{j-2}) \xrightarrow{p_{l_j+m}(P_{j-1})} J_{l_j+m}(B_{j-1}) \xrightarrow{p_m(P_j)} J_m(B_j)$$

are exact for all $m \geq 0$.

Example 1.10. Let l be an integer ≥ 2 for which R_{k+l-1} is formally integrable and g_{k+l-1} is involutive; then the hypotheses of Theorem 1.9,(ii) hold with $r_j = l + (j-1)$, for $j \geq 1$. In this case, one can give a more explicit description of a sequence (12) obtained from Theorem 1.9,(ii) and the above construction. First, let B_1'' be a vector bundle isomorphic to the cokernel of $p_{l-1}(\varphi)$ and P_1'' be a differential operator of order $l-1$ such that the sequence

$$J_{k+l-1}(E) \xrightarrow{p_{l-1}(D)} J_{l-1}(F) \xrightarrow{p(P_1'')} B_1'' \to 0$$

is exact. Since g_{k+l-1} is involutive, one defines vector bundles $B_1', \ldots, B_n'$ and morphisms of vector bundles

$$\sigma_1 : S^l T^* \otimes F \to B_1', \qquad \sigma_j : T^* \otimes B_{j-1}' \to B_j',$$

for $2 \leq j \leq n$, as follows. Let B_1' be a vector bundle isomorphic to the cokernel of $\sigma_l(\varphi)$ and σ_1 a morphism such that the sequence

$$S^{k+l} T^* \otimes E \xrightarrow{\sigma_l(\varphi)} S^l T^* \otimes F \xrightarrow{\sigma_1} B_1' \to 0$$

is exact. We set $B_0' = B_0 = F$; if $j \geq 2$, we obtain B_j' and σ_j from B_{j-2}', B_{j-1}' and σ_{j-1} by letting B_j' be a vector bundle isomorphic to the cokernel of $(\sigma_{j-1})_{+1}$ and σ_j be a morphism such that the sequence

$$S^{l_j-1+1} T^* \otimes B_{j-2}' \xrightarrow{(\sigma_{j-1})_{+1}} T^* \otimes B_{j-1}' \xrightarrow{\sigma_j} B_j' \to 0$$

is exact, where $l_1 = l$ and $l_j = 1$ for $2 \leq j \leq n$. In fact, if we write $\sigma_j^r = (\sigma_j)_{+r}$, the sequences

$$(14)$$
$$S^{k+l+m} T^* \otimes E \xrightarrow{\sigma_{l+m}(\varphi)} S^{l+m} T^* \otimes F \xrightarrow{\sigma_1^m} S^m T^* \otimes B_1' \xrightarrow{\sigma_2^{m-1}} S^{m-1} T^* \otimes B_2' \to$$

$$\cdots \to S^{m-n+2} T^* \otimes B_{n-1}' \xrightarrow{\sigma_n^{m-n+1}} S^{m-n+1} T^* \otimes B_n' \to 0$$

are exact for all $m \geq 0$. We set $B_{n+1}' = 0$ and

$$B_j = B_j' \oplus (\textstyle\bigwedge^{j-1} T^* \otimes B_1''),$$

for $1 \leq j \leq n+1$. We identify $T^* \otimes B_j$ with

$$(T^* \otimes B_j') \oplus (T^* \otimes \textstyle\bigwedge^{j-1} T^* \otimes B_1'').$$

Let

$$\mu_j : T^* \otimes \textstyle\bigwedge^{j-2} T^* \otimes B_1'' \to \textstyle\bigwedge^{j-1} T^* \otimes B_1''$$

be the morphism sending $\alpha \otimes \omega \otimes u$ into $(\alpha \wedge \omega) \otimes u$, for $\alpha \in T^*$, $\omega \in \bigwedge^{j-2} T^*$ and $u \in B_1''$. For $2 \leq j \leq n+1$, we write

$$\nu_j = (\sigma_j, \mu_j) : T^* \otimes B_{j-1} \to B_j.$$

Then there exists a differential operator $P_1' : \mathcal{F} \to \mathcal{B}_1'$ of order l such that $\sigma(P_1') = \sigma_1$ and $P_1' \cdot D = 0$, and we define $P_1 : \mathcal{F} \to \mathcal{B}_1$ by

$$P_1 f = (P_1' f, P_1'' f),$$

for $f \in \mathcal{F}$. If $2 \leq j \leq n+1$ and P_{j-1} is defined, then $P_j : \mathcal{B}_{j-1} \to \mathcal{B}_j$ is the unique first-order differential operator such that $\sigma(P_j) = \nu_j$ and $P_j \cdot P_{j-1} = 0$. We obtain a sequence

$$(15) \qquad \mathcal{E} \xrightarrow{D} \mathcal{B}_0 \xrightarrow{P_1} \mathcal{B}_1 \xrightarrow{P_2} \mathcal{B}_2 \to \cdots \to \mathcal{B}_n \xrightarrow{P_{n+1}} \mathcal{B}_{n+1} \to 0$$

such that the sequences
$$(16)$$
$$J_{k+l+m}(E) \xrightarrow{p_{l+m}(D)} J_{l+m}(B_0) \xrightarrow{p_m(P_1)} J_m(B_1) \xrightarrow{p_{m-1}(P_2)} J_{m-1}(B_2) \to \cdots$$
$$\to J_{m-n+1}(B_n) \xrightarrow{p_{m-n}(P_{n+1})} J_{m-n}(B_{n+1}) \to 0$$

are exact for all $m > 0$.

According to Lemma 1.3, the equation Ker $p(P_j)$ obtained from the sequence (12) given by Theorem 1.9,(ii) is formally integrable, for $j \geq 1$. If X is a real-analytic manifold, E, F are real-analytic vector bundles and $\varphi : J_k(E) \to F$ is an analytic morphism of vector bundles satisfying the hypotheses of Theorem 1.9,(ii), according to the above construction of the sequence (12), the vector bundles B_j and the operators P_j given by Theorem 1.9,(ii) can be chosen to be analytic. Then by Theorem 1.4, the sub-complex

$$\mathcal{E}_\omega \xrightarrow{D} (\mathcal{B}_0)_\omega \xrightarrow{P_1} (\mathcal{B}_1)_\omega \xrightarrow{P_2} (\mathcal{B}_2)_\omega \to \cdots \to (\mathcal{B}_{j-1})_\omega \xrightarrow{P_j} (\mathcal{B}_j)_\omega \xrightarrow{P_{j+1}} \cdots$$

of (12) is exact. If φ is analytic and satisfies the hypotheses of Theorem 1.9,(i), and if moreover $\pi_m : R_{m+r} \to R_m$ has constant rank for all $m \geq k + l_0$ and $r \geq 0$, then the vector bundles B_j and the operators P_j given by Theorem 1.9,(i) can also be chosen to be analytic, and the above sub-complex of (12) is exact (see Goldschmidt [1968a], Corollary 4). Goldschmidt [1970b] proves that, if X is connected and φ is analytic, then, outside an analytic set, φ satisfies all these regularity conditions: R_{k+l} is a vector bundle for all $l \geq 0$, and $\pi_m : R_{m+r} \to R_m$ has constant rank for all $m \geq k$ and $r \geq 0$.

If φ satisfies the hypotheses of Theorem 1.9,(ii) together with additional assumptions, the proof of Theorem 1.9,(ii) gives us the existence of a sequence of type (12) whose vector bundles can be explicitly described.

Theorem 1.11. *Let $\varphi : J_k(E) \to F$ be a morphism of vector bundles and let $D = \varphi \circ j_k$. Assume that $R_k = $ Ker φ is formally integrable and that $\sigma(D) : S^k T^* \otimes E \to F$ is surjective. Let $q \geq 0$ be an integer.*

(i) *Let*

$$0 = r_0 < r_1 < \cdots < r_q = r_{q+1}$$

be integers such that the cohomology groups

$$(17) \qquad H^{k+r_{j-1}+m-j-1,j+1}(g_k)$$

vanish for all $1 \le j \le q+1$ *and* $m \ge 1$, *with* $m \ne r_j - r_{j-1}$. *Then there exists a complex*

$$(18) \qquad \mathcal{E} \xrightarrow{\ D\ } \mathcal{B}_0 \xrightarrow{\ P_1\ } \mathcal{B}_1 \xrightarrow{\ P_2\ } \mathcal{B}_2 \to \cdots \to \mathcal{B}_{q-1} \xrightarrow{\ P_q\ } \mathcal{B}_q \to 0,$$

where B_j *is a vector bundle isomorphic to* $H^{k+r_j-j-1,j+1}(g_k)$ *for* $j \ge 0$ *and* $B_0 = F$, *and where* P_j *is a linear differential operator of order* $l_j = r_j - r_{j-1}$ *for* $1 \le j \le q$, *such that the sequences*
$$(19)$$
$$S^{k+m}T^* \otimes E \xrightarrow{\ \sigma_m(D)\ } S^m T^* \otimes B_0 \xrightarrow{\ \sigma_{m-l_1}(P_1)\ } S^{m-l_1}T^* \otimes B_1 \xrightarrow{\ \sigma_{m-l_1-l_2}(P_2)\ }$$

$$S^{m-l_1-l_2}T^* \otimes B_2 \to \cdots \to S^{m-l_1-\cdots-l_{q-1}}T^* \otimes B_{q-1}$$

$$\xrightarrow{\ \sigma_{m-l_1-\cdots-l_q}(P_q)\ } S^{m-l_1-\cdots-l_q}T^* \otimes B_q \to 0$$

are exact for all $m \ge 0$.

(ii) *Let* (18) *be a complex, where* B_j *is a vector bundle,* $B_0 = F$ *and* P_j *is a linear differential operator of order* $l_j \ge 1$. *If the sequences* (19) *are exact for all* $m \ge 0$, *the complex* (18) *is formally exact; moreover the sequences*

$$J_{k+m}(E) \xrightarrow{\ p_m(D)\ } J_m(F) \xrightarrow{\ p_{m-l_1}(P_1)\ } J_{m-l_1}(B_1) \xrightarrow{\ p_{m-l_1-l_2}(P_2)\ } J_{m-l_1-l_2}(B_2) \to$$

$$\cdots \to J_{m-l_1-\cdots-l_q}(B_q) \to 0$$

are exact for all $m \ge 0$. *Furthermore, if we set* $r_0 = 0$ *and* $r_j = l_1 + \cdots + l_j$, *for* $1 \le j \le q$, *and* $r_j = r_q$ *for* $j > q$, *then the cohomology groups* (17) *vanish for all* $j \ge 1$ *and* $m \ge 0$ *except possibly for* $m = l_j$ *and* $j = 1, \ldots, q$, *and* B_j *is isomorphic to* $H^{k+r_j-j-1,j+1}(g_k)$.

Example 1.12. If $k = 1$ and g_1 is involutive, and if $R_1 = \operatorname{Ker} \varphi$ is formally integrable and $\sigma(D) : T^* \otimes E \to F$ is surjective, then the hypotheses of Theorem 1.11,(i) hold, with $q = n - 1$ and $r_j = j$ for $0 \le j \le n - 1$. We thus obtain a complex (18), where B_j is equal to

$$H^{0,j+1}(g_1) = (\textstyle\bigwedge^{j+1}T^* \otimes E)/\delta(\textstyle\bigwedge^j T^* \otimes g_1),$$

for $1 \le j \le n - 1$, for which the sequences (19) are exact for all $m \ge 0$. In fact, since

$$\delta(\alpha \wedge u) = (-1)^i \alpha \wedge \delta u,$$

for $\alpha \in \bigwedge^i T^*$ and $u \in \bigwedge^j T^* \otimes S^m T^* \otimes E$, the morphism

$$(20) \qquad T^* \otimes \bigwedge^j T^* \otimes E \to \bigwedge^{j+1} T^* \otimes E,$$

sending $\alpha \otimes \omega \otimes u$ into $(\alpha \wedge \omega) \otimes u$, for $\alpha \in T^*$, $\omega \in \bigwedge^j T^*$, $u \in E$, induces by passage to the quotient a morphism of vector bundles

$$\sigma_j : T^* \otimes H^{0,j}(g_1) \to H^{0,j+1}(g_1).$$

Since $\sigma(D)$ is surjective, the diagram

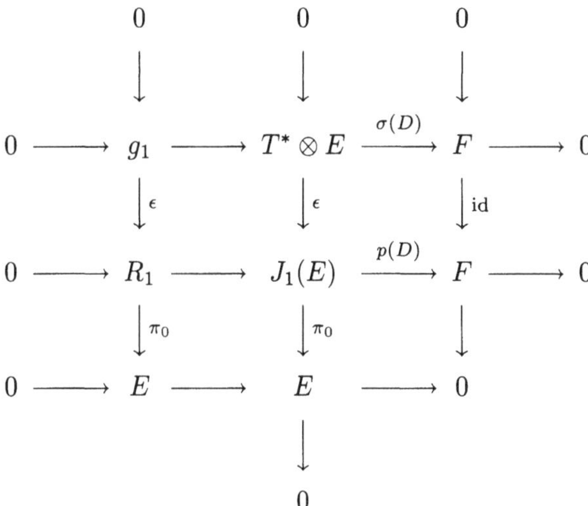

is commutative and exact; therefore $\pi_0 : R_1 \to E$ is surjective and we may identify $B_0 = F$ with $H^{0,1}(g_1)$ in such a way that $\sigma(D) = \sigma_0$. Then, if we write $P_0 = D$, the differential operators $P_j : B_{j-1} \to B_j$ are uniquely determined by the relations $\sigma(P_j) = \sigma_j$ and $P_j \cdot P_{j-1} = 0$, for $1 \le j \le n-1$ (see Goldschmidt [1967a], §5). The resulting complex (18) is called the sophisticated Spencer sequence of the first-order equation R_1.

Example 1.13. If there exists an integer $l \ge 0$ such that $R_{k+l} = \operatorname{Ker} p_l(\varphi)$ is formally integrable and g_{k+l+1} is involutive, the sophisticated Spencer sequence of R_{k+l} is a complex

$$(21) \qquad C^0 \xrightarrow{D_0} C^1 \xrightarrow{D_1} C^2 \xrightarrow{D_2} \cdots \to C^n \to 0,$$

where $C^0 = R_{k+l}$ and C^j is a vector bundle over X isomorphic to the cokernel of the composition

$$\bigwedge^{j-1} T^* \otimes g_{k+l+1} \xrightarrow{\delta} \bigwedge^j T^* \otimes g_{k+l} \xrightarrow{\mathrm{id} \otimes \epsilon} \bigwedge^j T^* \otimes R_{k+l},$$

and where D_j is a first-order linear differential operator such that:

(i)   Ker $p(D_0) = \lambda_1 R_{k+l+1}$ is formally integrable, its symbol is involutive, and the mapping $j_{k+l} : \mathcal{E} \to J_{k+l}(\mathcal{E})$ induces, by restriction, an isomorphism

$$\{u \in \mathcal{E} \mid Du = 0\} \to \{v \in \mathcal{C}^0 \mid D_0 v = 0\};$$

(ii)   the sequences

$$S^m T^* \otimes C^0 \xrightarrow{\sigma_{m-1}(D_0)} S^{m-1} T^* \otimes C^1 \xrightarrow{\sigma_{m-2}(D_1)} S^{m-2} T^* \otimes C^2 \to \cdots$$

$$\to S^{m-n} T^* \otimes C^n \to 0$$

are exact for all $m \geq 0$;

(iii)   if (12) is a complex, where $B_0 = F$ and where P_j is a differential operator of order l_j and $r_j = l_1 + \cdots + l_j$, for which the sequences (13) are exact at $J_{m-r_j}(B_j)$ for $m \geq r_{j+1}$ and $j \geq 0$, its cohomology at $\mathcal{B}_{j-1}$ is isomorphic to the cohomology of (21) at $\mathcal{C}^j$.

The sophisticated Spencer sequences were originally introduced by Spencer [1962]; other constructions are given in Bott [1963], Quillen [1964], Goldschmidt [1967a] and Spencer [1969].

Under certain regularity assumptions on the differential operator D, Theorem 1.14,(i) gives us the existence of a compatibility condition $D' : \mathcal{F} \to \mathcal{B}$ for D and the following theorem (see Goldschmidt [1968a], Theorem 1 and [1968b], Theorem 2) tells us that the solvability questions for the complex

$$\mathcal{E} \xrightarrow{D} \mathcal{F} \xrightarrow{D'} \mathcal{B}$$

can be reduced to those pertaining to a formally exact complex

$$\mathcal{E} \xrightarrow{D_1} \mathcal{F}_1 \xrightarrow{D'_1} \mathcal{B}_1$$

for which Ker $p(D_1)$ is formally integrable and $D_1 = P \cdot D$, where $P : \mathcal{F} \to \mathcal{F}_1$ is a differential operator obtained from D in finitely many steps. In particular, if f is a section of F, the solutions of the inhomogeneous equation $Du = f$ are the same as those of the equation $D_1 u = Pf$.

Theorem 1.14. *Let* $\varphi : J_k(E) \to F$ *be a morphism of vector bundles and let* $D = \varphi \circ j_k$. *Assume that* X *is connected and that there is an integer* $r_0 \geq 0$ *such that* R_{k+l} *is a vector bundle for all* $l \geq r_0$ *and that the mappings* $\pi_{k+l} : R_{k+l+r} \to R_{k+l}$ *have constant rank for all* $l \geq r_0$ *and* $r \geq 0$. *Then there exist integers* $m_0 \geq r_0$ *and* $l_0 \geq 0$, *a vector bundle* F_1 *and a differential operator* $P : \mathcal{F} \to \mathcal{F}_1$ *of order* $m_0 + l_0$ *such that the following assertions hold:*

(i)   *The differential operator* $D_1 = P \cdot D : \mathcal{E} \to \mathcal{F}_1$ *is of order* $k + m_0$ *and, if* $\psi = p(D_1) : J_{k+m_0}(E) \to F_1$, *the equation* Ker ψ *is equal* to $\pi_{k+m_0} R_{k+m_0+l_0}$ and is formally integrable; the solutions and formal solutions of this equation are exactly those of R_{k+r_0}.

(ii)   Let B, B_1 be vector bundles and $D' : \mathcal{F} \to \mathcal{B}$, $D_1' : \mathcal{F}_1 \to \mathcal{B}_1$ be differential operators of order l and q respectively. If the sequences

$$J_{k+l+r}(E) \xrightarrow{\ p_{l+r}(D)\ } J_{l+r}(F) \xrightarrow{\ p_r(D')\ } J_r(B),$$

$$J_{k+m_0+q+r}(E) \xrightarrow{\ p_{q+r}(D_1)\ } J_{q+r}(F_1) \xrightarrow{\ p_r(D_1')\ } J_r(B_1)$$

are exact for all $r \geq 0$, there exist an integer $m \geq 0$ and a differential operator $Q : \mathcal{B} \to J_m(\mathcal{B}_1)$ satisfying the following conditions:

(a)   the diagram

$$
\begin{array}{ccccc}
\mathcal{E} & \xrightarrow{\ D\ } & \mathcal{F} & \xrightarrow{\ D'\ } & \mathcal{B} \\
\downarrow{\scriptstyle \mathrm{id}} & & \downarrow{\scriptstyle P} & & \downarrow{\scriptstyle Q} \\
\mathcal{E} & \xrightarrow{\ D_1\ } & \mathcal{F}_1 & \xrightarrow{\ j_m \circ D_1'\ } & J_m(\mathcal{B}_1)
\end{array}
$$

commutes. Hence, if $f \in \mathcal{F}$ satisfies $D'f = 0$, then $D_1'(Pf) = 0$.

(b)   For $f \in \mathcal{F}$ satisfying $D'f = 0$, the solutions $u \in \mathcal{E}$ of the inhomogeneous equation $Du = f$ coincide with the solutions of the equation $D_1 u = Pf$. Moreover, P induces an isomorphism from the cohomology of the complex

$$\mathcal{E} \xrightarrow{\ D\ } \mathcal{F} \xrightarrow{\ D'\ } \mathcal{B}$$

to the cohomology of the complex

$$\mathcal{E} \xrightarrow{\ D_1\ } \mathcal{F}_1 \xrightarrow{\ D_1'\ } \mathcal{B}_1.$$

Hence, if $f_1 \in \mathcal{F}_1$ satisfies $D_1' f_1 = 0$, there exists $f \in \mathcal{F}$ satisfying $Pf = f_1$ and $D'f = 0$.

The first part of the above theorem is the prolongation theorem of Goldschmidt [1968a]; a result generalizing it together with the Cartan–Kuranishi prolongation theorem is given by Goldschmidt [1974], Theorem 1.

§2. Examples.

Example 2.1. Let E be a vector bundle over X. A connection on E is a linear differential operator

$$\nabla : \mathcal{E} \to T^* \otimes \mathcal{E}$$

such that

$$(22) \qquad \nabla(fs) = df \otimes s + f\nabla s,$$

where f is a real-valued function on X and s is a section of E. According to (1), the relation (22) is equivalent to the fact that ∇ is a first-order differential operator whose symbol

$$\sigma(\nabla) : T^* \otimes E \to T^* \otimes E$$

is the identity mapping.

Let ∇ be a connection on E. It determines a splitting $\chi_\nabla : E \to J_1(E)$ of the exact sequence of vector bundles

$$0 \to T^* \otimes E \xrightarrow{\epsilon} J_1(E) \xrightarrow{\pi_0} E \to 0$$

satisfying

$$\epsilon \circ p(\nabla) = \text{id} - \chi_\nabla \circ \pi_0$$

and a first-order differential equation

$$R_1 = \text{Ker } p(\nabla) = \chi_\nabla(E);$$

clearly $\pi_0 : R_1 \to E$ is an isomorphism and the symbol of R_1 is equal to 0.

By (22), the first-order linear differential operator

$$d^\nabla : \textstyle\bigwedge^j T^* \otimes \mathcal{E} \to \textstyle\bigwedge^{j+1} T^* \otimes \mathcal{E}$$

determined by

$$(23) \qquad d^\nabla(\omega \otimes s) = d\omega \otimes s + (-1)^j \omega \wedge \nabla s,$$

for $\omega \in \bigwedge^j T^*$ and $s \in \mathcal{E}$, is well-defined. According to (23), its symbol

$$(24) \qquad \sigma(d^\nabla) : T^* \otimes \textstyle\bigwedge^j T^* \otimes E \to \textstyle\bigwedge^{j+1} T^* \otimes E$$

is equal to (20). Moreover, the differential operator

$$d^\nabla \cdot \nabla : \mathcal{E} \to \textstyle\bigwedge^2 T^* \otimes \mathcal{E}$$

is of order zero and arises from a morphism of vector bundles

$$K : E \to \textstyle\bigwedge^2 T^* \otimes E,$$

the curvature of ∇, which we shall identify with a section of $\bigwedge^2 T^* \otimes E^* \otimes E$; in fact, we have

$$K(\xi,\eta)s = (\nabla_\xi \nabla_\eta - \nabla_\eta \nabla_\xi - \nabla_{[\xi,\eta]})s,$$

for $\xi,\eta \in \mathcal{T}$, $s \in \mathcal{E}$.

We denote by R_{l+1} the l-th prolongation of R_1. Using Proposition 2.8, Chapter IX, it can be shown that the sequence

$$R_2 \xrightarrow{\ \pi_1\ } R_1 \xrightarrow{\ K\circ\pi_0\ } \textstyle\bigwedge^2 T^* \otimes E$$

is exact; since the symbol of R_1 is equal to 0, by Theorem 1.6, we therefore see that R_1 is formally integrable if and only if the curvature K of ∇ vanishes (see Gasqui and Goldschmidt [1983], Theorem 2.1).

Suppose that the curvature K of ∇ vanishes. Then, according to (23), the sequence

$$(25) \qquad \mathcal{E} \xrightarrow{\ \nabla\ } T^* \otimes \mathcal{E} \xrightarrow{\ d^\nabla\ } \textstyle\bigwedge^2 T^* \otimes \mathcal{E} \xrightarrow{\ d^\nabla\ } \cdots \to \textstyle\bigwedge^n T^* \otimes \mathcal{E} \to 0$$

is a complex. Since the morphism (24) is equal to (20), and (25) is a complex, the construction given in Example 1.12 shows that (25) is the sophisticated Spencer sequence of R_1. Let U be a connected and simply-connected open subset of X; it is well-known that the mapping

$$(26) \qquad \{s \in C^\infty(U,E) \mid \nabla s = 0\} \to E_x,$$

sending s into $s(x)$, is an isomorphism of vector spaces. Let E' be the trivial vector bundle $U \times E_x$ and $\varphi : E|_U \to E'$ be the unique isomorphism of vector bundles sending $s \in C^\infty(U,E)$ satisfying $\nabla s = 0$ into the constant section $a \mapsto (a, s(x))$ of E' over U. Then the diagram

$$
\begin{array}{ccc}
\bigwedge^i T^* \otimes \mathcal{E} & \xrightarrow{\ \mathrm{id}\,\otimes\varphi\ } & \bigwedge^i T \otimes \mathcal{E}' \\
\Big\downarrow{\scriptstyle d^\nabla} & & \Big\downarrow{\scriptstyle d} \\
\bigwedge^{i+1} T^* \otimes \mathcal{E} & \xrightarrow{\ \mathrm{id}\,\otimes\varphi\ } & \bigwedge^{i+1} T^* \otimes \mathcal{E}'
\end{array}
$$

over U, where d is the exterior differential operator acting on functions with values in E_x, commutes. Therefore by the Poincaré lemma, the sequence (25) is exact and we have proved all but the last assertion of the next theorem, which follows from the fact that $H^1(U,\mathbb{R}) = 0$ (see Gasqui and Goldschmidt [1983], Proposition 2.1).

Theorem 2.2. *Let ∇ be a connection on E. Then the curvature K of ∇ vanishes if and only if $R_1 = \operatorname{Ker} p(\nabla)$ is formally integrable. If $K = 0$, then the complex (25) is the sophisticated Spencer sequence of R_1 and is exact; moreover, if U is a connected and simply-connected open subset of X, the mapping (26) is an isomorphism and the sequence*

$$
C^\infty(U, E) \xrightarrow{\nabla} C^\infty(U, T^* \otimes E) \xrightarrow{d^\nabla} C^\infty(U, \textstyle\bigwedge^2 T^* \otimes E)
$$

is exact.

Example 2.3. Let V be a finite-dimensional vector space and let E be the trivial vector bundle $X \times V$. The exterior differential operator $d : \mathcal{E} \to T^* \otimes \mathcal{E}$ for V-valued functions on X is a connection on E. The sequence (25) corresponding to d is the de Rham sequence

$$
(27) \qquad \mathcal{E} \xrightarrow{d} T^* \otimes \mathcal{E} \xrightarrow{d} \textstyle\bigwedge^2 T^* \otimes \mathcal{E} \xrightarrow{d} \cdots \to \textstyle\bigwedge^n T^* \otimes \mathcal{E} \to 0,
$$

which is exact by the Poincaré lemma. Thus the curvature of d vanishes and (27) is therefore the sophisticated Spencer sequence of the formally integrable equation on E equal to $\operatorname{Ker} p(d)$.

Example 2.4. We denote by $F_{\mathbb{C}}$ the complexification of a real vector bundle F. Assume that E is a complex vector bundle over X. Let $S^k E$ and $\bigwedge^k E$ denote the k-th symmetric and exterior powers (over $\mathbb{C}$) of the complex vector bundle E. The bundle $J_k(E)$ is a complex vector bundle if we set

$$
c \cdot j_k(s)(x) = j_k(c \cdot s)(x),
$$

for $x \in X$, $s \in \mathcal{E}_x$ and $c \in \mathbb{C}$. The real vector bundle $S^k T^* \otimes E$ is canonically isomorphic to the complex vector bundle $S^k T^*_{\mathbb{C}} \otimes_{\mathbb{C}} E$ and we shall identify these two vector bundles. The morphism

$$
\epsilon : S^k T^*_{\mathbb{C}} \otimes_{\mathbb{C}} E \to J_k(E)
$$

sends $((df_1 \cdot \ldots \cdot df_k) \otimes s)(x)$ into

$$
j_k\left(\left(\textstyle\prod_{i=1}^k f_i\right) \cdot s\right)(x),
$$

where $f_1, \ldots, f_k$ are complex-valued functions on X vanishing at x and s is a section of E over X. If F is also a complex vector bundle and $D : \mathcal{E} \to \mathcal{F}$ is a $\mathbb{C}$-linear differential operator of order k, then

$$
p_l(D) : J_{k+l}(E) \to J_l(F)
$$

is a morphism of complex vector bundles, as is the morphism

$$\sigma_l(D) : S^{k+l}T^*_{\mathbb{C}} \otimes_{\mathbb{C}} E \to S^l T^*_{\mathbb{C}} \otimes_{\mathbb{C}} F.$$

Assume now that $n = 2m + k$, where $m \geq 1$ and $k \geq 0$. An almost CR-structure (of codimension k) on X is a complex sub-bundle E'' of $T_{\mathbb{C}}$ of rank m (over $\mathbb{C}$) such that E'' and its complex conjugate $\overline{E}''$ have a zero intersection. This almost CR-structure is said to be a CR-structure if $\mathcal{E}''$ is stable under the Lie bracket, i.e. $[\mathcal{E}'', \mathcal{E}''] \subset \mathcal{E}''$.

Let $\mathcal{O}_X$ be the sheaf of complex-valued functions on X and let $\mathbf{1}_{\mathbb{C}}$ denote the trivial complex line bundle over X. If E'' is an almost CR-structure on X, let $\rho : T^*_{\mathbb{C}} \to E''^*$ be the projection induced by the inclusion $E'' \to T_{\mathbb{C}}$ and let

$$\overline{\partial}_b : \mathcal{O}_X \to \mathcal{E}''^*$$

be the first-order differential operator which is equal to the composition of the exterior differential operator $d : \mathcal{O}_X \to T^*_{\mathbb{C}}$ and $\rho : T^*_{\mathbb{C}} \to \mathcal{E}''^*$. The symbol $\sigma(\overline{\partial}_b) : T^*_{\mathbb{C}} \to E''^*$ of $\overline{\partial}_b$ is equal to ρ and so is surjective; therefore $R_1 = \mathrm{Ker}\ p(\overline{\partial}_b)$ is a first-order differential equation on $\mathbf{1}_{\mathbb{C}}$ and $\pi_0 : R_1 \to \mathbf{1}_{\mathbb{C}}$ is surjective (see Example 1.12). The symbol g_1 of R_1 is equal to the annihilator $E''^{\perp}$ of E''. The k-th prolongation g_{k+1} of g_1 is equal to the sub-bundle $S^{k+1}E''^{\perp}$ of $S^{k+1}T^*_{\mathbb{C}}$. Let $x \in X$; a basis $\{t_1, \ldots, t_n\}$ of $T_{\mathbb{C},x}$ over $\mathbb{C}$, for which the subspace of $T_{\mathbb{C}}$ spanned (over $\mathbb{C}$) by the vectors t_j, with $n - m + 1 \leq j \leq n$, is equal to E''_x, is quasi-regular for the sub-bundle $g_1 = E''^{\perp}$ of $T^*_{\mathbb{C}}$ at x in the sense that

$$\dim_{\mathbb{C}} g_{2,x} = \dim_{\mathbb{C}} g_{1,x} + \sum_{j=1}^{n-1} \dim_{\mathbb{C}}(T^*_{\mathbb{C},x,\{t_1,\ldots,t_j\}} \cap E''^{\perp}_x),$$

where $T^*_{\mathbb{C},x,\{t_1,\ldots,t_j\}}$ is the annihilator in $T^*_{\mathbb{C},x}$ of the subspace of $T_{\mathbb{C},x}$ spanned by $\{t_1, \ldots, t_j\}$ over $\mathbb{C}$. Then the arguments proving that condition (i) of Theorem 2.14, Chapter IX implies condition (ii) show that g_1 is involutive. Moreover, it is easily seen that

$$H^{0,j}(g_1) = \textstyle\bigwedge^j T^*_{\mathbb{C}} / (E''^{\perp} \wedge \textstyle\bigwedge^{j-1} T^*_{\mathbb{C}}).$$

Since the restriction mapping $\rho : \bigwedge^j T^*_{\mathbb{C}} \to \bigwedge^j E''^*$ induces an isomorphism of vector bundles

$$(28) \qquad \textstyle\bigwedge^j T^*_{\mathbb{C}} / (E''^{\perp} \wedge \textstyle\bigwedge^{j-1} T^*_{\mathbb{C}}) \to \textstyle\bigwedge^j E''^*,$$

we obtain a canonical isomorphism

$$\psi_j : H^{0,j}(g_1) \to \textstyle\bigwedge^j E''^*.$$

Clearly, the diagram

$$(29) \qquad \begin{array}{ccc} T^* \otimes H^{0,j}(g_1) & \xrightarrow{\ \sigma_j\ } & H^{0,j+1}(g_1) \\ \downarrow{\scriptstyle \mathrm{id}\,\otimes\psi_j} & & \downarrow{\scriptstyle \psi_{j+1}} \\ T^* \otimes \bigwedge^j E''^* & \xrightarrow{\ \gamma_j\ } & \bigwedge^{j+1} E''^* \end{array}$$

is commutative, where σ_j is defined in Example 1.12 and γ_j is the morphism $T_{\mathbb{C}}^* \otimes_{\mathbb{C}} \bigwedge^j E''^* \to \bigwedge^{j+1} E''^*$ sending $\alpha \otimes \beta$ into $\rho(\alpha) \wedge \beta$. It is easily seen that the sequence

$$0 \to S^2 E''^{\perp} \to S^2 T_{\mathbb{C}}^* \xrightarrow{\ \sigma_1(\bar{\partial}_b)\ } T_{\mathbb{C}}^* \otimes_{\mathbb{C}} E''^* \xrightarrow{\ \gamma_1\ } \bigwedge^2 E''^* \to 0$$

is exact. We now compute the first obstruction $\Omega : R_1 \to \bigwedge^2 E''^*$ to the integrability of R_1. If $p \in R_1$, then

$$\Omega(p) = \gamma_1 \epsilon^{-1} p_1(\bar{\partial}_b) q,$$

where $q \in J_2(\mathbf{1}_{\mathbb{C}})$ satisfies $\pi_1 q = p$. According to Proposition 2.8, Chapter IX, the sequence

$$(30) \qquad R_2 \xrightarrow{\ \pi_1\ } R_1 \xrightarrow{\ \Omega\ } \bigwedge^2 E''^*$$

is exact.

Lemma 2.5. *Let $x \in X$. If $f \in \mathcal{O}_{X,x}$ satisfies $(\bar{\partial}_b f)(x) = 0$ and u is the element $\epsilon^{-1} j_1(\bar{\partial}_b f)(x)$ of $T_{\mathbb{C}}^* \otimes_{\mathbb{C}} E''$, then*

$$(31) \qquad u(\zeta, \eta(x)) = \zeta \cdot (\eta \cdot f),$$

$$(32) \qquad \Omega(j_1(f)(x))(\xi(x), \eta(x)) = \langle [\xi, \eta], df \rangle(x),$$

for all $\xi, \eta \in \mathcal{E}_x''$, $\zeta \in T_{\mathbb{C},x}$.

Proof. We can write

$$\bar{\partial}_b f = \sum_{j=1}^{r} g_j \alpha_j,$$

where $\alpha_j \in \mathcal{E}_x''$, $g_j \in \mathcal{O}_{X,x}$ and $g_j(x) = 0$. Then

$$u = \sum_{j=1}^{r} (dg_j \otimes \alpha_j)(x)$$

and

$$u(\zeta, \eta(x)) = (\zeta \cdot g_j)\langle \eta, \alpha_j \rangle(x).$$

On the other hand,

$$\eta \cdot f = \langle \eta, \overline{\partial}_b f \rangle = \sum_{j=1}^{r} \langle \eta, g_j \alpha_j \rangle = \sum_{j=1}^{r} g_j \langle \eta, \alpha_j \rangle$$

and so

$$\zeta \cdot (\eta \cdot f) = \sum_{j=1}^{r} (\zeta \cdot g_j)\langle \eta, \alpha_j \rangle(x),$$

since $g_j(x) = 0$. Formula (32) is a direct consequence of (31).

Proposition 2.6. *Let E'' be an almost CR-structure on X. Then the following four statements are equivalent:*

(i) E'' is a CR-structure;

(ii) there exists a differential operator $\overline{\partial}_b : \mathcal{E}''^* \to \bigwedge^2 \mathcal{E}''^*$ such that the diagram

$$
\begin{array}{ccc}
\mathcal{T}_{\mathbb{C}}^* & \xrightarrow{\ d\ } & \bigwedge^2 \mathcal{T}_{\mathbb{C}}^* \\
\downarrow{\scriptstyle \rho} & & \downarrow{\scriptstyle \rho} \\
\mathcal{E}''^* & \xrightarrow{\ \overline{\partial}_b\ } & \bigwedge^2 \mathcal{E}''^*
\end{array}
$$

commutes;

(iii) $\pi_1 : R_2 \to R_1$ is surjective;

(iv) R_1 is a formally integrable differential equation.

Proof. The existence of the operator $\overline{\partial}_b$ of (ii) is easily seen to be equivalent to the following condition: if $\alpha \in \mathcal{E}''^\perp$, then $(d\alpha)(\xi, \eta) = 0$, for all $\xi, \eta \in \mathcal{E}''$. From the formula

$$(d\alpha)(\xi, \eta) = \xi \cdot \alpha(\eta) - \eta \cdot \alpha(\xi) - \alpha([\xi, \eta]),$$

we deduce the equivalence of (i) and (ii) (see Kuranishi [1977], Proposition 1). From the exactness of sequence (30), we see that (iii) is equivalent to the vanishing of Ω. If f is a constant function on X, then $\overline{\partial}_b f = 0$ and $\Omega(j_1(f)(x)) = 0$, for $x \in X$. Therefore (iii) holds if and only if the mapping

$$\Omega \circ \epsilon : E''^\perp \to \bigwedge^2 E''^*$$

is equal to zero. By (32), we have

$$(\Omega \circ \epsilon)(\alpha)(\xi, \eta) = \langle [\xi, \eta], \alpha \rangle,$$

for $\xi, \eta \in \mathcal{E}''$, $\alpha \in E''^{\perp}$. Thus $\Omega \circ \epsilon = 0$ if and only if E'' is a CR-structure, and so (i) is equivalent to (iii). Since $g_1 = E''^{\perp}$ is an involutive sub-bundle of $T_{\mathbb{C}}^*$ and g_2 is a vector bundle, the equivalence of (iii) and (iv) is provided by Theorem 1.6.

From the isomorphisms (28) and Proposition 2.6, it follows that, if E'' is a CR-structure, for all $j \geq 0$ there exists a differential operator

$$\overline{\partial}_b : \textstyle\bigwedge^j \mathcal{E}''^* \to \bigwedge^{j+1} \mathcal{E}''^*$$

such that the diagram

$$\begin{array}{ccc}
\bigwedge^j T_{\mathbb{C}}^* & \xrightarrow{\;\;d\;\;} & \bigwedge^{j+1} T_{\mathbb{C}}^* \\
\downarrow{\scriptstyle \rho} & & \downarrow{\scriptstyle \rho} \\
\bigwedge^j \mathcal{E}''^* & \xrightarrow{\;\overline{\partial}_b\;} & \bigwedge^{j+1} \mathcal{E}''^*
\end{array}$$

commutes. Clearly,

$$(33) \qquad \mathcal{O}_X \xrightarrow{\overline{\partial}_b} \mathcal{E}''^* \xrightarrow{\overline{\partial}_b} \textstyle\bigwedge^2 \mathcal{E}''^* \xrightarrow{\overline{\partial}_b} \cdots \to \bigwedge^m \mathcal{E}''^* \to 0$$

is a complex and

$$(34) \qquad \overline{\partial}_b(\alpha \wedge \beta) = \overline{\partial}_b \alpha \wedge \beta + (-1)^j \alpha \wedge \overline{\partial}_b \beta,$$

for all $\alpha \in \bigwedge^j \mathcal{E}''^*$, $\beta \in \bigwedge^r \mathcal{E}''^*$. According to (34) and (1),

$$\sigma(\overline{\partial}_b) : T_{\mathbb{C}}^* \otimes_{\mathbb{C}} \textstyle\bigwedge^j E''^* \to \bigwedge^{j+1} E''^*$$

is equal to γ_j. Since $\sigma(\overline{\partial}_b) : T_{\mathbb{C}}^* \to E''^*$ is surjective and g_1 is involutive, from the isomorphisms ψ_j and the commutativity of diagram (29), if E'' is a CR-structure, we see that (33) is isomorphic to the sophisticated Spencer sequence of the formally integrable first-order equation R_1, described in Example 1.12.

Example 2.7. We continue to use the notation and terminology introduced in Example 2.4. If X is a complex manifold, the sub-bundle T'' of $T_{\mathbb{C}}$ of tangent vectors of type $(0,1)$ satisfies the conditions $T'' \cap \overline{T}'' = 0$ and $[T'', T''] \subset T''$. The complex structure of X determined by T'' is a CR-structure (of codimension zero), and in this case the operator $\overline{\partial}_b$ is equal to the Cauchy–Riemann operator

$$\overline{\partial} : \mathcal{O}_X \to T''^*$$

and, if $n = 2m$, the sequence (33) is the Dolbeault sequence

$$(35) \qquad \mathcal{O}_X \xrightarrow{\overline{\partial}} T''^* \xrightarrow{\overline{\partial}} \textstyle\bigwedge^2 T''^* \xrightarrow{\overline{\partial}} \cdots \to \bigwedge^m T''^* \to 0$$

of X, which is always exact (see §3). More generally, let X be a real submanifold of a complex manifold Y of codimension k. If $i : X \to Y$ is the inclusion mapping and T_Y'' is the bundle of tangent vectors to Y of type $(0,1)$, let E'' be the sub-bundle of $T_{\mathbb{C}}$ with variable fiber determined by

$$i_*(E_x'') = i_*(T_{\mathbb{C},x}) \cap T_{Y,i(x)}'',$$

for $x \in X$. If E'' is a vector bundle, then clearly it is a CR-structure of codimension k on X; if $k = 1$, it is easily seen that this condition always holds. If E'' is a vector bundle, the operator $\overline{\partial}_b : \mathcal{O}_X \to \mathcal{E}''^*$ is called the tangential Cauchy–Riemann operator; its solutions include the restrictions to X of the holomorphic functions on Y. The complex (33) was first introduced in this case, with $k = 1$, by Kohn and Rossi [1965].

Example 2.8. In the preceding example, let $Y = \mathbb{C}^2$, with complex coordinates (z, w) and let X be the real hypersurface of Y defined by the equation $\mathrm{Im}\, w = |z|^2$. We consider the induced CR-structure E'' (of codimension 1) on X and its tangential Cauchy–Riemann operator $\overline{\partial}_b$, which is essentially the famous locally non-solvable example of H. Lewy [1957]: the sequence

$$(36) \qquad\qquad \mathcal{O}_X \xrightarrow{\;\overline{\partial}_b\;} \mathcal{E}''^* \to 0$$

given by (33) is not exact.

We now again consider the morphism of vector bundles $\varphi : J_k(E) \to F$ and the objects associated to it. If R_{k+l} is a vector bundle and there exists an integer $m \geq 0$ such that $g_{k+l+m} = 0$, we say that R_{k+l} is a differential equation of finite type. In this case, from Theorem 2.2 and the proof of Theorem 2.2 of Gasqui and Goldschmidt [1983], we obtain:

Theorem 2.9. *Let $\varphi : J_k(E) \to F$ be a morphism of vector bundles and let D be the differential operator $\varphi \circ j_k$ of order k. Let $l \geq 0$ be an integer such that $R_{k+l} = \mathrm{Ker}\, p_l(\varphi)$ is formally integrable and $g_{k+l+1} = 0$. Let (12) be a complex, where B_j is a vector bundle and P_j is a linear differential operator of order l_j. If $r_j = l_1 + \cdots + l_j$, suppose that the sequences (13) are exact at $J_{m-r_j}(B_j)$ for $m \geq r_{j+1}$ and $j \geq 0$. Then the complex (12) is exact and moreover, if U is a connected and simply-connected open subset of X and $x \in U$, the mapping*

$$\{s \in C^\infty(U, E) \mid Ds = 0\} \to R_{k+l,x},$$

sending s into $j_{k+l}(s)(x)$, is an isomorphism of vector spaces and the sequence

$$C^\infty(U, E) \xrightarrow{\;D\;} C^\infty(U, F) \xrightarrow{\;P_1\;} C^\infty(U, B_1)$$

is exact.

To prove Theorem 2.9, one shows that the cohomology of the sequence (12) is isomorphic to the cohomology of the complex (25), where $E = R_{k+l}$ and ∇ is the connection on R_{k+l} corresponding to the unique morphism $\chi_\nabla : R_{k+l} \to J_1(R_{k+l})$ for which the diagram

$$
\begin{array}{ccc}
R_{k+l+1} & \xrightarrow{\ \mathrm{id}\ } & R_{k+l+1} \\
\downarrow{\scriptstyle \pi_{k+l}} & & \downarrow{\scriptstyle \lambda_1} \\
R_{k+l} & \xrightarrow{\ \chi_\nabla\ } & J_1(R_{k+l})
\end{array}
$$

commutes, where π_{k+l} is an isomorphism. Using Theorem 2.2, one sees that the curvature of ∇ vanishes and that the desired assertions hold.

Example 2.10. If β is a section of $\bigotimes^q T^*$, we denote by $\mathcal{L}_\xi \beta$ the Lie derivative of β along a vector field ξ on X. Let g be a Riemannian metric on X. We wish to describe the compatibility condition for the first-order linear differential operator

$$
D : \mathcal{T} \to S^2 \mathcal{T}^*,
$$

sending ξ into $\mathcal{L}_\xi g$, under various assumptions on g. We consider the morphism $\varphi = p(D)$, with $E = T$ and $F = S^2 T^*$ and some of the objects introduced in Example 2.3, Chapter IX. The mapping $\sigma(\varphi) : T^* \otimes T \to S^2 T^*$ is surjective and the fiber $g_{1,x}$ of its kernel g_1 at $x \in X$ is equal to the Lie algebra of the orthogonal group of the Euclidean vector space $(T_x, g(x))$; moreover $g_2 = 0$ (see Gasqui and Goldschmidt [1983], §3). Thus we see that $R_1 = \mathrm{Ker}\, p(\varphi)$ is a differential equation of finite type. The solutions of the equation R_1 or of the homogeneous equation $D\xi = 0$ are the Killing vector fields of (X, g). We denote by E_j the sub-bundle of $\bigwedge^{j+1} T^* \otimes \bigwedge^2 T^*$, which is the kernel of the morphism of vector bundles

$$
\mu : \bigwedge^{j+1} T^* \otimes \bigwedge^2 T^* \to \bigwedge^{j+2} T^* \otimes T^*
$$

determined by

$$
\mu(\omega \otimes (\alpha \wedge \beta)) = (\alpha \wedge \omega) \otimes \beta - (\beta \wedge \omega) \otimes \alpha,
$$

for $\omega \in \bigwedge^{j+1} T^*$, $\alpha, \beta \in T^*$; then $E_1 = G$. The Spencer cohomology of g_1 is given by:

$$
\begin{aligned}
&H^{0,0}(g_1) = T, \quad H^{0,1}(g_1) \simeq S^2 T^*, \quad H^{0,j}(g_1) = 0, \\
&H^{1,0}(g_1) = H^{1,1}(g_1) = 0, \quad H^{1,j}(g_1) \simeq E_{j-1},
\end{aligned}
\tag{37}
$$

for $j > 1$, and $H^{m,i}(g_1) = 0$, for $m \geq 2$, $i \geq 0$, where the isomorphisms depend only on g (see Gasqui and Goldschmidt [1983], §3).

Let H be the sub-bundle of $T^* \otimes G$ consisting of those elements ω of $T^* \otimes G$ which satisfy the relation

$$\omega(\xi_1, \xi_2, \xi_3, \xi_4, \xi_5) + \omega(\xi_2, \xi_3, \xi_1, \xi_4, \xi_5) + \omega(\xi_3, \xi_1, \xi_2, \xi_4, \xi_5) = 0,$$

for all $\xi_1, \xi_2, \xi_3, \xi_4, \xi_5 \in T$. In fact, according to the second Bianchi identity $(\mathcal{DR})(g) = \nabla^g \mathcal{R}(g)$ is a section of H. Let $\delta : H \to T^* \otimes G$ be the inclusion mapping; the image B'_j of the morphism

$$\delta : \textstyle\bigwedge^{j-1} T^* \otimes H \to \bigwedge^j T^* \otimes G,$$

sending $\omega \otimes u$ into $(-1)^{j-1} \omega \wedge \delta u$, for $\omega \in \bigwedge^{j-1} T^*$, $u \in H$, is a sub-bundle of $\bigwedge^j T^* \otimes G$. According to the exactness of the sequences (3.25_l) and (3.31_l) of Gasqui and Goldschmidt [1983], there exist morphisms $\sigma_1 : S^3 T^* \otimes S^2 T^* \to H$ and $\sigma_j : T^* \otimes B'_{j-1} \to B'_j$, for $2 \leq j \leq n$, such that the sequences (14) are exact for all $m \geq 0$; in fact, σ_1 is the first prolongation of τ.

We set $\nabla = \nabla^g$ and $R = \mathcal{R}(g)$. Let

$$\mathcal{R}'_g : S^2 T^* \to \mathcal{G}, \quad (\mathcal{DR})'_g : S^2 T^* \to \mathcal{H}$$

be the linear differential operators of order 2 and 3 which are the linearizations along g of the non-linear operators $h \mapsto \mathcal{R}(h)$ and $h \mapsto (\mathcal{DR})(h)$ respectively, where h is a Riemannian metric on X. If $h \in S^2 T^*$, we therefore have

$$\mathcal{R}'_g(h) = \frac{d}{dt} \mathcal{R}(g + th)|_{t=0}, \qquad (\mathcal{DR})'_g(h) = \frac{d}{dt} (\mathcal{DR})(g + th)|_{t=0}.$$

The invariance of the operators $\mathcal{R}$ and $\mathcal{DR}$ gives us the formulas

$$(38) \qquad \mathcal{R}'_g(\mathcal{L}_\xi g) = \mathcal{L}_\xi R, \quad (\mathcal{DR})'_g(\mathcal{L}_\xi g) = \mathcal{L}_\xi \nabla R,$$

for all $\xi \in T$ (see Gasqui and Goldschmidt [1983], Lemma 4.4). By formula (4.30) of Gasqui and Goldschmidt [1983], we have

$$\sigma((\mathcal{DR})'_g) = \sigma_1.$$

Let $\tilde{G}$ be the sub-bundle of G with variable fiber, whose fiber at $x \in X$ is

$$\tilde{G}_x = \{ (\mathcal{L}_\xi R)(x) \mid \xi \in T_x \text{ with } (\mathcal{L}_\xi g)(x) = 0 \},$$

and let $\alpha : G \to G/\tilde{G}$ be the canonical projection. If $\tilde{G}$ is a vector bundle, we define a second-order linear differential operator

$$D_1 : S^2 T^* \to \mathcal{G}/\tilde{\mathcal{G}}$$

by setting

$$(D_1 h)(x) = \alpha(\mathcal{R}'_g(h - \mathcal{L}_\xi g))(x),$$

for $x \in X$ and $h \in S^2 T_x^*$, where ξ is an element of T_x satisfying $h(x) = (\mathcal{L}_\xi g)(x)$ whose existence is guaranteed by the surjectivity of $\sigma(D)$. By the first relation of (38), we see that this operator is well-defined; clearly we have $D_1 \cdot D = 0$. By Proposition 5.1 of Gasqui and Goldschmidt [1983], the sequence

$$J_3(T) \xrightarrow{p_2(D)} J_2(S^2 T^*) \xrightarrow{p(D_1)} G/\tilde{G} \to 0$$

is exact. For $j \geq 1$, we consider the vector bundle

$$B_j = B'_j \oplus (\textstyle\bigwedge^{j-1} T^* \otimes G/\tilde{G})$$

and we let

$$P_1 : S^2 T^* \to \mathcal{B}_1$$

be the third-order linear differential operator given by

$$P_1 h = ((\mathcal{DR})'_g(h), D_1 h),$$

for $h \in S^2 T^*$.

For $x \in X$, let ρ denote the representation of $g_{1,x}$ on T_x and also on $\bigotimes^q T_x^*$. We say that (X, g) is a locally symmetric space if $\nabla R = 0$. According to Theorem 7.1 and §5 of Gasqui and Goldschmidt [1983], we have:

Theorem 2.11. *If (X, g) is a locally symmetric space, $R_3 = \operatorname{Ker} p_2(\varphi)$ is a formally integrable differential equation and $\tilde{G}$ is a vector bundle equal to the "infinitesimal orbit of the curvature"*

$$\{\rho(u)R \mid u \in g_1\}.$$

We now suppose that $\nabla R = 0$. From the second relation of (38), we deduce that $P_1 \cdot D = 0$. Since $g_2 = 0$, the hypotheses of Theorem 1.9,(ii) hold for φ with $k = 1$, $r_j = j + 2$, for $j \geq 1$. Therefore, if $\nu_j : T^* \otimes B_{j-1} \to B_j$ is the morphism defined in Example 1.10, for $2 \leq j \leq n + 1$, according to the construction given there, there exists a complex

$$(39) \qquad \mathcal{T} \xrightarrow{D} S^2 T^* \xrightarrow{P_1} \mathcal{B}_1 \xrightarrow{P_2} \mathcal{B}_2 \to \cdots \to \mathcal{B}_n \xrightarrow{P_{n+1}} \mathcal{B}_{n+1} \to 0,$$

with $\sigma(P_j) = \nu_j$, for $2 \leq j \leq n + 1$, of the type (15) of Example 1.10 such that the sequences (16) are exact for all $m \geq 0$, with $E = T$, $B_0 = S^2 T^*$. In Gasqui and Goldschmidt [1983], §7, the operators $P_2, \ldots, P_{n+1}$ are determined explicitly (see also Gasqui and Goldschmidt [1988a]).

From Theorem 2.9, we now deduce the following result of Gasqui and Goldschmidt [1983] (Theorem 7.2):

Theorem 2.12. *If $\nabla R = 0$, the complex (39) is exact and, if U is a simply-connected open subset of X, the sequence*

$$C^\infty(U, T) \xrightarrow{\ D\ } C^\infty(U, S^2 T^*) \xrightarrow{\ P_1\ } C^\infty(U, H \oplus G/\tilde{G})$$

is exact.

According to Lemma 5.3 of Gasqui and Goldschmidt [1983], the assumption

$$H \cap (T^* \otimes \tilde{G}) = 0$$

implies that (X, g) is a locally symmetric space; if this condition is satisfied, there is an exact sequence

$$\mathcal{T} \xrightarrow{\ D\ } S^2\mathcal{T}^* \xrightarrow{\ D_1\ } \mathcal{G}/\tilde{\mathcal{G}} \xrightarrow{\ P_2'\ } \mathcal{B}_2' \to \cdots \to \mathcal{B}_{n+1}' \xrightarrow{\ P_{n+2}'\ } \mathcal{B}_{n+2}' \to 0$$

obtained from the complex of type (15) corresponding to the differential operator D_1 (see Gasqui and Goldschmidt [1988a], Theorem 2.3).

Proposition 6.1 of Gasqui and Goldschmidt [1983] tells us that

Proposition 2.13. *If X is connected, the differential equation R_1 is formally integrable if and only if (X, g) has constant curvature.*

Consider the connection ∇ in $\bigwedge^l T^*$ and the corresponding differential operator

$$d^\nabla : \bigwedge^{j+1} \mathcal{T}^* \otimes \bigwedge^l \mathcal{T}^* \to \bigwedge^{j+2} \mathcal{T}^* \otimes \bigwedge^l \mathcal{T}^*$$

of Example 2.1; according to Gasqui and Goldschmidt [1983], §6, if $l = 2$, it induces by restriction a first-order linear differential operator

$$d^\nabla : \mathcal{E}_j \to \mathcal{E}_{j+1}, \quad \text{for } j \geq 1.$$

Suppose that (X, g) has constant curvature K; then $R = -K\tau(g \otimes g)$. Consider the first-order linear differential operator

$$D_g : S^2 \mathcal{T}^* \to \mathcal{G}$$

defined by

$$D_g h = \mathcal{R}_g'(h) + 2K\tau(h \otimes g),$$

for $h \in S^2 \mathcal{T}^*$. According to Gasqui and Goldschmidt [1983], §6 and [1984a], §16, the Calabi sequence

$$(40) \qquad \mathcal{T} \xrightarrow{\ D\ } S^2 \mathcal{T}^* \xrightarrow{\ D_g\ } \mathcal{E}_1 \xrightarrow{\ d^\nabla\ } \mathcal{E}_2 \xrightarrow{\ d^\nabla\ } \cdots \to \mathcal{E}_{n-1} \to 0$$

is a complex. By (37), the hypotheses of Theorem 1.11,(i) are satisfied with $k = 1$, $q = n - 1$, $r_j = j + 1$, for $1 \leq j \leq n - 1$; it is shown in Gasqui and Goldschmidt [1983] that (40) is a complex of the type (18), for which the sequences (19) are exact. From Theorem 2.9, we obtain:

Theorem 2.14. *If (X, g) has constant curvature, the complex (40) is exact and, if U is a simply-connected open subset of X, the sequence*

$$C^\infty(U, T) \xrightarrow{D} C^\infty(U, S^2 T^*) \xrightarrow{D_g} C^\infty(U, G)$$

is exact.

The sequence (40) is the resolution of the sheaf of Killing vector fields of the space of constant curvature (X, g) introduced by Calabi [1961]. The sequences of Example 2.10 have been used to study infinitesimal rigidity questions for symmetric spaces related to the Blaschke conjecture, namely by Bourguignon for the real projective spaces $\mathbb{RP}^n$ (see Besse [1978]), and by Gasqui and Goldschmidt [1983, 1984b, 1988b, 1989a, 1989b] for the complex projective spaces $\mathbb{CP}^n$, with $n \geq 2$, the complex quadrics of dimension ≥ 5, and for arbitrary products of these spaces with flat tori. In particular, the infinitesimal orbit of the curvature is determined explicitly for these spaces.

Example 2.15. Let g be a Riemannian metric on X. We say that the Riemannian manifold (X, g) is conformally flat if, for every $x \in X$, there is a diffeomorphism φ of a neighborhood U of x onto an open subset of $\mathbb{R}^n$ and a real-valued function u on U such that

$$\varphi^* g' = e^u g,$$

where g' is the Euclidean metric on $\mathbb{R}^n$. We consider some of the objects introduced in Example 2.10. Let

$$\mathrm{Tr} : S^2 T^* \to \mathbb{R},$$
$$\mathrm{Tr}^j : \textstyle\bigwedge^{j+1} T^* \otimes T^* \otimes T^* \to \bigwedge^j T^* \otimes T^*$$

be the trace mappings defined by

$$\mathrm{Tr}\, h = \sum_{i=1}^{r} h(t_i, t_i),$$
$$(\mathrm{Tr}^j u)(\xi_1, \ldots, \xi_j, \eta) = \sum_{i=1}^{n} u(t_i, \xi_1, \ldots, \xi_j, t_i, \eta),$$

for all $h \in S^2 T_x^*$, $u \in (\bigwedge^{j+1} T^* \otimes T^* \otimes T^*)_x$, $\xi_1, \ldots, \xi_j, \eta \in T_x$, with $x \in X$, where $\{t_1, \ldots, t_n\}$ is an orthonormal basis of T_x. We denote by $S_0^2 T^*$ the kernel of Tr and we set

$$E_j^0 = E_j \cap \mathrm{Ker}\, \mathrm{Tr}^j.$$

If $(\ ,\ )$ is the scalar product on T^* induced by g, we consider the scalar product $(\ ,\ )$ on $\bigwedge^l T^*$ determined by

$$(41) \qquad (\alpha_1 \wedge \cdots \wedge \alpha_l, \beta_1 \wedge \cdots \wedge \beta_l) = \det(\alpha_i, \beta_j),$$

for $\alpha_1, \ldots, \alpha_l, \beta_1, \ldots, \beta_l \in T^*$. This scalar product in turn induces a scalar product on $\bigwedge^{j+1} T^* \otimes \bigwedge^2 T^*$ and hence on its sub-bundle E_j. We denote by $\rho_j : E_j \to E_j^0$ the orthogonal projection onto E_j^0.

We suppose that $n \geq 3$; we wish to determine the compatibility condition for the first-order linear differential operator

$$D : T \to S_0^2 T^*$$

sending ξ into $\frac{1}{2}\left(\mathcal{L}_\xi g - \frac{1}{n}\operatorname{Tr}(\mathcal{L}_\xi g)g\right)$, when (X, g) is conformally flat. We consider the morphism $\varphi = p(D)$ with $E = T$ and $F = S_0^2 T^*$. The mapping $\sigma(\varphi) : T^* \otimes T \to S_0^2 T^*$ is surjective. We have $g_3 = 0$ and so $R_1 = \operatorname{Ker} p(D)$ is a differential equation of finite type; the solutions of R_1 or of the homogeneous equation $D\xi = 0$ are the conformal Killing vector fields of (X, g). The only non-zero Spencer cohomology groups of g_1 are given by:

$$
\begin{aligned}
& H^{0,0}(g_1) = T, \quad H^{0,1}(g_1) \simeq S_0^2 T^*, \\
(42) \qquad & H^{1,j}(g_1) \simeq E_{j-1}^0, \quad \text{for } 2 \leq j \leq n-2, \\
& H^{2,n-1}(g_1) \simeq \textstyle\bigwedge^n T^* \otimes S_0^2 T^*, \quad H^{2,n}(g_1) = \textstyle\bigwedge^n T^* \otimes T^*,
\end{aligned}
$$

where the isomorphisms depend only on g (see Gasqui and Goldschmidt [1984a], §2).

The Weyl tensor $\mathcal{W}(g)$ of (X, g) is the section $\rho_1 R$ of E_1^0. If $n \geq 4$, a result of H. Weyl asserts that the Riemannian manifold (X, g) is conformally flat if and only if its Weyl tensor vanishes. If $n = 3$, we have $E_1^0 = 0$ and the Weyl tensor always vanishes.

Proposition 5.1 of Gasqui and Goldschmidt [1984a] tells us that

Proposition 2.16. *If $n \geq 4$, the differential equation R_1 is formally integrable if and only if $W = 0$.*

The linearization along g of the non-linear differential operator $h \mapsto \mathcal{W}(h)$, where h is a Riemannian metric on X, is the second-order linear differential operator

$$\mathcal{W}_g' : S^2 T^* \to \mathcal{E}_1$$

defined by

$$\mathcal{W}_g'(h) = \frac{d}{dt}\mathcal{W}(g + th)\big|_{t=0}.$$

If $W = \mathcal{W}(g) = 0$, the operator $\mathcal{W}'_g$ takes its values in $\mathcal{E}^0_1$. We consider the first-order differential operators

$$P_{j+1} = \rho_{j+1} d^\nabla : \mathcal{E}^0_j \to \mathcal{E}^0_{j+1},$$

for $1 \leq j \leq n - 4$, and

$$P_{n-1} = d^\nabla \cdot \mathrm{Tr}^{n-1} : \textstyle\bigwedge^n T^* \otimes S^2_0 T^* \to \bigwedge^n T^* \otimes T^*.$$

Let

$$\phi : \textstyle\bigwedge^{n-2} T^* \otimes T^* \otimes T^* \otimes T^* \to \bigwedge^{n-1} T^* \otimes \bigwedge^2 T^*$$

be the morphism sending $\beta_1 \otimes \beta_2 \otimes \alpha_1 \otimes \alpha_2$ into $(\beta_1 \wedge \alpha_1) \otimes (\beta_2 \wedge \alpha_2)$, for $\beta_1 \in \bigwedge^{n-2} T^*$, $\beta_2, \alpha_1, \alpha_2 \in T^*$. The mapping

$$\hat{\phi} : \textstyle\bigwedge^{n-2} T^* \otimes T^* \to \bigwedge^{n-1} T^* \otimes \bigwedge^2 T^*$$

sending u into $\phi(u \otimes g)$ is an isomorphism. Let R^0 be the section of $T^* \otimes T$ determined by

$$g(R^0(\xi), \eta) = \mathrm{Ric}(\xi, \eta),$$

where Ric is the Ricci curvature of g, and $\xi, \eta \in T$. We define a morphism

$$\theta : \textstyle\bigwedge^{n-2} T^* \otimes \bigwedge^2 T^* \to \bigwedge^{n-1} T^* \otimes T^*$$

by

$$\theta(u)(\xi_1, \ldots, \xi_{n-1}, \eta) = \frac{1}{n-2} \sum_{l=1}^{n-1} (-1)^{l+n-1} u(\xi_1, \ldots, \hat{\xi}_l, \ldots, \xi_{n-1}, R^0(\xi_l), \eta),$$

for $u \in \bigwedge^{n-2} T^* \otimes \bigwedge^2 T^*$, $\xi_1, \ldots, \xi_{n-1}, \eta \in T$.

Assume that $n \geq 4$ and that $W = 0$. Then there exists a unique second-order differential operator

$$P_{n-2} : \mathcal{E}^0_{n-3} \to \textstyle\bigwedge^n T^* \otimes S^2_0 T^*$$

such that

$$\mathrm{Tr}^{n-1} P_{n-2} = \frac{(-1)^n}{2} (d^\nabla \cdot \hat{\phi}^{-1} \cdot d^\nabla - \theta) : \mathcal{E}^0_{n-3} \to \textstyle\bigwedge^{n-1} T^* \otimes T^*$$

(see Gasqui and Goldschmidt [1984a], Chapter I).

By (42), the hypotheses of Theorem 1.11,(i) are satisfied with $k = 1$, $q = n - 1$, $r_j = j + 1$, for $1 \leq j \leq n - 3$, and $r_{n-2} = n$, $r_{n-1} = n + 1$. Gasqui and Goldschmidt [1984a] show that the sequence

(43)

$$T \xrightarrow{D} S^2_0 T^* \xrightarrow{\mathcal{W}'_g} \mathcal{E}^0_1 \xrightarrow{P_2} \cdots \to \mathcal{E}^0_{n-3} \xrightarrow{P_{n-2}} \textstyle\bigwedge^n T^* \otimes S^2_0 T^* \xrightarrow{P_{n-1}} \bigwedge^n T^* \otimes T^* \to 0$$

is a complex of type (18), for which the sequences (19) are exact. From Theorem 2.9, we obtain the following result of Gasqui and Goldschmidt [1984a]:

Theorem 2.17. *If $n \geq 4$ and (X, g) is conformally flat, the sequence (43) is exact and, if U is a simply-connected open subset of X, the sequence*

$$C^\infty(U, T) \xrightarrow{\ D\ } C^\infty(U, S_0^2 T^*) \xrightarrow{\ W_g'\ } C^\infty(U, E_1^0)$$

is exact.

Gasqui and Goldschmidt [1984a] also construct a resolution of type (18) of the sheaf of conformal Killing vector fields of a conformally flat space when $n = 3$.

We now return to the morphism φ of §1, the corresponding differential operator $D : \mathcal{E} \to \mathcal{F}$ and the objects associated to them. If U is an open subset of X, let $C_0^\infty(U, E)$ denote the space of sections of E with compact support contained in U. Suppose that E and F are endowed with scalar products $(\ ,\ )$. The formal adjoint $D^* : \mathcal{F} \to \mathcal{E}$ of D is the unique differential operator such that

$$\int_U (Du, v) = \int_U (u, D^* v),$$

for any oriented subset U of X and for all $u \in C_0^\infty(U, E)$, $v \in C^\infty(U, F)$; it is of order k.

Let (12) be a complex of differential operators for which the sequences (13) are exact at $J_{m-r_j}(B_j)$ for $m \geq r_{j+1}$ and $j \geq 0$. If the vector bundles E, F, B_j are endowed with scalar products, in general the adjoint complex

$$\cdots \to \mathcal{B}_j \xrightarrow{\ P_j^*\ } \mathcal{B}_{j-1} \xrightarrow{\ P_{j-1}^*\ } \cdots \to \mathcal{B}_2 \xrightarrow{\ P_2^*\ } \mathcal{B}_1 \xrightarrow{\ P_1^*\ } \mathcal{B}_0 \xrightarrow{\ D^*\ } \mathcal{E},$$

with $B_0 = F$, does not necessarily satisfy the analogous condition. However, there are just a few examples for which it does, namely:

Example 2.3 (continued). Let g be a Riemannian metric on X and fix a scalar product on the vector space V. Then we obtain scalar products on the vector bundles $\bigwedge^j T^* \otimes E$; the adjoint sequence

$$\bigwedge^n T^* \otimes \mathcal{E} \xrightarrow{\ d^*\ } \bigwedge^{n-1} T^* \otimes \mathcal{E} \xrightarrow{\ d^*\ } \cdots \to \bigwedge^2 T^* \otimes \mathcal{E} \xrightarrow{\ d^*\ } T^* \otimes \mathcal{E} \xrightarrow{\ d^*\ } \mathcal{E} \to 0$$

of (27) is the sophisticated Spencer sequence of the equation Ker $p(d^*)$, where $d^* : \bigwedge^n T^* \otimes \mathcal{E} \to \bigwedge^{n-1} T^* \otimes \mathcal{E}$, and is exact.

Example 2.15 (continued). The vector bundles of the sequence (43) all inherit scalar products from the metric g. If $n \geq 4$ and $W = 0$, the adjoint sequence

$$\bigwedge^n T^* \otimes T^* \xrightarrow{\ P_{n-1}^*\ } \bigwedge^n T^* \otimes S_0^2 T^* \xrightarrow{\ P_{n-2}^*\ } \mathcal{E}_{n-3}^0 \to \cdots \to \mathcal{E}_1^0 \xrightarrow{\ P_1^*\ } S_0^2 T^* \xrightarrow{\ D^*\ } T^* \to 0$$

of (43), where $P_1 = W_g'$, is again a complex of type (18), for which the sequences (19) are exact; moreover it is exact (see Gasqui and Goldschmidt [1984a]).

§3. Existence Theorems for Elliptic Equations.

We again consider the morphism $\varphi : J_k(E) \to F$, the differential operator $D = \varphi \circ j_k : \mathcal{E} \to \mathcal{F}$ and the objects of §1 associated to them. If $x \in X$ and $\alpha \in T_x^*$, let

$$\sigma_\alpha(D) : E_x \to F_x$$

be the linear mapping defined by

$$\sigma_\alpha(D)u = \frac{1}{k!}\sigma(D)(\alpha^k \otimes u),$$

where $u \in E_x$ and α^k denotes the k-th symmetric product of α. If E and F are endowed with scalar products and $D^* : \mathcal{F} \to \mathcal{E}$ is the formal adjoint of D, then we have

$$(44) \qquad\qquad \sigma_\alpha(D^*) = (-1)^k \sigma_\alpha(D)^*.$$

If B is another vector bundle and $D' : \mathcal{F} \to \mathcal{B}$ is a differential operator of order l, then, for all $x \in X$ and $\alpha \in T_x^*$, it is easily verified that

$$(45) \qquad\qquad \sigma_\alpha(D' \cdot D) = \sigma_\alpha(D') \cdot \sigma_\alpha(D) : E_x \to B_x.$$

We say that $\alpha \in T_x^*$, with $x \in X$, is non-characteristic for D if the mapping $\sigma_\alpha(D) : E_x \to F_x$ is injective. Thus α is non-characteristic for D if and only if there are no non-zero elements e of E_x such that $\alpha^k \otimes e \in g_k$.

Assume now that $k = 1$ and thus that D is a first-order operator. We say that a subspace U of T_x^*, with $x \in X$, is non-characteristic for D if

$$(U \otimes E) \cap g_1 = 0.$$

Let U be a sub-bundle of T^*; if $x_0 \in X$ and g_{1,x_0} is involutive and if U_{x_0} is a maximal non-characteristic subspace of $T_{x_0}^*$ for D, then, for all $x \in X$ belonging to a neighborhood of x_0, the subspace U_x of T_x^* is maximal non-characteristic.

Assume that D is a first-order differential operator satisfying the following conditions:
  (i) $R_1 = \text{Ker } p(D)$ is formally integrable;
  (ii) $\sigma(D) : T^* \otimes E \to F$ is surjective;
  (iii) g_1 is involutive.
We consider the initial part

$$(46) \qquad\qquad \mathcal{E} \xrightarrow{\ D\ } \mathcal{F} \xrightarrow{\ D'\ } \mathcal{B}$$

of the sophisticated Spencer sequence of the first-order equation R_1 given by Example 1.12, where D' is a differential operator of order 1. We now describe the normal forms for D and for the complex (46) introduced by Guillemin [1968] (see also Spencer [1969]).

Given a point $x_0 \in X$, let U_{x_0} be a maximal non-characteristic subspace of $T^*_{x_0}$. We choose a local coordinate system $(x^1, \ldots, x^n)$ for X on a neighborhood of x_0 such that $dx^1, \ldots, dx^k$ forms a basis for U_{x_0} at x_0. The subspace U_x of T^*_x spanned by $dx^1, \ldots, dx^k$ at x is maximal non-characteristic for D for all x belonging to a neighborhood V of x_0. Let U be the sub-bundle of $T^*|_V$ spanned by the sections $dx^1, \ldots, dx^k$. We consider the complex (46) restricted to V. For $i = 1, \ldots, k$, the morphism

$$\sigma_{dx^i}(D) : E \to F$$

is injective and the vector bundle

$$E_i = \sigma_{dx^i}(D)(E)$$

is isomorphic to E. It is easily seen that the sum $E_1 + \cdots + E_k$ is direct, and we choose a complement E_0 to this sum in F; thus we have

$$F = E_0 \oplus E_1 \oplus \cdots \oplus E_k.$$

For $i = 0, 1, \ldots, k$, we let $\pi_i : F \to E_i$ be the projection and we set

$$D_0 = \pi_0 D, \qquad D_i = \sigma_{dx^i}(D)^{-1} \cdot \pi_i \cdot D,$$

for $i = 1, \ldots, k$. Then in terms of the differential operators

$$D_0 : \mathcal{E} \to \mathcal{E}_0, \qquad D_i : \mathcal{E} \to \mathcal{E},$$

we have

$$D = D_0 + \sum_{i=1}^{k} \sigma_{dx^i}(D)D_i.$$

From the definitions, for $i = 1, \ldots, k$, we see that

$$D_i = \frac{\partial}{\partial x^i} + L_i,$$

and that

$$L_i = L_i\left(\frac{\partial}{\partial x^{k+1}}, \ldots, \frac{\partial}{\partial x^n}\right), \qquad D_0 = D_0\left(\frac{\partial}{\partial x^{k+1}}, \ldots, \frac{\partial}{\partial x^n}\right)$$

are differential operators involving differentiations only along the sheets of the foliation $x^1 = $ constant, $x^2 = $ constant$, \ldots, x^k = $ constant.

The main results of Guillemin [1968] are summarized in

Theorem 3.1. *For $1 \leq i,\, j \leq k$, there exist first-order differential operators*

$$D'_{ij} : \mathcal{E}_0 \to \mathcal{E}, \quad D'_i : \mathcal{E}_0 \to \mathcal{E}_0$$

such that

$$[D_i, D_j] = D'_{ij} D_0, \quad D_0 D_i = D'_i D_0,$$

where $[D_i, D_j] = D_i D_j - D_j D_i$. Moreover, the differential operator D_0 also satisfies conditions (i), (ii) and (iii); if $D'_0 : \mathcal{E}_0 \to \mathcal{B}_0$ is the first-order differential operator which is the compatibility condition for D_0 whose symbol $\sigma(D'_0) : T^* \otimes E_0 \to B_0$ is surjective, the complex (46) can be described as follows:

We may identify F with $E_0 \oplus (U \otimes E)$ and B with

$$B_0 \oplus (U \otimes E_0) \oplus (\textstyle\bigwedge^2 U \otimes E)$$

and the operators D and D' are given by

$$Du = D_0 u + \sum_{i=1}^{k} dx^i \otimes D_i u,$$

$$D'\Big(f_0 + \sum_{i=1}^{k} dx^i \otimes f_i\Big)$$

$$= D'_0 f_0 + \sum_{i=1}^{k} dx^i \otimes (D_0 f_i - D'_i f_0)$$

$$+ \sum_{1 \leq i < j \leq k} dx^i \wedge dx^j \otimes (D_i f_j - D_j f_i - D'_{ij} f_0),$$

for $u \in \mathcal{E}$, $f_0 \in \mathcal{E}_0$, $f_1, \ldots, f_k \in \mathcal{E}$.

We no longer make any assumptions on the operator D. We say that D is a determined (resp. an underdetermined) differential operator if, for all $x \in X$, there exists a non-zero cotangent vector $\alpha \in T_x^*$ such that $\sigma_\alpha(D)$ is an isomorphism (resp. is surjective). The following result is due to Quillen [1964] (see also Gasqui [1976]).

Proposition 3.2. *Let $\varphi : J_k(E) \to F$ be a morphism of vector bundles and let $D = \varphi \circ j_k$. Then the differential operator D is underdetermined if and only if the morphisms*

$$\sigma_l(\varphi) : S^{k+l} T^* \otimes E \to S^l T^* \otimes F$$

are surjective for all $l \geq 0$. If D is underdetermined, then $R_k = \operatorname{Ker} \varphi$ is formally integrable and the morphisms

$$p_l(\varphi) : J_{k+l}(E) \to J_l(F)$$

are surjective for all $l \geq 0$.

From the proof of Theorem 1.6, it follows that the second assertion of this proposition is an easy consequence of the first. From Proposition 3.2, we infer that, if D is underdetermined, then g_1 is involutive.

We say that D is elliptic (resp. determined elliptic, underdetermined elliptic) if for all $x \in X$ and $\alpha \in T_x^*$, with $x \in X$, the mapping $\sigma_\alpha(D)$ is injective (resp. is an isomorphism, is surjective). If there exists an integer $l \geq 0$ such that $g_{k+l} = 0$, then D is elliptic (see Goldschmidt [1967a], Proposition 6.2).

Lemma 3.3. *Let B be a vector bundle over X and $D' : \mathcal{F} \to \mathcal{B}$ be a differential operator of order k. Assume that E, F, B are endowed with scalar products. If for all $x \in X$ and $\alpha \in T_x^*$, with $\alpha \neq 0$, the sequence*

$$(47) \qquad E_x \xrightarrow{\;\sigma_\alpha(D)\;} F_x \xrightarrow{\;\sigma_\alpha(D')\;} B_x$$

is exact, then the differential operator

$$\Box = DD^* + D'^*D' : \mathcal{F} \to \mathcal{F}$$

of order $2k$ is determined elliptic.

Proof. Let $x \in X$ and $\alpha \in T_x^*$, with $\alpha \neq 0$. If $u \in E_x$, by (44) and (45), we have

$$(\sigma_\alpha(\Box)u, u) = (-1)^k((\sigma_\alpha(D)\sigma_\alpha(D)^* + \sigma_\alpha(D')^*\sigma_\alpha(D'))u, u)$$

$$= (-1)^k\{(\sigma_\alpha(D)^*u, \sigma_\alpha(D)^*u) + (\sigma_\alpha(D')u, \sigma_\alpha(D')u)\}.$$

If $\sigma_\alpha(\Box)u = 0$, we deduce that $\sigma_\alpha(D')u = 0$ and $\sigma_\alpha(D)^*u = 0$. From the exactness of the sequence (47), we infer that $u = 0$ and so $\Box$ is determined elliptic.

Theorem 3.4. *If $D : \mathcal{E} \to \mathcal{F}$ is a determined elliptic or underdetermined elliptic operator, then the sequence*

$$\mathcal{E} \xrightarrow{\;D\;} \mathcal{F} \to 0$$

is exact.

If D is determined elliptic, this result is classic (see Hörmander [1963], Theorem 7.5.1). If D is underdetermined elliptic, one chooses scalar products on the vector bundles E and F; by Lemma 3.3, the operator $DD^* : \mathcal{F} \to \mathcal{F}$ is determined elliptic and the theorem for D follows from the previous case.

Let

$$(48) \qquad \mathcal{B}_0 \xrightarrow{Q_0} \mathcal{B}_1 \xrightarrow{Q_1} \mathcal{B}_2 \xrightarrow{Q_2} \cdots \to \mathcal{B}_{r-1} \xrightarrow{Q_{r-1}} \mathcal{B}_r \to 0$$

be a complex, where B_j is a vector bundle over X and Q_j is a linear differential operator of order $l_j \geq 1$. We say that (48) is an elliptic complex if, for all $x \in X$ and $\alpha \in T_x^*$, with $\alpha \neq 0$, the sequence

$$0 \to B_{0,x} \xrightarrow{\sigma_\alpha(Q_0)} B_{1,x} \xrightarrow{\sigma_\alpha(Q_1)} B_{2,x} \to \cdots \to B_{r-1,x} \xrightarrow{\sigma_\alpha(Q_{r-1})} B_{r,x} \to 0$$

is exact.

Examples 2.7 and 2.8 (continued). If X is a complex manifold, the Cauchy–Riemann operator $\bar{\partial} : \mathcal{O}_X \to T''^*$ is elliptic and the Dolbeault sequence (35) is an elliptic complex. On the other hand, if E'' is a CR-structure on X of codimension > 0, the tangential Cauchy–Riemann operator $\bar{\partial}_b : \mathcal{O}_X \to \mathcal{E}''^*$ fails to be elliptic. The operator $\bar{\partial}_b$ of Example 2.8 is determined.

The following theorem is a direct consequence of the proof of Quillen's theorem (see Quillen [1964] and Goldschmidt [1967a], Proposition 6.5; see also Proposition 6.2, Chapter VIII).

Theorem 3.5. *Suppose that the complex* (48) *satisfies the following condition: for all $m \geq 0$, the sequences*

$$S^{m+l_0}T^* \otimes B_0 \xrightarrow{\sigma_m(Q_0)} S^m T^* \otimes B_1 \xrightarrow{\sigma_{m-l_1}(Q_1)} S^{m-l_1}T^* \otimes B_2 \xrightarrow{\sigma_{m-l_1-l_2}(Q_2)} \cdots$$

$$\to S^{m-l_1-\cdots-l_{r-2}}T^* \otimes B_{r-1} \xrightarrow{\sigma_{m-l_1-\cdots-l_{r-1}}(Q_{r-1})} S^{m-l_1-\cdots-l_{r-1}}T^* \otimes B_r \to 0$$

are exact. If Q_0 is elliptic, then (48) *is an elliptic complex.*

If D is elliptic and if there exists an integer $l \geq 0$ such that R_{k+l} is formally integrable and g_{k+l+1} is involutive, then, by Proposition 6.4 of Goldschmidt [1967a], the operator D_0 of the sophisticated Spencer sequence (21) of R_{k+l} is also elliptic, and so from Theorem 3.5 we deduce that (21) is an elliptic complex.

We are interested in knowing under which conditions an elliptic complex is exact. Proposition 3.2 asserts that an elliptic complex (48) with $r = 1$ is a complex of type (12) and Theorem 3.4 tells us that it is always exact. If $r > 1$, the following example shows that a formally exact elliptic complex is not necessarily exact, and hence that in this case we must restrict our attention to complexes of type (12).

Example 3.6. Suppose that $X = \mathbb{C}$ with its complex coordinate $z = x + \sqrt{-1}\, y$. We set

$$\frac{\partial}{\partial z} = \frac{1}{2}\left(\frac{\partial}{\partial x} - \sqrt{-1}\frac{\partial}{\partial y}\right), \quad \frac{\partial}{\partial \bar{z}} = \frac{1}{2}\left(\frac{\partial}{\partial x} + \sqrt{-1}\frac{\partial}{\partial y}\right).$$

Let E be the trivial complex line bundle over X and $F = E \oplus E$. Consider the first-order differential operator $D : \mathcal{E} \to \mathcal{F}$ given by

$$Df = \left(z^2 \frac{\partial f}{\partial z} - f, \frac{\partial f}{\partial \bar{z}} \right),$$

for $f \in \mathcal{E}$. Let $B = E$ and $D' : \mathcal{F} \to \mathcal{B}$ be the first-order differential operator defined by

$$D'(u, v) = \frac{\partial u}{\partial \bar{z}} - z^2 \frac{\partial v}{\partial z} + v,$$

for $u, v \in \mathcal{E}$. Then we obtain a complex

$$(49) \qquad\qquad \mathcal{E} \xrightarrow{\;D\;} \mathcal{F} \xrightarrow{\;D'\;} \mathcal{B} \to 0.$$

Since $\sigma_\alpha \left(\frac{\partial}{\partial \bar{z}} \right) : E_x \to E_x$ is an isomorphism, for $x \in X$ and $\alpha \in T_x^*$, with $\alpha \neq 0$, we easily see that (49) is an elliptic complex. If $R_{l+1} = \operatorname{Ker} p_l(D)$, the sequence

$$0 \to R_{l+1} \to J_{l+1}(E) \xrightarrow{\;p_l(D)\;} J_l(F) \xrightarrow{\;p_{l-1}(D')\;} J_{l-1}(B) \to 0$$

is exact for $l \geq 0$, and the sequence (49) is formally exact by Lemma 1.3. Indeed, since

$$p_l \left(\frac{\partial}{\partial \bar{z}} \right) : J_{l+1}(E) \to J_l(E)$$

is surjective for all $l \geq 0$, we see that $p_{l-1}(D')$ is surjective for $l \geq 1$. Moreover, let $a \in X$ and $u, v \in \mathcal{E}_a$ satisfy

$$(50) \qquad\qquad p_{l-1}(D')(j_l(u)(a), j_l(v)(a)) = 0;$$

there exists $f_1 \in \mathcal{E}_a$ such that $j_l \left(\frac{\partial f}{\partial \bar{z}} \right)(a) = j_l(v)(a)$. Set

$$u' = u - z^2 \frac{\partial f_1}{\partial z} + f_1;$$

because of (50), we obtain $j_{l-1} \left(\frac{\partial u'}{\partial \bar{z}} \right)(a) = 0$. Then according to the complex analytic analogue of Example 1.5, we can choose a holomorphic function f_2 on a neighborhood of a such that

$$j_l \left(z^2 \frac{\partial f_2}{\partial z} - f_2 - u' \right)(a) = 0;$$

if $f = f_1 + f_2$, we have

$$j_l(Df - (u,v))(a) = 0.$$

Therefore R_{l+1} is a vector bundle for all $l \geq 0$. As in Example 1.5, it is easily seen that $\pi_0 : R_{l+1} \to E$ is an isomorphism on $X - \{0\}$ and hence that R_1 is formally integrable on $X - \{0\}$. However, $\dim_{\mathbb{C}} g_{l+1,a} = 1$ for $a = 0$, and so $\pi_l : R_{l+1} \to R_l$ is not surjective at $a = 0$ and does not have constant rank on X. On the other hand, the complex (49) is not exact. In fact, consider $(-z,0) \in C^\infty(F)$; we have $D'(-z,0) = 0$ and a solution of the equation $Df = (-z,0)$ is a holomorphic function f satisfying $z^2 \frac{\partial f}{\partial z} - f = -z$. According to the argument given in Example 1.5, there does not exist such a function f on a neighborhood of 0.

We no longer impose any of the above restrictions on the operator D. Assume that X is connected and that there is an integer $r_0 \geq 0$ such that R_{k+l} is a vector bundle for all $l \geq r_0$ and that the mappings $\pi_{k+l} : R_{k+l+r} \to R_{k+l}$ have constant rank for all $l \geq r_0$ and $r \geq 0$. Then according to Theorem 1.9,(i), we may consider a complex (12), where B_j is a vector bundle over X and $B_0 = F$, and where P_j is a differential operator of order l_j; if $r_j = l_1 + \cdots + l_j$, we may suppose that the sequences (13) are exact at $J_{m-r_j}(B_j)$ for $m \geq r_{j+1}$ and $j \geq 0$.

If D is not elliptic, in general the complex (12) will not be exact; indeed, we have seen that the complex (36), which is the Spencer sequence associated to the first-order operator $\overline{\partial}_b$ of Example 2.8, is not exact. We now state

The Spencer Conjecture. *If D is elliptic, the complex (12) is exact.*

The main cases when this conjecture is known to be true for overdetermined operators are described in the following two theorems; the first one is due to Spencer and the second one to MacKichan, Sweeney and Rockland.

Theorem 3.7. *Suppose that X is a real-analytic manifold, E, F are real-analytic and D is an analytic differential operator. If D is elliptic, the complex (12) is exact.*

Theorem 3.8. *Suppose either that E is a real line bundle, or that E is a complex line bundle, F is a complex vector bundle and that D is $\mathbb{C}$-linear. If D is an elliptic first-order operator and if R_{l+1} is formally integrable for some $l \geq 0$, the complex (12) is exact.*

The important special case of Theorem 3.8, namely that of a first-order elliptic operator, acting on the trivial complex line bundle and determined by complex vector fields, for which R_1 is formally integrable, can be treated using the Newlander–Nirenberg theorem (see Trèves [1981], Theorem 1.1, Chapter I).

To verify the Spencer Conjecture, it suffices to consider the case of operators D for which $R_k = \text{Ker } p(D)$ is formally integrable. Indeed, according to Goldschmidt [1968a, 1968b], we may replace the differential operator D by the operator $D_1 = P \cdot D$ given by Theorem 1.14,(i); then D_1 is elliptic, $\text{Ker } p(D_1)$ is formally integrable and the cohomology of the sequence (12) is isomorphic to that of a sequence (12) corresponding to D_1.

We now suppose that $R_k = \text{Ker } p(D)$ is formally integrable. Let $l \geq 0$ be an integer for which g_{k+l+1} is involutive. According to Example 1.13, to prove the Spencer Conjecture for D, it suffices to show that the sophisticated Spencer sequence (21) of R_{k+l} is exact at $\mathcal{C}^j$, for $j > 0$. By Theorem 3.5 and the remark which follows it, if D is elliptic, (21) is an elliptic complex. The first-order operator D_0 of (21) satisfies the assumptions considered above for the existence of Guillemin normal forms.

We now give a proof due to Spencer [1962] of the exactness of the complex (21) under the hypotheses of Theorem 3.7; it consists of an adaptation of H. Cartan's argument for the exactness of the Dolbeault sequence (see also Spencer [1969]). As we have seen above, Theorem 3.7 is a consequence of this result.

Assume that the hypotheses of Theorem 3.7 hold. Then the vector bundles and the differential operators D_j of (21) are real-analytic. Let x_0 be an arbitrary point of X. We choose real-analytic scalar products on the vector bundles C^j over a neighborhood U of x_0 and consider the complex (21) restricted to U. The Laplacian

$$\square = D_{j-1}D_{j-1}^* + D_j^* D_j : \mathcal{C}^j \to \mathcal{C}^j,$$

with $D_j = 0$ for $j = -1$ and n, is real-analytic. As (21) is an elliptic complex, according to Lemma 3.3, $\square$ is a determined elliptic operator. Now let u be a section of C^j over a neighborhood of x_0 satisfying $D_j u = 0$. By Theorem 3.4, there exists a section v of C^j over a neighborhood $V \subset U$ of x_0 such that $\square v = u$ on V. We set $w = u - D_{j-1}D_{j-1}^* v$; since $D_j u = 0$, we have $D_j w = 0$ on V. On the other hand,

$$D_{j-1}^* w = D_{j-1}^* u - D_{j-1}^* \square v = 0,$$

and so $\square w = 0$ on V. Since $\square$ is a determined elliptic, analytic operator, it follows that w is real-analytic (see Hörmander [1963], Chapter X). According to Theorem 1.11,(ii) and Theorem 1.4, there exists an analytic section w' of C^{j-1} over a neighborhood V' of x_0 such that $D_{j-1}w' = w$ on V'; thus

$$u = D_{j-1}(D_{j-1}^* v + w')$$

on a neighborhood of x_0, and this proves Theorem 3.7.

Example 2.7 (continued). The Cauchy–Riemann operator $\bar{\partial} : \mathcal{O}_X \to T''^*$ of a complex manifold X is real-analytic (with respect to the structure of real-analytic manifold on X determined by its complex structure). Since the first-order equation $R_1 = \operatorname{Ker} p(\bar{\partial})$ is formally integrable and involutive and (35) is the Spencer sequence of R_1 described in Example 1.12, Theorem 3.7 implies that the sequence (35) is exact.

We now describe the δ-estimate discovered by Singer. For the moment, we no longer make any of the above assumptions on E and the operator D. We use some of the notation introduced in Example 2.4, and suppose that E and F are complex vector bundles and that $D : \mathcal{E} \to \mathcal{F}$ is a $\mathbb{C}$-linear differential operator. Hermitian scalar products $(\, , \,)$ on E and F determine a Hermitian scalar product on $E \otimes_{\mathbb{C}} F$ by setting

$$(e_1 \otimes f_1, e_2 \otimes f_2) = (e_1, e_2)(f_1, f_2),$$

for $e_1, e_2 \in E$, $f_1, f_2 \in F$. We fix Hermitian scalar products $(\, , \,)$ on $T_{\mathbb{C}}^*$ and E. We obtain a Hermitian scalar product $(\, , \,)$ on $\bigwedge^l T_{\mathbb{C}}^*$ for which (41) holds for all $\alpha_1, \ldots, \alpha_l, \beta_1, \ldots, \beta_l \in T_{\mathbb{C}}^*$. The imbedding of $S^l T^*$ into $\bigotimes^l T^*$ defined in §1, Chapter IX extends uniquely to a $\mathbb{C}$-linear imbedding $S^l T_{\mathbb{C}}^* \to \bigotimes^l T_{\mathbb{C}}^*$; we identify $S^l T_{\mathbb{C}}^*$ with its image in $\bigotimes^l T_{\mathbb{C}}^*$ under this mapping. Thus the Hermitian scalar product on $\bigotimes^l T_{\mathbb{C}}^*$ induces a scalar product on $S^l T_{\mathbb{C}}^*$, and the Hermitian scalar products on $T_{\mathbb{C}}^*$ and E induce a scalar product on

$$\bigwedge^j T^* \otimes S^{k+l} T^* \otimes E = \bigwedge^j T_{\mathbb{C}}^* \otimes_{\mathbb{C}} S^{k+l} T_{\mathbb{C}}^* \otimes_{\mathbb{C}} E.$$

We consider the restriction $(\, , \,)$ of this Hermitian scalar product to $\bigwedge^j T^* \otimes g_{k+l}$ and set $\|u\|^2 = (u, u)$, for $u \in \bigwedge^j T^* \otimes g_{k+l}$.

Definition. We say that the $\mathbb{C}$-linear differential operator $D : \mathcal{E} \to \mathcal{F}$ of order k satisfies the δ-estimate if there exist Hermitian scalar products on $T_{\mathbb{C}}^*$ and E such that
$$\|\delta u\|^2 \geq \|u\|^2,$$

for all $u \in (T^* \otimes g_{k+1}) \cap \operatorname{Ker} \delta^*$, where $\delta^* : T^* \otimes g_{k+1} \to g_{k+2}$ is the adjoint of δ.

The following result of Rockland [1972] (Appendix A) is verified by adapting the proof of MacKichan [1971] (Example 4.3) that the Cauchy–Riemann operator $\bar{\partial} : \mathcal{O}_X \to T''^*$ of Example 2.7 satisfies the δ-estimate.

Proposition 3.9. *If E is a complex line bundle, then a first-order $\mathbb{C}$-linear differential operator $D : \mathcal{E} \to \mathcal{F}$ satisfies the δ-estimate.*

Spencer [1962] proposed a generalization of the $\bar{\partial}$-Neumann problem for overdetermined linear elliptic systems on small convex domains. This

Spencer–Neumann problem has been studied by Sweeney [1968, 1976] and has been solved only in the case of operators satisfying the δ-estimate by MacKichan [1971], using the work of Kohn and Nirenberg [1965]. An estimate of Sweeney [1969] implies that the corresponding harmonic spaces vanish and gives us the following result due to MacKichan [1971] and Sweeney [1969].

Theorem 3.10. *Suppose that D is a $\mathbb{C}$-linear elliptic differential operator satisfying the δ-estimate. If there exists an integer $l \geq 0$ such that R_{k+l} is formally integrable, then the complex (12) is exact.*

Theorem 3.8 now follows from Proposition 3.9 and the above theorem.

The δ-estimate also gives a generalization of symmetric hyperbolic operators, in conjunction with the Guillemin normal form of Theorem 3.1 (see MacKichan [1975]).

Example 3.11. Let X be an open subset of $\mathbb{R}^k \times \mathbb{C}^r$, with coordinates $(t, z) = (t^1, \ldots, t^k, z^1, \ldots, z^r)$. We write $z^j = x^j + \sqrt{-1}\, y^j$ and

$$\frac{\partial}{\partial z^j} = \frac{1}{2}\left(\frac{\partial}{\partial x^j} - \sqrt{-1}\frac{\partial}{\partial y^j}\right), \quad \frac{\partial}{\partial \bar{z}^j} = \frac{1}{2}\left(\frac{\partial}{\partial x^j} + \sqrt{-1}\frac{\partial}{\partial y^j}\right).$$

Let E be the trivial complex line bundle of rank m over X and F be the direct sum of $(r+1)$-copies of E. Let

$$P_1 : \mathcal{E} \to \mathcal{E}, \quad P_2 : \mathcal{E} \to \mathcal{E}$$

be the first-order differential operators given by

$$P_1 u = \sum_{i=1}^{k} a_i(t, z)\frac{\partial u}{\partial t^i}, \quad P_2 u = \sum_{j=1}^{r} b_j(t, z)\frac{\partial u}{\partial z^j},$$

for $u \in \mathcal{E}$, where $a_j(t, z)$, $b_l(t, z)$ are $m \times m$ matrices which are holomorphic in z. Assume that $k \geq 1$ and that P_1 is determined in the t-variables, i.e. for all $x \in X$, there exists

$$\tag{51} \alpha = \sum_{j=1}^{k} \alpha_j dt^j \in T_x^* - \{0\}$$

for which $\sigma_\alpha(P_1) : E_x \to E_x$ is an isomorphism. Consider the first-order differential operator $D : \mathcal{E} \to \mathcal{F}$ defined by

$$Du = \left(P_1 u + P_2 u, \frac{\partial u}{\partial \bar{z}^1}, \ldots, \frac{\partial u}{\partial \bar{z}^r}\right),$$

for $u \in \mathcal{E}$. Then using a variant of Proposition 3.2, one verifies that $R_1 = \operatorname{Ker} p(D)$ is a formally integrable differential equation, that $\sigma(D) : T^* \otimes E \to F$ is surjective and that g_1 is involutive. Moreover, if B is the direct sum of $\binom{r+1}{2}$-copies of E, the differential operator $D' : \mathcal{F} \to B$ defined by

$$D'(u_0, u_1, \ldots, u_r) = \left(\frac{\partial u_0}{\partial \bar{z}^j} - (P_1 + P_2)u_j, \frac{\partial u_q}{\partial \bar{z}^l} - \frac{\partial u_l}{\partial \bar{z}^q} \right)_{\substack{j,l,q=1,\ldots,r \\ l<q}},$$

for $u_0, u_1, \ldots, u_r \in \mathcal{E}$, is the compatibility condition for D; in fact,

$$\text{(52)} \qquad\qquad \mathcal{E} \xrightarrow{D} \mathcal{F} \xrightarrow{D'} B$$

is the initial part of the sophisticated Spencer sequence of the first-order equation R_1. If P_1 is elliptic in the t-variables, i.e. for all $x \in X$ and $\alpha \in T_x^* - \{0\}$ of the form (51), the mapping $\sigma_\alpha(P_1) : E_x \to E_x$ is an isomorphism, then D is elliptic and, by Theorem 3.5, the sequence (52) is the initial part of an elliptic complex of type (12). It is easily seen that the excactness of the complex (52) is equivalent to the solvability of the system

$$P_1 u + P_2 u = f,$$

$$\frac{\partial u}{\partial \bar{z}^j} = 0, \quad 1 \leq j \leq r,$$

for $u \in \mathcal{E}$, where $f \in \mathcal{E}$ satisfies $\frac{\partial f}{\partial \bar{z}^j} = 0$, for $1 \leq j \leq r$. This inhomogeneous system was first considered by Nirenberg [1974]; the exactness of the complex (52) was proved by Menikoff [1977] for a class of such systems.

Example 3.12. Let X be an open subset of $\mathbb{R}^{r+k}$, with coordinates $(x, y) = (x^1, \ldots, x^r, y^1, \ldots, y^k)$, and $k \geq 1$. Let E be a trivial vector bundle over X and F be the direct sum of $(k+1)$-copies of E. Let

$$P_0 : \mathcal{E} \to \mathcal{E}, \quad P_j : \mathcal{E} \to \mathcal{E},$$

with $j = 1, \ldots, k$, be first-order differential operators, where

$$P_j = \frac{\partial}{\partial y^j} + L_j$$

and where

$$P_0 = P_0 \left(\frac{\partial}{\partial x^1}, \ldots, \frac{\partial}{\partial x^r} \right), \quad L_j = L_j \left(\frac{\partial}{\partial x^1}, \ldots, \frac{\partial}{\partial x^r} \right)$$

are operators which involve differentiations only along the sheets of the foliations $y^1 = \text{constant}, \ldots, y^k = \text{constant}$. Assume that P_0 is determined and that

$$[P_0, P_j] = [P_j, P_l] = 0,$$

for $j, l = 1, \ldots, k$. Consider the first-order differential operator $D : \mathcal{E} \to \mathcal{F}$ defined by

$$Du = (P_0 u, P_1 u, \ldots, P_k u),$$

for $u \in \mathcal{E}$. Then using Proposition 3.2, one verifies that $R_1 = \text{Ker } p(D)$ is a formally integrable differential equation, that $\sigma(D) : T^* \otimes E \to F$ is surjective and that g_1 is involutive. Set $y^0 = x^1$ and let U be the sub-bundle of T^* spanned by $dy^0, dy^1, \ldots, dy^k$. We identify $F = B_0$ with $U \otimes E$ and set $B_j = \bigwedge^{j+1} U \otimes E$, for $j \geq 0$. We consider the first-order differential operator $D_j : \mathcal{B}_{j-1} \to \mathcal{B}_j$, with $j \geq 1$, determined by

$$D_j(dy^{\alpha_1} \wedge \cdots \wedge dy^{\alpha_j} \otimes u) = \sum_{l=0}^{k} dy^l \wedge dy^{\alpha_1} \wedge \cdots \wedge dy^{\alpha_j} \otimes P_l u,$$

where $0 \leq \alpha_1, \ldots, \alpha_j \leq k$ and $u \in \mathcal{E}$. We obtain a complex

$$(53) \qquad \mathcal{E} \xrightarrow{D} \mathcal{F} \xrightarrow{D_1} \mathcal{B}_1 \xrightarrow{D_2} \mathcal{B}_2 \to \cdots \to \mathcal{B}_{k+1} \to 0$$

which is the sophisticated Spencer sequence of the first-order equation R_1. In view of Theorem 3.1, if the D_0-component of the Guillemin decomposition of a first-order operator vanishes, then this operator is isomorphic to one of the type considered here. We write $U = X \cap (\mathbb{R}^r \times \{0\})$. If the restriction of P_0 to U is elliptic, then D is elliptic and, by Theorem 3.5, the sequence (53) is an elliptic complex of type (12).

In Dencker [1982], the notion of operators $\mathcal{E} \to \mathcal{E}$ of real principal type is defined; determined elliptic operators belong to this class. The following result is due to D. Yang [1986].

Theorem 3.13. *If the restriction of the differential operator P_0 to U is of real principal type and if $r \geq 3$, then the complex (53) is exact.*

Thus if the restriction of P_0 to U is elliptic, the above theorem gives examples of elliptic complexes of type (12) which are exact.

BIBLIOGRAPHY

ALLARD, W. K.

1989 An improvement of Cartan's test for ordinarity, *in* "Partial differential equations and the calculus of variations, Volume I, Essays in honor of Ennio De Giorgi," edited by F. Colombini et al., Birkhäuser, Boston, Basel, Berlin, 1989, 1–8.

BERGER, E., BRYANT, R. L. AND GRIFFITHS, P. A.

1983 The Gauss equations and rigidity of isometric embeddings, *Duke Math. J.*, **50** (1983), 803–892.

BESSE, A. L.

1978 "Manifolds all of whose geodesics are closed," Ergeb. Math. Grenzgeb., Bd. 93, Springer-Verlag, Berlin, Heidelberg, New York, 1978.

1987 "Einstein manifolds," Ergeb. Math. Grenzgeb., 3 Folge, Bd. 10, Springer-Verlag, Berlin, Heidelberg, New York, 1987.

BOTT, R.

1963 "Notes on the Spencer resolution," Harvard University, Cambridge, Mass., 1963.

BOURBAKI, N.

1961 "Éléments de mathématique, Algèbre commutative; Chapitre 4: Idéaux premiers associés et décomposition primaire," Hermann, Paris, 1961.

1965 "Éléments de mathématique, Topologie générale; Chapitre 2: Structures uniformes," 4e édition, Hermann, Paris, 1965.

BRYANT, R. L.

1979 "Some aspects of the local and global theory of Pfaffian systems," Ph.D. thesis, University of North Carolina, Chapel Hill, 1979.

BRYANT, R. L., GRIFFITHS, P. A. AND YANG, D.

1983 Characteristics and existence of isometric embeddings, *Duke Math. J.*, **50** (1983), 893–994.

CALABI, E.

1961 On compact, Riemannian manifolds with constant curvature. I, *in* "Differential geometry," Proc. Sympos. Pure Math., Vol. III, Amer. Math. Soc., Providence, R.I., 1961, 155–180.

CARATHEODORY, C.

1909 Untersuchungen über die Grundlagen der Thermodynamik, *Math. Ann.*, **67** (1909), 355–386.

CARTAN, E.

1899 Sur certaines expressions différentielles et le problème de Pfaff, *Ann. Sci. École Norm.*, (3) **16** (1899), 239–332.

1901 Sur l'intégration des systèmes d'équations aux différentielles totales, *Ann. Sci. École Norm.*, (3) **18** (1901), 241–311.

1910 Les systèmes de Pfaff à cinq variables et les équations aux dérivées partielles du second ordre, *Ann. Sci. École Norm.*, (3) **27** (1910), 109–192.

1922 "Leçons sur les invariants intégraux," Hermann, Paris, 1922.

1931 Sur la théorie des systèmes en involution et ses applications à la relativité, *Bull. Soc. Math. France*, **59** (1931), 88–118.

1946 "Les systèmes différentiels extérieurs et leurs applications géométriques," Hermann, Paris, 1945.

1953 "Œuvres complètes," Gauthier-Villars, Paris, 1953.

CARTAN, E. AND EINSTEIN, A.

1979 "Letters on absolute parallelism, 1929–1932," English translation, edited by R. Debever, Princeton University Press, Princeton, N.J., 1979.

CHERN, S. S.

1985 Deformation of surfaces preserving principal curvatures, *in* "Differential geometry and complex analysis, A volume dedicated to the memory of Harry Ernest Rauch," edited by I. Chavel and H. M. Farkas, Springer-Verlag, Berlin, Heidelberg, New York, Tokyo, 1985, 155–163.

CHERN, S. S. AND GRIFFITHS, P. A.

1978 Abel's theorem and webs, *Jahresber. Deutsch. Math.-Verein*, **80** (1978), 13–110.

CHERN, S. S., GRIFFITHS, P. A. AND BRYANT, R. L.

1982 "Exterior differential systems," China Scientific Press, Beijing, 1982.

CHOW, W. L.

1940 Über Systeme von linearen partiellen Differentialgleichungen erster Ordnung, *Math. Ann.*, **117** (1940), 98–105.

DARBOUX, G.

1870 Sur les équations aux dérivées partielles du second ordre, *Ann. Sci. École Norm. Sup.*, (1) **7** (1870), 163–173.

1882 Sur le problème de Pfaff, *Bull. Sci. Math.*, (2) **6** (1882), 14–36; 49–68.

DENCKER, N.

1982 On the propagation of polarization sets for systems of real principal type, *J. Funct. Anal.*, **46** (1982), 351–372.

DETURCK, D. M.

1981 Existence of metrics with prescribed Ricci curvature: local theory, *Invent. Math.*, **65** (1981), 179–207.

1982   Metrics with prescribed Ricci curvature, *in* "Seminar on differential geometry," edited by S. T. Yau, Ann. of Math. Studies, No. 102, Princeton University Press, Princeton, N.J., University of Tokyo Press, Tokyo, 1982, 525–537.

DeTurck, D. M. and Yang, D.

1984   Existence of elastic deformations with prescribed principal strains and triply orthogonal systems, *Duke Math. J.*, **51** (1984), 243–260.

1986   Local existence of smooth metrics with prescribed curvature, *Contemp. Math.*, **51** (1986), 37–43.

Douady, A.

1966   Le problème des modules pour les sous-espaces analytiques compacts d'un espace analytique donné, *Ann. Inst. Fourier (Grenoble)*, **16**, 1 (1966), 1–95.

Ehrenpreis, L.

1970   "Fourier analysis in several complex variables," Wiley-Interscience, New York, London, Sydney, Toronto, 1970.

Ehrenpreis, L., Guillemin, V. W. and Sternberg, S.

1965   On Spencer's estimate for δ-Poincaré, *Ann. of Math.*, **82** (1965), 128–138.

Fulton, W. and Lang, S.

1985   "Riemann–Roch algebra," Grundlehren Math. Wiss., Vol. 277, Springer-Verlag, Berlin, Heidelberg, New York, Tokyo, 1985.

Gabber, O.

1981   The integrability of the characteristic variety, *Amer. J. Math.*, **103** (1981), 445–468.

Gardner, R. B.

1967   Invariants of Pfaffian systems, *Trans. Amer. Math. Soc.*, **126** (1967), 514–533.

1989   "Method of equivalence and its applications," CBMS-NSF Regional Conf. Ser. in Appl. Math., Vol. 58, SIAM, Philadelphia, 1989.

Gardner, R. B. and Shadwick, W. F.

1987   A simple characterization of the contact system on $J^k(E)$, *Rocky Mountain J. Math.*, **17** (1987), 19–21.

Gasqui, J.

1975   Sur l'existence d'immersions isométriques locales pour les variétés riemanniennes, *J. Differential Geom.*, **10** (1975), 61–84.

1979a   Equivalence projective et équivalence conforme, *Ann. Sci. École Norm. Sup.*, (4) **12** (1979), 101–134.

1979b   Sur l'existence locale de certaines métriques riemanniennes plates, *Duke Math. J.*, **46** (1979), 109–118.

1982   Sur la résolubilité locale des équations d'Einstein, *Compositio Math.*, **47** (1982), 43–69.

GASQUI, J. AND GOLDSCHMIDT, H.

1983   Déformations infinitésimales des espaces riemanniens localement symétriques. I, *Adv. in Math.*, **48** (1983), 205–285.

1984a   "Déformations infinitésimales des structures conformes plates," Progress in Mathematics, Vol. 52, Birkhäuser, Boston, Basel, Stuttgart, 1984.

1984b   Infinitesimal rigidity of $S^1 \times \mathbb{RP}^n$, *Duke Math. J.*, **51** (1984), 675–690.

1988a   Complexes of differential operators and symmetric spaces, *in* "Deformation theory of algebras and structures and applications," edited by M. Hazewinkel and M. Gerstenhaber, Kluwer Academic Pub., Dordrecht, Boston, London, 1988, 797–827.

1988b   Some rigidity results in the deformation theory of symmetric spaces, *in* "Deformation theory of algebras and structures and applications," edited by M. Hazewinkel and M. Gerstenhaber, Kluwer Academic Pub., Dordrecht, Boston, London, 1988, 839–851.

1989a   Rigidité infinitésimale des espaces projectifs et des quadriques complexes, *J. Reine Angew. Math.*, **396** (1989), 87–121.

1989b   Infinitesimal rigidity of products of symmetric spaces, *Illinois J. Math.*, **33** (1989), 310–332.

GOLDSCHMIDT, H.

1967a   Existence theorems for analytic linear partial differential equations, *Ann. of Math.*, **86** (1967), 246–270.

1967b   Integrability criteria for systems of non-linear partial differential equations, *J. Differential Geom.*, **1** (1967), 269–307.

1968a   Prolongations of linear partial differential equations: I. A conjecture of Élie Cartan, *Ann. Sci. École Norm. Sup.*, (4) **1** (1968), 417–444.

1968b   Prolongations of linear partial differential equations: II. Inhomogeneous equations, *Ann. Sci. École Norm. Sup.*, (4) **1** (1968), 617–625.

1970a   Formal theory of overdetermined linear partial differential equations, *in* "Global analysis," Proc. Sympos. Pure Math., Vol. XVI, Amer. Math. Soc., Providence, R.I., 1970, 187–194.

1970b   Points singuliers d'un opérateur différentiel analytique, *Invent. Math.*, **9** (1970), 165–174.

1972a   Sur la structure des équations de Lie: I. Le troisième théorème
             fondamental, *J. Differential Geom.*, **6** (1972), 357–373.
1972b   Sur la structure des équations de Lie: II. Equations formelle-
             ment transitives, *J. Differential Geom.*, **7** (1972), 67–95.
1974   Prolongements d'équations différentielles linéaires. III. La suite
             exacte de cohomologie de Spencer, *Ann. Sci. École Norm. Sup.*,
             (4) **7** (1974), 5–28.

GOLDSCHMIDT, H. AND STENBERG, S.
1973   The Hamilton–Cartan formalism in the calculus of variations,
             Ann. Inst. Fourier (Grenoble), **23**, 1 (1973), 203–267.

GOODMAN, J. AND YANG, D.
1990   Local solvability of non-linear partial differential equations of
             real principal type, preprint.

GRAUERT, H.
1960   Ein Theorem der analytischen Garbentheorie und die Mod-
             ulräume komplexer Structuren, *Inst. Hautes Études Sci. Publ.
             Math.*, No. 5 (1960), 64 pp.

GREEN, M. L.
1989a   Koszul cohomology and geometry, *in* "Lectures on Riemann
             surfaces," edited by M. Cornalba, X. Gomez-Mont and A. Ver-
             jovsky, World Scientific, Singapore, 1989, 177–200.
1989b   Restrictions of linear series to hyperplanes, and some results
             of Macauley and Gotzmann, *in* "Algebraic curves and projec-
             tive geometry, Proceedings Trento 1988," edited by E. Ballico
             and C. Ciliberto, Lecture Notes in Math., Vol. 1389, Springer-
             Verlag, Berlin, Heidelberg, New York, 1989, 76–86.

GRIFFIN, M.
1933   Invariants of pfaffian systems, *Trans. Amer. Math. Soc.*, **35**
             (1933), 929–939.

GRIFFITHS, P. A.
1983   "Exterior differential systems and the calculus of variations,"
             Progress in Mathematics, Vol. 25, Birkhäuser, Boston, Basel,
             Stuttgart, 1983.

GRIFFITHS, P. A. AND JENSEN, G. R.
1987   "Differential systems and isometric embeddings," Ann. of Math.
             Studies, No. 114, Princeton University Press, Princeton, N.J.,
             1987.

GROMOV, M.
1985   Pseudo holomorphic curves in symplectic manifolds, *Invent.
             Math.*, **82** (1985), 307–347.

GUILLEMIN, V. W.

1968   Some algebraic results concerning the characteristics of overdetermined partial differential equations, *Amer. J. Math.*, **90** (1968), 270–284.

GUILLEMIN, V. W., QUILLEN, D. AND STERNBERG, S.

1970   The integrability of characteristics, *Comm. Pure Appl. Math.*, **23** (1970), 39–77.

GUILLEMIN, V. W. AND STERNBERG, S.

1964   An algebraic model of transitive differential geometry, *Bull. Amer. Math. Soc.*, **70** (1964), 16–47.

HARTSHORNE, R.

1977   "Algebraic geometry," Graduate Texts in Math., Vol. 52, Springer-Verlag, Berlin, Heidelberg, New York, 1977.

HARVEY, R. AND LAWSON, H. B.

1982   Calibrated geometries, *Acta. Math.*, **148** (1982), 47–157.

HÖRMANDER, L.

1963   "Linear partial differential operators," Academic Press, New York, Springer-Verlag, Berlin, Göttingen, Heidelberg, 1963.

1973   "An introduction to complex analysis in several variables," North-Holland Pub. Co., Amsterdam, London, American Elsevier Pub. Co., New York, 1973.

JACOBOWITZ, H.

1978   The Poincaré lemma for $d\omega = F(x, \omega)$, *J. Differential Geom.*, **13** (1978), 361–371.

KÄHLER, E.

1934   "Einfürhung in die Theorie der Systeme von Differentialgleichungen," B. G. Teubner, Leipzig, 1934.

KOHN, J. J. AND NIRENBERG, L.

1965   Non-coercive boundary value problems, *Comm. Pure Appl. Math.*, **18** (1965), 443–492.

KOHN, J. J. AND ROSSI, H.

1965   On the extension of holomorphic functions from the boundary of a complex manifold, *Ann. of Math.*, **81** (1965), 451–472.

KURANISHI, M.

1957   On E. Cartan's prolongation theorem of exterior differential systems, *Amer. J. Math.*, **79** (1957), 1–47.

1961   On the local theory of continuous infinite pseudo groups, II, *Nagoya Math. J.*, **19** (1961), 55–91.

1962   "Lectures on exterior differential systems," Tata Institute of Fundamental Research, Bombay, 1962.

1964   Sheaves defined by differential equations and application to deformation theory of pseudo-group structures, *Amer. J. Math.*, **86** (1964), 379–391.

1967   "Lectures on involutive systems of partial differential equations," Publicações da Sociedade de Matemática de São Paulo, São Paulo, 1967.

1977   The integrability condition of deformations of CR structures, *in* "Contributions to algebra, a collection of papers dedicated to Ellis Kolchin," edited by H. Bass, P. J. Cassidy and J. Kovacic, Academic Press, New York, San Francisco, London, 1977, 267–278.

LEWY, H.

1957   An example of a smooth linear partial differential equation without solution, *Ann. of Math.*, **66** (1957), 155–158.

MACKICHAN, B.

1971   A generalization to overdetermined systems of the notion of diagonal operators, I. Elliptic operators, *Acta Math.*, **126** (1971), 83–119.

1975   A generalization to overdetermined systems of the notion of diagonal operators, II. Hyperbolic operators, *Acta Math.*, **134** (1975), 239–274.

MALGRANGE, B.

1963   Sur les systèmes différentiels à coefficients constants, *in* "Les équations aux dérivées partielles," Colloques internationaux du C.N.R.S., No. 117, Editions du C.N.R.S., Paris, 1963, 113–122.

1966   "Cohomologie de Spencer (d'après Quillen)," Secrétariat mathématique d'Orsay, 1966.

1966-67   "Théorie analytique des équations différentielles," Séminaire Bourbaki, 1966–67, Exposé No. 329, 13 pp.

1972   Equations de Lie. I, II, *J. Differential Geom.*, **6** (1972), 503–522; **7** (1972), 117–141.

MATSUSHIMA, Y.

1953   On a theorem concerning the prolongation of a differential system, *Nagoya Math. J.*, **6** (1953), 1–16.

1954-55   Sur les algèbres de Lie linéaires semi-involutives, *in* "Colloque de topologie de Strasbourg," Université de Strasbourg, 1954–55, 17 pp.

MENIKOFF, A.

1977   The local solvability of a special class of overdetermined systems of partial differential equations, *Amer. J. Math.*, **99** (1977), 699–711.

MOORE, J. D.

1972   Isometric immersions of space forms into space forms, *Pacific J. Math.*, **40** (1972), 157–166.

MUMFORD, D.

1966   "Lectures on curves on an algebraic surface," Ann. of Math.
            Studies, No. 59, Princeton University Press, Princeton, N.J.,
            1966.

NIRENBERG, L.

1973   "Lectures on linear partial differential equations," CBMS Re-
            gional Conf. Ser. in Math., No. 17, Amer. Math. Soc., Provi-
            dence, R.I., 1973.

1974   On a question of Hans Lewy, *Uspekhi Mat. Nauk.*, **29** (1974),
            2, 241–251.

PFAFF, J. F.

1814-1815   Methodus generalis, aequationes differentiarum partialium nec
            non aequationes differentiales vulgates, ultrasque primi ordi-
            nis, inter quotcunque variables, complete integrandi, *Abh. kgl.
            Akad. d. Wiss. Berlin*, (1814-1815), 76–136.

QUILLEN, D. G.

1964   "Formal properties of over-determined systems of linear par-
            tial differential equations," Ph.D. thesis, Harvard University,
            Cambridge, Mass., 1964.

RIQUIER, C.

1910   "Les systèmes d'équations aux dérivées partielles," Gauthier-
            Villars, Paris, 1910.

ROCKLAND, C.

1972   "A new construction of the Guillemin Poincaré complex," Ph.D.
            thesis, Princeton University, Princeton, N.J., 1972.

SERRE, J.-P.

1955   Faisceaux algébriques coherents, *Ann. of Math.*, **61** (1955),
            197–278.

SINGER, I. M. AND STERNBERG, S.

1965   The infinite groups of Lie and Cartan, Part I, (The transitive
            groups), *J. d'Analyse Math.*, **15** (1965), 1–114.

SPENCER, D. C.

1961   A type of formal exterior differentiation associated with pseu-
            dogroups, *Scripta Math.*, **26** (1961), 101–106.

1962   Deformations of structures on manifolds defined by transi-
            tive, continuous pseudogroups. I, II, *Ann. of Math.*, **76** (1962),
            306–445.

1969   Overdetermined systems of linear partial differential equations,
            Bull. Amer. Math. Soc., **75** (1969), 179–239.

SPIVAK, M.

1979   "A comprehensive introduction to differential geometry," Vols. I-
            V, 2nd edition, Publish or Perish, Houston, Texas, 1979.

SWEENEY, W. J.

1967   The δ-Poincaré estimate, *Pacific J. Math.*, **20** (1967), 559–570.

1968   The *D*-Neumann problem, *Acta Math.*, **120** (1968), 223–277.

1969   A uniqueness theorem for the Neumann problem, *Ann. of Math.*, **90** (1969), 353–360.

1976   A condition for subellipticity in Spencer's Neumann problem, *J. Differential Equations*, **21** (1976), 316–362.

TANAKA, N.

1973   Rigidity for elliptic isometric imbeddings, *Nagoya Math. J.*, **51** (1973), 137–160.

TRÊVES, F.

1975   "Basic linear partial differential equations," Academic Press, New York, San Francisco, London, 1975.

1981   "Approximation and representation of functions and distributions annihilated by a system of complex vector fields," Ecole Polytechnique, Centre de Mathématique, 1981.

WARNER, F.

1971   "Foundations of differentiable manifolds and Lie groups," Scott, Foreman, Glenview, Ill., London, 1971.

WEINSTEIN, A.

1971   Symplectic manifolds and their Lagrangian submanifolds, *Adv. in Math.*, **6** (1971) 329–346.

YANG, D.

1986   Local solvability of overdetermined systems defined by commuting first order differential operators, *Comm. Pure Appl. Math.*, **39** (1986), 401–421.

1987   Involutive hyperbolic differential systems, *Mem. Amer. Math. Soc.*, **68** (1987), No. 370.

INDEX

GPSR Compliance
The European Union's (EU) General Product Safety Regulation (GPSR) is a set
of rules that requires consumer products to be safe and our obligations to
ensure this.

If you have any concerns about our products, you can contact us on

ProductSafety@springernature.com

In case Publisher is established outside the EU, the EU authorized
representative is:

Springer Nature Customer Service Center GmbH
Europaplatz 3
69115 Heidelberg, Germany